69th Highway Geology Symposium (HGS 2018)

Better Highways Through Applied Geology

Portland, Maine, USA
10-13 September 2018

ISBN: 978-1-5108-8659-9

Printed from e-media with permission by:

Curran Associates, Inc.
57 Morehouse Lane
Red Hook, NY 12571

Some format issues inherent in the e-media version may also appear in this print version.

Printed by Curran Associates, Inc. (2019)

Highway Geology Symposium (HGS)
C/O John Pilipchuk
1589 Mail Service Center
Raleigh, NC, 27699-1589

Phone: (919) 707-6851
Fax: (919) 250-4237

jpilipchuk@ncdot.gov

Additional copies of this publication are available from:

Curran Associates, Inc.
57 Morehouse Lane
Red Hook, NY 12571 USA
Phone: 845-758-0400
Fax: 845-758-2633
Email: curran@proceedings.com
Web: www.proceedings.com

69th Highway Geology Symposium (HGS 2018)

Better Highways Through Applied Geology

Portland, Maine, USA
10-13 September 2018

Tuesday, September 11

Technical Session 1

Location: Massachusetts/New Hampshire

Tuesday, September 11(continued)

10:10 AM – 10:40 AM
Morning Coffee Break—Sponsored by Haley and Aldrich, Inc.
Location: Vermont/Connecticut/Rhode Island (Exhibitor Area)

10:40 AM – 11:00 AM
110 **Young Author Presentation: Laboratory Investigations of the Oldest Concrete Pavement in America – Applied Geology in Civil Engineering**
Author: Blake Lemcke

11:00 AM – 11:20 AM
128 **Young Author Presentation: From Field Data Collection to Soils Analysis in A Few Mouse Clicks – Going (Even More) Digital at North Dakota DOT**
Authors: Jesse Greenwald, Colter Schwagler

11:20 AM – 11:40 AM
N/A **PhotoMonitoring of Landslides**
Authors: Paolo Caprossi, Paolo Mazzanti

11:40 PM – 1:00 PM
Lunch—Sponsored by Ameritech Slope Constructors, Inc.
Location: Vermont/Connecticut/Rhode Island (Exhibitor Area)

Technical Session 2

Location: Massachusetts/New Hampshire
Pete Ingraham, Moderator

1:00 PM – 1:20 PM
145 **Rattlesnake Hills Landslide: Overview and Monitoring**
Authors: George Machan, Charlie Hammond

1:20 PM – 1:40 PM
165 **Adding Another Dimension to Rock Cut Slope Evaluations: Looking Out as Well as Up**
Authors: Chris, Ruppen, Don Gaffney, Joel Borrelli

1:40 PM – 2:00 PM
177 **Practical Aspects of Using Structure from Motion Photogrammetry Techniques for Characterizing and Monitoring of Rock Slopes**
Authors: Randy Post, Roger Pihl, Alex Brown, Ty Ortiz

2:00 PM – 2:20 PM
197 **The use of Google Earth/ Google Street View Combined with High Resolution Digital Surface Models (DSMs) for Rockfall Hazard Rating**
Author: Yonathan Admassu

2:20 PM – 2:40 PM
213 **Application of High-Speed Photogrammetry for Rock Cut Assessment**
Authors: Angus MacPhail, Dave Gauthier, D. Jean Hutchinson

Tuesday, September 11(continued)

2:40 PM – 3:00 PM
232 **3-D Geo-View of Subsurface Conditions for Rapid Roadway Stability Assessment**
Author: Joel Daniel

3:00 PM – 3:30 PM
Afternoon Break—Sponsored by Hager-Richter Geoscience, Inc. and Gilson Company, Inc.
Location: Vermont/Connecticut/Rhode Island (Exhibitor Area)

Technical Session 3

Location: Massachusetts/New Hampshire
Bob Henthorne, Moderator

3:30 PM – 3:50 PM
251 **What are the Benefits of Geotechnical Data Interchange?**
Author: Scott L. Deaton

3:50 PM – 4:10 PM
269 **Geotechnical Solutions for the I-95 Betsy Ross Bridge Interchange Structure Alternatives over Soft Soils in Spaghetti Junction**
Authors: Sarah McInnes, Geoff Stryker, John Pizzi

4:10 PM – 4:30 PM
288 **Risk Assessment for Landslides on the Last Chance Grade, Crescent City, California**
Authors: Scott Anderson, Cole Christiansen, Dave Gauthier, Sebastian Cohen

4:30 PM – 4:50 PM
303 **Landslide Applications of the Geotechnical Observational Approach**
Authors: George Machan, Wade Osborne, Chris Carpenter, Charles M. Hammond, Philip Wurst

4:50 PM – 5:20 PM
547 **Highway Geology Symposium Field Trip Preview**
Presenter: Bob Marvinney
Location: Massachusetts/New Hampshire

Free evening to explore and dine in the Old Port

Wednesday, September 12

6:00 AM – 7:00 AM
To-Go Continental Breakfast—Sponsored by Maccaferri, Inc. and Precision Blasting Services
Location: Highway Geology Symposium Registration Area

Highway Geology Symposium Field Trip

6:45 AM – 7:15 AM
Load buses for Field Trip
Pick-up Location: Meet in Hotel Front Lobby

7:15 AM – 5:00 PM
Field Trip
Lunch—Sponsored by Geobrugg
Afternoon Beverages—Sponsored by Golder Associates
(NO GLASS ALLOWED INSIDE BUSES)

5:30 PM – 6:30 PM
Highway Geology Symposium Social Hour—Sponsored by Access Limited Construction
Location: Vermont/Connecticut/Rhode Island (Exhibitor Area)

Highway Geology Symposium Banquet Dinner

6:30 PM – 9:30 PM
Highway Geology Symposium Banquet Keynote Speaker —Sponsored by IDS GeoRadar
Entertainment by Tim Sample:
Location: Massachusetts/New Hampshire

Thursday, September 13

6:30 AM – 9:00 AM
Continental Breakfast—Sponsored by Scarptec, Inc. and Rocscience, Inc.
Location: Vermont/Connecticut/Rhode Island (Exhibitor Area)

8:00 AM – 10:00 AM
N/A **Highway Geology Symposium Exhibitor Area Open**
Exhibitors can break down after morning coffee break

Technical Session 4

Location: Massachusetts/New Hampshire
Steve Sweeney, Moderator

7:30 AM – 7:50 AM
323 **Soil Mixing: An Innovative Solution for Resiliency in a Flood-Prone Canyon**
Author: Todd Schlittenhart

Thursday, September 13 (continued)

7:50 AM – 8:10 AM
340 **Innovative Use of Horizontal Directional Continuous Rock Coring For The Design of The LSIORB East End Tunnels**
Author: Craig S. Lee

8:10 AM – 8:30 AM
355 **Geotechnical Seismic Design in New England**
Author: Craig Coolidge

8:30 AM – 8:50 AM
375 **Geotechnical risks from abandoned coal mines to transportation infrastructure and mitigation – an overview**
Authors: David Knott, Athena Livesey, Robert Kingsland, Thomas Lefchik, Elizabeth Dwyre

8:50 AM – 9:10 AM
398 **Icefall Hazard Predictive Indicators + Mitigation Techniques – Results Of A 3-YR. Research Study In Alaska**
Authors: David J. Scarpato, Matt Murphy

9:10 AM – 9:30 AM
421 **Rock Slope Scaling Investigative Approach and Volume Estimation Method**
Author: John Duffy

9:30 AM – 10:00 AM
Morning Coffee Break—Sponsored by HI-TECH Rockfall Construction, Inc.
Location: Vermont/Connecticut/Rhode Island (Exhibitor Area)

10:00 AM – 10:20 AM
433 **What Were We Thinking A case History of Extreme Slope Scaling in Washington**
Authors: Marc Fish, Jim Struthers, Mike Mullhern

10:20 AM – 10:40 AM
453 **Slope Access Safety Evaluation (SASE) Form**
Author: William CB Gates

10:40 AM – 11:00 AM
466 **Rockfall in New Jersey: A Proactive and Collaborative Approach**
Authors: Amber B. Granger, John Jamerson, Scott J. Deeck, Edward M. Zamiskie Jr.

11:00 AM – 11:20 AM
482 **Development of a Modular Rockfall Protection Wall to Mitigate Earthquake-Induced Slope Hazards Along a Coastal Transportation Corridor**
Authors: Rori Green, Cedric Lambert, Charlie Watts, Daniel Kennett, Eric Ewe, Emerson Ryder, Michal Tutko

11:20 AM – 11:40 AM
500 **Rock Slope Remediation at the Penobscot Narrows Bridge**
Authors: Bryan C. Steinert, Amber Granger, Laura Krusinski, Wayne Chadbourne

Thursday, September 13 (continued)

11:40 AM – 12:00 PM

516 **Advantages of Using A Downhole Optical Televiewer For Rock Cut Slope Design—An Example In Central Pennsylvania**
Authors: Jeremy Robinson, Andrew Smithmyer

12:00 PM – 12:20 PM

532 **Design of Pinned Drapery Systems for Rockfall Protection**
Author: Mike Koutsourais

12:20 PM

Closing Remarks and Adjournment

69TH HIGHWAY GEOLOGY SYMPOSIUM

September 10-13, 2018

Holiday Inn By The Bay
Portland, Maine

2018 Proceedings

69TH HIGHWAY GEOLOGY SYMPOSIUM
September 10-13, 2018

Holiday Inn By The Bay
Portland, Maine

Grateful Acknowledgments

We would like to thank the following people who helped make made this Symposium possible.

Krystle Pelham
John Pilipchuk
Peter Ingraham
Chris Ruppen
Richard Lane
Jim Coffin
Jeff Dean
HGS Steering Committee

Maine Geological Survey
Robert Marvinney
Maine Department of Transportation
Laura Krusinski
Kate Maguire
Delaney Meeting and Event Management

Picture of Portland Head Light.
Located in Fort Williams Park, Cape Elizabeth, Maine.

69th ANNUAL
HIGHWAY GEOLOGY SYMPOSIUM
Table of Contents

The Proceedings of the
69th Highway Geology Symposium
are dedicated to
Dave Bingham and Joe Gutierrez

Photo of Dave Bingham and Joe Gutierrez taken by Joe Jennings at the 1999 Southeastern Transportation Geotechnical Engineering Conference (STGEC)

Joe Gutierrez
1926-2018

Joe Gutierrez was born on August 12, 1926, in in Lafayette, IN. He obtained his Bachelor of Science in Geology from the University of North Carolina, Chapel Hill in1956. Joe proudly served in the United States Army Air Force during WWII. He started his career working for the North Carolina Department of Transportation in 1956 as a Highway Geologist, shortly after that he went to work for W.E. Graham & Sons, which eventually became Vulcan Materials Mideast Division, where he retired in 1992. Joe continued working for Vulcan as a temporary part-time employee leading school groups through the Joseph Andres Gutierrez Earth Science Museum. This museum is Joe's legacy for teaching children that "Everything comes from the Earth, and if you can't grow it, you've got to mine it."

Joe was always trying to advance the fields of Mining and Geology through outreach classes. The Joseph Andres Gutierrez Earth Science Museum was opened and dedicated to Joe at his retirement in 1992, due to his outreach programs. This museum has won numerous state and national awards over many years and is still active. Joe taught children and adults about the importance of mining with over 3000 students per year touring the facility and Vulcan's North Quarry. Joe was instrumental for thousands of scouts obtaining their geology badges. He was an important part of Vulcan's and the Nation Stone Association's outreach program, which developed nationwide into educating people about the importance for mining. This played a key role in getting new properties rezoned and permitted for mining. Joe remained involved as a temporary part-time employee at Vulcan Materials leading tours through the museum until 2013 at 86 years of age, which was 21 years after he retired. He became a Vulcan Mideast Division icon and was loved by teachers from across NC, SC, and VA. Joe received the Non-School Teachers Award in 1999 from the North Carolina Science Teachers Association (NCSTA).
Joe was a member of the HGS Steering Committee for approximately 30 years and served in several officer positions during his attendance. He served as Secretary and Treasurer in the early 70's for many years and Vice Chairman in late 70's. He was a mentor to many younger

geologists that followed in his footsteps over the years and will always be remembered for his generosity.

Joe enjoyed helping others. He will be remembered for his dedication to educating youth about geology and his fun wit with such a positive attitude.

Dave Bingham
1932-2018

Dave Bingham was born on May 2, 1932 in Wake County, North Carolina. After graduating from high school, he served in the U.S. Coast Guard from 1952 – 1956, then went on to earn his Bachelor of Science degree in Geology from the University of North Carolina, Chapel Hill in 1959. Dave worked with the NC Department of Transportation as the State Geotechnical Engineer. At one time or another, he worked in all four offices across the state, from coastal plain to mountains. After retiring from the DOT in 1989, he worked at Law Engineering and other geotechnical consulting firms. Dave was extremely active in the Highway Geology Symposium. He was treasurer of the HGS from 1977 thru 1990, and took over that role from one of the founding members, A. Carter Dodson. Dave encouraged as many employees as he could to attend the HGS and other organizations as a means to staying abreast of new technologies and products. He was a thoughtful manager, doing his best to look out for employees in tight times in North Carolina. He loved hunting, being outdoors, and being a geologist. David was married to Peggy Perry Haithcock for 62 years, and had two daughters, Lynnette and Dana, and four grandchildren.

At a Glance Schedule of Events

69th Highway Geology Symposium
Portland, Maine
September 10-13, 2018

Monday, September 10

11:00 AM – 5:00 PM
Highway Geology Symposium Registration Open in Hotel Lobby

1:00 PM – 5:00 PM
Transportation Research Board Technical Session: "Geotechnical Asset Management: Implementation of Programs and Advances in Technology"
Location: Massachusetts

5:00 PM – 8:30 PM
Highway Geology Symposium Exhibitor Area Open
Location: Vermont/Connecticut/Rhode Island

5:00 PM – 6:15 PM
HGS National Steering Committee Meeting
Location: New Hampshire

6:30 PM – 8:30 PM
Ice Breaker Social—Sponsored by BGC Engineering and Hager-Richter Geoscience, Inc.
Location: Vermont/Connecticut/Rhode Island (Exhibitor Area)

Tuesday, September 11

6:30 AM – 9:00 AM
Continental Breakfast —Sponsored by Terracon Consultants, Inc.
Location: Vermont/Connecticut/Rhode Island (Exhibitor Area)

6:30 AM – 5:00 PM
Highway Geology Symposium Registration Open in Hotel Lobby

8:00 AM – 5:00 PM
Highway Geology Symposium Exhibitor Area Open

Tuesday, September 11(continued)

7:30 AM – 8:10 AM
Welcome and Opening Remarks
Krystle Pelham, HGS Organizing Committee Chair
Dedication of Proceedings - John Pilipchuk
Joyce Taylor, P.E., Chief Engineer, Maine Department of Transportation
Location: Massachusetts/New Hampshire

GEOTECHNICAL ASSET MANAGEMENT (GAM) SUBCOMMITTEE AFP00(1)

2018 TRB Midyear Meeting at the 69th Highway Geology Symposium (HGS), Portland, Maine

Date: Monday, September 10, 2018, 12:30 PM – 5:00 PM

Location: Massachusetts Room

Session Theme: Geotechnical Asset Management: Implementation of Programs and Advances in Technology

Time	Topic	Discussion Lead/Presenter
12:30 – 12:35	Welcome and Introductions	Darren Beckstrand, Landslide Technology
12:35 – 12:50	GAM Subcommittee Business	Scott Anderson, BGC Engineering
Presentations		
12:50 – 1:25	The Unstable Slope Management Program: A Tool for Federal Land Management Agencies and Beyond *The Federal Highway Administration has completed an unstable slope inventory and assessment tool for Federal Land Management Agencies, including comprehensive rating criteria, digital applications, and online mapping tools.*	Doug Anderson, Western Federal Lands Division of the Federal Highway Administration
1:25 – 2:00	Montana's Rock Slope Asset Management Program (RAMP) *MDT's comprehensive RAMP Program combines TAM principles (i.e. deterioration, Return-on-investment and other fiscal modeling) with technical decision support tools to assist policy makers with setting budgets, planners to group rock slope improvements with nearby projects, and geotechnical personnel with reducing user and Department risk due to rock slopes.*	Jeff Jackson, Montana Department of Transportation
2:00 – 2:35	Applications of Remote Monitoring Technologies to GAM *A review of various remote sensing and monitoring methods and techniques for managing geotechnical assets.*	Jean Hutchinson, Univ. of Queens
2:35 – 3:00	Break	
3:00 – 3:35	Legislating Geotechnical Asset Management: Lessons Learned *An accounting of efforts to include management of geotechnical assets into a Minnesota House of Representatives Bill and the lessons learned.*	John Siekmeier, Minnesota House of Representatives
3:35 – 4:10	Update on the NCHRP GAM Implementation Manual *The implementation process in the manual is intended to be simple and practical to enable broad adoption across the nation for all types of geotechnical assets. The recommended GAM processes also were developed to facilitate the integration of geotechnical assets into the broader asset and performance management programs in a DOT. The Manual includes a Microsoft Excel based tool, the GAM Planner, to enable agencies to start GAM now without needing additional specialized resources.*	Mark Vessely, BGC Engineering
4:10 – 5:00	Discussion	Group

Highway Geology Symposium History, Organization, and Function

Inaugural Meeting

Established to foster a better understanding and closer cooperation between geologists and civil engineers in the highway industry, the Highway Geology Symposium (HGS) was organized and held its first meeting on March 14, 1950, in Richmond Virginia. Attending the inaugural meeting were representatives from state highway departments (as referred to at that time) from Georgia, South Carolina, North Carolina, Virginia, Kentucky, West Virginia, Maryland, and Pennsylvania. In addition, a number of federal agencies and universities were represented. A total of nine technical papers were presented.

W.T. Parrott, an engineering geologist with the Virginia Department of Highways, chaired the first meeting. It was Mr. Parrott who originated the Highway Geology Symposium.

It was at the 1956 meeting that future HGS leader, A.C. Dodson, began his active role in participating in the Symposium. Mr. Dodson was the Chief Geologist for the North Carolina State Highway and Public Works Commission, which sponsored the 7th HGS meeting.

Symposium Locations

Since the initial meeting, 69 consecutive annual meetings have been held in 33 different states. Between 1950 and 1962, the meetings were east of the Mississippi River, with Virginia, West Virginia, Ohio, Maryland, North Carolina, Pennsylvania, Georgia, Florida, and Tennessee serving as host state.

In 1962, the symposium moved west for the first time to Phoenix, Arizona where the 13th annual HGS meeting was held. Since then it has alternated, for the most part, back and forth from the east to the west. The Annual Symposium has moved to different location as shown on the next page.

Organization

Unlike most groups and organizations that meet on a regular basis, the Highway Geology Symposium has no central headquarters, no annual dues and no formal membership requirements. The governing body of the Symposium is a steering committee composed of approximately 20 - 25 engineering geologist and geotechnical engineers from state and federal agencies, colleges and universities, as well as private service companies and consulting firms throughout the country. Steering committee members are elected for three-year terms, with their elections and re-elections being determined principally by their interests and participation in and contribution to the Symposium. The officers include a chairman, vice chairman, secretary, and treasurer. all of whom are elected for a two-year term. Officers, except for the treasurer, may only succeed themselves for one additional term.

A number of three-member standing committees conduct the affairs of the organization. The lack of rigid requirements, routing and relatively relaxed overall functioning of the organization is what attracts many participants.

List of Highway Geology Symposium Meetings

No.	Year	HGS Location	No.	Year	HGS Location
1st	1950	Richmond, VA	2nd	1951	Richmond, VA
3rd	1952	Lexington, VA	4th	1953	Charleston, WV
5th	1954	Columbus, OH	6th	1955	Baltimore, MD
7th	1956	Raleigh, NC	8th	1957	State College, PA
9th	1958	Charlottesville, VA	10th	1959	Atlanta, GA
11th	1960	Tallahassee, FL	12th	1961	Knoxville, TN
13th	1962	Phoenix, AZ	14th	1963	College Station, TX
15th	1964	Rolla, MO	16th	1965	Lexington, KY
17th	1966	Ames, IA	18th	1967	Lafayette, IN
19th	1968	Morgantown, WV	20th	1969	Urbana, IL
21st	1970	Lawrence, KS	22nd	1971	Norman, OK
23rd	1972	Old Point Comfort, VA	24th	1973	Sheridan, WY
25th	1974	Raleigh, NC	26th	1975	Coeur d'Alene, ID
27th	1976	Orlando, FL	28th	1977	Rapid City, SD
29th	1978	Annapolis, MD	30th	1979	Portland, OR
31st	1980	Austin, TX	32nd	1981	Gatlinburg, TN
33rd	1982	Vail, CO	34th	1983	Stone Mountain, GA
35th	1984	San Jose, CA	36th	1985	Clarksville, TN
37th	1986	Helena, MT	38th	1987	Pittsburg, PA
39th	1988	Park City, UT	40th	1989	Birmingham, AL
41st	1990	Albuquerque, NM	41st	1991	Albany, NY
43rd	1992	Fayetteville AR	44rd	1993	Tampa, FL
45th	1994	Portland, OR	46th	1995	Charleston, WV
47th	1996	Cody, WY	48th	1997	Knoxville, TN
49th	1998	Prescott, AZ	50th	1999	Roanoke, VA
51st	2000	Seattle, WA	52nd	2001	Cumberland, MD
53rd	2002	San Luis Obispo, CA	54th	2003	Burlington, VT
55th	2004	Kansas City, MO	56th	2005	Wilmington, NC
57th	2006	Breckinridge, CO	58th	2007	Pocono Manor, PA
59th	2008	Santa Fe, NM	60th	2009	Buffalo, NY
61st	2010	Oklahoma City, OK	62nd	2011	Lexington, KY
63rd	2012	Redding, CA	64th	2013	North Conway, NH
65th	2014	Laramie, WY	66th	2015	Sturbridge, MA
67th	2016	Colorado Springs	68th	2017	Marietta, GA
69th	2018	Portland, ME			

Meeting sites are chosen two to four years in advance and are selected by the Steering Committee following presentations made by representatives of potential host states. These presentations are usually made at the steering committee meeting, which is held during the Annual Symposium. Upon selection, the state representative becomes the state chairman and a member of the Steering Committee.

HGS History, Organization, and Function (continued)

The symposia are generally scheduled for two and one-half days, with a day-and-a-half for technical papers plus a full day for the field trip. The Symposium usually begins with a TRB session and an evening Ice-Breaker the first day, a full day of technical presentations the second day, a field trip on the third day followed by the annual banquet that evening, and a half day of technical presentations on the final day.

The Field Trip

The field trip is the focus of the meeting. In most cases, the trips cover approximately 150 to 200 miles, provide for six to eight scheduled stops, and require about eight hours. Occasionally, cultural stops are scheduled around geological and geotechnical points of interests. To cite a few examples: in Wyoming (1973), the group viewed landslides in the Big Horn Mountains; Florida's trip (1976) included a tour of Cape Canaveral and the NASA space installation; the Idaho and South Dakota trips dealt principally with mining activities; North Carolina provided stops at a quarry site, a dam construction site, and a nuclear generation site; in Maryland, the group visited the Chesapeake Bay hydraulic model and the Goddard Space Center. The Oregon trip included visits to the Columbia River Gorge and Mount Hood; the Central mine region was visited in Texas; and the Tennessee meeting in 1981 provided stops at several repaired landslide in Appalachia regions of East Tennessee.

In Utah (1988) the field trip visited sites in Provo Canyon and stopped at the famous Thistle Landslide, while in New Mexico, in 1990, the emphasis was on rockfall treatments in the Rio Grande River canyon and included a stop at the Brugg Wire Rope headquarters in Santa Fe.

Mount St, Helens was visited by the field trip in 1994 when the meeting was in Portland, Oregon, while in 1995 the West Virginia meeting took us to the New River Gorge Bridge that has a deck elevation of 876 feet above the water.

In Cody, Wyoming the 1996 field trip visited the Chief Joseph Scenic Highway and the Beartooth Uplift in northwest Wyoming. In 1997 the meeting in Tennessee visited the newly constructed future I-26 highway in the Blue Ridge of East Tennessee. The Arizona meeting in 1998 visited the Oak Creek Canyon near Sedona and a mining ghost town at Jerome, Arizona. The Virginia meeting in 1999 visited the "Smart Road" Project that was under construction. This was a joint research project of the Virginia Department of Transportation and Virginia Tech University. The Seattle Washington meeting in 2000 visited an ancient lahar in the Mount Rainier area. A stop during the Maryland meeting in 2001 was the Sideling Hill road cut for I-68 which displayed a tightly folded syncline in the Allegheny Mountains.

The California field trip in 2002 provided a field demonstration of the effectiveness of rock netting against rock falls along the Pacific Coast Highway. The Kansas City meeting in 2004 visited the Hunt Subtropolis which is said to be the "world's largest underground business complex". It was created through the mining of limestone by way of the room and pillar method. The Rocky Point Quarry provided an opportunity to search for fossils at the North Carolina meeting in 2005. The group also visited the US-17 Wilmington Bypass Bridge which was under construction. Among the stops at the Pennsylvania meeting were the Hickory Run Boulder Field, the No.9 Mine and Wash Shanty Museum, and the Lehigh Tunnel.

HGS History, Organization, and Function (continued)

The New Mexico field trip in 2008 included stops at a soil nailed wall along US-285/84 north of Santa Fe and a road cut through the Bandelier Tuff on highway 502 near Los Alamos where rockfall mesh was used to protect against rockfalls. The New York field trip in 2009 included the Niagara Falls Gorge and the Devil's Hole Trail. The Oklahoma field trip in 2010 toured the complex geology of the Arbuckle Mountains in the southern part of the state along with stops at Tucker's Tower and Turner Falls.

In the bluegrass state of Kentucky, the 2011 HGS field trip included stops at Camp Nelson which is the site of the oldest exposed rocks in Kentucky near the Lexington and Kentucky River Fault Zones. Additional stops at the Darby Dan Farm and the Woodford Reserve Distillery illustrated how the local geology has played such a large part in the success of breeding prized Thoroughbred horses and made Kentucky the "Birthplace of Bourbon".

In Redding, California, the 2012 field trip included stops at the Whiskeytown Lake, which is one in a series of lakes that provide water and power to northern California. Additional stops included Rocky Point, a roadway construction site containing Naturally Occurring Asbestos (NOA), and Oregon Mountain where the geology and high rainfall amounts have caused Hwy 299 to experience local and global instabilities since first constructed in 1920.

The 2013 field trip of New Hampshire highlighted the topography and geologic remnants left by the Pleistocene glaciation that fully retreated approximately 12,000 years ago. The field rip included stops at various overlooks of glacially-carved valleys and ranges; the Old Man of the Mountain Memorial Plaza, which is a tribute to the famous cantilevered rock mass in the Franconia Notch that collapsed on May 3, 2003; the lacustrine deposits and features of the Glacial Lake Ammonoosuc; views of the Presidential Range; bridges damaged during Tropical Storm Irene in August 2011; and the Willey Slide, located in the Crawford Notch where all members of the Willey family were buried by a landslide in 1826.

The 2014 field trip presented a breathtaking tour of the geology and history of southeast Wyoming, ascending from the high plains surrounding Laramie at 7000 feet to the Medicine Bow Mountains along the Snowy Range Scenic Byway. Visible along the way were a Precambrian shear zone, and glacial deposits and features. From the glacially carved Mirror Lake and the Snowy Range Ski Area, the path wound east to the Laramie Mountains and the Vedauwoo Recreational Area, a popular rock climbing and hiking area before returning to Laramie.

In Sturbridge, MA, the 2015 field trip focused on the Connecticut Valley, a Mesozoic rift basin that signaled the breakup of Pangea, and the Berkshires, which represents the collision and amalgamation of an island arc system with the North American Laurentian margin.

The field trip in 2016 was an urban setting along the western edge of Colorado Springs and around Manitou Springs. Stops included the Pikeview Quarry, Garden of the Gods Visitor Center, and several other locations where rockfall and debris flow mitigation, post-flooding highway embankment repair, and a nonconformity in the rock records that spans 1.3 billion years were observed.

The 2017 field trip provided an opportunity to view the geology of northern Georgia. Stops included the Bellwood Quarry, which, at one time was run by the City of Atlanta and also served as a prison labor camp. It will eventually serve as a 2.4 billion-gallon water storage facility for the City of Atlanta upon completion of a tunnel to connect the quarry to two water treatment plans and three pump stations. Additional stops included the Buzzi Unicem Cement Plant to get a close up view of the Clairmont Melange, The Cooper Furnace near the Allatoona Dam, and the New Riverside Ochre-Emerson Barite mine.

At the technical sessions, case histories and applied state-of-the-art papers are most common; with highly theoretical papers the exception. The papers presented at the technical sessions are published in the annual proceedings. All proceedings are available to download from www.HighwayGeologySymposium.org.

Banquet speakers are also a highlight and have been varied through the years.

A Medallion Award was initiated in 1970 to honor those persons who have made significant contributions to the Highway Geology Symposium. The selection was- and is currently made from the members of the national steering committee of the HGS.

A number of past members of the national steering committee have been granted Emeritus status. These individuals, usually retired, resigned from the HGS Steering Committee, or are deceased, have made significant contributions to the Highway Geology Symposium. A total of 38 persons have been granted Emeritus status.

Several Proceedings volumes have been dedicated to past HGS Steering Committee members who have passed away. The 36th HGS Proceedings were dedicated to David L. Royster (1931 - 1985, Tennessee) at the Clarksville, Indiana Meeting in 1985. In 1991 the Proceedings of the 42nd HGS held in Albany, New York were dedicated to Burrell S. Whitlow (1929 - 1990, Virginia). The 64th HGS Proceedings were dedicated to Earl Wright (1931 – 2012) at the North Conway, New Hampshire meeting. The 65th proceedings were dedicated to Nicholas Priznar (1952 – 2014) at the Laramie, Wyoming meeting. The 76th HGS held at Colorado Springs, Colorado dedicated the proceedings to Vern McGuffy (1934 – 2016). The proceedings for the 68th HGS held in Marietta, Georgia were dedicated to Richard (Dick) Cross (1944 – 2016). The proceedings for the 69th HGS are dedicated to Dave Bingham (1932-2018) and Joe Gutierrez (1926-2018).

Young Author Award Winners

2014 Simon Boone, "Performance of Flexible Debris Flow Barriers in a Narrow Canyon"

2015 Cory Rinehart, "High Quality H20: Utilizing Horizontal Drains for Landslide Stabilization"

2016 Todd Hansen, "Geologic Exploration for Ground Classification: Widening of the I-70 Veterans Memorial Tunnels"

2017 James Arthurs, "Construction of Transportation Infrastructure in Weathered Volcanic Ash Soils"

HGS Medallion Award Recipients

Hugh Chase	1970
Tom Parrott	1970
Paul Price	1970
K.B. Woods	1971
R.J. Edmondson	1972
C.S. Mullin	1974
A.C. Dodson	1975
Burrell Whitlow	1978
Bill Sherman	1980
Virgil Burgat	1981
Henry Mathis	1982
David Royster	1982
Terry West	1983
Dave Bingham	1984
Vernon Bump	1986
C.W. "Bill" Lovell	1989
Joseph A. Gutierrez	1990
Willard McCasland	1990
W.A. "Bill" Wisner	1991
David Mitchell	1993
Harry Moore	1996
Earl Wright	1997
Russell Glass	1998
Harry Ludowise	2000
Sam Thornton	2000
Bob Henthorne	2004
Mike Hager	2005
Joseph A. Fischer	2007
Ken Ashton	2008
A. David Martin	2008
Michael Vierling	2009
Dick Cross	2009
John F. Szturo	2010
Christopher Ruppen	2012
Jeff Dean	2012
John Pilipchuk	2015
Peter Ingraham	2016

Emeritus Members of the Steering Committee

Emeritus Status is granted by the Steering Committee

R.F. Baker
John Baldwin
David Bingham
Vernon Bump
Virgil E. Burgat
Robert G. Charboneau
Hugh Chase
Richard Cross
A.C. Dodson
Walter F. Fredericksen
Brandy Gilmore
Robert Goddard
Joseph Gutierrez
G. Michael Hager
Rich Humphries
Charles T. Janik
John Lemish
Bill Lovell
A. David Martin
Henry Mathis
Willard McCasland
George S. Meadors, Jr.
David Mitchell
Harry Moore
W.T. Parrot
Paul Price
Nick Priznar
David L. Royster
Mitchell Smith
Willard L. Sitz
Bill Sherman
Jim Stroud
Berke Thompson
Sam Thornton
Burrell Whitlow
W. A. “Bill” Wisner
Earl Wright
Ed J. Zeigler

HGS National Steering Committee Officers

HGS National Steering Committee Members

HGS National Steering Committee Members (continued)

Kyle Halverson
Chief Geologist
Kansas Department of Transportation
Bureau of Structures and Geotechnical Services
700 SW Harrison St.
Topeka, KS 66603
Office: 785-291-3860
Cell: 785-845-4332
Email: kyle.halverson@ks.gov

Bob Henthorne
Mid-States Materials
1800 Brickyard Road
Topeka, KS 66618
Phone: (785) 640-2477
Email:
bhenthorne@midstatesmaterials.com

Peter Ingraham
Golder Associates Inc.
670 North Commercial Street, Suite 103
Manchester, NH 03101-1146
Phone: (603) 668-0880
Fax: (603) 668-1199
Email peter_ingraham@golder.com

Richard Lane
NHDOT (Retired)
213 Pembroke Hill Rd.
Pembroke, NH 03275
Phone: (603) 485-3202
Email: lanetrisbr@hotmail.com

Sarah McInnes
PA DOT
District 6-0
7000 Geerdes Blvd.
King of Prussia, PA 19406
Phone: (610) 205-6544
FAX: (610) 205-6599
Email: smcinnes@pa.gov

Victoria Porto
PA DOT (Retired)
10 Pine Lake Drive
Carlisle, PA 17015
Phone: (717) 805-5941
Email: vamporto@aol.com

Erik Rorem
Geobrugg North America, LLC
Phone: (505) 771-4080
Fax: (505) 771-4081
Email: erik.rorem@geobrugg.com

Christopher A. Ruppen
Michael Baker Jr., Inc.
100 Airside Drive
Moon Township, PA 15108
Phone: (724) 495-4079
Cell: (412) 848-2305
Fax: (724) 495-4017
Email: cruppen@mbakerintl.com

Stephen Senior
Ontario Min of Trans. (Retired)
11 Dewbourne Ave.
Richmond Hill, ON L4B 3G7 Canada
Phone: (416) 235-3734
Fax: (416) 235-4101
Email: sa.senior@rogers.com

Deana Sneyd
Petrologic Solutions, Inc.
3997 Oak Hill Road
Douglasville, GA 30135
Phone: (678) 313-4147
Email: dsneyd@gmail.com

HGS National Steering Committee Members (continued)

Steven Sweeney
NY Thruway (Retired)
105 Albert Rd.
Delanson, NY 12053
Email: 2ssweeney@gmail.com

John F. Szturo
HNTB Corporation
715 Kirk Drive
Kansas City, MO 64105
Phone: (816) 527-2275 (Direct Line)
Cell: (913) 530-2579
Fax: (816) 472-5013
Email: jszturo@hntb.com

Michael P. Vierling
NY Thruway (Retired)
323 Boght Road
Watervliet, NY 12189-1106
Phone: (518) 233-1197
Email: rocdoc1956@gmail.com

Terry West
Earth and Atmospheric Science Dept.
Purdue University
West Lafayette, IN 47907-1297 Phone:
(765) 494-3296
Fax: (765)496-1210
Email: trwest@purdue.edu

Richard Wilson
S&ME, Inc.
2020 Liberty Road, Suite 105
Lexington, KY 40505
Phone: (859) 293-5518
Cell: (502) 682-1203
Email: rwilson@smeinc.com

HIGHWAY GEOLOGY SYMPOSIUM
Past, Present, and Future Symposium Contact List

2013	New Hampshire	Krystle Pelham	603-271-1657	Krystle.Pelham@dot.state.nh.us
2014	Wyoming	Jim Coffin	307-777-4205	Jim.coffin@wyo.gov
2015	Massachusetts	Peter Ingraham	603-688-0880	peter_ingraham@golder.com
2016	Colorado	Ty Ortiz	303-921-2634	Ty.ortiz@state.co.us
2017	Georgia	Deana Sneyd	678-313-4147	Dsneyd61@gmail.com
2018	Maine	Krystle Pelham	603-271-1657	Krystle.Pelham@dot.state.nh.us
2019	Oregon	Scott Burns	503-725-3389	BurnsS@pdx.edu

69th ANNUAL HIGHWAY GEOLOGY SYMPOSIUM
Sponsors

The following companies have graciously contributed toward the sponsorship of the Symposium. The HGS relies on sponsor contributions for refreshment breaks, field trip lunches and other activities. We gratefully appreciate the contributions made by these sponsors.

Platinum Sponsors

Geobrugg North America, LLC.
22 Centro Algodones
Algodones, New Mexico 87001
P: (505) 771 4080
Fax: (505) 771 4081
www.geobrugg.com

Geobrugg supplies safety nets and meshes of high-tensile steel wire. Many years of experience and a global network with branches and partners in over 50 countries ensures fast, thorough, and cost-effective solutions for customer requirements. We are partners, consultants, developers, and project managers for our customers.

Geotechnical Contractors

Ameritech Slope Constructors, Inc
PO Box 2702
Asheville, North Carolina 28802
P: 828-633-6352
www.ameritech.pro

Ameritech Slope Constructors, Inc. is a specialty contracting company specializing in Civil/Geotechnical Construction projects, including rock and soil slope stabilization, rock scaling, rock bolting, high strength steel mesh drapes and barriers as well as dry mix shotcrete. We also drill and break large boulders and overhanding ledges using nonexplosive rock removal methods and mechanical rock splitters.

69th ANNUAL HIGHWAY GEOLOGY SYMPOSIUM
Gold Sponsors

BGC Engineering, Inc.
Suite 211
701 - 12th Street
Golden, Colorado
USA, 80401
P: (720)-598-5982
info@bgcengineering.ca

BGC Engineering Inc. is an international applied earth science consulting firm of over 400 engineers, geoscientists, technicians and computer scientists. Our strength is in developing new solutions, and informing and communicating risk-based decisions in complex environments.

Terracon Consultants, Inc.
77 Sundial Avenue, Suite 401W
Manchester, New Hampshire 031031
P: (603) 647-9700
www.terracon.com

Terracon is a national, 100 percent employee-owned, consulting engineering firm providing quality services to clients. Since 1965, Terracon has evolved into a successful multi-discipline firm specializing in:

- Environmental
- Facilities
- Geotechnical
- Materials

69th ANNUAL HIGHWAY GEOLOGY SYMPOSIUM

Silver Sponsors

Access Limited Construction
1102 Pike Lane
Oceano, California 93445
P: (805)-592-2230
www.accesslimitedconstruction.com

Access Limited Construction is a General Contractor located in San Luis Obispo, California. An industry leader, we provide rockfall mitigation, slope stabilization, and difficult drilling services for transportation, energy, mining and private sector clients. With our fleet of Spyder Excavators, we can access steep terrain and hard to reach projects throughout the United States from the East Coast to Hawaii.

Haley & Aldrich, Inc.
75 Washington Avenue, Suite 1A
Portland, Maine 04101
P: 207-482-4607
www.haleyaldrich.com

The first geotechnical engineering firm in New England, **Haley & Aldrich, Inc**. has been working on transportation projects for more than 60 years. We are familiar with the unique subsurface conditions and environmental concerns of the region and we have worked with the region's many transportation agencies.

69th ANNUAL HIGHWAY GEOLOGY SYMPOSIUM

Silver Sponsors (continued)

IDS GeoRadar NORTH AMERICA

IDS GeoRadar
14828 w. 6th Avenue, Unit 12-B
Golden, Colorado 80401
P: (303)-726-6024
www.idsgeoradar.com

IDS GeoRadar manufactures, sells, and services radar instruments. IBIS mine and geohazard monitors, and GPR tools, lead the sector in features, affordability and utility.

Bronze Sponsors

Atlas Pipe Piles

Atlas Pipe Piles
1855 East 122nd Street
Chicago, Illinois 60633
P: (312)-275-1608
www.atlaspipepiles.com

Atlas Pipe Piles manufactures new ERW straight seam steel pipe piling up to 20" diameter and .750" wall that can be found in multiple projects.

69th ANNUAL HIGHWAY GEOLOGY SYMPOSIUM

Bronze Sponsors (continued)

GeoStabilization International®

GeoStabilization International
543 31 Road
Grand Junction, Colorado 81504
P: (855)-579-0536
www.geostabilization.com

GeoStabilization International® is the leading geohazard mitigation firm operating throughout the United States and Canada. Our passion is to develop and install innovative solutions that protect people and infrastructure from the dangers of geohazards. We specialize in emergency landslide repairs, rockfall mitigation, and grouting using design/build and design/build/warranty contracting. GeoStabilization's team includes some of the brightest and most dedicated professionals in the geohazard mitigation industry. Our expertise, proprietary tools, and worldwide partnerships allow us to repair virtually any slope stability or foundation problem in any geologic setting

GOLDER

Golder Associates
670 North Commercial Street, Suite 103
Manchester, New Hampshire 03101-1146
Phone: (603) 668-0880
Fax: (603) 668-1199
www.golder.com

Golder is respected across the globe for providing consulting, design and construction services in our specialist areas of earth and environment. Our highly skilled engineers, scientists, project managers and other technical specialists are committed to helping clients achieve project success on projects around the globe.

69th ANNUAL HIGHWAY GEOLOGY SYMPOSIUM

Bronze Sponsors (continued)

GILSON COMPANY, INC.

Gilson Company, Inc.
PO Box 200
7975 N Central Drive
Lewis Center, Ohio 43035
P: (800)-444-1508
www.globalgilson.com

In 1939, **Gilson** introduced our Testing Screen in response to Mining and Highway Construction industry demands for greater control of material quality. The Testing Screen is still a center piece in most modern aggregate labs, but Gilson has not been standing still. As the company has grown, so has our reputation for superior expertise, innovation, and development of particle size analysis and sample dividing equipment.

HAGER-RICHTER GEOSCIENCE, INC.

Hager-Richter Geoscience, Inc.
8 Industrial Way – D10
Salem, NH 03079
P: (603) 893-9944
www.hager-richter.com

HAGER-RICHTER GEOSCIENCE, INC. is a well-established small business that specializes in Surface and Borehole Geophysics for Engineering applications. The firm has been in business since 1984, has earned a national reputation, and has a nationwide practice. Hager-Richter is headquartered in Salem, New Hampshire and has had a fully staffed and equipped New York/New Jersey Regional Office in New Jersey since 2001. Hager-Richter has extensive experience in providing high resolution geophysical services to support transportation infrastructure projects in the Northeastern, Southeastern, and Midwestern sections of the U.S.

69th ANNUAL HIGHWAY GEOLOGY SYMPOSIUM

Bronze Sponsors (continued)

HI-TECH Rockfall Construction, Inc.
PO Box 674
Forest Grove, Oregon 97116
P: (503)-357-6508
www.hitechrockfall.com

HI-TECH Rockfall is a General Contractor that specializes in Rockfall Mitigation and has been in the industry for over 22 years.

Maccaferri Inc
9210 Corporate Blvd., Suite 220
Rockville, Maryland 20850
P: (240)-818-4116
www.maccaferri.com/us

With over 60 years' experience in rockfall and geohazard mitigation, **Maccaferri** offers a wide range of systems to stabilize rock faces, soil slopes, shallow landslides and debris flow, reducing risk to people, buildings, and infrastructure. We offer a range of engineered systems, certified and tested by leading institutes in accordance with the latest standards. Maccaferri solutions are designed using state-of-the-art modeling software and techniques.

69th ANNUAL HIGHWAY GEOLOGY SYMPOSIUM

Bronze Sponsors (continued)

Precision Blasting Services
6990 Summers Road
Montville, Ohio 44064
P: (440)-823-2263
www.AcademyBlasting.com

Precision Blasting Services is a rock blasting and overbreak control consulting firm, specification writing, and training provider for DOTs, consulting firms, and drill and blast companies.

Rocscience Inc
54 St. Patrick Street
Toronto, Ontario M5T 1V1
P: 416-698-8217
www.rocscience.com

Rocscience is a world leader in developing 2D & 3D software for civil, mining, and geotechnical engineers. For over 20 years, we've used leading-edge research to build geotechnical tools used by over 7,000 engineers around the world for slope stability, excavation design, and geotechnical analysis.

69th ANNUAL HIGHWAY GEOLOGY SYMPOSIUM

Bronze Sponsors (continued)

ROCK ENGINEERING SOLUTIONS

Scarptec, Inc.
PO Box 326
Monument Beach, Massachusetts 02553
P: (603)-361-0397
www.scarptec.com

Scarptec, Inc. provides geological engineering and design solutions for challenging site conditions, with a laser-sharp focus on constructability. Whether it's a deep excavation or falling rock, ice, snow, and soil, Scarptec works collaboratively with our clients and other stakeholders to help mitigate the forces of Mother Nature.

Ameritech
Slope
Constructors
INC
Geotechnical Contractors
Provides:
Rock Scaling, Slope Stabilization Using Soil Nails and High Strength Mesh, Rock Bolting, Rockfall Drape and Barriers, Rock Drains, and Shotcrete.
Contact Information:
Office Address:
21 Overland Blvd.
Asheville, NC 28806
Mailing Address:
P.O. Box 2702
Asheville, NC 28802
Phone: 828-633-6352
Fax: 828-633-6353
Website: www.ameritech.pro

Call Us - (805) 592-2230
1102 Pike Lane, Oceano, California 93445
info@accesslimitedconstruction.com
ACCESS LIMITED
CONSTRUCTION
SPECIALIZED IN:
ROCKFALL MITIGATION
SLOPE STABILIZATION
DIFFICULT DRILLING
www.AccessLimitedConstruction.com

Atlas Pipe Piles
Makers of Straight Seam
ERW Pipe Piles
Chicago, IL
Blytheville, AR
Sizes up to 20" x .750"
Booth # 3
AMERICAN METAL

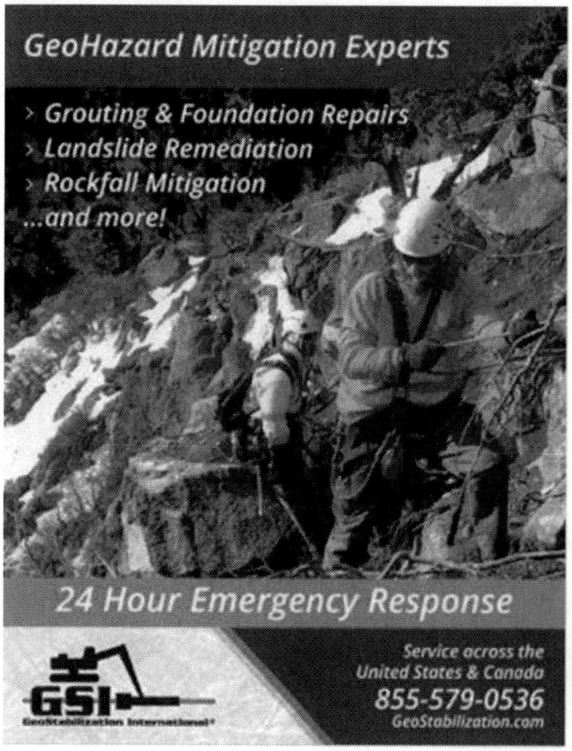
GeoHazard Mitigation Experts
> Grouting & Foundation Repairs
> Landslide Remediation
> Rockfall Mitigation
...and more!
24 Hour Emergency Response
GSI
Service across the
United States & Canada
855-579-0536
GeoStabilization.com

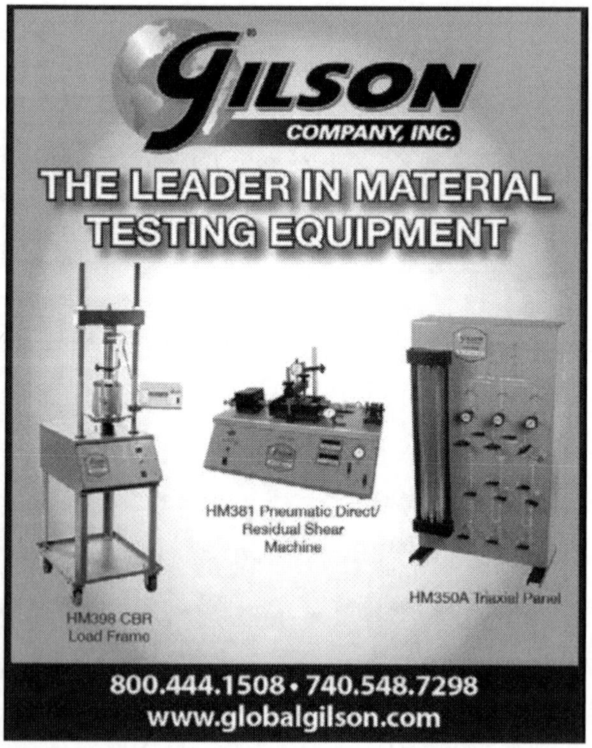
GILSON
COMPANY, INC.
THE LEADER IN MATERIAL
TESTING EQUIPMENT
HM381 Pneumatic Direct/
Residual Shear
Machine
HM398 CBR
Load Frame
HM350A Triaxial Panel
800.444.1508 • 740.548.7298
www.globalgilson.com

SCARP LLC
ROCK ENGINEERING SOLUTIONS
*Difficult Access Rockfall Mitigation Solutions
*Ice and Falling Debris Hazard Evaluations
*Rock Slope Engineering Services During Construction
*Anchorage Design and Testing
*Slope Stabilization Feasibility and Design

69th ANNUAL HIGHWAY GEOLOGY SYMPOSIUM
Exhibitor Display Locations

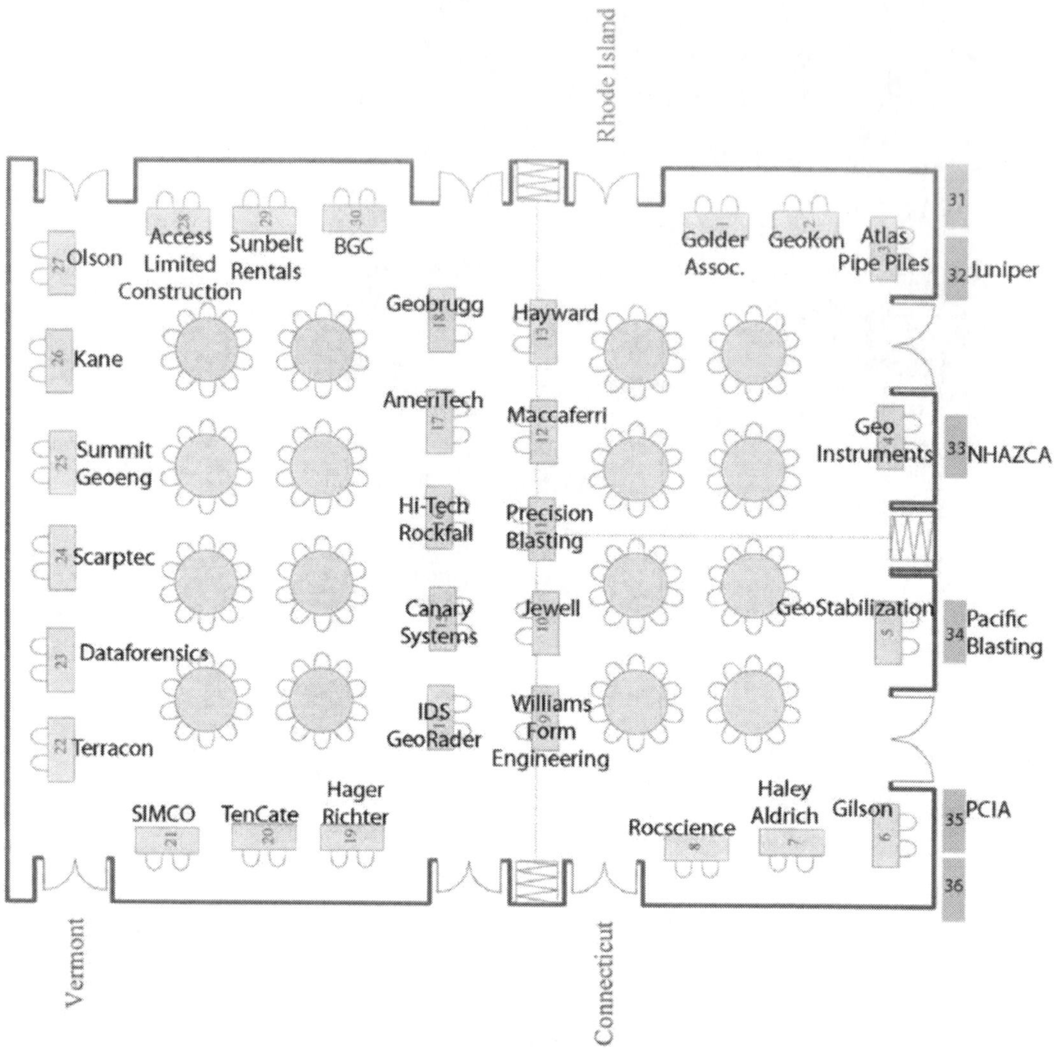

69th ANNUAL HIGHWAY GEOLOGY SYMPOSIUM
Exhibitors

Thank you to all participating exhibitors. The exhibit booths are in the Vermont/Connecticut/Rhode Island rooms.

ACCESS LIMITED CONSTRUCTION

Access Limited Construction
1102 Pike Lane
Oceano, California 93445
P: (805)-592-2230
www.accesslimitedconstruction.com

Ameritech Slope Constructors INC
Geotechnical Contractors

Ameritech Slope Constructors, Inc.
PO Box 2702
Asheville NC 28802
P: (828) 633-6352
Fax: (828) 398-2041
www.ameritech.pro

Atlas Pipe Piles

Atlas Pipe Piles
1855 East 122nd Street
Chicago, Illinois 60633
P: (312)-275-1608
www.atlaspipepiles.com

BGC BGC ENGINEERING INC.
AN APPLIED EARTH SCIENCES COMPANY

BGC Engineering, Inc.
Suite 211
701 - 12th Street
Golden, Colorado
USA, 80401
Tel: (720)-598-5982
info@bgcengineering.ca

CANARY SYSTEMS®

Canary Systems, Inc.
5 Gould Road
New London, New Hampshire 03257
P: 603-526-9800
www.canarysystems.com

69th ANNUAL HIGHWAY GEOLOGY SYMPOSIUM
Exhibitors

Dataforensics
2310 Parklake Drive, Suite 525
Atlanta, Georgia 30345
P: 678-406-0106
www.dataforensics.net

Geobrugg North America, LLC.
22 Centro Algodones
Algodones, New Mexico 87001
P: (505) 771 4080
Fax: (505) 771 4081
www.geobrugg.com

Geo-Instruments
24B Celestial Drive
Narragansett, Rhode Island 02882
P: 800-477-2506
www.geo-instruments.com

GeoKon, Inc.
48 Speicer Street
Lebanon, New Hampshire 03766
P: 603-448-3216
www.geokon.com

GeoStabilization International
543 31 Road
Grand Junction, Colorado 81504
P: (855)-579-0536
www.geostabilization.com

Gilson Company, Inc.
PO Box 200
7975 N Central Drive
Lewis Center, Ohio 43035
P: (800)-444-1508
www.globalgilson.com

69th ANNUAL HIGHWAY GEOLOGY SYMPOSIUM
Exhibitors

GOLDER

Golder Associates
670 North Commercial Street, Suite 103
Manchester, New Hampshire 03101-1146
Phone: (603) 668-0880
Fax: (603) 668-1199
www.golder.com

HAGER-RICHTER GEOSCIENCE, INC.

Hager-Richter Geoscience, Inc.
8 Industrial Way – D10
Salem, NH 03079
P: (603) 893-9944
www.hager-richter.com

HALEY ALDRICH

Haley & Aldrich, Inc.
75 Washington Avenue, Suite 1A
Portland, Maine 04101
P: 207-482-4607
www.haleyaldrich.com

HAYWARD BAKER
A KELLER COMPANY

Hayward Baker
7550 Teague Road, Suite 300
Hanover, Maryland 21076
P: 410-551-8200
www.haywardbaker.com

HI-TECH ROCKFALL CONSTRUCTION
ROCKS

Hi-Tech Rockfall Construction, Inc.
PO Box 674
Forest Grove, Oregon 97116
P: (503) 357-6508
Fax: (503) 357-7323
www.hitechrockfall.com

IDS GeoRadar
NORTH AMERICA

IDS GeoRadar
14828 w. 6th Avenue, Unit 12-B
Golden, Colorado 80401
P: (303)-726-6024
www.idsgeoradar.com

69th ANNUAL HIGHWAY GEOLOGY SYMPOSIUM
Exhibitors

Jewell Instruments
850 Perimeter Road
Manchester, New Hampshire 03103
P: 603-621-6034
www.jewellinstruments.com

Juniper Unmanned
15000 W 6th Ave, Suite 330
Golden, Colorado 80401
P: 720-440-9960
www.JuniperUnmanned.com

KANE GeoTech, Inc.
7400 Shoreline Drive, Suite 6
Stockton, California 95219
P: 209-472-1822
www.kanegeotech.com

Maccaferri Inc
9210 Corporate Blvd., Suite 220
Rockville, Maryland 20850
P: (240)-818-4116
www.maccaferri.com/us

NHAZCA S.r.l
Via Vittorio Bachelet 12
Rome, Italy 00185
P: +39-0695065820
www.nhazca.it

69th ANNUAL HIGHWAY GEOLOGY SYMPOSIUM

Exhibitors

Olson Engineering, Inc.
12401 W. 49th Avenue
Wheat Ridge, Colorado 80033
P: 303-423-1212
www.olsonengineering.com

Pacific Blasting
3183 Norland Avenue
Burnaby, British Columbia V5B 3A9
P: 604-291-1255
www.pacificblasting.com

Professional Climbing Instructors Association
PO Box 121
Orono, Maine 04473
P: 207-866-7562
www.pcia.us

Precision Blasting Services
6990 Summers Road
Montville, Ohio 44064
P: (440)-823-2263
www.AcademyBlasting.com

Rocscience Inc
54 St. Patrick Street
Toronto, Ontario M5T 1V1
P: 416-698-8217
www.rocscience.com

Scarptec, Inc.
PO Box 326
Monument Beach, Massachusetts 02553
P: (603)-361-0397
www.scarptec.com

69th ANNUAL HIGHWAY GEOLOGY SYMPOSIUM

Exhibitors

SIMCO Drilling Equipment, Inc.
802 Furnas Drive
Osceola, Iowa 50213
P: 641-342-2166
www.simcodrill.com

Summit Geoengineering
P.O. Box 7216
Lewiston, Maine 04243
P: 207-576-3313
www.summitgeoeng.com

Sunbelt Rentals
6770 Dorsey Road
Elkridge, Maryland 21075
P: 410-379-2800
www.mabey.com

TenCate Geosynthetics
365 South Holland Drive
Pendergrass, Georgia 30567
P: 706-693-1803
www.tencategeo.com

Terracon Consultants, Inc
77 Sundial Avenue, Suite 401W
Manchester, New Hampshire 03103
P: 603-647-9700
www.terracon.com

Williams Form Engineering Corp.
8165 Graphic Drive
Belmont, Michigan 49306
P: 616-866-0815
www.williamsform.com

69th ANNUAL HIGHWAY GEOLOGY SYMPOSIUM
Holiday Inn By The Bay Hotel and Convention Center
Floor Plan

LOBBY & CONVENTION HALL LEVEL

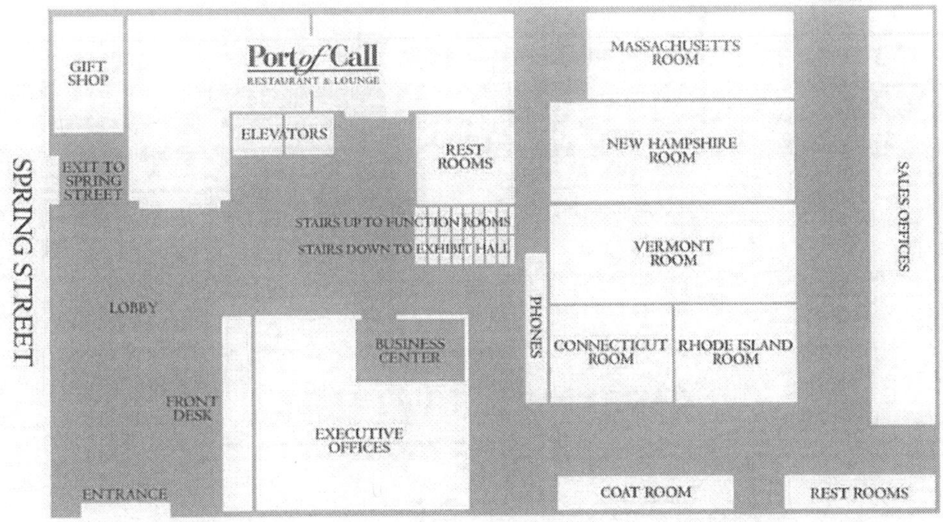

Highway Geology Symposium

Thank you for attending the 69th

Highway Geology S*ymposium*

69th ANNUAL HIGHWAY GEOLOGY SYMPOSIUM

Papers

Assessment and Mitigation Process for Bridge Foundation Reuse – Case Study

James Arthurs

Federal Highway Administration – Central Federal Lands Highway Division, 12300 W. Dakota Ave., Lakewood, CO 80228, (720) 963-3633, james.arthurs@dot.gov

Khamis Haramy

Federal Highway Administration – Central Federal Lands Highway Division, 12300 W. Dakota Ave., Lakewood, CO 80228, (720) 963-3521, khamis.haramy@dot.gov

Prepared for the 69th Highway Geology Symposium, September, 2018

Acknowledgements

The authors wish to thank Mr. Frank Jalinoos from Federal Highway Administration (FHWA) Turner-Fairbanks Highway Research Center for providing resources to make this study possible; Mr. Ken Southall, FHWA, Federal Lands Highway (FLH) project engineer, for his record of observations and photos during construction; and Mr. Karl Eikerman, FHWA, FLH bridge engineer, for help in evaluating the bridge structural capacity.

Disclaimer

Statements and views presented in this paper are strictly those of the authors and do not necessarily reflect positions held by their affiliations, the Highway Geology Symposium (HGS), or others acknowledged above. The mention of trade names for commercial products does not imply the approval of endorsement by HGS.

ABSTRACT

Aging bridges and transportation infrastructure in the U.S. present a challenge to Federal and State DOTs, local governments, and other transportation agencies. Reuse of bridge foundations is a viable option for bridge replacement and rehabilitation efforts. Foundation reuse may be a practical technique to reduce project costs, schedule, and environmental and mobility impacts; however, this is not without challenges, including uncertainties in existing foundation geometry and conditions, remaining service life of the substructure elements, and load carrying capacity. These challenges can be managed with appropriate investigations and mitigative efforts.

The Willow Valley Creek Bridge located on County Road 209, southeast of Flagstaff in Coconino County, New Mexico was investigated, designed and constructed with reused foundations. The bridge, constructed in 1934 and widened in 1964, was a 104-foot long, 34-foot wide, 3-span structure supported on mortared cut-stone masonry abutments and piers founded on shallow foundations. Original bridge and as-built plans of the widening project are available. Core sampling of the structure supporting elements and foundation strata, and comprehensive non-destructive borehole and surface geophysical testing was conducted to evaluate the foundation system for reuse. Investigations indicated the bridge was founded on limestone bedrock; areas of potential voids or weak material were noted in the substructure supporting elements. To address concerns about these deficiencies, a grouting program was implemented to improve the strength of the bridge substructure. Grouting was monitored during construction to ensure the quality of the construction.

This manuscript discusses the foundation assessment, mitigation plan, and construction methods employed to meet design requirements for reuse.

INTRODUCTION

Bridge infrastructure in the United States is gradually deteriorating, prompting a high demand for bridge replacement while total transportation funding is limited. Bridge substructure and foundation elements typically comprise a significant fraction of bridge construction costs (*1*). Therefore, the potential for reusing bridge foundation elements will substantially reduce costs relative to a conventional bridge replacement project.

The key factors influencing the decision to reuse bridge foundations are economics, environment, and construction schedule. Specifically, reuse of bridge foundation elements can reduce project costs, environmental impacts, and road closures, and can expedite construction. However, certain conditions must be met prior to implementing foundation reuse. Existing foundations must provide sufficient capacity to resist applied bridge loads. Determining the load carrying capacity of an existing bridge typically requires access to the original bridge plans, geotechnical and structural design calculations, subsurface information, and as-built construction drawings. Also, the proposed bridge deck geometry must be relatively similar to the original bridge. Finally, the existing condition and remaining service life of the bridge components to be reused must meet the AASHTO LRFD requirements and design life of the replacement bridge. Availability of bridge inspection reports and detailed assessment of the bridge structure also helps during the evaluation process.

For many Federal Lands Highway projects, full road closures to replace bridges are not typically feasible. Many of these roads represent a single access route to an area and occasionally no detour is available or if available could be very long. In both cases, requirements for public and emergency vehicle access may dictate that temporary crossings and detours be constructed to provide continuous access during construction. Construction of temporary detours is costly, both in economic and environmental terms. Reuse of bridge foundations reduces road closure times and requirements for temporary detours.

Willow Valley Creek Bridge

The Willow Valley Creek Bridge was identified as a candidate for foundation reuse and is located on the Lake Mary Road (Coconino County Road 209), approximately 40 miles southeast of Flagstaff and 50 miles north of Payson, Arizona (latitude 34° 38' 11.8" North and longitude 111° 21' 38.6" West, Figure 1). The FHWA is also rehabilitating a segment of this low volume road that provide access to private residences, Lake Mary and Mormon Lake within the Coconino National Forest, and regional commercial traffic including logging trucks. Due to the high elevation and rolling terrain, this road is popular with bicyclists and runners, with several bicycling events held each year. The

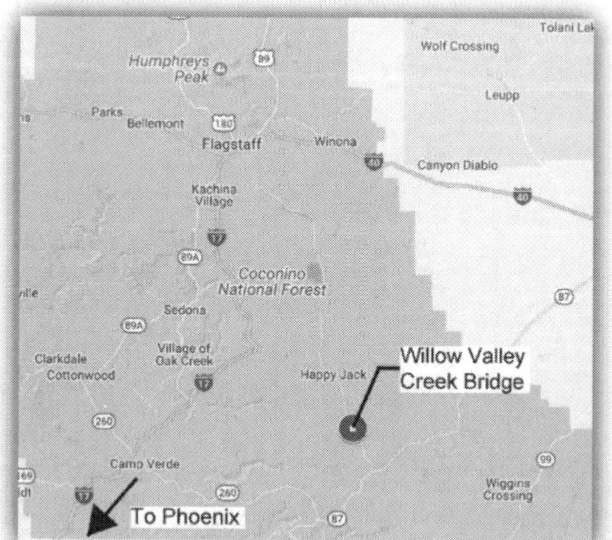

Figure 1 – Willow Valley Creek bridge location

purpose of the bridge rehabilitation was to widen the roadway and bridge to provide 5-foot shoulders along each side to improve the entire corridor for multimodal use, which may create an opportunity for the route to become a world-class training and event course. The rehabilitation plan also emphasized retaining the existing stone masonry piers and abutments and provide a steel girder superstructure that retains the same visual appearance as the existing bridge, blending well within the forest environment of the project site.

The 3-span bridge, constructed in 1934 and later widened in 1964, is 104-feet long by 34-feet wide providing a total 31-foot clear width. The bearing-to-bearing distances for the center and exterior spans are 50 feet and 25 feet, respectively. Constructed as near vertical gravity walls of cut-stone masonry, the two bridge piers and abutments are founded on spread footings bearing on bedrock. The wing-walls extend about 10 feet at 45 degrees from the abutment walls. Pier height ranged from 14 to 19 feet and abutments between 13 and 24 feet (Figure 2).

The plans depict approximately 4 feet of soil above bedrock at the bridge piers and 15 feet of soil above rock at the abutments. Due to variance in overburden/rock interface and scour depth, the actual embedded depths of each pier and abutment also varied along the lengths.

Figure 2 –Willow Valley Creek Bridge (left) and close up photo of bridge pier (right); note the near vertical joint near middle of photo separating the original bridge structure on the right from the 1964 widening on the left.

Geological Setting

The Willow Valley Creek Bridge is located southeast of Flagstaff, Arizona, within the Mormon Lake Area of the Mongollon Slope Section, Colorado Plateau Province (*2*). The site is mapped as underlain by recent alluvial soils and Permian Kaibab Formation which includes silty to sandy dolomite, limestone, and fine-grained sandstone (*3*). Tertiary basaltic rocks are also mapped as outcropping near the project area. The project site is located near the Mormon Mountain Anticline, a broad regional structure whose east flank is characterized by low northeasterly dips.

As described in the "Evaluation" section below, the bridge site is underlain by sandy clay with local gravel and cobble residual and alluvial soils. These soils are typically medium stiff, moist, and moderately plastic. Underlying these surficial soils are interbedded limestone and calcareous shale bedrock, interpreted to belong to the Kaibab Formation.

PROPOSED REHABILITATION

This project proposed bridge widening with additional 5-foot wide shoulders on each side of the bridge for pedestrian and cyclist use. Widening across the bridge would therefore include complete replacement of the superstructure. The proposed replacement consists of an approximately 107-foot long, 37-foot wide new bridge structure. This would provide a clear width of 34 feet with two 12-foot lanes, each with 5-foot paved shoulders.

Due to the relatively limited widening required and desire to reduce the bridge closure time, foundation reuse was proposed for this project. To verify that the bridge substructure and foundations have sufficient capacity to support the added superstructure loads, a detailed conventional boring and nondestructive investigation program was deployed to evaluate the condition of the existing structure and the soil and rock supporting it. The investigations and evaluation are presented below.

FIELD INVESTIGATIONS

Conventional Geotechnical Methods

To ensure a stable and safe structural design and to characterize the integrity of the bridge substructure walls and foundations for reuse, the existing foundation was investigated by advancing ten boreholes within or near the existing bridge structure. Two boreholes were advanced in the roadway behind the bridge abutments using a combination of augur drilling and HQ-wireline coring. The other eight boreholes (B1 through B8) shown in Figure 3 were advanced from the bridge deck with two borings at each foundation element at locations where gaps between the girders exist and into the supporting geological strata using HQ-wireline coring.

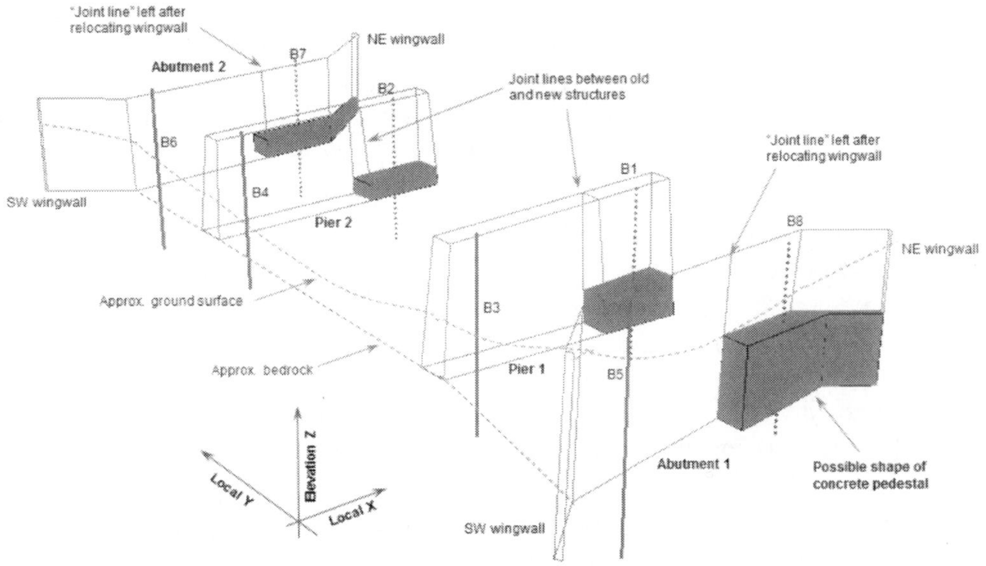

Figure 3 – Borehole locations

Continuous concrete and rock core samples of the bridge sub-structural elements and bedrock were collected and characterized by visual classification in the field. Percent core recovery and Rock Quality Designation (RQD) were determined in the field for each core run.

Unconfined compressive strength (UCS) tests were conducted on select core samples retrieved from the limestone, concrete, and masonry materials obtained from the boreholes.

Non-destructive Geophysical Methods

Various nondestructive testing methods including borehole-geophysical logging, three-dimensional seismic tomography, and ground penetrating radar surveys were used to image the interior of the substructure. Boreholes through the existing pier and abutment cut-stone masonry walls and the foundation bedrock (B1 through B8) were logged with various downhole geophysical methods (*4*). Specifically, the purpose of the downhole geophysical logging is to characterize the masonry and concrete integrity near the drill hole walls by detecting fractures, cracks, and defects within the walls and foundation elements, as well as to investigate the limestone bedrock density and competency. Downhole methods included caliper, compensated density, electric log, natural gamma, optical televiewer, full waveform sonic, and velocity (compression and shear) measurements.

Other nondestructive testing techniques conducted through the sub-structural elements were also used to characterize the bridge structural elements and provide additional information about the condition of the foundation materials (*5*, *6*, *7*). These techniques included sonic echo/impulse response, ultra-seismic, spectral analysis of surface waves, multichannel reflection survey, impact echo, ground penetrating radar (GPR), sonic pulse velocity, electrical resistivity imaging, seismic refraction tomography, and multichannel analysis of surface wave (MASW). Three-dimensional tomographic imaging using direct and reflected seismic waves was completed to evaluate foundation element integrity and determine the volume of areas that will require grouting at both abutments and piers of the bridge.

FINDINGS

Boreholes advanced behind the bridge abutments, encountered fill and natural sandy clay soils. These soils were generally medium stiff, moist, and medium plastic. Gravel, cobble, and boulders were encountered locally. The soils were underlain by limestone and calcareous shale bedrock extending to the maximum depths explored. The bedrock is medium strong to strong, based on uniaxial compressive strength laboratory testing. Core recovered from the bedrock indicated rock quality designation (RQD) values between 80 and 100, and unconfined compressive strength (UCS) between 5,400 and 9,200 pounds per square inch (psi). Based on these measurements, and information collected from the as-built plans, the geotechnical capacity of the foundation was deemed to be sufficient to support the proposed bridge loads.

Boreholes advanced through the bridge structure encountered asphalt pavement, concrete, and stone masonry of the existing bridge structure, deck, and pavement. These artificial materials were underlain by similar limestone and shale bedrock, confirming that the bridge structure is supported by spread footings bearing on bedrock. Core recovered from the stone masonry had RQD values between 16 and 100, with lower values generally encountered in the original portion of the structure. UCS testing on stone/mortar samples indicated intact strengths between 4,300 and 7,900 psi. Various anomalies, interpreted to be either weak masonry/mortar or void space, were identified in the bridge abutments and piers from the 3-D geophysical data. Void space was also recorded using the optical televiewer (Figure 4). These voids were interpreted to be related to the original construction methods. Results of the borehole geophysics were used to inform and calibrate the 3-D tomography and set the velocity threshold for void areas. In general, the borehole geophysics indicated that the masonry and concrete within the newer bridge section is more competent with less voids than that in the older section. Both compressional and shear waves measured higher velocities within the newer section, also indicating more competent materials.

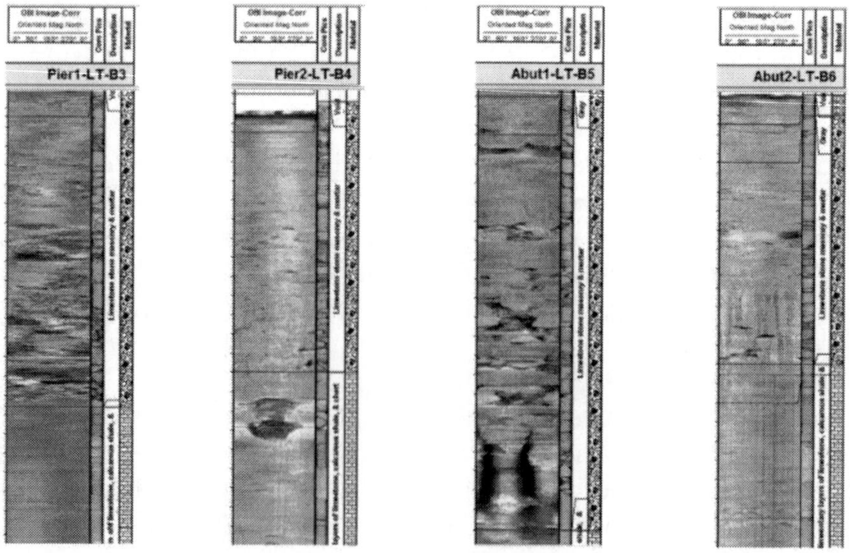

Figure 4 – Example downhole image showing void space within the mortared cut-stone masonry B3-B4-B5-B6. Old section (B3 and B5) versus newer section (B4 and B6).

The ground penetrating radar test method was used on the lower sides of both piers and both abutments, testing horizontally into the structures to look for unusual features. There was also a single GPR scan taken vertically into the soil next to Pier 1, as well as two vertical scans from the roadway deck top taken along the centerlines of Piers 1 and 2. All GPR scans were done with a 400 MHz antenna to allow for penetration through the entire horizontal width of the piers and abutments coupled with relatively good resolution of features. A distinct change in the reflections from the interior of the pier at about 22 feet from the start of the scan was observed in the data. This change occurred where the original pier meets the newer portion of the pier and is likely representative of the various stone and mortar interfaces in the pier. Although GPR surveys were used at this site this type of data was not useful in determining bridge subsurface element integrity and in developing a plan for improving the structural integrity of the bridge elements.

The 3-D volumetric distribution of seismic velocity within each foundation element, reconstructed through 3-D tomographic inversion contours using the measured travel times and distances between sources and receivers. Figure 5 shows an example of an isometric projection of generated velocity contours at a cross-section (tomogram) at the center-line through a reconstructed velocity distribution image for Abutment 1.

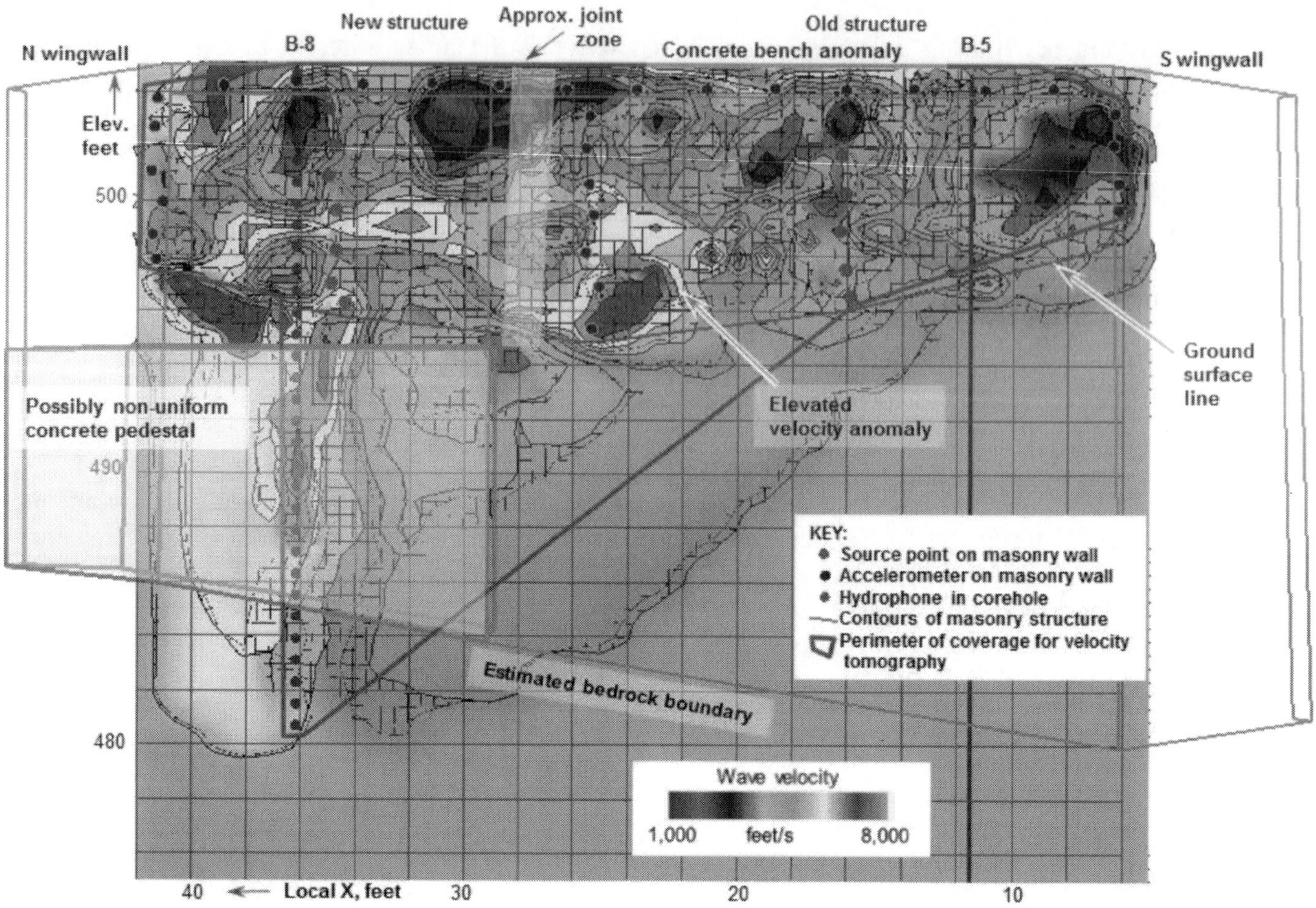

Figure 5 – Tomograph combined with the volumetric contour image of velocity distribution reconstructed along Abutment 1.

The color velocity distribution was obtained from measurements at a series of acoustic transmitters and receivers placed along the vertical walls of the structures above the ground level. In Figure 5, cooler colors (i.e. purple/blue) indicate areas with lower velocities, interpreted as areas of weak/voided space within the structural elements and were utilized to guide the development of the foundation repair plan. Note that as shown on Figure 5, limitations in the tomography method controlled which portions of the foundation elements were imaged.

Based on all geophysical investigation and conventional coring results, a velocity below 3,000 feet per second (ft/s) was considered as a general indicator of deficiency for structural integrity. Figure 6 shows plan and profile views of the deficient zones (velocity < 3,000 ft/s) and their approximate volume. The total volume for Pier 1 was estimated at 292 cubic feet (cu ft) significantly larger than Pier 2 which was estimated at 68.9 cu ft. While significantly smaller than for Pier 1, the anomalies at Pier 2 still tend to occupy the central part of the Pier profile, with much larger anomalies in the old structure, and with the largest anomaly near the top of the structure expanding toward the west wall. Abutments 1 and 2 indicated much smaller volume of deficient masonry with about 40 and 11 cu ft, respectively. Due to both the original construction methods and the limitations of the investigation methods, the contrast in velocity between parts of the structure, rather than the absolute value of velocity, were emphasized in assessments of structural integrity.

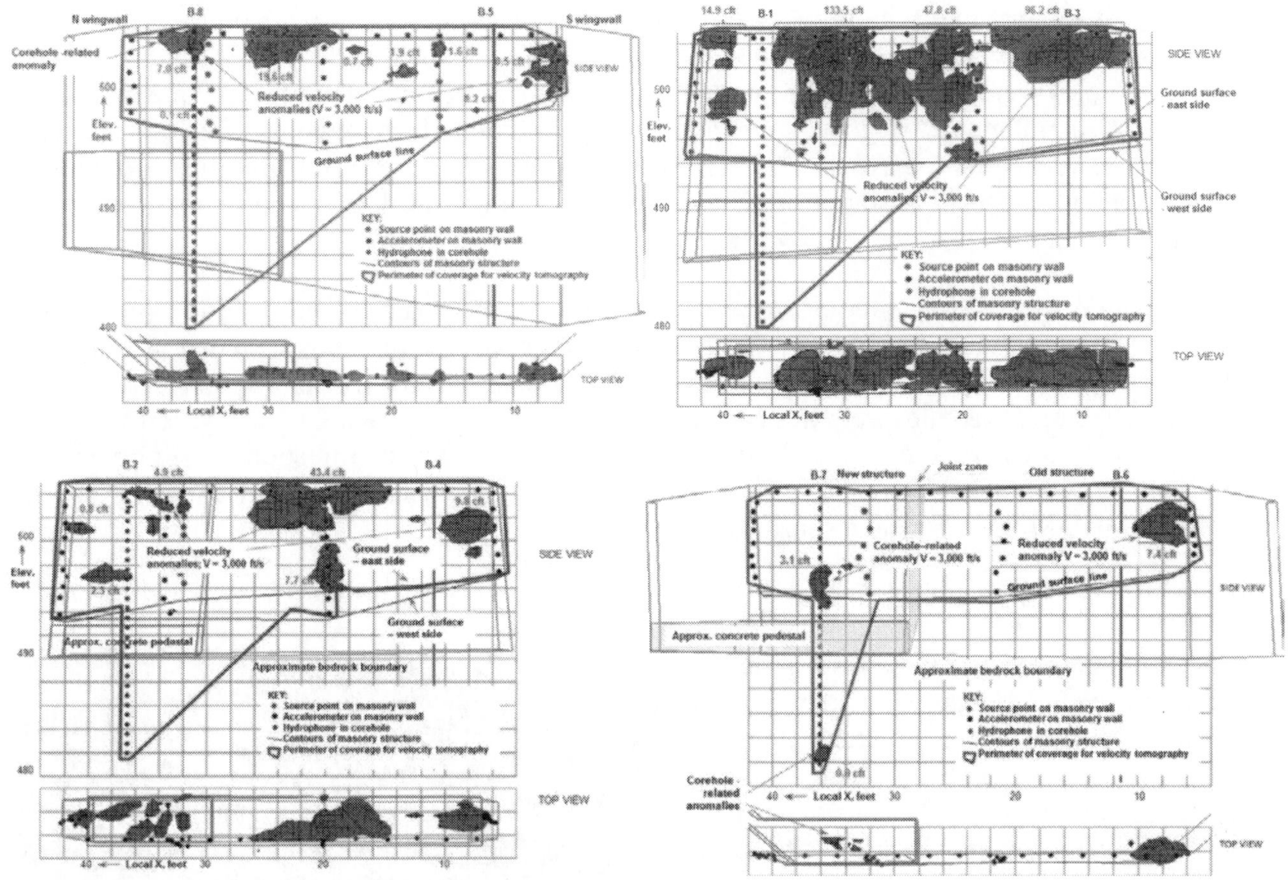

Figure 6 – Tomographic representation of possible compromised structural integrity volumes for Abutment 1 (upper left), Pier 1 (upper right), Pier 2 (lower left) and Abutment 2 (lower right)

STRUCTURAL REPAIR

Analysis of geotechnical investigations and the three-dimensional seismic tomographic imaging of the foundation elements indicated that the masonry structures are in generally good condition for substructure elements/foundation reuse and should be able to support the higher live loads as required by AASHTO with minor improvements through grouting. Caliper data from downhole geophysics generally indicated higher estimated grout volume than the three dimensional tomographic images. This data represents more accurate spot location estimates of volume, but it is projected across a significantly larger area without consideration of lateral or vertical variability in the concrete. Therefore, relying on caliper data alone may result in an unrealistically high estimation of the required grout volume compared to the total percent of concrete in each abutment or pier.

Grouting Plan

Due to weak areas and void space identified by the geotechnical and geophysical methods, a grouting mitigation program was developed and incorporated into the project plans to improve the overall integrity and stability of the bridge substructure. The grout volume required was calculated based on the aforementioned results of the 3-D tomography survey (Figure 6). Grout injection holes were strategically located at asymmetrical intervals along the abutments and piers to maximize potential for grout to penetrate the interpreted weak/void spaces. An example injection hole layout for Abutment 1 is illustrated in Figure 8. The grouting program was recommended under the assumption that the bridge deck and girders would be removed prior to grout placement.

Construction specifications required the contractor to monitor and record drilling conditions and grout injection volumes per hole. Refusal criteria for grout injection was included in the contract documents and specified maximum values for: grout returned to surface, displacement of bridge structure facing, injection pressure, and grout volume. The contractor was required to evaluate these criteria at each stage of grout injection. It was critical that the injection was pumped at low pressure to avoid deflection of the substructure elements. Displacement monitoring was essential to assure stability of the grouted substructure elements.

Execution

The contractor elected to perform grouting operations prior to removal of the existing bridge deck and super structure elements as shown in Figure 7. The original injection hole layout only considered how to best fill the potential voids with grout and did not consider the location of various superstructure members with respect to hole location. Therefore, many of the injection points required adjustment during construction. Hole relocations were kept to the minimum feasible to avoid the steel girders of the bridge structure.

Figure 7 – Injection grouting at Abutment 1.

Initial injection points had very high grout takes, with injection typically being terminated by reaching the specified maximum volume criteria. This raised concerns for the contractor and the FHWA construction staff that the actual quantity placed would be considerably higher than the estimated bid quantity. As the grout injection progressed, the volume of grout injected at subsequent, adjacent injection points was lower than the anticipated values. This was interpreted to be related to communication of grout between the injection points.

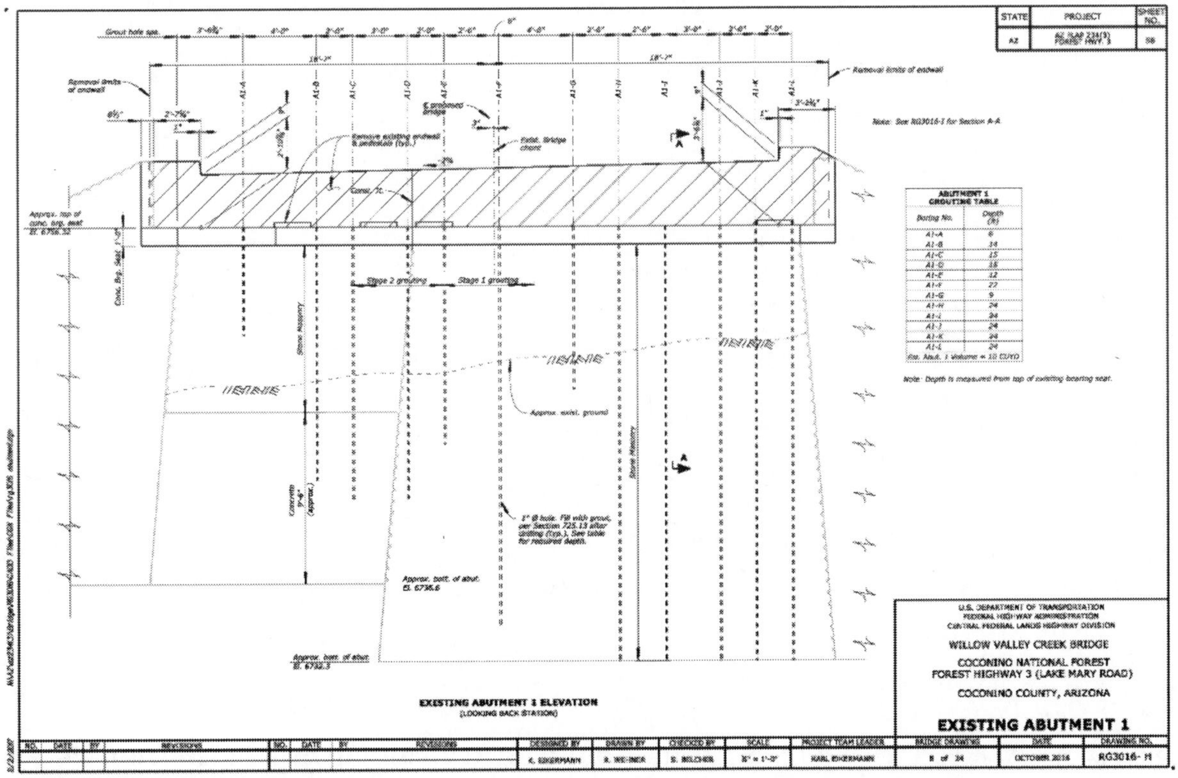

Figure 8 – Abutment 1 (East) grouting plan.

Three example grout injection logs are presented in Figure 9. These logs represent grout injection at three holes located on along the Abutment 1 substructure element. Of these three points, A1-D was injected first, followed by A1-F, and then A1-G. Note that as injection proceeded to subsequent points, the total injection quantity decreased. This indicates substantial interconnectedness between the voids in the abutment. Communication between grout injection points was not anticipated by FHWA, and concerns were raised that grout was potentially migrating out of the bridge structure, either through the face of the walls, into the abutment fill, or into the foundation strata, raising a concern about contract grout quantity overruns. The contractor continued to place grout and monitor the bridge structure for movement and any signs of seepage. No seepage of grout was noted in the wall face or in the ground near the bridge. For the first several holes, the maximum grout volume was the limiting refusal criteria. After several holes in Abutment 2 were grouted in this manner, the following holes took considerably less grout. For these injection points, maximum pressure was the limiting refusal criteria. Figure 10 shows the estimated and actual total volume of grout placed at each injection point in Abutment 1. Although a large discrepancy between the estimated and actual grout volumes existed at each point, the total estimated and actual grout volumes for Abutment 1 are almost identical: 266 cubic feet and 264 cubic feet, respectively. A similar pattern continued for the other structural elements. The total project grout volumes were 657 cubic feet (approximately 25 cubic yards) estimated and 560 cubic feet (approximately 21 cubic yards) placed mostly attributed to interconnectivity of the void spaces within each structural element.

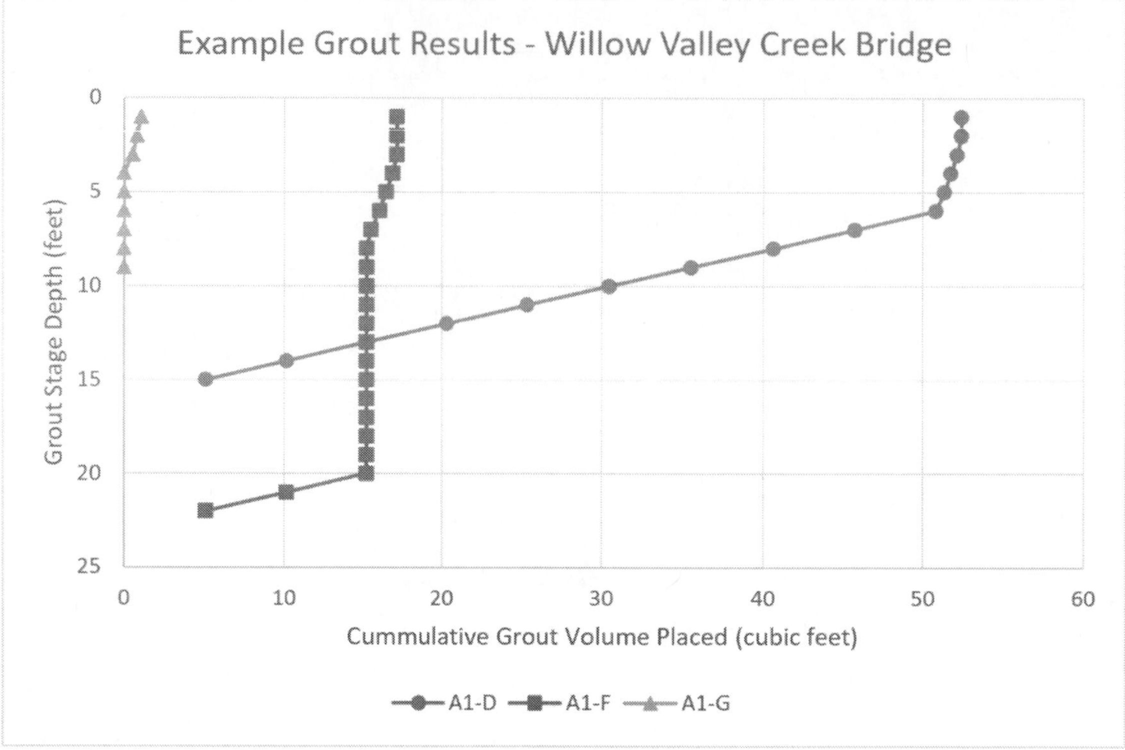

Figure 9 – Example grouting logs from Abutment 1. Grouting proceeded from Hole A1-D, then A1-F, and last, A1-G. Note that total grout volume decreases as the program progresses.

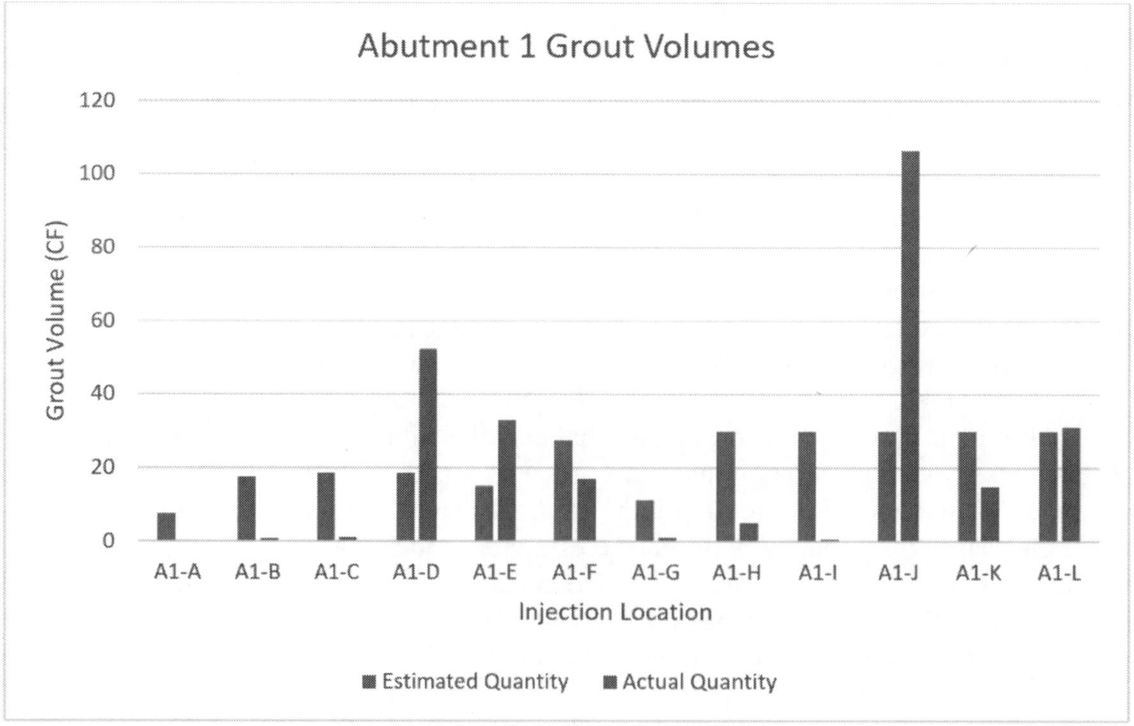

Figure 10 – Total volumes of grout placed at each injection point in Abutment 1.

Once the grouting was successfully completed and cured, the bridge deck was removed and new caps were installed on top of the existing abutment and pier walls. Steel girders were installed and deck was placed as shown in Figure 11. Staged construction was used to maintain public access across the bridge during the deck replacement rather than temporary detours or alternate routes.

Figure 11 – Bridge construction photos.

SUMMARY AND CONCLUSIONS

The bridge foundations were investigated for reuse and deemed sufficient to safely support the proposed structure loads in accordance with the AASHTO guidelines with minor repairs. By reusing the existing foundations an approximate savings of 25 to 30 percent of the total actual construction cost was achieved. In addition, indirect costs by reducing the construction duration, improving traffic control, preserving historic cut masonry stone walls, and improving productivity were also realized. This project was deemed successful.

Three distinct methods of investigation, namely, rock coring, downhole geophysics, and 3-D seismic tomography of the existing bridge structure were particularly useful in guiding decision making for foundation reuse and developing a mitigation program. The combination of these methods provided the most reliable indication of void space within the bridge substructure. Rock coring and downhole geophysics provided information useful to selecting an appropriate critical velocity for interpretation of the tomography data. Developing estimated grout volumes from these exploration results was challenging, requiring multiple iterations and input from several geotechnical professionals within FHWA. In the end, the estimated contract quantity of 25 cubic yards was only 4 cubic yards greater than the placed grout volume.

Logs of the grouting operation were essential to evaluate the efficacy of the grouting program. Records of the grouting operation included drilling logs, which were useful for identifying actual void areas encountered, and grouting logs, which recorded grout take at each interval and identified where grout was placed. The grouting logs indicated significant connectivity between the injection points that was not originally anticipated in the design. During placement of grout in the first few injection points, FHWA field personnel were concerned that the contract quantity may have been significantly underestimated. As previously discussed, grout take at subsequent injection points was considerably lower, with total volumes at each structure element relatively close to the contract amount.

REFERENCES

1. Boeckmann, A.Z. and J.E. Loehr. *Current Practices and Guidelines for the Reuse of Bridge Foundations*. Project 20-05, Topic 47-03. National Cooperative Highway Reasearch Program (NCHRP), Transportation Research Board (TRB). 2017.
2. Trapp, R.A. and S.J. Reynolds. *Map Showing Names and Outlines of Physiographic Areas in Arizona used by the Arizona Geological Survey with Comprehensive Base Map*. Open File Report OFR 95-2a. Arizona Geological Survey, 1995.
3. Weir, G.W., G.E. Ulrich., and L.D. Nealey. *Geologic Map of the Sedona 30' X 60' Quadrangle, Yavapai and Coconino Counties, Arizona*. Miscellaneous Investigations Series Map I-1896. United States Geological Survey, 1989.
4. Colog Borehole Geophysics/Hydrophysics. *Willow Valley Bridge* (unpublished field data). 2014.
5. Knox, J.M., D. Sack, J. Sheehan, P. Sirles, and L.D. Olson. *Nondestructive Evaluation and Geophysical Investigation Report, Lake Mary Road – Willow Valley Bridge Rehabilitation Project, FHWA Re-use of Existing Foundations Research Project DTFH61-14-D-00010 TOPR 1: Characterization of Bridge Foundation (CBF) Research Program, Near Flagstaff, Arizona*. Olson Engineering, unpublished report, Job No. 4811A, 2015.
6. Descour, J.M. *Seismic Evaluation of Foundation Integrity, Willow Valley Bridge, Lake Mary Road, Coconino County, AZ*. C-Thru Ground, Inc., unpublished report, 2015.
7. Sitek, M.A., C. Bojanowski, and S. Lottes. *Numerical Modeling of Lake Mary Bridge*. Transportation Research and Analysis Computing Center, Energy Systems Division, Argonne National Laboratory, 2015.

Rock Slope Failure Mode Analysis and Mitigation Using LiDAR Data in Fairlee, Vermont

Erik D. Friede, E.I.T
GZA GeoEnvironmental, Inc.
477 Congress Street
Suite 700
Portland, Maine 04101
(207) 358-5131
Erik.Friede@gza.com

Andrew R. Blaisdell, P.E.
GZA GeoEnvironmental, Inc.
477 Congress Street
Suite 700
Portland, Maine 04101
(207) 358-5117
Andrew.Blaisdell@gza.com

Christopher L. Snow, P.E.
GZA GeoEnvironmental, Inc.
477 Congress Street
Suite 700
Portland, Maine 04101
(207) 358-5117
Christopher.Snow@gza.com

Prepared for the 69th Highway Geology Symposium, September, 2018

Acknowledgements

The authors would like to thank the following individuals/entities for their contributions to this project:

Ethan Thomas – Vermont Agency of Transportation
Callie Ewald – Vermont Agency of Transportation
Michael Carter – Doucet Survey, Inc.
David Lamothe – GZA GeoEnvironmental, Inc.

Disclaimer

Statements and views presented in this paper are strictly those of the authors(s), and do not necessarily reflect positions held by their affiliations, the Highway Geology Symposium, or others acknowledged above. The mention of trade names for commercial products does not imply the approval or endorsement by HGS.

ABSTRACT

In May 2017, the Vermont Agency of Transportation (VTrans) asked GZA GeoEnvironmental, Inc. (GZA) to assess a recent rockfall that had left a mass of failed material above Interstate Route 91 (I-91) in Fairlee, Vermont. VTrans' initial assessment identified the risk that release of the failed material could overwhelm the existing catchment and impact the interstate highway.

GZA developed a comprehensive assessment that included LiDAR survey and hand measurements to characterize the failure area. The LiDAR data allowed characterization of the scarp behind and the base plane beneath the failed material while avoiding direct access on or beneath the unstable mass. Split-FX® software was used to create a mesh from the point cloud, assess bedrock structure and dimensions of key features, and develop detailed cross-sections. Most significantly, the Split-FX® mesh helped reveal the location and orientation of the scarp for use in kinematic analyses.

Evaluations confirmed toppling instability as the predominant failure mode, and the orientation of the base plane indicated a combined sliding and toppling instability mode was the likely cause of instability of the Displaced Mass. GZA recommended scaling to remove the failed material, combined with passive dowels to stabilize the remaining scarp face against continued toppling. Catchment performance was evaluated for the proposed scaling, and for the proposed final slope configuration, using RocFall® software. LiDAR data was used extensively in these evaluations, as input to Split-FX®, which was used to evaluate the geometry of the failed rock mass, thickness of individual blocks, and cross-section dimensions for rockfall evaluations. The value of LiDAR for rock slope mapping was illustrated throughout the course of the project.

INTRODUCTION

On May 8, 2017, a rockfall event occurred, originating from the toe of a potentially unstable rock mass (herein referred to as "the Displaced Mass"), located along the southbound lanes of Interstate Route 91 (I-91 SB) near Fairlee, Vermont. The small rockslide that occurred in May 2017 did not enter the travelway. However, the Vermont Agency of Transportation (VTrans) observed a scarp behind the Displaced Mass, and a wide tension crack between the scarp and the Displaced Mass as shown in **Figure 1.** The toe of the Displaced Mass was observed to be approximately 30 feet above pavement grade, and the overall Rock Mass is approximately 25 feet high and 240 feet wide. Therefore, despite the presence of a catchment ditch greater than 25 feet in width, concern was expressed regarding the potential for the ditch capacity to be exceeded if the entire Rock Mass were to release at once, allowing rockfall to enter the travelway.

Figure 1 – Photographs of the Displaced Mass, Front (Left) and Side showing Tension Crack and Scarp (Right)

The following sections of this paper address pertinent aspects of the geologic setting, field investigation, and rock slope engineering. The benefits of utilizing LiDAR data to assess the stability and mitigation of the Displaced Mass, as well as the stability of toppling and sliding modes from the cut slope, are highlighted throughout the following sections.

PROJECT AREA

I-91 serves as a major artery providing north-south access through the heart of New England. I-91 generally follows the Vermont-New Hampshire border and extends north into Canada and south through Massachusetts and Connecticut (*1*). The site is located along I-91 SB at mile marker 94.5, near the base of Sawyer Mountain, and about 1 mile north of Lake Morey, as shown on **Figure 2.** The overall length of the rock slope along the east side of the SB barrel is approximately 0.3 miles.

The area of investigation was approximately 240 feet long, extending roughly 100 feet north and south of the Displaced Mass. GZA established a baseline with stationing to identify measurement locations and features along the slope, shown on **Figure 2.**

Figure 2 – Project Location on Aerial Photograph (left) and Project Area with GZA Baseline (right)

This rock slope was previously identified and characterized as part of VTrans' Rockfall Hazard Rating System (RHRS). The slope held a 'B' rating, which corresponds to the third-from-highest risk level in the VTrans system and therefore would not typically be programmed for mitigation. The scarp existed in a condition similar to the current condition in 2012, at the time of the last rating, and the relatively low hazard rating was attributed primarily to a catchment width greater than 20 feet. However, concern that a much larger volume of the Displaced Mass could become destabilized and fail at one time was the driver for this assessment.

OBJECTIVES AND APPROACH

The primary objectives of the project were to characterize the bedrock structure, provide geological engineering recommendations to stabilize the Displaced Mass, and mitigate future rockfall risk to the travelling public. The challenge was to gather field data to characterize the Displaced Mass and the nearby rock slope, while avoiding direct access above or beneath potentially loose, unstable material.

To accomplish this, GZA proceeded with a multi-phased approach that included a field mapping program to characterize the primary structure of the Displaced Mass using hand measurements and a terrestrial LiDAR survey. GZA utilized Split-FX® analytical software, developed by Split Engineering, LLC, to create a model from the LiDAR survey and extract joint orientation and slope geometric measurements from the point cloud. The data collected from the field mapping and LiDAR survey were used to conduct our analyses and develop recommendations to stabilize the Displaced Mass.

After the primary structures were characterized, GZA evaluated the performance of the existing catchment using RocFall® software, developed by Rocscience, Inc. Catchment performance was evaluated for two conditions: during scaling of the existing loose material and for the proposed final configuration.

GEOLOGIC CONDITIONS

Available bedrock mapping published by the U.S. Geological Survey indicates the bedrock at the site is part of the Sawyer Mountain Formation (*2*). Mapped rocks of this unit are described as greenish gray to dark gray, pyritic locally calcareous phyllite and light gray, locally pyritic and calcareous, fine- to medium-grained, feldspar-rich metasandstone. The site is also mapped between two northeast-southwest trending thrust faults. The mapped bedrock geology in the site vicinity is shown in **Figure 3.**

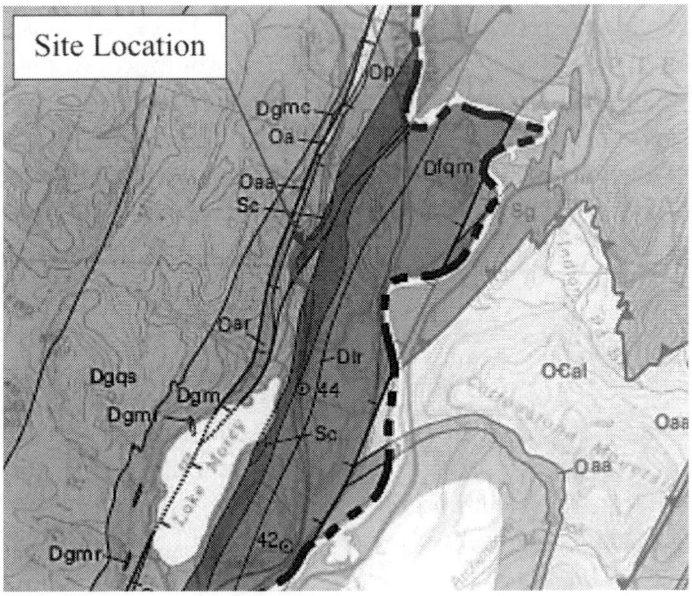

Figure 3 – Bedrock Geology

Field observations by GZA and VTrans geologists were generally consistent with the bedrock mapping in the project area. The rock exposed in the existing slope was interpreted to consist of medium hard, dark gray (where fresh) to light gray and rusty brown (where weathered), fine- to medium-grained, Phyllite. Distinct fault zones or planes were not observed in the study area. Typically, rock fragments could be removed from the rock slope by hand with little effort where they could be reached near the base of the slope.

The overall cut sloped down to the west and the cut slope aspect was slightly north of due west. The average cut slope inclination is approximately 60 degrees. The typical rock slope height is approximately 90 feet. The catchment distance between the toe of the rock slope and edge of paved shoulder is approximately 28 feet, and the bottom of the catchment ditch is approximately 4 feet below the adjacent roadway elevation. The slope contained sparse vegetation growing on lower angle areas.

An apparent continuous, approximately 65-foot-long scarp had formed along the steeply dipping foliation over the full length of the back of the Displaced Mass, and a apparent displacement was observed between the Displaced Mass and scarp with a maximum open width of approximately 11 feet. The foliation within the Displaced Mass had rotated approximately 50 degrees out of the slope, returning to the foliation orientation of the intact rock at the apparent bottom of the scarp. Therefore, the entire Displaced Mass appeared to have translated and rotated-out away from the scarp, consistent with toppling instability. The bottom and top of the Displaced Mass were approximately 21 and 55 feet above pavement level, respectively.

FIELD INVESTIGATIONS

Hand Measurements and Visual Observations

Field measurements were made by hand and with laser scanning on June 22, 2017. The hand measurements were made with a Brunton® compass, and the GeoID mobile application. Ropes access techniques were not used to map this slope in consideration of the unstable nature of the Displaced Mass and the ability to obtain the desired data using LiDAR scanning. A total of 22 features were hand-mapped using the GeoID V1.8 application, and approximately 40 percent of these readings were field-checked for accuracy against readings taken with a Brunton® compass. Field mapping performed by GZA included a visual assessment of general rock type and measurements of joint characteristics including dip, dip direction, spacing, continuity, roughness, aperture, filling, and seepage.

Terrestrial LiDAR Scanning

GZA subcontracted Doucet Survey, Inc. (Doucet) of Newmarket, New Hampshire to conduct the LiDAR scanning. A Leica C10 High Definition Laser Scanner was used at three scan locations. Traditional survey techniques were used to set permanent control points along the west side of the SB barrel using a 3-second Trimble Robotic Total Station and Automatic Level.

The LiDAR data set consisted of a point cloud with a typical spacing of ½ to 1 inch between survey points, each with unique x, y, and z coordinates and intensity. The individual scans were registered to form a continuous point cloud model covering the study area and portions of the catchment and roadway below. Target locations were georeferenced to a 6-millimeter (mm) accuracy based on least squares analysis. Doucet provided a raw point cloud of the registered and georeferenced data in *.PTS format. An image of the point cloud is presented in **Figure 4.**

Figure 4 – LiDAR Point Cloud Image of the Displaced Mass

Point Cloud Data Interpretation

The method used by Split-FX® to assess structural geology involves creating a "mesh" and "patches" based on the point cloud data. A mesh is a polygonal surface model generated using the point cloud data, and it represents a reconstruction of the surface geometry from the densely-sampled points. The mesh is created based on an average number of points per triangle, or by defining a uniform triangle size. A comparison of the point cloud and the generated mesh in the vicinity of the Displaced Mass is shown in **Figure 5.**

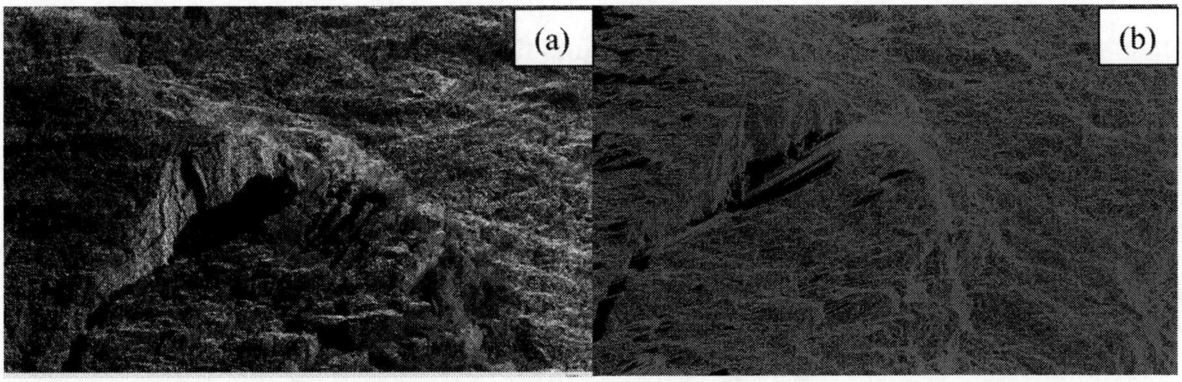

Figure 5 – Images of Point Cloud (a) and the Generated Mesh (b)

After the mesh was created, patches were created. Patches are planes fitted to real discontinuity surfaces present in a cloud. Patches are created by grouping adjacent mesh triangles based on similarity of the orientation. The pole (i.e., dip and dip direction) of a patch is developed using least squares to fit a plane through the points bounded by the grouped triangles (not the mesh triangles themselves). The orientation of a patch developed in Split-FX® is comparable to a hand-measurement taken with a Brunton® compass or GeoID. Patches can either be found automatically, based on user-selected values of variation in orientation between adjacent triangles and minimum patch size, or they can be selected for a specific set of mesh triangles.

The mesh was also utilized for creation of cross-sections. A cross-section line can be selected at any location and orientation within the rock slope, and the cross-section is developed as a series of segments corresponding to the coordinates of each mesh triangle that is crossed by the section line. At locations where the LiDAR line of sight is "shadowed" and there is a hole in the mesh, the cross-section line will show a gap. Cross-section coordinates can be inputted directly into rockfall catchment modeling software.

Split-FX® also allows measurement of joint spacing and continuity within the point cloud. The point cloud is oriented so that the desired measurement is normal to the screen for the measurement. Measurement units correspond to the units embedded in the point cloud data, in this case, feet (*3*).

Comparison of Brunton® Compass Measurements, GeoID Measurements, and Split-FX® Results

During field mapping, some of the discontinuities of interest were out of reach to allow measurement with the Brunton® compass, and/or were on surfaces small enough that use of the Brunton® would be difficult. Therefore, we utilized the GeoID application on an iPhone to collect strike and dip readings; this method was shown to be within 1 degree of a Brunton compass measurement for half of the readings taken with GeoID and the Brunton® compass, with a maximum variation of 6 and 10 degrees for dip and dip direction, respectively. Given the good agreement, GeoID was considered to be validated for discontinuity measurements.

We used the discontinuity measurements collected with the GeoID to compare dip and dip direction to the patches created from the mesh generated in Split-FX®. Due to the clarity of the LiDAR scan, GZA was able to directly compare hand readings to individual patches at the same location by locating the spray-painted points within the cloud, an example of which is shown in **Figure 6**.

Figure 6 – Comparison of Patches to Known Hand-Reading Locations

The results indicated that the average absolute values of variation in dip angle and dip direction between the hand readings and Split-FX® measurements were 2.3 degrees and 7.3 degrees, respectively. These variations were considered within the margin of variation that would be typical for a Brunton® compass, and the overall range of variation in dip direction would not significantly alter the interpreted bedrock structure as it relates to kinematic analyses. The comparison did not show a significant directional trend to the variations for either dip or dip direction. Therefore, the Split-FX® mesh and the patches were considered to be validated, and suitable for use in our kinematic analyses.

The largest variations in dip direction were on planes that are orthogonal to the face, which were typically stepped and variable over small distances, and the orientation also made them the most likely to be shadowed from the scan. Typically, planes with a dip direction perpendicular to the scan direction are the most difficult to interpret in Split-FX®. However, this shortcoming of LiDAR point cloud interpretation was anticipated, and the orientation of these planes was characterized adequately by hand readings.

ROCK SLOPE STABILITY

Bedrock Structure

We developed a lower-hemisphere pole plot from hand measurements collected along the base of the cut and developed a pole concentration contour plot to develop four preliminary joint set groupings. The automatically-generated patch data from Split-FX® were also plotted and evaluated to assess consistency with the hand measurements. The automatically-generated poles from Split-FX® patches were generally consistent with the hand measurements. However, a fifth joint set was recognized in the patch data that was found to be persistent at higher elevations in slope, as shown in **Figure 7.**

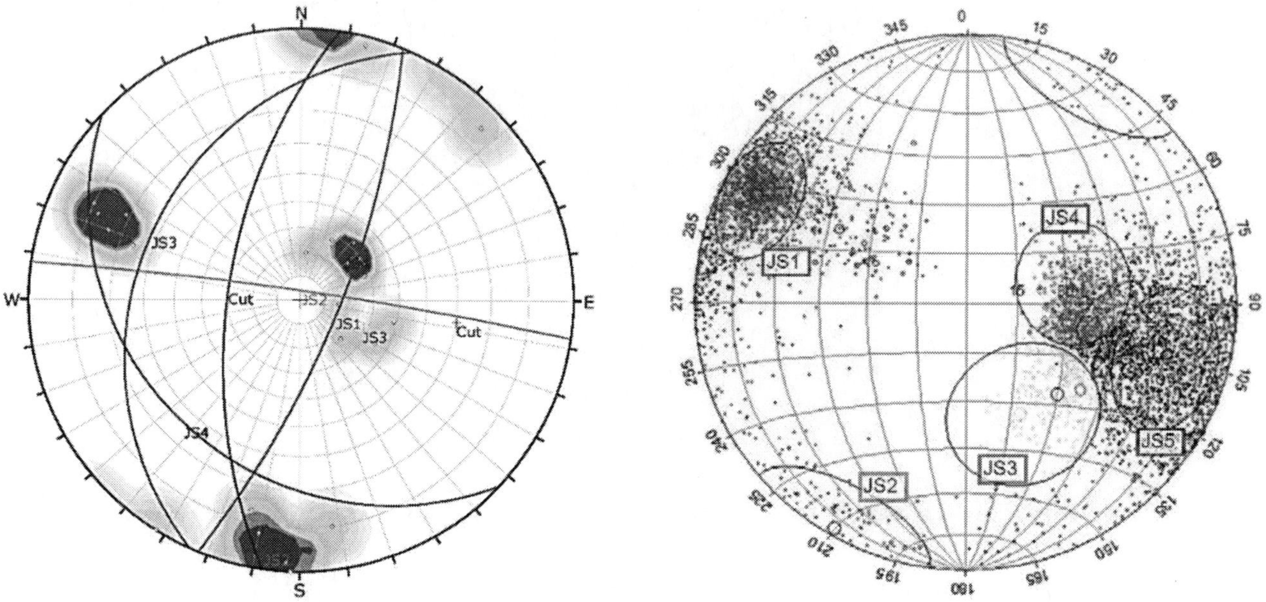

Figure 7 – Joint Set Groupings: Poles of Hand Measurements (left) and Poles of Split-FX® Patches (right)

A total of thirty (30) joint observations were used in our kinematic evaluations, including 22 hand measurements taken near the base of the slope of JS1 through JS4 and 8 Split-FX® patches on planes with varying orientation higher on the slope, including JS5 and the scarp orientation. The engineering analysis was based on representative rock structure orientations consisting of 5 joint sets (**Figure 8**).

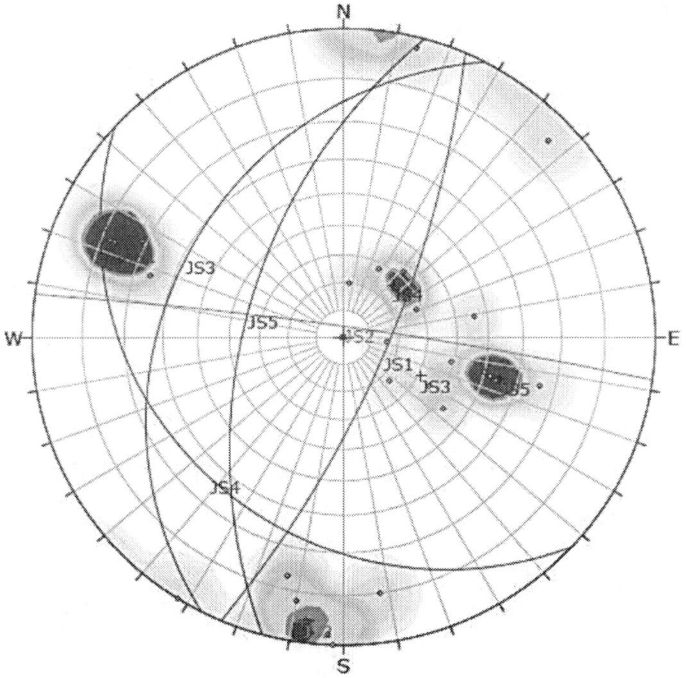

	Color	Dip	Dip Direction	Label
User Planes				
5		60	278	Cut
Mean Set Planes				
1m		74	113	JS1
2m		86	8	JS2
3m		31	297	JS3
4m		26	223	JS4
5m		55	285	JS5

Figure 8 – Pole Plot of Final Joint Set Groupings

The structure was observed to be relatively consistent across the study area, except for the rotated material within the Displaced Mass. JS1 is the foliation set that is sub-parallel to the rock slope and the highway, and JS2 is the near-vertical set that is orthogonal to the foliation set. JS3, JS4 and JS5 are moderately dipping sets that dip out of the slope. The orientations of these sets have similarities, and they may be local variations of the same overall structural feature, but they were treated separately for kinematic analyses. JS3, JS4 and JS5 were frequently stepped between JS1 planes.

Kinematic Stability Assessment

Based on observation of the foliation orientation and slope performance, toppling was confirmed as the most apparent instability mode. Toppling instability can occur where elongated blocks form along near-vertical discontinuities that dip into the slope (*4*). The toppling conditions for the Displaced Mass includes two intersecting joint sets with a near-vertical intersection line dipping into the slope, which forms the sides of discrete toppling blocks. This condition is represented on the stereonet by two great circles intersecting in Zone 1, 2, or 3, as shown on **Figure 9.** The second condition is represented by third joint set with poles in Zone 1 that acts as a release plane or sliding plane, allowing the blocks to topple (*5*). The limits of these zones are defined by the orientation and slope of the rock cut and the estimated friction angle along the base plane, which was assumed to be 30 degrees for this analysis.

JS1 and JS2 intersected in Zone 2, indicating the first condition for toppling is met. Poles in the JS4 set and some poles in the JS3 set fell in Zones 2 and 3, and poles in the JS5 set and some poles in the JS3 set fell in Zone 1. Two planes exposed beneath the Displaced Mass were mapped in JS3 and JS4, which possibly allowed the initial blocks to be released by sliding along these planes. These conditions resulted in a combined sliding and toppling mode. The stepped and/or discontinuous nature of the release planes was concluded to increase the effective friction angle of the material at the base of the Displaced Mass, providing the resistance that resulted in the observed rotation of the blocks about the base, and preventing the release of the entire unstable Rock Mass.

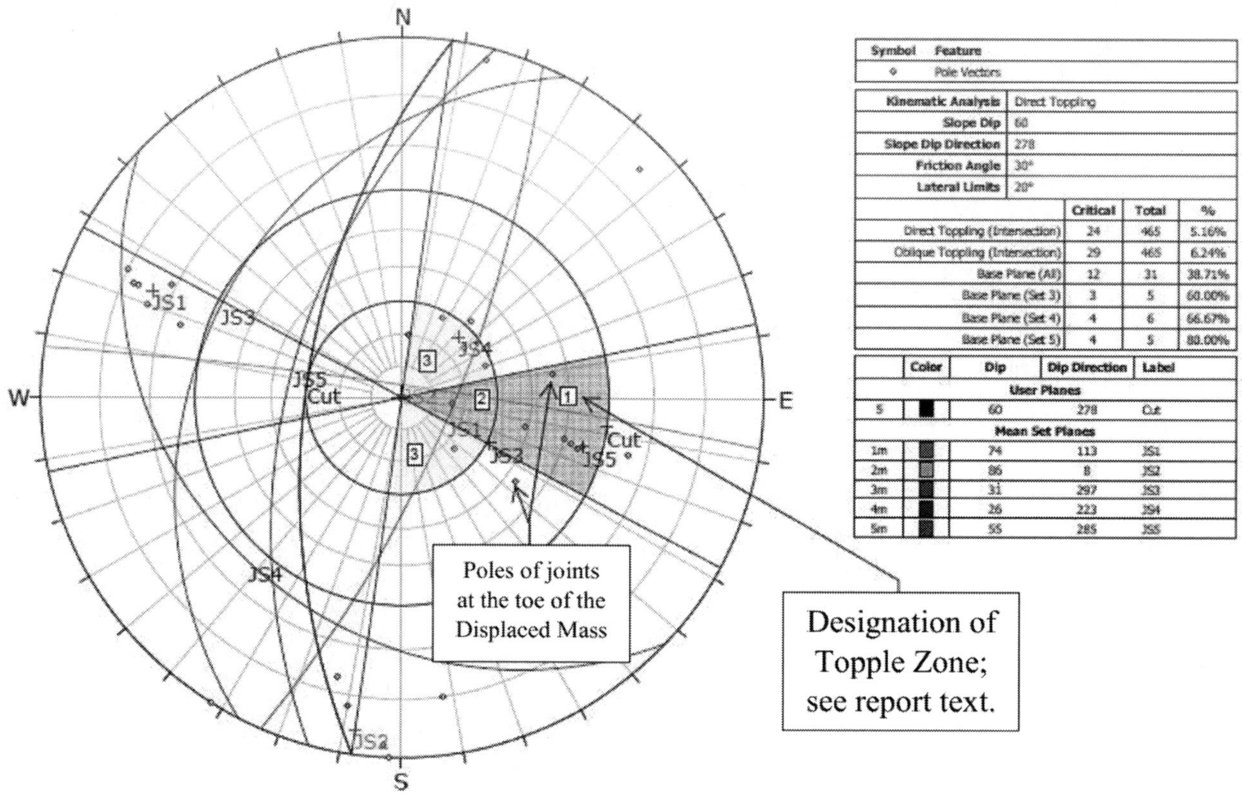

Figure 9 – Toppling Instability Evaluation

EVALUATION OF TOPPLING MITIGATION

The stepped nature of the release planes at the bottom of the toppled blocks was a likely source of sliding resistance for the Displaced Mass, but water pressure and/or ice-jacking were anticipated to force the failed mass to slide in the future. Therefore, we concluded the Displaced Mass should be removed by scaling to expose the intact rock mass believed to form the bottom and back of the Displaced Mass.

After scaling back to the face, the exposed scarp is expected to be 15 feet high and is considered susceptible to future toppling. The areas shown in pink in **Figure 10** are interpreted to represent the likely base of the Displaced Mass. In order to limit the risk of future toppling, we evaluated mechanical stabilization of the near-vertical foliation plane to be exposed after scaling.

Figure 10 – Patches (Pink) Representing Base Planes in Beneath the Displaced Mass

The geometric parameters required to analyze toppling included:

- The height and angle of the slope face
- The anticipated base inclination
- The block base angle
- The upper slope angle
- The bench width

These parameters were estimated using the orientation of patches and the measurement tool in Split-FX®. The overall base inclination was taken as the line of intersection of two planes underlying the Displaced Mass and below the scarp, and the inclination of the toppling blocks to be stabilized was taken as the dip of the foliation plane at the back of the tension crack, 77 degrees. The upper slope angle, bench length, and final slope height were interpreted using measurements in Split-FX®. The typical block width JS1 joint spacing estimated using Split-FX®. A representative spacing of 1 foot was selected.

RocTopple was used to evaluate the factor of safety against toppling and to design the reinforcement necessary to achieve suitable safety factors. The toppling analysis in RocTopple is conducted using a two-dimensional model based on the analytical method of Goodman and Bray (*5*). An overhanging face geometry cannot be modeled in RocTopple; therefore, the face slope angle was modeled as 89 degrees, understanding that the front few blocks modeled in RocTopple were previously released during the initial rockfall event in May 2017. The factor of safety against toppling instability, assuming the joints are 50 percent filled with water, was calculated for the scaled slope configuration, without reinforcement, to be approximately 0.3.

Given that toppling has not yet occurred above the scarp, the results indicate that the modelled conditions were more conservative than the field conditions. The differences are believed to be that the foliation planes are not as closely spaced or continuous as assumed, or that the rock mass has not sustained the assumed hydrostatic pressure.

Having shown the model to be conservative, GZA evaluated the scarp stabilization using this base model and a design factor of safety of 1.5. Reinforcement was added in the model to achieve the desired factor of safety. The results of the RocTopple evaluation show that two rows of rock dowels, spaced at 6 feet on center along the length of the final exposed face, provided a factor of safety greater than 1.5 for the 50 percent pressure condition, selected as the design basis, and greater than 1.0 for the 80 percent pressure condition, selected to represent an extreme event. Results for the reinforced slope in the 50 percent pressure condition are shown in **Figure 11.**

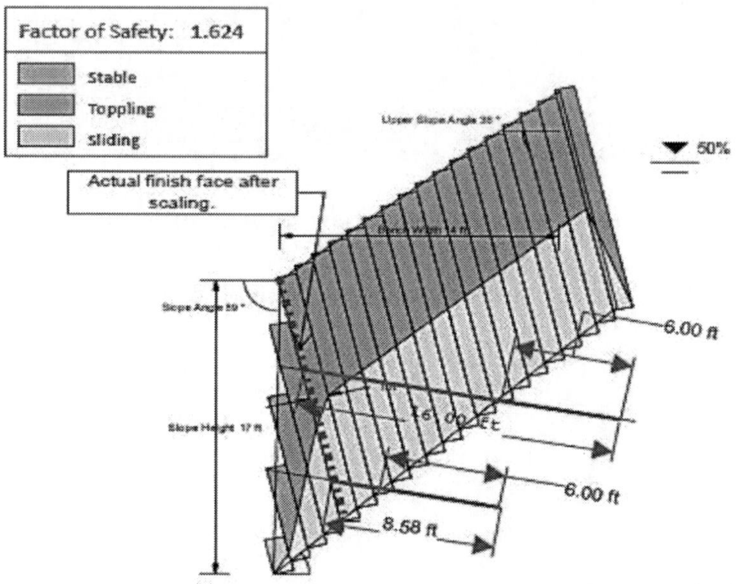

Bolt Details

Installed Block	Bolt Name	Length (ft)	Angle (°)	Tip Location	Anchor Location	Capacity (kips)	Out-of-Plane Spacing (ft)
2	Bolt 1	11	-10	-0.36, 6.00	10.47, 4.09	30	6
4	Bolt 2	18	-10	0.31, 12.00	18.04, 8.87	30	6

Required Dowel Capacity = 30 kips
Spaced 6' O.C. along exposed face

Figure 11 – 50 Percent Water Pressure Model Evaluated in RocTopple with Reinforcement

A dowel capacity of 30 kips was required to achieve stability for this dowel configuration, which was to be developed in the portion of the dowels behind the base failure plane. Based on the geometry and the use of 75 ksi No. 8 bars installed in a 3-inch-diameter hole, the minimum required bond length was 6 feet, and the minimum required dowel lengths were 16 and 9 feet for the top and bottom row of dowels, respectively, as shown on **Figure 11**.

EVALUATION OF CATCHMENT PERFORMANCE

We evaluated rockfall catchment in the vicinity of the Displaced Mass considering two scenarios. The first scenario was during scaling, at which time the Displaced Mass would be removed, and we expected that large volumes of rock could be released concurrently. This scenario was developed to evaluate the safety of the travelling public during scaling, attempting to show that one lane of traffic could safely remain open during construction, while the remainder of the east side of the SB barrel and shoulder would be closed. The second scenario evaluated the final slope configuration after completion of scaling and stabilization based on the existing catchment geometry. This condition considered the top of the existing rock slope extending above the near-vertical stabilized face and the moderately-dipping plane anticipated to be exposed after scaling is completed. Smaller blocks released along the upper slope will have the potential to fall a relatively long distance, impact the exposed plane, and be launched toward the SB barrel in a way that is not feasible for the existing slope.

Computer-based catchment evaluations were conducted using the analytical software RocFall by Rocscience to analyze the scaling scenario and the final condition described above. RocFall employs user-defined slope and catchment geometries and a series of input parameters to simulate the rockfall behavior for a given slope. Rockfall parameters include:

- The size and shape of the rocks that compose a rockfall event
- The surface roughness
- The coefficient of friction of the slope
- The coefficients of restitution of the slope and catchment

The restitution coefficients have a significant impact on rockfall modelling. Three surface materials were modeled, include the rock slope, the catchment area and the pavement. Considering the moderately hard rock at the site and the likelihood of falling rocks to fracture upon impact with the rock surface, restitution coefficients selected for rock were in the lower to middle range of values reported in the references. The surface roughness for bedrock was modeled with an average height of 6 inches and an average spacing of 1 foot. These parameters were selected based on the spacing and continuity of observed joint sets, and visual observation of the asperity of the surfaces developed in individual rockfall animations that appeared to be consistent with field conditions. The catchment area was modeled as a talus cover, with properties consistent with rock fill or hard soil.

The output from the RocFall analysis includes the stopping point of each block dropped. The typical criterion for acceptable rockfall catchment design used by VTrans is that at least 95 percent of the modeled rockfall is contained in the catchment, in this case, the outside edge of the paved shoulder.

Slope geometry was based on a representative cross-section through the highest portion of the scarp developed using Split-FX®. For both analysis scenarios, we assumed that the critical geometry for rockfall catchment would be the final slope geometry, with rocks falling either from the back of the Displaced Mass, just below the scarp for the scaling analysis, or from above the scarp for the final configuration analysis. Therefore, the section exported from Split-FX® was modified by removing the displaced material, leaving the anticipated final geometry, shown in **Figure 12.**

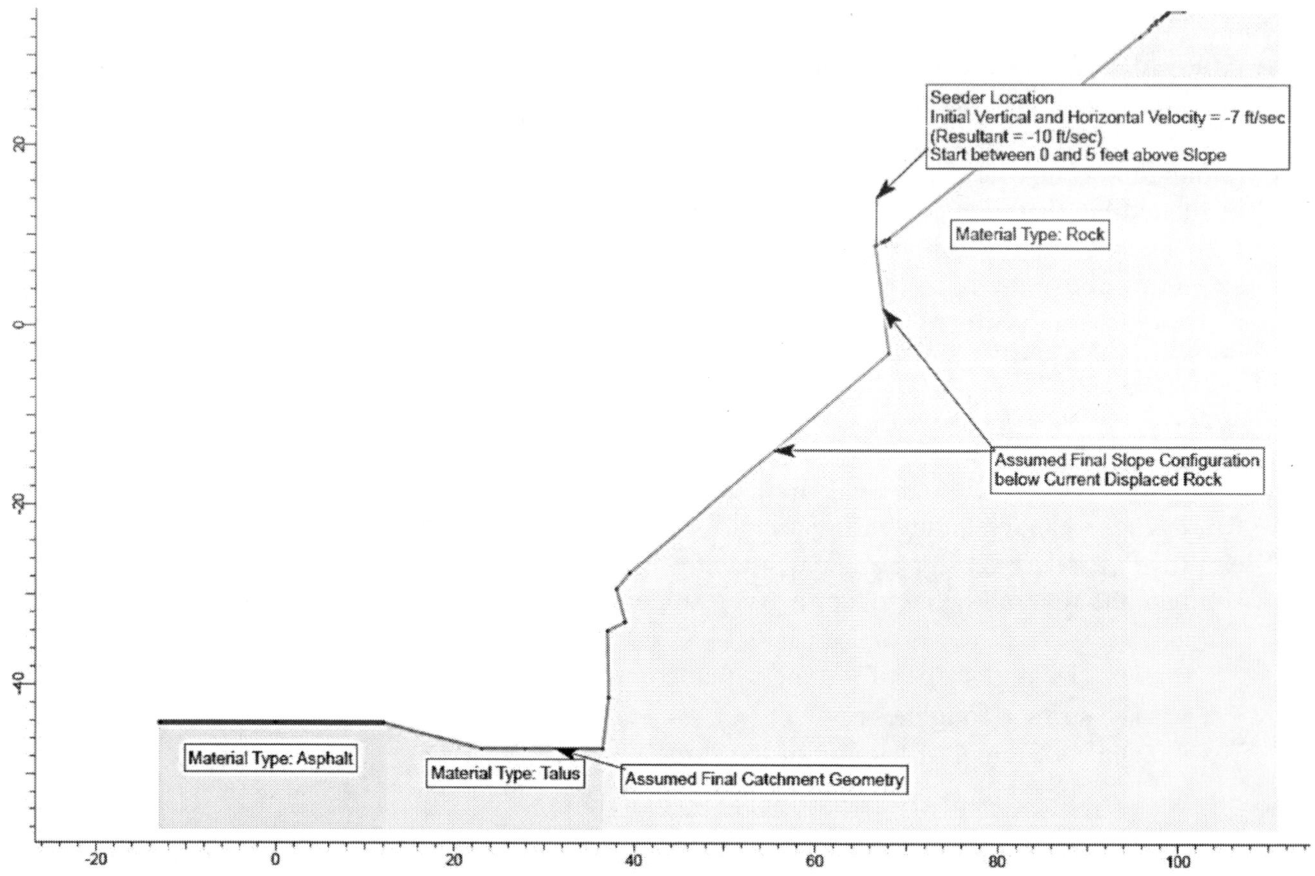

Figure 12 – Final Slope Configuration for Catchment Performance Evaluation

Based on the bedrock structure and observations of past rockfall, we anticipated that individual falling rocks would initially have an elongated shape. RocFall allows the user to select the rock shape from several standard shapes and aspect ratios to best represent the actual mode of rolling, which is an advantage over the modeling ability of the Colorado Rockfall Simulation Program (CRSP), which only allows selection of spherical or square rocks. A super ellipse shape with an aspect ratio of 1:2 was selected, as this is the most elongated rock type with the sharpest corners available in the program.

Scaling Scenario

We recommended that loose soil and rock currently in the ditch be moved from the base and be placed along the pavement side of the catchment to create a temporary berm to enhance the effectiveness of the catchment during scaling. The temporary berm was modeled in RocFall with 1H:1V side slopes, approximately 2 feet above the roadway. The slope and catchment geometry used as the basis for rockfall evaluations during scaling can be seen in **Figure 13.**

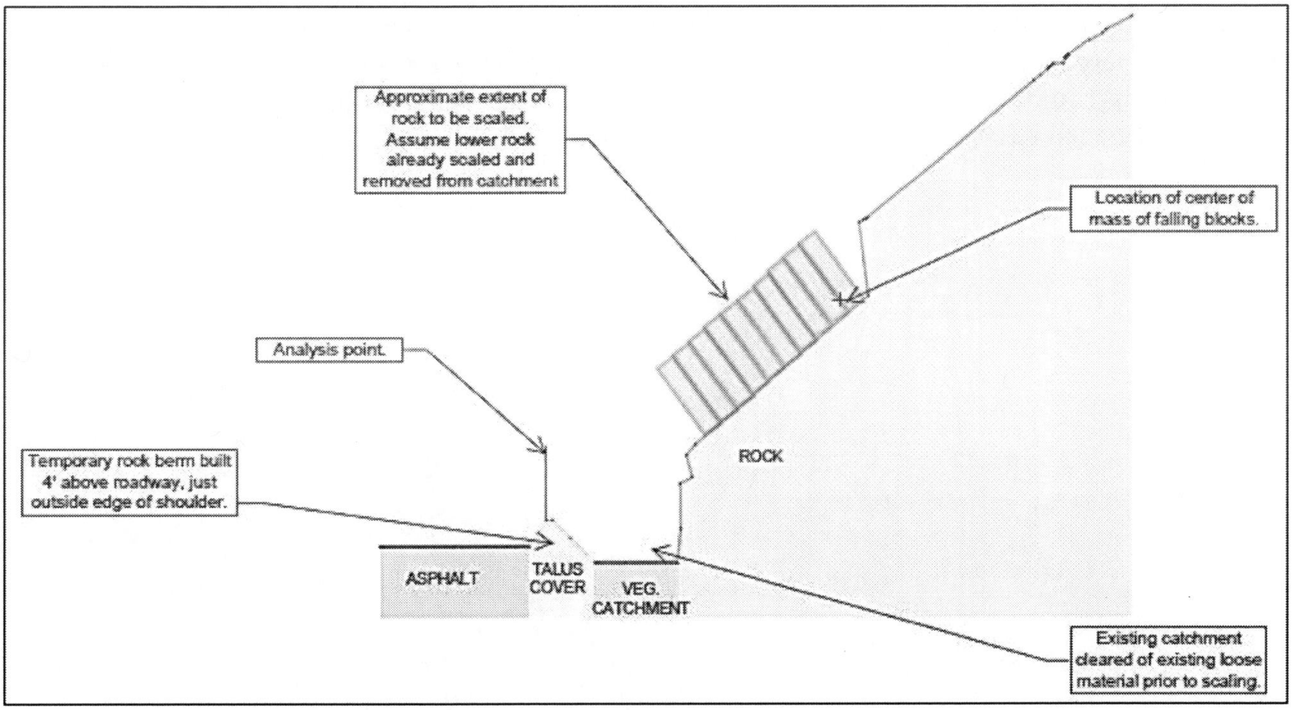

Figure 13 – Initial Slope Configuration for Rockfall Evaluation During Scaling

A combination of measurements of loose blocks observed in the Displaced Mass and an assumption of potential larger intact rocks that could fall during scaling were used as a basis for the modeled block sizes for this evaluation. Blocks were typically observed to be elongated slabs, and the spacing of the JS1 foliation joints in the displaced zone was typically less than 2 feet. A summary of the modeled block sizes is shown in the table below. The block dimensions below were inputted to RocFall, where they were used to determine the block weight using a rock unit weight of approximately 170 pounds per cubic foot, and the calculated weight was then used to create a "Super Ellipse" shape in RocFall with the same weight.

Table 1 – Catchment Evaluation Block Dimensions

Block Designation	Block Dimensions (ft x ft x ft)
Large	10 x 10 x 2
Medium	4 x 4 x 2
Small	2 x 1.5 x 1

Three seeder lines were used as source locations for the falling rocks. The seeder line lengths were selected to model each rock size falling from a range of heights varying from a rock falling from the finish slope surface up to a rock falling from the top of the displaced zone. 1,000 blocks of each size were modeled, resulting in a total of 3,000 blocks modeled for the scaling scenario. The rocks were modeled with an initial velocity of 7 feet per second to account for the effort required to remove the blocks from the slope.

Two catchment conditions were modeled for the scaling scenario. The first consisted of an empty catchment (**Figure 13**), assuming the condition was the same as that created immediately prior to scaling. For the second condition, we assumed that a large volume of rock would be released simultaneously, resulting in about 4 feet of rock filling the entire catchment, followed immediately by release of the design block sizes from the back of the Displaced Mass before the catchment could be cleared of rockfall debris. The considered cross-section and the RocFall-generated block paths are shown as **Figure 14.**

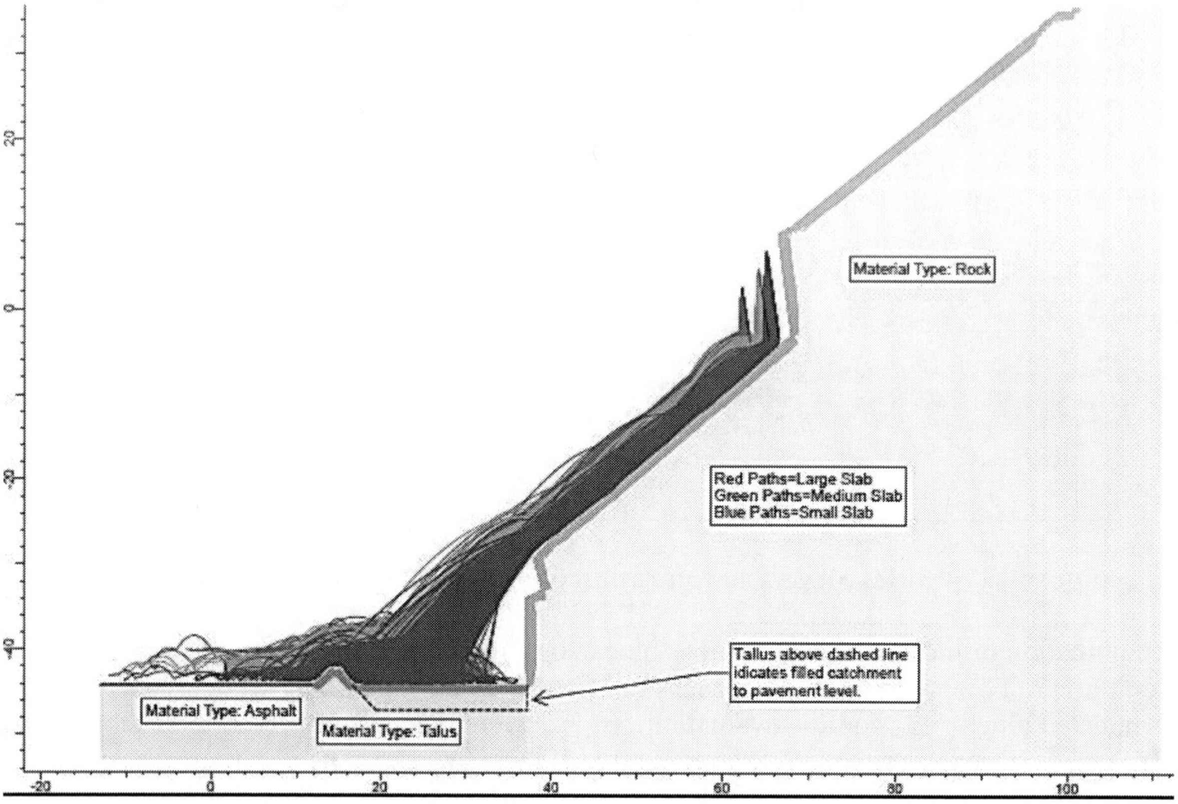

Figure 14 – Filled Catchment Configuration and Rock Paths

The percent retained in the catchment for both scenarios is plotted on **Figure 15**. The results indicate that greater than 95 percent of the falling blocks would be retained inside of the temporary berm if the catchment remained empty for the duration of the scaling efforts. The percent retained behind the berm for the partially-filled catchment case was 91 percent, and 95 percent retention was achieved about 4 feet into the pavement, which was anticipated to be well within the proposed lane closure during scaling.

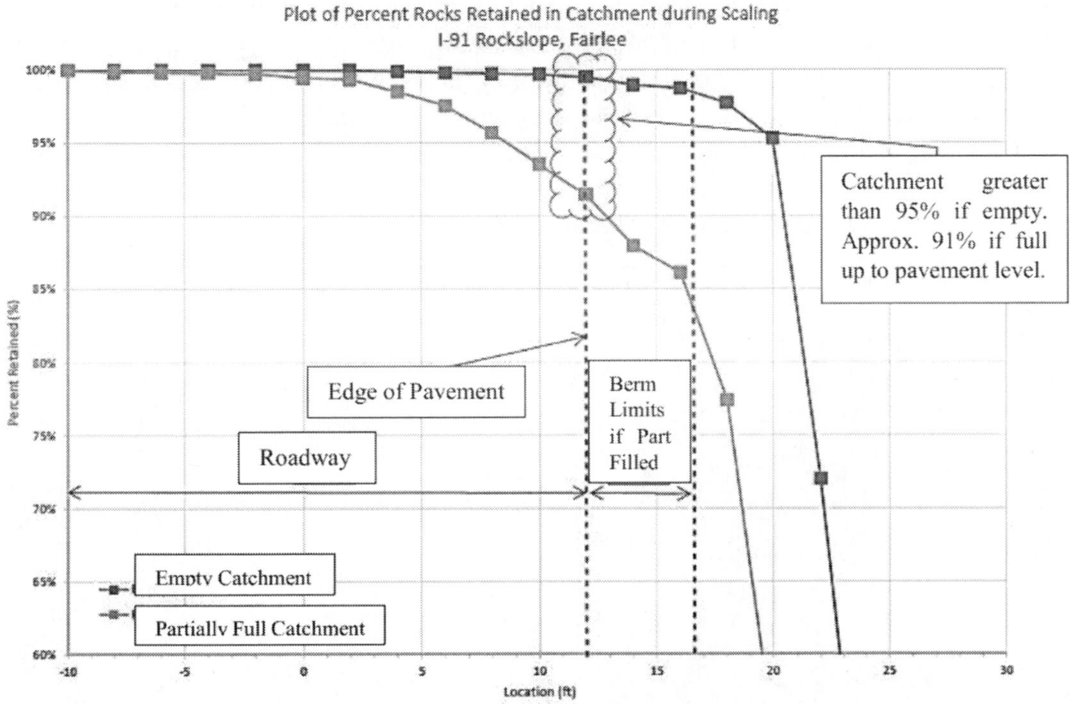

Figure 15 – Catchment Performance Results for Both Scaling Scenarios

Based on these results, we concluded that scaling could be controlled such that the rockfall retention within the enhanced catchment and the closed lane met VTrans catchment criteria of at least 95 percent retained, promoting safety for passing vehicles.

Final Slope Geometry

Previously-fallen blocks in the catchment, observed during the field visit, were used as a basis for the modeled block sizes for the final slope evaluation. Block dimensions of 2 by 2 by 0.5 feet and 1 by 1 by 0.1 feet were used to model anticipated "medium" and "small" blocks, respectively. These block sizes were used to account for the fracturing of falling rock fragments into smaller pieces.

It is anticipated that rocks of this dimension will be loosened over time by environmental factors, such as frost-jacking or hydrostatic pressure, and the source location could be anywhere between the top of the rock slope, about 90 feet above the road, to just above the stabilized scarp face, approximately 50 feet above the road. Therefore, the seeder line extended over most of the rock slope above the stabilized area, modeling rocks to be released with an equal distribution from the area above the face. 1,000 blocks were modeled of each size, resulting in a total of 2,000 blocks modeled for this scenario.

Our analysis assumed that all fallen rock from the scaling activities would be removed. The results indicate that approximately 89 percent of the assumed blocks falling from above the crest of the stabilized face would be retained in the catchment, which did not meet the 95 percent catchment criteria.

GZA plotted the bounce height of rocks at the edge of the paved shoulder, as shown in **Figure 16**. This function is not available in CRSP, and it allows the user to assess the number of rocks passing a point that would impact a barrier of a given height. The results indicated that approximately 72 percent of the rocks passing this location were predicted to be at or below 2 feet from the ground. Therefore, the addition of a 2-foot-high barrier was recommended to provide rockfall retention to greater than 95 percent.

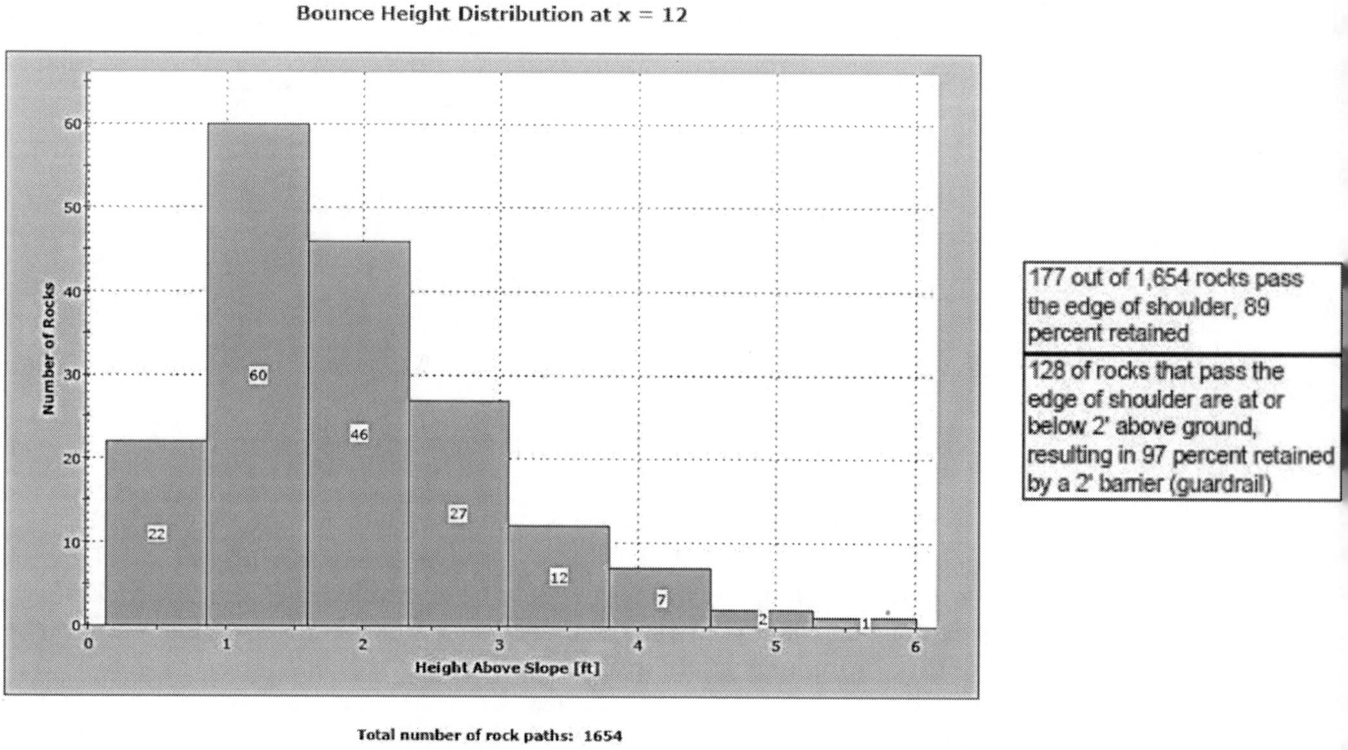

Figure 16 – Summary of Bounce Height for Blocks Entering Roadway

DISCUSSION AND CONCLUSIONS

VTrans approached GZA with a request for a rock slope assessment with a number of challenges, including a potentially hazardous work environment with difficult access to the primary area of concern. Terrestrial LiDAR, processed using Split-FX® in conjunction with a targeted hand-mapping program, was identified as an excellent solution to these challenges. Data generated using Split-FX® was used directly in a suite of engineering software packages developed by RocScience, Inc., including Dips®, RocFall® and RocTopple. These programs were used to evaluate the conditions driving the observed failure, and to assist in GZA's design remediation to limit future rockfall during and post-construction from impacting the travelway. GZA recommended that a combination of slope clearing, scaling, mechanical stabilization and potential catchment enhancement be completed for the project.

Some of the advantages gained from using the LiDAR survey in comparison to conventional hand-readings and survey include the following:

- The use of LiDAR survey allowed evaluation of the unstable mass to be completed with minimal risk to GZA and VTrans engineers, geologists, subcontractors, and the travelling public. Direct access above and beneath the Displaced Mass would have been required if hand measurements and traditional survey were used, and this access could have jeopardized stability of the Displaced Mass.

- The use of LiDAR survey allowed for detailed measurements of the Displaced Mass to be collected from the point cloud, which would not have been possible using traditional survey. The ability to rotate the point cloud in space and take a variety of accurate measurements of the Displaced Mass allowed GZA to develop a detailed characterization for development of design recommendations for scaling and stabilization. Having the ability to complete additional detailed measurements as engineering evaluations develop is a unique advantage of using LiDAR, and it was shown to be an invaluable tool for the project.

- Split-FX®-generated discontinuity data developed in the study area gave insight into localized variations in bedrock structure that were not prevalent along the base of the cut. Identifying this subtle variation may have been more difficult using the smaller data set typically attainable via hand measurements.

- Highly-detailed rock slope sections were extracted from the Split-FX®-generated mesh. These sections are more accurate representations of the critical slope areas than sections derived from typical roadway survey or by collecting limited optical survey points along a section. Consequently, the reliability of the catchment evaluation was enhanced using the LiDAR-based data.

- Doucet established permanent control points at the site. These can be used for future scans to assess changes in the slope. VTrans will have the ability to monitor the slope going forward and identify the nature and magnitude of displacements.

The next phase of the project is planned for summer 2019 and will consist of scaling and stabilization of the rock slope. The remediation process is expected to include removal of additional vegetation that could worsen stability, scaling of the Displaced Mass remaining on the slope, and placement of rock dowels to limit the potential for large rock releases.

REFERENCES

1. Nitzman, A. *Interstate-Guide I-91,* AA Roads, July 14, 2016.

2. Ratcliffe, N.M., Stanley, R.S., Gale, M.H., Thompson, P.J., and Walsh, G.J. (2011). Bedrock geologic map of Vermont: U.S. Geological Survey Scientific Investigations Map 3184, 3 Sheets, scale 1:100,000. USGS.

3. Split Engineering, LLC., *Split-FX® User's Manual, Software Version 2.1*. Tucson, AZ, 2011.

4. Hoek, E., and Bray, J., *Rock Slope Engineering, Revised Third Edition*, Spon Press, New York, NY, 1981.

5. Goodman, R.E. and Bray, J.W. (1976). "Toppling of Rock Slopes," *Proceedings of the Specialty Conference on Rock Engineering for Foundations and Slopes*, ASCE, Boulder, CO, August 15-18, 1976.

Geotechnical Challenges for Bridge Foundation & Roadway Embankment Design in Peats and Deep Glacial Lake Deposits

Brian T. Felber, PE
HNTB Corporation
9 Entin Road, Suite 202
Parsippany, NJ 07054
(973)-800-5502
bfelber@hntb.com

Prepared for the 69th Highway Geology Symposium, September, 2018

Acknowledgements

The author would like to thank the individuals/entities for their contributions in the work described:

Matthew Riegel, PE – HNTB Corporation, review
John Schuring, PE – HNTB Corporation, review
John Szturo, PG – HNTB Corporation, suggestion to prepare paper

Disclaimer

Statements and views presented in this paper are strictly those of the author(s), and do not necessarily reflect positions held by their affiliations, the Highway Geology Symposium (HGS), or others acknowledged above. The mention of trade names for commercial products does not imply the approval or endorsement by HGS.

Copyright Notice

ABSTRACT

Design of bridge foundations and roadway embankments in the NJ Hackensack Meadowlands is complicated by low strength compressible organic silts and clays, peats, and glacial lake clays, which result in low lateral resistances, global stability concerns, and significant consolidation and secondary settlements. This paper will describe experiences from five projects in this region, but mainly the most recent, design of a roughly $300 million two-mile long new highway, in this challenging setting.

This paper will include description of the estimation of soil properties for peats and glacial lake clays in this region and the benefits that Cone Penetration Testing with pore water dissipation testing and shear wave velocity testing can provide in similar deposits. A discussion is also provided of ground improvement alternatives considered and those selected for roadway embankment, including lightweight soil aggregate, expanded polystyrene (EPS), cellular concrete, surcharge with prefabricated vertical drains (PVD), and timber pile supported embankment relief platforms. Accelerated Bridge Construction (ABC) techniques were used to replace two 100+ year-old rail bridges with limited track outage time using tied back micropile and lagging walls and temporary jump span bridges. This paper will also describe on the benefits of vibration and displacement monitoring to reduce risks associated with impacting adjacent facilities.

INTRODUCTION

This paper describes challenges and solutions considered for bridge foundation and roadway embankment design in the NJ Hackensack Meadowlands region. Geotechnical design in this region is complicated by low strength compressible organic silts and clays, peats, and glacial lake clays, which result in low lateral resistances, global stability concerns, and significant consolidation and secondary settlements.

The Hackensack Meadowlands is located in northeastern New Jersey flanking both sides of the Hackensack River, and surrounding the Passaic River at Newark Bay. The Hackensack Meadowlands generally includes undeveloped natural wetlands, rail, highway, and utilities infrastructure, landfills, superfund sites, and industrial land uses, as well as the Meadowlands Sports Complex (Met Life Stadium).

This paper describes experiences from design of a roughly $300 million two-mile long new highway, in this challenging setting, which crosses facilities for three railroads and a major United States Postal Service (USPS) distribution hub, and is also based on four other significant transportation infrastructure projects in the region with similar lithology.

Numerous project constraints influenced the selection of the preferred alternative for the example project:

- Low undrained shear strengths in the peat and to a lesser extent in the glacial lake clays, resulting in low lateral resistance and global stability concerns
- Susceptibility of peat and glacial lake clays to significant time dependent consolidation and secondary settlement
- Depth to rock greater than 150 feet for portions of the alignment
- Shallow groundwater and the need to dewater, and also the and risk that treatment of groundwater may be required
- Extremely aggressive corrosion rate inferred from electrochemical testing
- Right of entry agreements prohibited environmental testing prior to the property acquisition, which introduces risk given the past land use and known contaminated sites within the region
- Protection of existing utilities, including twin 72-inch water aqueducts, which have been in service for more than 100 years and are the primary water supply for a major city, several sewer force mains as large as 54-inch in diameter, two sewer screening facilities, electrical duct banks from an adjacent power plant, and a petroleum pipeline
- Protection of existing structures, including nine passenger rail bridge abutments and two roadway bridge piers, with the rail bridges more than 100 years old
- Limited footprint available for embankment and foundations due to three existing railroads including one freight transfer yard facility
- Right-of-way limited to minimize loss of USPS parking spaces

Objective

The objective of this paper is to provide practical observations related to the following:

- Estimation of soil properties for peat and glaciolacustrian varved clay, which may be useful given that correlations and typical properties of peats are less readily available
- Benefits of including Cone Penetration Testing (CPT) with pore water pressure dissipation testing and shear wave velocity tests as part of a subsurface exploration
- Comparison of ground improvement alternatives considered, including lightweight soil aggregate, expanded polystyrene (EPS), cellular (foamed) concrete, surcharge with prefabricated vertical drains (PVD), and timber pile column supported embankment, as they relate to settlement, stability, constructability, and cost
- Accelerated Bridge Construction (ABC) techniques proposed to replace two over 100+ year-old rail bridges with limited track outage time using tied back micropile and lagging walls and jump spans
- Aggressive corrosion mitigation strategy
- Benefits of vibration and displacement monitoring to reduce risk to existing foundations and utilities

GEOLOGY

The Hackensack Meadowlands and example project sites are situated in the Piedmont Physiographic Province. Surficial geology and geomorphology of the region have been dominated by a series of glacial advances and retreats over northern New Jersey during three glaciations, the pre-Illinoian, Illinoian, and late Wisconsinan, listed from oldest to youngest, with the most recent occurring approximately 12,000 years ago.

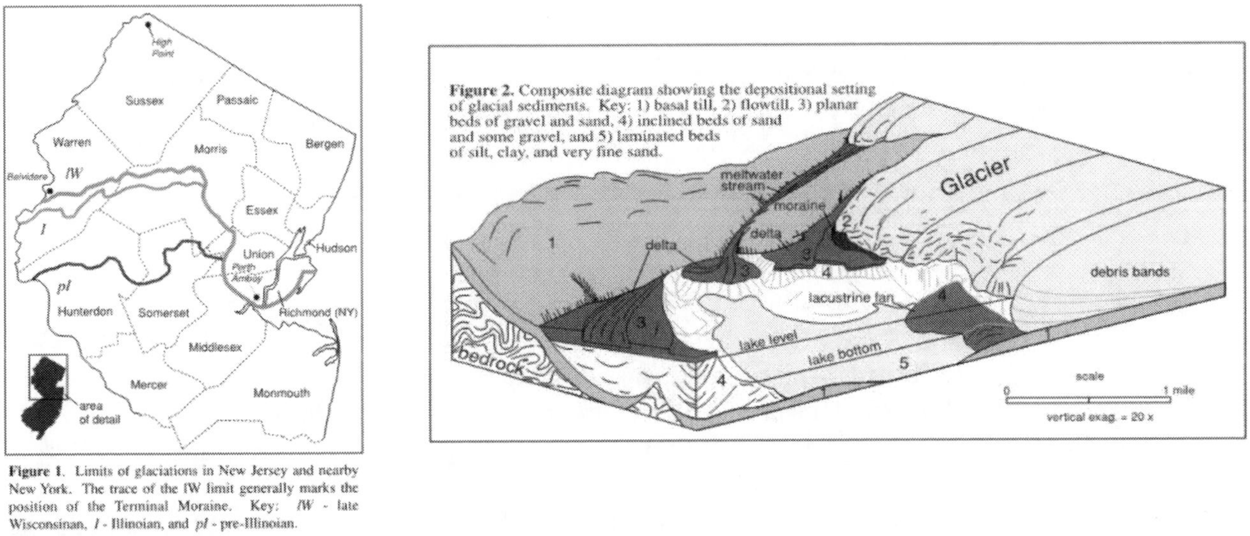

Figure 1 – Limits of NJ Glaciation and Composite Diagram of Glacial Sediment Deposition (Ref. 17)

The Palisades Sill is the dominant geologic feature located to the east of the site. The Palisades were formed from an intrusion of magma, which cooled to form diabase bedrock when the North American and African Plates began to separate, roughly 200 million years ago in the Jurassic Period. Subsequent glacial and erosional processes along the Hudson River have exposed the outcrop as a north to south feature, which projects as much as 500 feet above the river level.

Bedrock geology of the Hackensack Meadowlands to the west of the sill is dominated by sedimentary rocks of the Passaic Formation, Newark Supergroup, generally siltstone and shale. Bedrock of the Lockatong Formation, generally including arkosic sandstone, siltstone, mudstone, argillite, and hornfels are common in the region. It is common to encounter bedrock influenced by contact metamorphism from the diabase intrusion. Bedrock is generally deeper than 90 feet below the ground surface through the alignment of the example project, and is deeper than 150 feet for portions of the example project alignment. Decomposed rock is present above the competent bedrock in isolated locations, generally less than ten feet in thickness.

Figure 2 – Bedrock and Surficial Geology Maps
Left–Bedrock, Jd = Palisade Diabase, TRl, TRla, & TRpg = Lockatong Formation (Ref. 15)
Right–Surficial, Qm = Tidal Marsh and Estuarine Deposits (Meadow Mat) (Ref. 19)

As the glaciers advanced and subsequently retreated they left in their path ice and debris creating a dam which formed Glacial Lakes Hackensack and Bayonne, which once encompassed the area. Over time the lakes were filled as slow-moving waters deposited their sediment loads of fine silts and clays eventually filling in the lake. At some point, the terminal moraine "dam" broke and Glacial Lakes Hackensack and Bayonne emptied, at which time deltaic deposits, were deposited along the example project alignment at its southern extents. Sea level rose as water once trapped in the glacial ice melted and returned to the ocean, and the region gradually became the estuarine tidal meadows that exists today.

Overburden soil deposits within former Glacial Lake Hackensack, which encompassed much of the alignment, are associated with these glacial events. Beginning with the deepest materials above the bedrock are ablation glacial till composed of very dense sands and gravels. These granular deposits are overlain by discontinuous ridges of lacustrian fan deposits, primarily consisting of sand with varying and lesser amounts of silt. The lacustrian fan deposits and till are overlain by glaciolacustrian varved silt and clay, which can be as thick as 200 feet. Situated above the soft and compressible glaciolacustrian materials are more recent alluvial sands and outwash deposits. This veneer of granular soils was laid down by meandering streams on the lowlands and in fan deposits associated with upland waterways. The final natural deposits are organic materials, including peat and organic silts, also known as meadow mat; these highly compressible materials are often unstable when subjected to loading. Certain portions of the alignment have been reclaimed as usable land; various thicknesses of manmade fill are nearly continuous throughout the region.

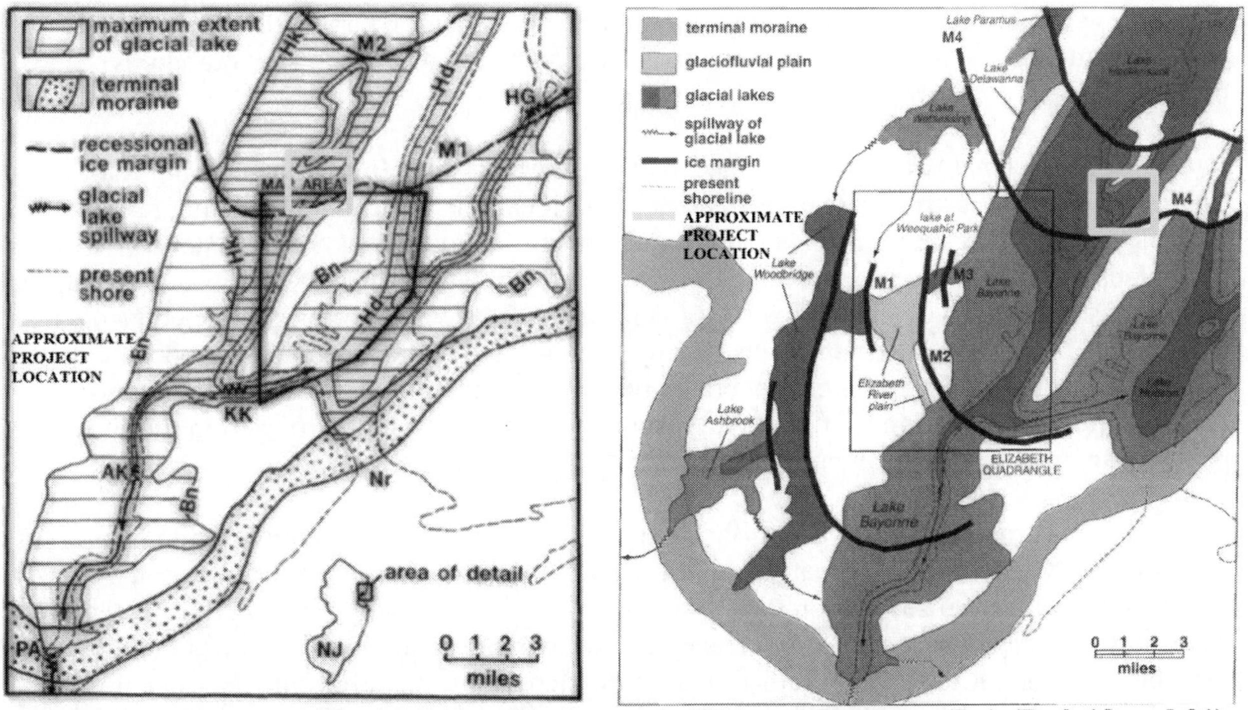

Figure 3 – Extents of Glacial Lakes Hackensack (Hk) and Bayonne (Bn) (Ref. 18 and 21)

SUBSURFACE EXPLORATION

Existing soil boring data taken between 1975 and 1999 for past projects adjacent to the example project's proposed alignment were available, including 56 Standard Penetration Test (SPT) soil borings. Multiple phases of subsurface exploration were performed for this project between November 2013 and September 2017, with a total of 96 SPT borings.

Soils were classified in accordance with the Burmister Soil Classification System, a system commonly used locally in the NY metropolitan region (Ref. 4). Standard Penetration Testing (SPT) was performed at each boring, in general accordance with ASTM D1586. Soil samples were retrieved using a 24-inch-long split-spoon sampler (2-inch O.D., 1-1.375-inch I.D.), driven by a 140-lb hammer free falling 30 inches. Groundwater levels were recorded when encountered. Thin walled tube undisturbed samples were advanced in general accordance with ASTM D1587, in the organics and lake bottom deposits.

The subsurface exploration also included 45 Piezocone Penetration Test (CPTu or CPT) and Seismic Piezocone Penetration Test (SCPTu or SCPT) soundings, advanced in general accordance with ASTM D5778. The penetrometers featured equal end area friction sleeves (i.e. not tapered), a net end area ratio of 0.8, and cone tips with a 60-degree apex angle. The pore pressure filter was located directly behind the cone tip in the "U2" position, and was 6-millimeter-thick made of porous polyethylene with an average pore size of 125 microns. The filter was saturated with silicon oil or glycerin under vacuum pressure before being used. The CPTu was conducted at a constant rate of 2 centimeter per second and rod inclination was measured. Tip and Sleeve offset were accounted for in the results. Pore pressure dissipation tests, hold periods to determine the time required for pore water to return to its equilibrium pressure, were performed to aid in the estimation of consolidation parameters. Seismic testing was performed to aid in establishing shear wave velocity and maximum shear modulus values for design.

Multi-channel Analysis of Surface Waves (MASW) and Ground Penetrating Radar (GPR) geophysical testing were performed at select locations adjacent the existing railroad bridges to verify that the location and limits of existing foundations were consistent with the as-built plans. MASW uses shear wave velocity to identify relative stiffness of subsurface materials, with high shear wave velocity indicative of dense materials. MASW was conducted using a Geometrics Stratavisor 24-channel seismograph and 4.5 Hz vertical geophones spaced 2 feet apart. Seismic surface waves were generated by striking an aluminum plate on the ground surface with a twelve-pound hammer. High shear wave velocities identified from the shear wave profile were inferred to be concrete from the existing shallow foundation.

Upon completion of the subsurface exploration, a laboratory testing program was performed to verify the visual-manual field classifications and to aid in determination of the engineering soil properties. Laboratory testing included water content, Atterberg limits, grain size analysis, percent passing no. 200 sieve, unit weight determination, organic content by ignition, direct shear testing, undrained unconsolidated (UU) triaxial testing, isotopically consolidated and undrained (CIU) triaxial testing, compressive strength and elastic moduli of intact rock, pH, resistivity, sulfate ion concentration, and chloride ion concentration.

SPT $N1_{60}$ values per stratum vs. elevation and equivalent SPT $N1_{60}$ values correlated from Cone Penetrations Testing per stratum vs. elevation are presented on Figure 4 below. This yielded the following correlation, which shows reasonably close agreement;

SPT $N1_{60} = C_n (q_c / p_a) / [8.5 (1 - (I_c / 4.6))]$, where

C_n is the overburden correction factor, q_c is the measured cone tip resistance in kips per square foot (ksf), p_a is atmospheric pressure in ksf, and I_c is the CPT material index. The close interval continuous data collection of the CPT results in a significantly greater data set than the SPT test.

87 SPT soil borings taken as part of this project yielded 1471 SPT N_{60} values as a result of the sampling interval. This data was obtained at a cost of approximately $382,523 (in addition to drilling and sampling this cost includes mobilization, Shelby tubes, well, etc.). This corresponds to $260 per SPT N_{60} value. 45 CPT's taken as part of this project yielded 17,672 equivalent SPT N_{60} values. This data was obtained at a cost of $144,517 (in addition to pushing the CPT, this cost includes mobilization, shear wave velocity tests, pore water pressure dissipation tests, etc.). This results in $8 per equivalent SPT N_{60} value.

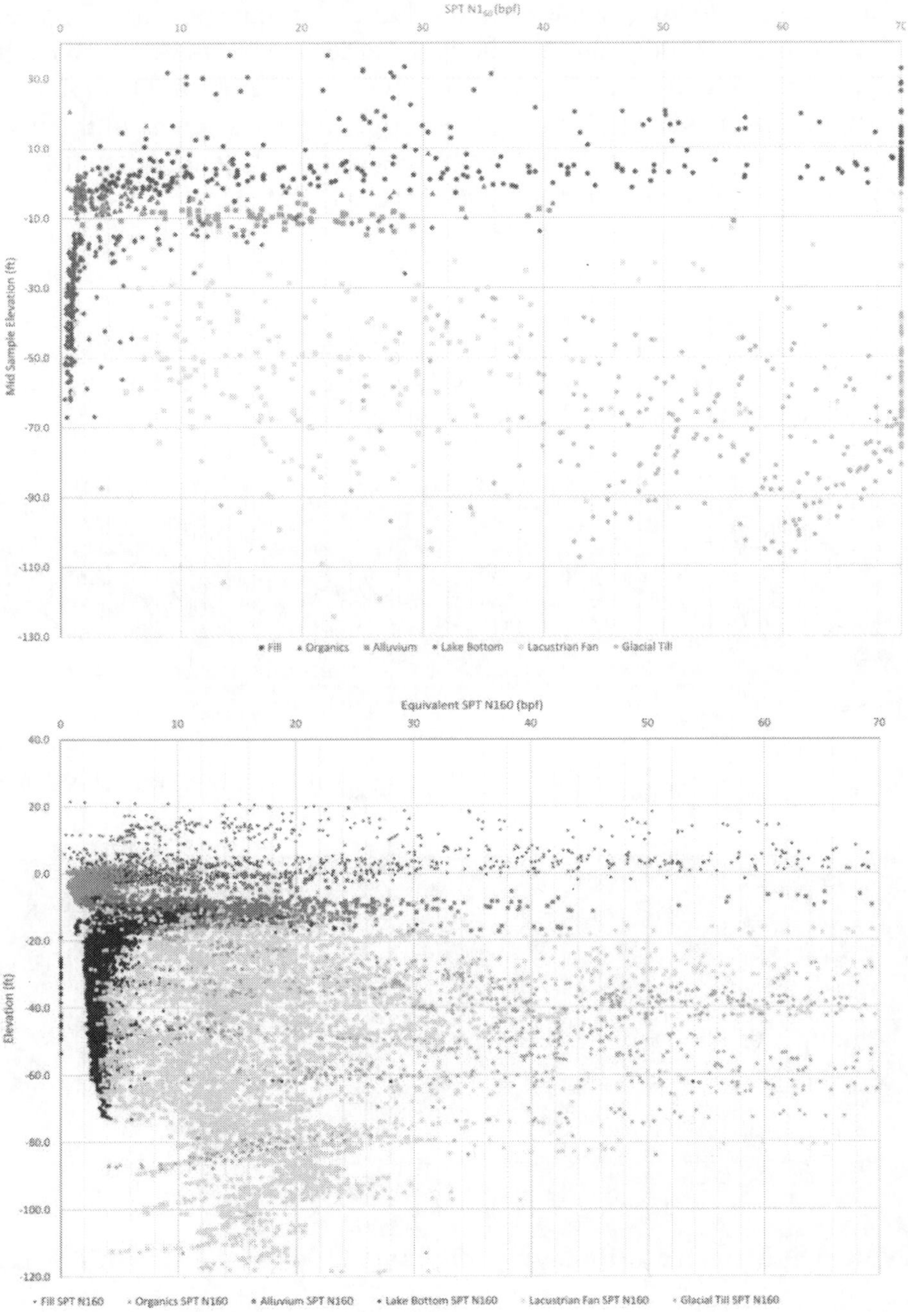

Figure 4 – Comparison Between SPT $N1_{60}$ and CPT Equivalent SPT $N1_{60}$

This example illustrates the economy and benefit of improved data resolution by using CPT's, as shown in Fig. 4. The author is not recommending SPT borings be entirely replaced, as the ability to collect physical samples for laboratory testing to calibrate CPT's results is essential. Even greater benefits are realized when the additional information obtained from the porewater pressure dissipation test data and shear wave velocity test data are considered, and the ability to reliably correlate to virtually all soil parameters.

SUBSURFACE CONDITIONS

Seismicity

CPT shear wave velocity test data was used to determine site class. The Peak Ground Acceleration for the site is 0.100g, and the site is Seismic Site Class E, which results in a site adjusted peak ground acceleration of 0.249g. The southern portion of the alignment, including the two southern most proposed rail bridges, is Seismic Site Class D, resulting in an adjusted peak ground acceleration of 0.159g. Tolerable factors of safety against liquefaction were calculated using the Boulanger and Idriss (2006) Method (Ref. 3).

Lithology

Much of the example project alignment is underlain by fill, organic soils, alluvium, lake bottom deposits, and lacustrian fan deposits overlying till, which is common throughout the Hackensack Meadowlands region, as shown in Figure 5, below.

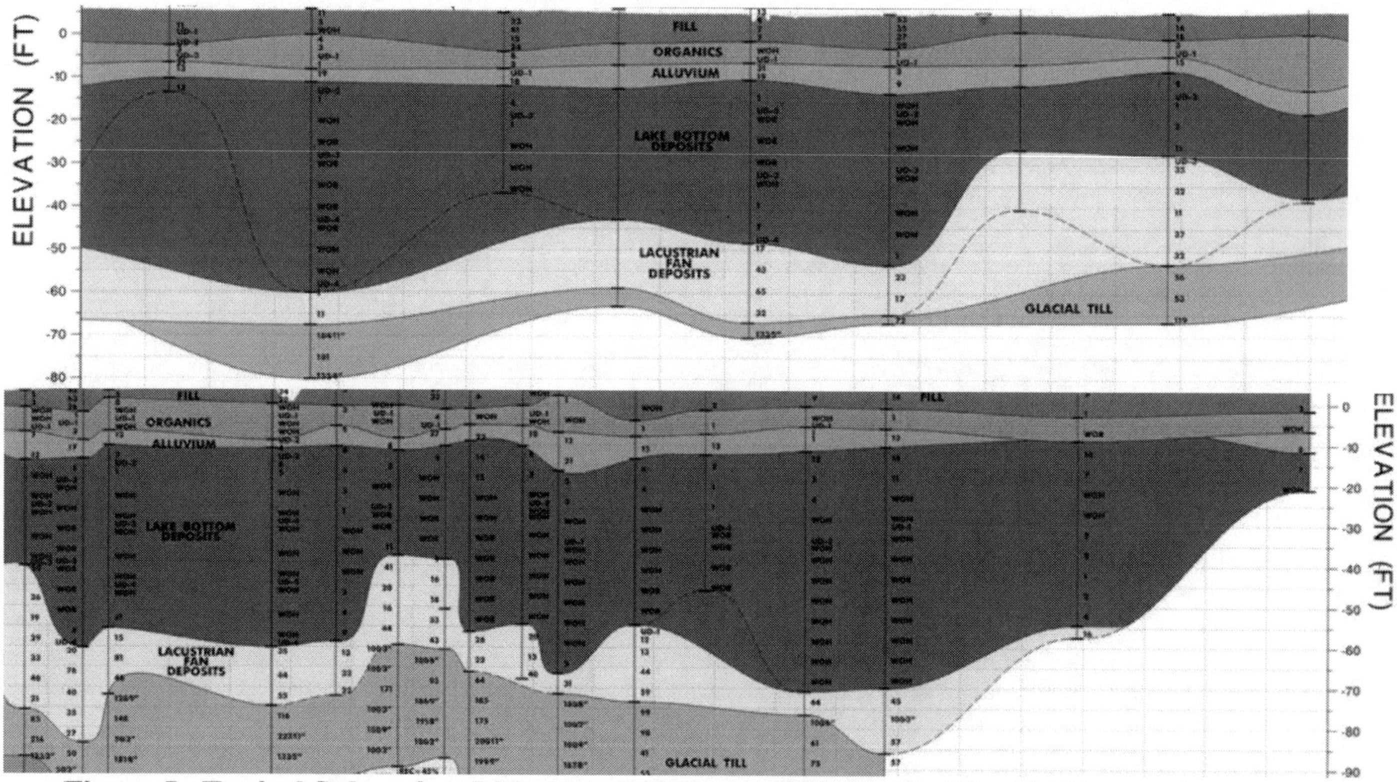

Figure 5– Typical Subsurface Lithology (1000' and 1100' Typical Profile Excerpts)

Fill

Fill was generally granular in nature and comprised of varying proportions of sand, gravel, and fines. At some locations, brick, glass, coal, porcelain, wood, and cinders were also encountered in the fill. The thickness of the unit ranged between 0 and 20 feet thick, but was generally less than ten feet thick. The fill's relative density ranged between very loose and very dense, with SPT N_{60} values ranging from one and refusal, with an average of 21 blows per foot (bpf).

Organics

Underlying the fill, and nearly continuously present throughout the site, exists dark brown to black organic soils, deposited post glacially in tidal marshes and brackish estuaries. The organics were generally amorphous or fibrous peat, however, organic clays and silts, and varying amounts of sand may be intermittent throughout this layer, which is known locally as Meadow Mat. Fibrous peat is most commonly identified in the stratum. The SPT N_{60} values generally ranged between weight of rods and 10, with an average of 3 bpf and median of 1 bpf, which corresponds to a consistency of very soft to soft. This material is highly compressible, and susceptible to consolidation and secondary settlements when loaded. This stratum varied in thickness from 0 to 10 feet, with an average of 6.5 feet. Plasticity index ranged from 2 to 564 with an average of 166. Natural moisture content ranged from 11% to 742% with an average of 271%. Typical sample photos are shown below in Figure 6.

Alluvium

Deposited by post glacial streams, the Holocene age alluvial deposits of the Hackensack Meadowlands generally consist of sand with varying amounts of silt and clay or low plasticity fines. Alluvium was encountered between the organics and lake bottom deposits. The thickness of this stratum varied from 0 to 8 feet, with an average of 4 feet, where present. SPT N_{60} values ranged from 3 to 31 bpf, with an average of 12 bpf, corresponding to a relative density of medium dense.

Lake Bottom

Glaciolacustrian brown and red cohesive fine-grained soils generally with $^1/_{16}$ to ¼ inch thick varves of sand and silt were encountered throughout much of the alignment. This stratum ranges from 10 to greater than 60 feet thick, when present. SPT N_{60} values generally ranged from weight of rods to generally less than 13 bpf and an average of 1 bpf, with several outliers excluded. It is not uncommon for 30 feet in this material to consecutively have less than or equal to one blow per foot. This stratum is believed to be combination of Glacial Lake Hackensack deposits underlain by Glacial Lake Bayonne deposits, both of the late Wisconsinan stage of the Pleistocene Epoch. Plasticity index ranged from 3 to 30 with an average of 17. Natural moisture content ranged from 17% to 62% with an average of 37%. Typical sample photos are shown below in Figure 6.

Lacustrian Fan

Lake bottom deposits were underlain by lacustrian fan deposits, which generally varied from 5to 65 feet thick. SPT N_{60} values generally ranged between 3 bpf and refusal, with an average of 32 bpf, which corresponded to a relative density of dense. The gradation primarily consisted of sand with varying amounts of gravel and silt.

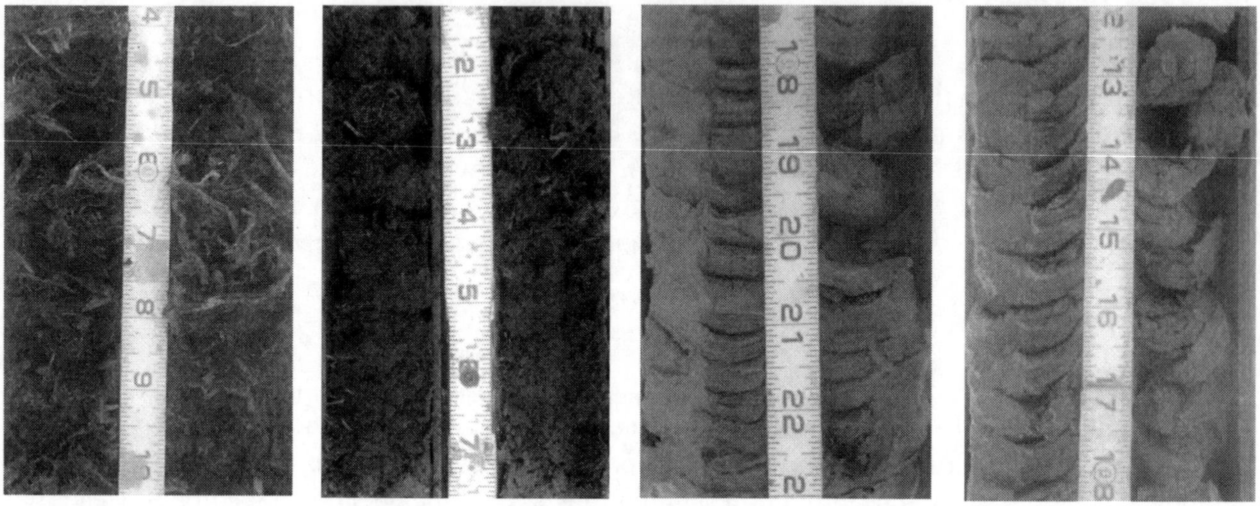

Figure 6 – Typical Split Spoon Samples
Left - Fibrous and Amorphous Peat (Meadow Mat)
Right - Varved Glaciolacustrian Silty Clay (Lake Bottom)

Glacial Till

Rahway till was deposited as the glaciers advanced and retreated, and scraped away rock more susceptible to erosion, creating a well sorted, very dense stratum. SPT N_{60} values generally ranged from 20 bpf to refusal, with an average of 82 bpf. Till may be 40 feet thick or more, but many borings were terminated after several consecutive refusal samples in till.

Decomposed Rock

Decomposed rock was encountered in several borings. The thickness of decomposed rock ranged from 0 to 15 feet thick. SPT testing consistently encountered refusal in this material, which generally consists of brown, red, and white silt, with varying amount of gravel and sand. The soil particles were cemented and exhibited rock like structure.

Bedrock

Bedrock was only encountered in several of the soil boring, with top of bedrock encountered between elevation -85 feet and elevation -125 feet. Hornfels and Diabase bedrock were encountered. Given the depth of rock, it is out of the zone of interest for this project, as driven piles could achieve the required resistance in the overlying till.

Groundwater

Groundwater observation wells were installed in 21 soil borings, upon completion of drilling. The wells were constructed of 2 inches or 4-inch nominal diameter PVC casing with a 5 or 10-foot perforated screen. The wells were equipped with Onset HOBO® MX2001water level data loggers to record drilling to monitor groundwater elevations over time, which was useful for obtaining seasonal and tidal fluctuations, as well as artesian conditions. Groundwater is commonly at or shallowly below the ground surface in this region.

SOIL PROPERTIES

The geotechnical analysis was most sensitive to the properties of the peat (organics) and the varved glacial lake clay (lake bottom deposits). This section describes the multifaceted approach to estimate critical parameters in these two strata and along with the various methods used to estimate these parameters. Soil properties were determined by comparison and interpretation of SPT and CPT in-situ test results, laboratory test results, published correlations, and typical values. This section also discusses the corrosion testing results.

Undrained Shear Strength

Two common published correlations for undrained shear strength, Stroud and Butler 1974 (Ref. 9) and Sowers 1979 (Ref. 22), were used to estimate the undrained shear strength of the lake bottom deposits based on SPT N_{60} and plasticity index. SPT N_{60} correlation to undrained shear strength was relied upon for the selection of undrained shear strength as the results did not correlate well to the laboratory measured undrained shear strength.

Laboratory measurement of undrained shear strength from unconsolidated undrained (UU) triaxial tests were performed on undisturbed samples in accordance with ASTM D2850. These test results were weighted most heavily in selection of undrained shear strength as they are direct measurements. Care was taken to minimize sample disturbance and properly saturated samples before running the test. The data sample included:

- 50 UU triaxial tests on organics
- 137 UU triaxial tests on lake bottom deposits

Undrained shear strength was correlated from in-situ CPT results. The following two CPT correlations were also used to estimate undrained shear strength:

1) Total cone tip resistance, total stress, and bearing factor = $(q_t - \sigma_{v0})/N_{kt}$, where $N_{kt} = 15$
2) Porewater pressure = $(u_2 - u_0) / N_{Du}$, where $N_{Du} = 10$

The variation of N_{kt} and N_{Du} from within the recommended range of these parameters results in a range of undrained shear strengths. Selection of N_{kt} and N_{Du} from within the recommended range was based on correlation to UU triaxial test results.

For the peat, a correlation to effective stress based on data published for peats from Holland (Ref. 5) was also found to yield similar results to the UU triaxial tests:

$$S_u = 2.1 + 0.62\,\sigma'_{vo},$$

where S_u is the undrained shear strength in kpa and σ'_{vo} is the vertical effective stress in kpa. Being the peat is normally consolidated, this results in high c/p ratios, which is supported by research for peats from Holland, which suggest the c/p ratio for peat may be 0.62 for slightly overconsolidated peats (Ref. 6).

For the lake bottom deposits, the undrained shear strength UU triaxial test results were comparable to 0.22 Pc, where Pc is the preconsolidation pressure.

For organic soils encountered on the project site, strength parameters were based on statistical evaluation of laboratory testing results, as correlations for highly variable materials like peat are less reliable. Based on these findings an undrained shear strength of 250 psf was estimated for the organics. For the lake bottom deposits, statistical evaluation of UU triaxial test results, the CPT porewater pressure correlation, and the 0.22 Pc correlation were found to be in close agreement, and an undrained shear strength of 500 psf was estimated.

Internal Effective Drained Friction Angle

Internal effective drained friction angles for the organics and lake bottom deposits were estimated based on laboratory direct shear tests performed in accordance with ASTM D3080 and laboratory isotopically consolidated and undrained (CIU) triaxial testing performed in accordance with ASTM D4767. The data sample included:

- 6 Direct Shear Tests in Organics
- 8 Direct Shear Tests in Lake Bottom Deposits
- 15 CIU Triaxial Tests in Lake Bottom Deposits

These results were compared to a CPT correlation for internal effective drained friction angle = $29.5\ B_q^{\ 0.121}[0.256 + 0.336\ B_q + \log Q]$, where Bq is the normalized pore water pressure = $(u_2-u_0)/(q_t-\sigma_{vo})$, and Q is a stress normalized CPT parameter to account for depth = $(q_t-\sigma_{vo})/\sigma'_{vo}$.

These results were compared to the following published correlations of internal effective drained friction angle of clays to plasticity index:

1) Adapted from Terzaghi et.al. 1996 (Ref. 9, Figure 7-45)
2) Bjerrum and Simons 1960 (Ref. 7, Page 74, Table 5.16)
3) Louisiana Alluvial Clays (Ref. 7, Figure 5.44)

The CIU triaxial tests, direct shear tests, and CPT correlation resulted in considerably greater friction angles for the organics than the correlations to plasticity index. It is believed the fibrous nature of the peat contribute to the high strength parameter. This notion is supported by data from peats in Holland (Ref. 6), which documents internal effective drained friction angles in organic soils and peats ranging from 35 to 83 degrees and attributes these high values to fibers.

The CIU triaxial tests, direct shear tests, and CPT correlations resulted in slightly greater friction angles than the correlations to plasticity index. It is believed the presence of silt and sand varves in the lake bottom deposits contributed to the high friction angles measured.

Internal effective drained frictions angles of 33 and 26 degrees, were assigned for the organics and lake bottom deposits, respectively.

Consolidation Parameters & CPT Porewater Pressure Dissipation Tests

Laboratory incremental consolidation tests were conducted in accordance with ASTM D2435, and they served as the primary method to determine the compression index, Cc, recompression index, Cr, secondary compression index, $C\alpha$, initial void ratio e_0, preconsolidation pressure, Pc, and the vertical coefficient of consolidation, c_v. The data sample included:

- 19 consolidation tests in the organics
- 42 consolidation tests in the lake bottom deposits

These values were compared with published correlations to moisture content and Atterberg limits, and typical values published for similar materials, but the laboratory test results were weighted most heavily in the selection of the consolidation parameters.

Both the organics and lake bottom deposits were evaluated as normally consolidated soils (OCR = 1), although some OCR's were greater than 1.0 based on unit weight assumptions and estimated preconsolidation pressures, both of which influence OCR.

Based on statistical considerations of the lab test data per stratum, the following properties were estimated:

Table 1 – Consolidation Properties of Hackensack Meadowlands Data Set

Stratum	Initial Void Ratio, e_0	Compression Index, Cc	Recompression Index, Cr	Coefficient of Vertical Consolidation, C_v (ft^2/yr)
Organics	6.5	4.0	0.55	30
Lake Bottom	1.08	0.3	0.03	175

The compression index values fall within the wide range of typical values for organic soils provided by Holtz and Kovacs (Ref. 12) of 1.5 to 15.

Mesri and Godlewski 1977 (Ref. 9) found the ratio of the secondary compression index to the consolidation index ($C\alpha/Cc$) to typically be 0.05 ± 0.01 for organic clays and silts and 0.075 ± 1 for peats. Consolidation tests from the Hackensack Meadowland data set, suggest the $C\alpha/Cc$ ratio may be considerably lower. It is recommended site-specific consolidation testing be performed or test embankments instrumented and monitored to better estimate the secondary settlement index.

Cone Penetrating Testing porewater pressure dissipation tests were conducted to obtain the horizontal coefficient of consolidation, to aid in estimating time rate of consolidation settlement. The coefficient of horizontal consolidation, c_h, in the organics varied from 380 ft^2/yr to 25,700 ft^2/yr, with an average of 7,400 ft^2/yr. The coefficient of horizontal consolidation in the lake bottom deposits varied from 480 ft^2/yr to 29,600 ft^2/yr with an average of 7,000 $ft^2/year$. Typically, the ratio of c_h/c_v is 1.2 to 1.5 for clays and 2 to 10 for varved clays (Ref. 11). The coefficients of vertical consolidation were estimated using the minimum coefficients of horizontal coefficient divided by the c_h/c_v ratio. These in-situ test results were supported by the coefficient of vertical consolidation from laboratory consolidation tests, however, the resulting coefficient of vertical consolidation was greater than typical values from Holtz and Kovacs (Ref. 12). Elias et. al. reports, "even with proper laboratory techniques and high-quality samples the designer is fortunate to be within 50 percent of the actual coefficient of consolidation" (Ref. 11).

Although the results were highly variable, the CPT's porewater dissipation tests are beneficial in providing in-situ properties to better estimate the time rate of settlement, especially in varved soils such as the lake bottom, but also in fibrous peats given the high void ratio and horizontal drainage.

Shear Wave Velocity and Elastic Modulus

Shear wave velocity tests from the seismic CPT's were used to evaluate the seismic site class and liquefaction potential. Shear wave velocity was also used to estimate maximum shear modulus, using the following equation:

$G_0 = \rho\, v_s^2$, where

G_0 is maximum shear modulus,
ρ is soil density,
v_s is shear wave velocity

This allowed shear modulus to be calculated using a published shear modulus reduction curves for sand as a function of shear strain (Ref. 8, Figures 4-14 and 4-15), and elastic modulus to be estimate based using the following equation:

$E = G\, 2(1+\nu)$, where

E is the elastic modulus,
G is the shear modulus,
ν is the Poisson's ratio

The resulting elastic modulus values were considerably higher than those estimated using SPT correlations, and allowed for refined elastic settlement magnitudes in cohesionless soils.

Corrosion

Laboratory testing, included pH (ASTM G51), resistivity (ASTM G57), sulfate content (ASTM C1580), chloride content (ASTM D4237), and organic content (ASTM D2974), was performed to assess the aggressive nature of the soil and groundwater found on-site to aid in estimating the corrosion rate, service life of steel elements, and mitigation strategies. The electrochemical testing results document the site is an aggressive corrosive environment. Due to the laboratory testing results and the historic land uses of the Hackensack Meadowlands, the consultant has assumed 0.003 inches of corrosion loss will occur per year which is based on the FHWA's GEC No. 12 - Design and Construction of Driven Piles (Ref. 10).

The following results are based on approximately 25 suites of electrochemical testing:

- Resistivity (ohm-cm) 149 to 10,200, with an average of 2129 and median of 1300
- pH 4.8 to 10.0, with an average of 6.9 and median of 6.8
- Sulfates (ppm) 30 to 7,268, with an average of 911 and median of 242
- Chlorides (ppm) 181 to 1293, with an average of 737 and median of 737
- Organic Content (%) 0 to 72, with an average of 11 and median of 5

Aggressive thresholds based on AASHTO LRFD Bridge Design Specifications (Ref. 1, 10.7.5) are less than 2,000 ohm-cm resistivity, greater than 1,000 ppm sulfates, greater than 1,000 ppm chlorides, or pH less than 5.5 or between 5.5 and 8.5 with a high organic content.

To mitigate this risk the consultant minimized the use of steel foundation elements, and employed mitigative measures using multiple levels of corrosion protection ASTM A690 Marine Grade Steel, inorganic zinc primer, and coal tar epoxy coatings.

ALTERNATIVES AND RECOMMENDATIONS

Roadway Embankment Alternatives

Construction of roadway embankment was influenced by right-of-way, global stability, settlement, protection of adjacent structures and utilities, desired construction duration, and cost. Alternatives considered included surcharged embankment with wick drains, lightweight embankment using expanded polystyrene (EPS), cellular (foamed) concrete, or lightweight soil aggregate, anchored sheetpile supported embankment, and column supported embankment with prefabricated modular walls.

The surcharged embankment alternative included installing prefabricated vertical wick drains for consolidation acceleration, installing settlement platforms and vibrating wire piezometers, constructing embankment with sloped sides, and placing temporary surcharge. For this approach, settlement platform and vibrating wire piezometer data would be monitored and used to establish the completion of consolidation settlement, at which time settlement platforms will reach a plateau and increased porewater pressure, associated with the added embankment stress, which will dissipate to an equilibrium condition. This solution is preferred where sufficient right-of-way exists given its simplicity and lower cost. This alternative posed a risk to existing structures and utilities at some locations and would have required multiple stages in coordination with utilities relocation. Time available for surcharging is also a key factor in the desirability of this alternative.

The lightweight embankment alternative consisted of over excavating and replacing existing soils with lightweight materials such as expanded polystyrene (EPS), cellular (foamed) concrete, or lightweight soils aggregates. The intent of this alternative is to yield no net stress increase to prevent global stability issues and minimize settlement. This alternative was generally not desirable because much of the alignment had limited proposed fill heights to balance buoyant forces on EPS and because existing soils being over excavated would be mainly the organics, which have a low unit weight of 65 pcf. Additional concerns with this approach are potential degradation of EPS from petroleum products and other potential chemical attack. This alternative would also require excavation and disposal in environmentally regulated soils, potential dewatering, and potential treatment of groundwater removed. This alternative is desirable as it may eliminate time for surcharging.

The anchored sheetpile supported embankment alternative included installation of two parallel rows of permanent steel sheetpiles, installation of settlement platforms and vibrating wire piezometers, placement of proposed fill between the two sheetpile walls, installation of walers, installation of encapsulated threaded bar tie rods between the two walls within an isolation casing, additional placement of proposed fill and surcharge. For this approach, settlement platform and vibrating wire piezometer data would be monitored and used to document the completion of consolidation settlement, at which time settlement platforms will reach a plateau and increased porewater pressure, associated with the added embankment stress, will dissipate to an equilibrium condition. After the completion of settlement, the surcharge would be removed and excavation would be performed to install proposed drainage pipes. Isolation of the tie rods from large magnitudes and excavation for the drainage complicate this alternative and make it labor intensive. Corrosion mitigation also increased the cost of this alternative, as well as needs to install facing for esthetics.

The column supported embankment and prefabricated modular wall alternative consisted of installing timber pile or continuous modulus columns, placing a load transfer platform consisting of aggregate and geogrid, constructing prefabricated modular walls and filling between the walls. This alternative did not require a surcharge. Pile were proposed to be driven to dense underlying granular soils. This alternative was implemented through a performance based specification, to allow the Contractor's engineer to optimize column materials, column spacing, and load transfer platform design to achieve tolerable factors of safety against global stability and tolerable settlement.

Table 2 – Roadway Embankment Alternatives

Alternative	Right-of-way	Global Stability	Settlement	Nearby Utilities or Structures	Construction Duration	Cost
Surcharged Embankment	Inadequate	Inadequate for Maximum Height	Risk Takes Longer Than Estimate	Problematic at Select Locations	Moderate	Low
Lightweight Embankment	Inadequate	Adequate	Still Susceptible to Secondary Settlements	Acceptable	Short	Moderate
Anchored Sheetpile Supported Embankment	Adequate	Requires Anchored Walls	Risk Tie Rods will be Stressed if Magnitude is Larger than Estimated	Requires Multiple Stages and Relocation	Long	High
Column Supported Embankment & Prefabricated Modular Wall	Adequate	Adequate	Low Risk	Restricts Access Requiring Utilities Relocation	Moderate	High

Green cells are desirable, yellow cells are neutral, and red cells are undesirable

Roadway Embankment and Retaining Wall Design

Limit equilibrium global stability analysis for the alternatives listed above was performed using Geo-Slope Geostudio 2016 Slope W software. Stability was analyzed for the undrained condition and drained (effective stress) condition. The undrained shear strength of the peat layer was controlling the design, and resulted in lower factors of safety for stability than the drained analysis, because of the high friction angle used for peat. Rocscience's Settle 3D software and hand calculations were used to estimate the magnitude of settlement, the time rate of settlement, the time rate of settlement for the proposed surcharge height and wick drain spacing.

Geotechnical and structural axial resistances for timber piles were calculated, and pile group settlement was check and required piles be driven to underlying dense granular soils. The load transfer platform was designed using the beam method (Ref. #) and geogrid properties were adjusted by a creep reduction factor, installation damage reduction factor, deterioration reduction factor, and an overall factor of safety of 2. A combination of uniaxial and biaxial geogrids was utilized. The cost of timber piles is anticipated to range between $20 and $25 per liner foot for furnishing and driving, which is believed to be cheaper than alternative

Jump Span Design

Accelerated Bridge Construction techniques are proposed for the replacement of two100+ year-old rail bridges due to limited track outage time. To install a full height cast in place concrete abutment founded on micropiles below an existing 3 track rail embankment, while maintaining rail service, tied back micropile and lagging walls and temporary jump span bridges are proposed. This work includes installing two rows of nine 12.75-inch OD x 0.5-inch wall micropiles from top of rail tracks during limited nighttime track closures. Upon completion of the micropile installation, during a weekend track outage, a temporary jump span and cap will be set on the micropiles. While trains remain in service, excavation will occur sequentially below the embankment and as lagging is placed between the micropiles and tiebacks and struts will be installed. This work is proposed to be performed near several existing shallow foundation rail abutments. The temporary jump span wall micropiles are proposed to be ASTM A252 Grade 3 steel to allow them to be weldable to install channels to hold the lagging. Upon completion of the micropile and lagging walls and excavation, 13.375-inch OD x 0.514-inch walls will be installed for the permanent abutment below the active tracks. The system was designed for apparent earth pressure and cooper E-80 live load surcharge. Tension load testing will be performed on a sacrificial micropile and a sacrificial tie back to determine the nominal unit grout-to-ground bond resistance. All tiebacks will be load testing.

CONCLUSIONS

Conclusions from this project include:

- CPT's add value - low cost greater resolution of data and ability to obtain many parameters
- CPT porewater dissipation testing is valuable in estimating the time rate of settlement, particularly in varved clays, as they yield the horizontal coefficient of consolidation.
- CPT shear wave velocity tests aid in evaluating site class, liquefaction, and maximum shear modulus, which may be beneficial in estimating refined elastic settlement magnitudes.
- SPT borings to obtain undisturbed samples for UU triaxial testing should be performed in conjunction with CPT's to aid in fitting N_{kt} and N_{Du} factors for estimating undrained shear strength, as well as more accurate estimates of total unit weight from lab measurement.
- The relationship between undrained shear strength of peat and vertical effective stress from the data in Holland closely matched the NJ Meadowlands peat undrained shear strengths
- The measured drained friction angle for organics from CIU triaxial and CPT correlation was surprisingly high, likely due to the fibrous nature.

- The measured drained friction angle from CIU triaxial and CPT correlation for the lake bottom deposits was slightly higher than published correlations and typical values.
- Timber piles may be more economical than other types of column to support embankment.
- Obtain enough right-of-way to construct sloped embankments with surcharge, which will be considerably cheaper than retaining wall supported embankment.
- Wick drains are relatively inexpensive and may be beneficial to include to reduce risk of delay of construction claims given the uncertainty in estimating the time rate of settlement.
- Robust pre- and post-construction inspection of existing structures, vibration monitoring, and tilt monitoring are valuable in reducing risk to the Contractor and Owner.
- Highly aggressive electrochemical properties should be anticipated

REFERENCES:

1. AASHTO. *LRFD Bridge Design Specifications*. 2014.
2. ASTM. *STP 820 – Testing of Peats and Organic Soils*. 1982.
3. Boulanger, R.W., Idriss, I.M. *Liquefaction Susceptibility Criteria for Silts and Clays*. Journal of Geotechnical and Geoenvironmental Engineering. November 2006.
4. Burmister, D.M. *Suggested Methods of Test for Identification of Soils*. American Society for Testing Materials. 1958.
5. Den Haan, E., Kruse, G. *Characterization and Engineering Properties of Dutch Peats*. 2007.
6. Den Haan, E., Feddema, A. *Deformation and Strength of Embankments on Soft Dutch Soil*. Proceedings of the Institution of Civil Engineers Geotechnical Engineering Volume 166 Issue GE3. June 2013.
7. Duncan, J.M., Wright, S.G., Brandon, T. *Soil Strength and Slope Stability*. 2nd Ed.
8. FHWA. GEC No. 3. *LRFD Seismic Analysis and Design of Transportation Geotechnical Features and Structural Foundations*. August 2011.
9. FHWA. GEC No. 5 – Evaluation of Soil and Rock Properties. 2002
10. FHWA. GEC No. 12 – Design and Construction of Driven Pile Foundations September 2016.
11. FHWA. *Ground Improvement Methods*. October 2004.
12. Holtz, R.D., Kovacs, W.D. An Introduction to Geotechnical Engineering. 1st Edition. 1981.
13. Lunne, T., Robertson, P.K., Powell, J.J.M. *Cone Penetration Testing in Geotechnical Practice*. 1997.
14. Mayne, P.W. Engineering Design Using the Cone Penetration Test - Geotechnical Applications Guide. 7th Printing - January 2017
15. NJGS. Bedrock Geologic Map of Northern New Jersey. 1996.
16. NJGS. Bedrock Map of the Hackensack Meadowlands. 1959.
17. NJGS. Glacial Sediment and the Ice Age in New Jersey. www.state.nj.us/dep/njgs/enviroed/infocirc/glacial.pdf
18. NJGS. Surficial Geologic Map of the Jersey City Quadrangle 1995.
19. NJGS. Surficial Geologic Map of Northern New Jersey. 2002.
20. NJGS. Surficial Geologic Map of the Weehawken and Central Park Quadrangles. 1993.
21. NJGS. Surficial Geologic Map of the Elizabeth Quadrangle. 2002.
22. Sowers, G.B., Sowers, G.F. Introductory Soil Mechanics and Foundations. 3rd ed. 1970.
23. US Army Corps of Engineers. New York District. Meadowlands Environmental Site Investigation. May 2004.

Slope Stability Analysis for TH53 Relocation, Virginia, MN

Anya Brose
Itasca Consulting Group
111 Third Avenue South
Suite 450
Minneapolis, MN 55401
612-371-4711
abrose@itascacg.com

Lee Petersen
Itasca Consulting Group
111 Third Avenue South
Suite 450
Minneapolis, MN 55401
612-371-4711
lpetersen@itascacg.com

Ryan Peterson
Itasca Consulting Group
111 Third Avenue South
Suite 450
Minneapolis, MN 55401
612-371-4711
rpeterson@itascacg.com

Derrick Dasenbrock
Minnesota Department of Transportation
1400 Gervais Ave
St Paul, MN 55109
651-366-5597
derrick.dasenbrock@state.mn.us

Gary Person
Minnesota Department of Transportation
1400 Gervais Ave
St Paul, MN 55109
651-366-5598
gary.person@state.mn.us

Andrew Shinnefield
Minnesota Department of Transportation
1400 Gervais Ave
St Paul, MN 55109
651-366-5431
andrew.shinnefield@state.mn.us

Luigi Cotesta
Itasca Canada
166 Douglas Street
Sudbury, Ontario P3E 1G1
1-705-522-2697
lcotesta@itasca.ca

Prepared for the 69th Highway Geology Symposium, September, 2018

Acknowledgements

For example: The author(s) would like to thank the individuals/entities for their contributions in the work described:

Derrick Dasenbrock– MnDOT
Andrew Shinnefeld – MnDOT
Gary Person – MnDOT
Luigi Cotesta- Itasca Canada

Disclaimer

Statements and views presented in this paper are strictly those of the author(s), and do not necessarily reflect positions held by their affiliations, the Highway Geology Symposium (HGS), or others acknowledged above. The mention of trade names for commercial products does not imply the approval or endorsement by HGS.

ABSTRACT

In September 2017, the Minnesota Department of Transportation (MnDOT) completed the relocation of a portion of T.H. 53 between Eveleth and Virginia, MN. The relocation project included the construction of a 1,100-foot bridge crossing the existing Rouchleau iron ore pit. The relocation of T.H. 53, plus subsequent mining, will eventually create a so-called isthmus that carries the roadway from the current alignment to the east end of the bridge. Once mining occurs, the isthmus cross-section will be a trapezoid 300 ft wide at the top, with downslopes of 53 degrees, and up to about 500 ft deep. Information regarding rock bedding, jointing, and faulting was collected using three methods: down-the-hole televiewer logging; photogrammetry; and geomechanical core logging.

The presence of subvertical joints indicated the possibility of flexural toppling along the isthmus slope. Using the collected discontinuity data, a kinematic analysis was performed and the resulting factor of safety was found acceptable. In addition to a kinematic analysis, multiple two-dimensional stability analyses were performed using *UDEC*, a discrete element modeling software. The *UDEC* models showed an unacceptable factor of safety. Consequently, a three-dimensional *3DEC* analysis was performed. The *3DEC* analysis was necessary to fully capture behavior of isthmus geometry and jointing, and to show that flexural toppling wasn't a concern.

INTRODUCTION

This rock discontinuity characterization and slope stability assessment was part of the Highway 53 (T.H. 53) re-location project (completed September 2017). MnDOT built the existing alignment in 1960 on land owned by iron mining interests (currently held by Cliffs Natural Resources and RGGS). The easement that the agency signed in order to build the road included a requirement that MnDOT would move the road with three years notice if the mining company needed to get to the ore underneath the highway. In 2010, the mining interests notified MnDOT that the road would need to be moved. Later that year, MnDOT and the mining company agreed to a 2017 date for the road move.

The location of the new alignment with respect to the existing Rouchleau iron ore pit is shown in Figure 1. The east pier and abutment and the west abutment will be founded on bedrock. The roadway was placed at approximately the current ground surface. Future mining southeast of the east abutment will create slopes up to 500 ft high. After mining, the new roadway will be atop a trapezoidal cross-section 300 ft wide at the top and with down slopes of about 53 degrees, as shown in Figure 2. MnDOT wished to ensure that the future rock slopes will be stable. A key component of rock slope design is the character of the discontinuities present in the rock. The existing rock exposure in the Rouchleau pit provides a valuable source of information. In addition, coreholes drilled along the alignment to the southeast will also provide discontinuity information.

Using collected rock mass and discontinuity data, a kinematic analysis was performed that resulted in an acceptable factor of safety. A two-dimensional numerical model using *UDEC* was created to assess slope failure modes not within the capabilities of the kinematic analysis. When the factor of safety resulting from the *UDEC* model was lower than the kinematic analysis, a three-dimensional *3DEC* model was created to better represent the interaction of the joints and the isthmus geometry.

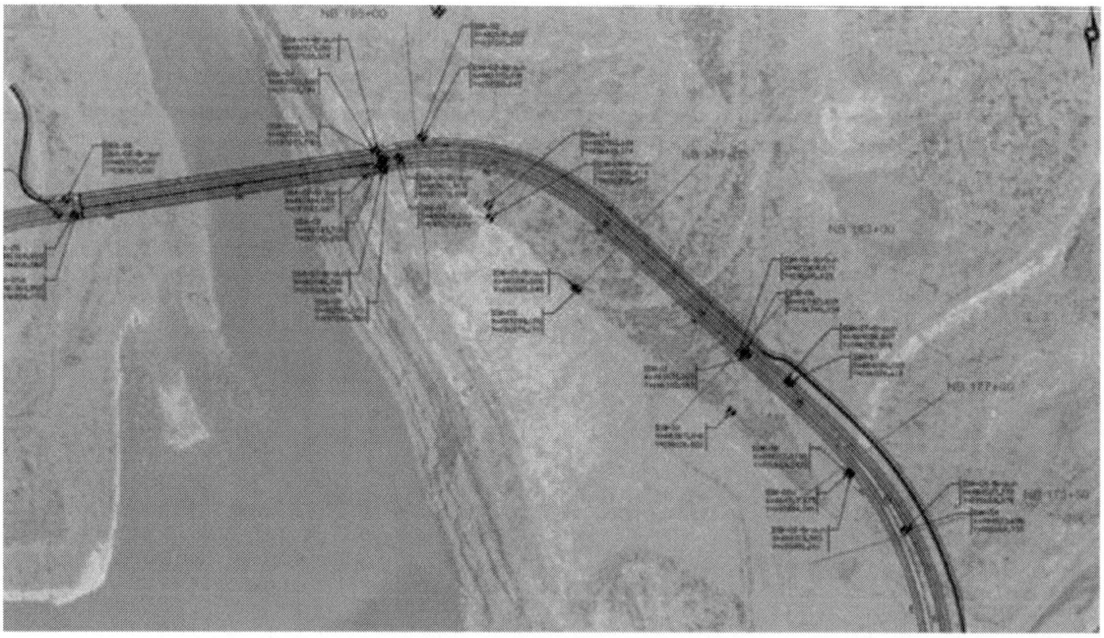

Figure 1 – T.H. 53 alignment, showing Rochleau iron ore pit.

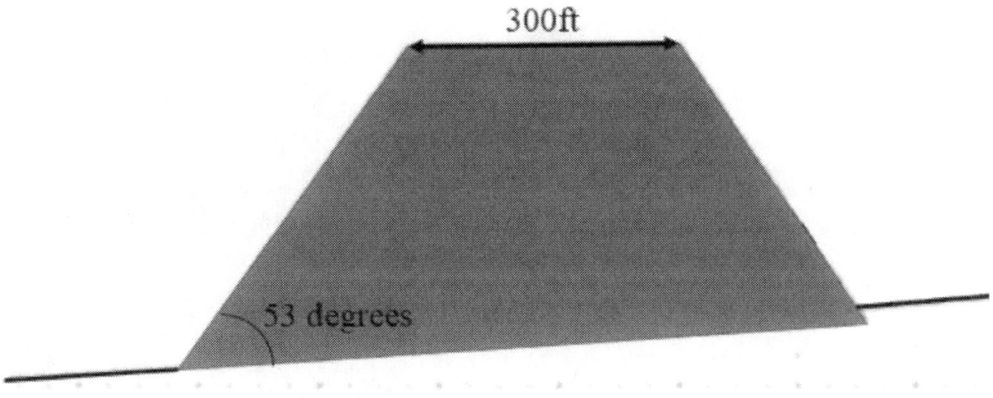

Figure 2 – Isthmus cross-section at Station 183+00.

ROCK MASS AND DISCONTINUITY DATA COLLECTED

In order to accurately model the isthmus, a field program was developed to collect rock mass and discontinuity information. The field program included the following.

- Coring, and geological and geotechnical core logging (22 coreholes with approximately 7,000 ft of core)
- Laboratory testing
- Point load testing
- Down-the-hole logging
- Photogrammetry

Rock Mass Data

Rock mass properties were determined from geotechnical core logging and laboratory testing. Lithology, weathering, RQD, and core recovery were recorded during core logging and used in rock classification. Over 1600 point load tests were performed over 10 holes and 54 uniaxial tests were performed over 15 holes to compare rock strength across the isthmus. Eventually, the isthmus was divided into five groups of similar strength and properties, shown in Figure 3 and Table 1. (Note: the “Boundary” group was assigned artificial properties to represent boundary conditions.)

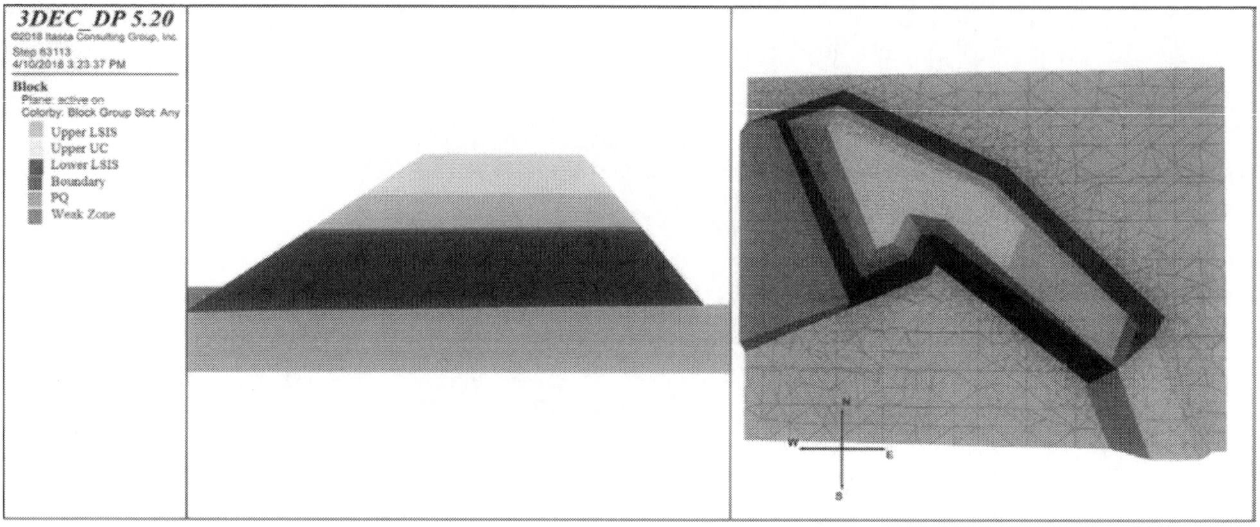

Figure 3 – Geologic units in section view (left) and in plan view (right).

Table 1 – Rock Mass Properties							
Material	Constitutive model	Density [pcf]	Shear modulus [psf]	Bulk modulus [psf]	Cohesion [psf]	Tension [psf]	Friction [deg.]
Upper UC	Ubiquitous Joint	190	3.69e8	7.99e8	2.96e4	3.74e3	44.2
Upper LSIS	Ubiquitous Joint	190	3.94e8	8.54e8	3.71e4	6.17e3	47
Weak zone	Ubiquitous Joint	190	3.88e8	8.41e8	3.86e3	3.18e2	20.9
Lower LSIS	Ubiquitous Joint	190	3.46e8	7.48e8	3.96e4	7.48e3	47.3
PQ	Ubiquitous Joint	190	4.23e8	9.17e8	3.56e4	5.42e3	46.8

Discontinuity Data

Discontinuity information was collected using three methods: geotechnical core logging, down-the-hole televiewer logging, and photogrammetry. Down-the-hole logging provided discontinuity orientation and spacing of the in-situ rock, but it did not provide persistence information. To obtain joint persistence information and additional orientation data, remotely piloted drones were used to obtain photos for photogrammetry, which allowed for data collection along the rock slopes despite the large spans and dangerous conditions. Geotechnical core logging provided information on joint condition, spacing, infill, roughness, etc., which was later used in determining discontinuity properties for the models.

Four major joint sets were identified in using the above methods, excluding bedding planes. All joints sets were subvertical with the average dip ranging from 83-90 degrees, while bedding planes were subhorizontal (Figure 4). A stereonet showing the four joint sets, with bedding planes filtered out, is shown in Figure 5.

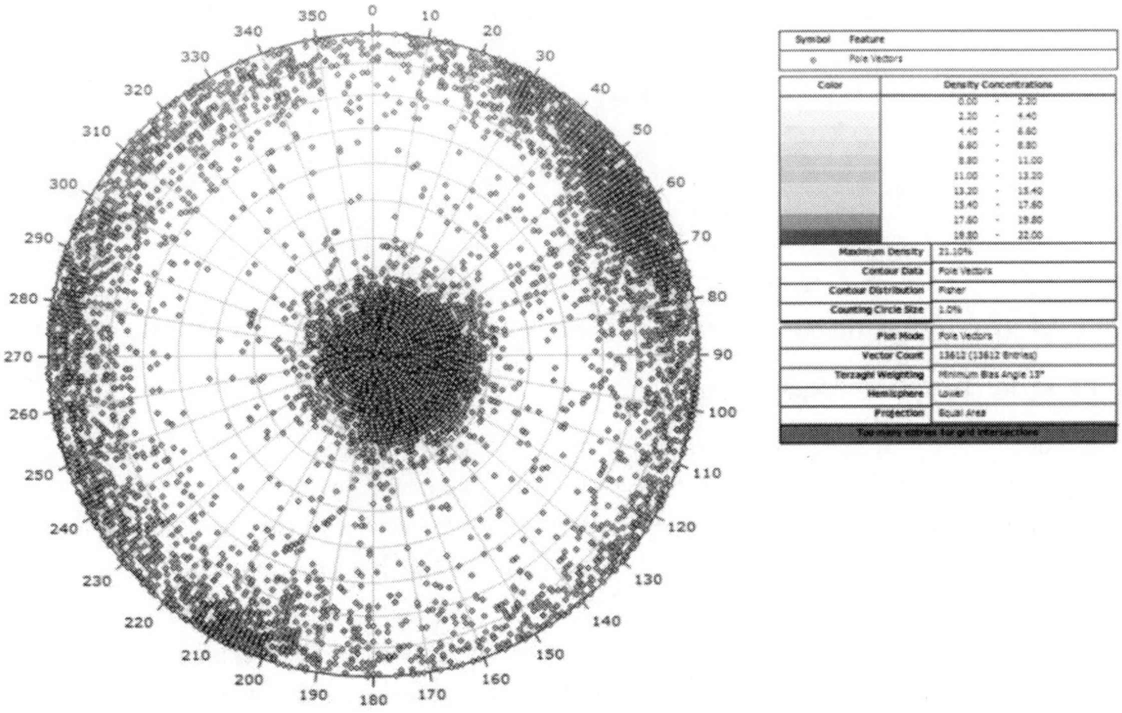

Figure 4 – Lower hemisphere stereonet of all joint and bedding orientations for the isthmus rocks.

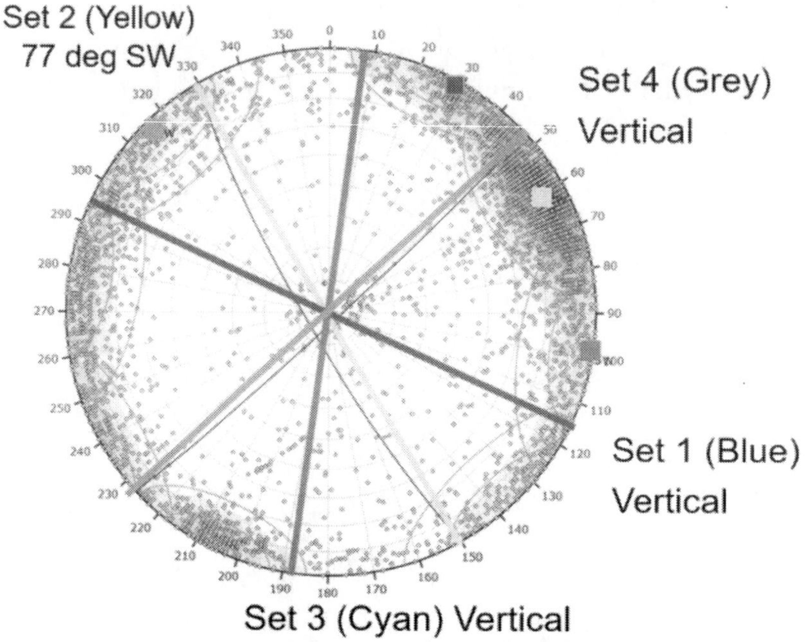

Figure 5 – Lower hemisphere stereonet of joint set orientations for the isthmus rocks.

Description of the Toppling Mechanism

There are two toppling mechanisms of slope instability: (1) block toppling, a relatively shallow mechanism in which rock blocks tip over and tumble down the slope; and (2) more deep-seated flexural toppling in which shear movement along joints and flexure of the resulting rock columns lead to slope movement (Nichol, Hungr, & Evans, 2002).

Flexural toppling is a ductile mechanism, as opposed to a brittle mechanism (Nichol, Hungr, & Evans, 2002). A ductile mechanism will move relatively slowly, instead of the rapid, catastrophic movements associated with brittle mechanisms. Flexural toppling generally occurs along joints steeply dipping into a slope, with relatively close spacing (Nichol, Hungr, & Evans, 2002). An example of flexural toppling is shown in Figure 6.

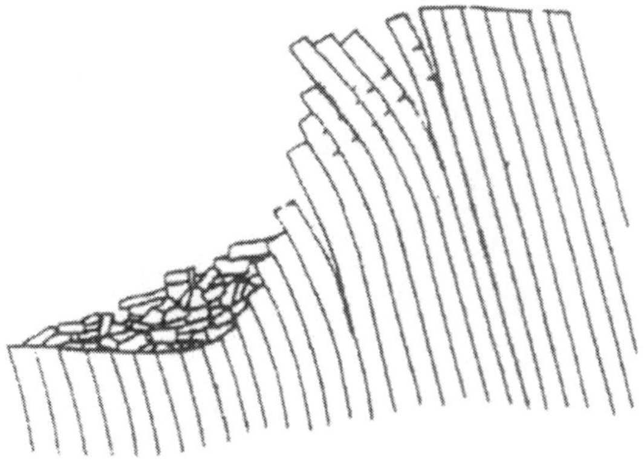

Figure 6 – Examples of rock movement caused by flexural toppling (Hittinger & Goodman, 1978).

Figure 7 shows the future isthmus intersected by the average orientation of the four joint sets. The rules of thumb listed above generally are met for the north-, northeast-, and east-facing slopes created by future mining. This caused concern that the slopes will be susceptible to flexural toppling.

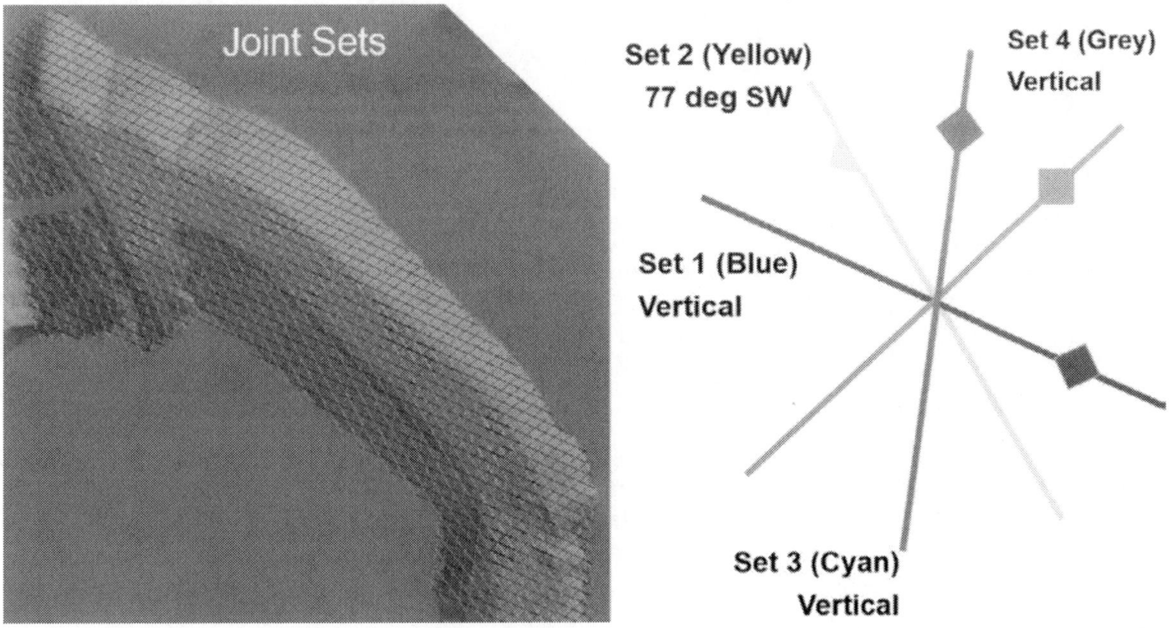

Figure 7 – Average joint set orientations intersected with the isthmus.

TWO-DIMENSIONAL STABILITY ASSESSMENT

Following a kinematic analysis which produced an acceptable safety factor, *UDEC*, a two-dimensional discrete element modeling software, was used to model multiple sections of the isthmus (Figure 8) in order to gain further insight into the failure mechanism. Six cross sections along the isthmus were chosen (Figure 9), and two joint sets were modeled: the subhorizontal bedding planes and Joint Set 2, which is defined by joints dipping 75-degrees southwest.

Properties of the bedding planes and joints were determined from geotechnical core logging, down-the-hole logging, and photogrammetry data. *UDEC* models were run both deterministically and stochastically. In the deterministic model, the friction angle of the joints and bedding planes are equal to the mean values 42.7° and 41.6°, respectively. In the stochastic models, the properties were sampled from the distributions illustrated in Figure 10.

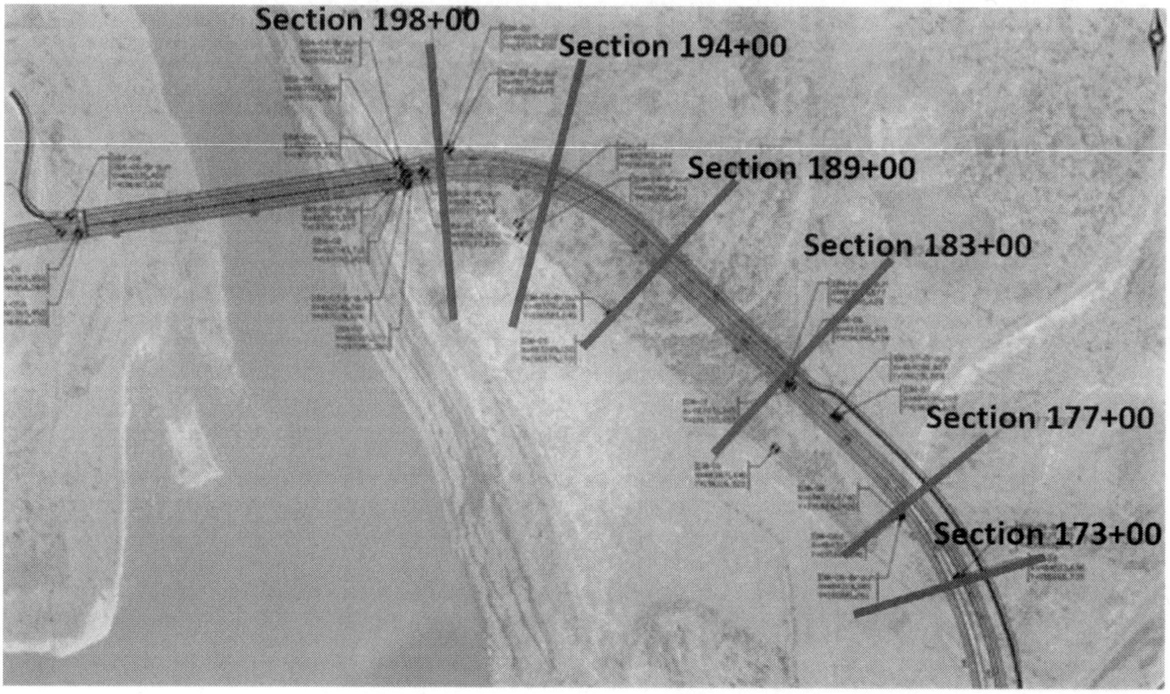

Figure 8 – Aerial image of the bridge alignment prior to construction, showing sections used for *UDEC* modeling.

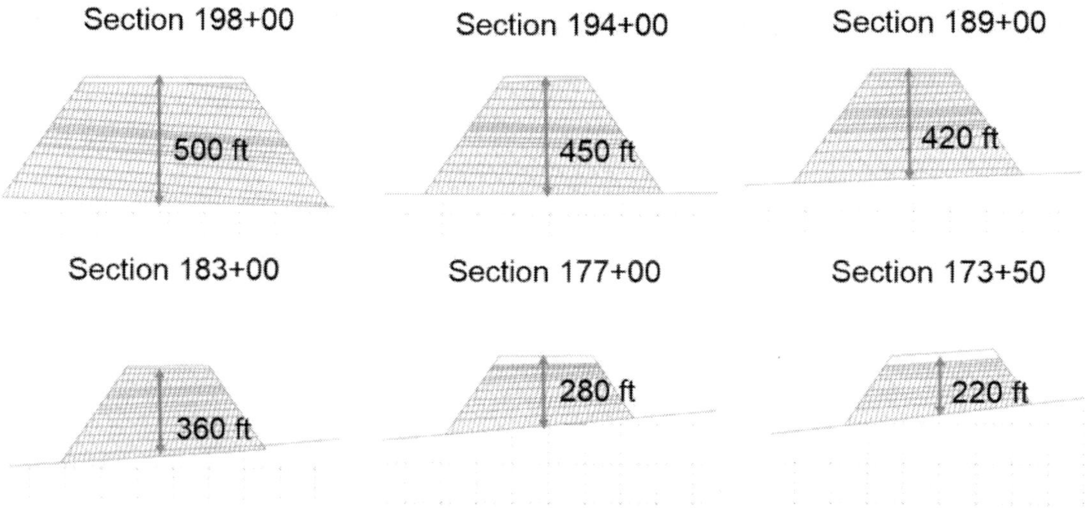

Figure 9 – Cross-sections modeled using *UDEC* (sections looking toward bridge, for joints dipping 75 degrees left).

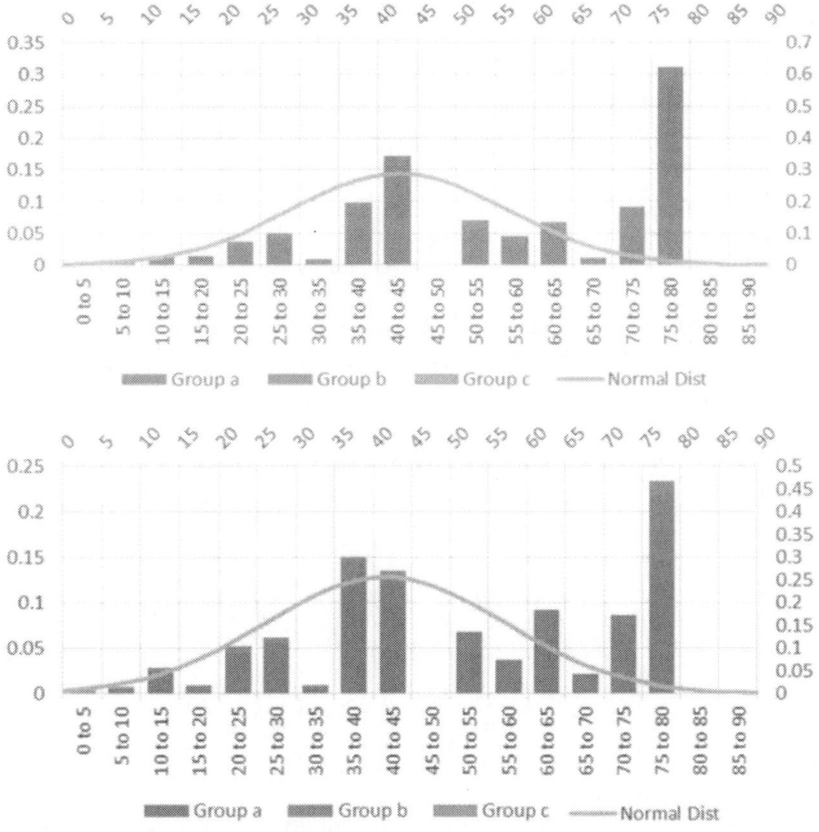

Figure 10 – Joint frictional strength properties (top) and bedding frictional strength properties (bottom).

The factor of safety for each deterministic model was calculated using the strength reduction method (SRM), a function which is built into UDEC. The SRM involved reducing strength properties until failure to obtain a factor of safety for the slope.

Displacement vectors and failure locations in the *UDEC* models were used to evaluate the failure mechanism of the slope. The left image in Figure 11 shows the displacement vectors (red is the greatest magnitude, cyan is the smallest) and the right image shows the tensile and shear failure. Failure extends from the upper left slope, daylighting on the lower right slope and the failure location corresponds to the displacement vectors in the upper right section of the slope. Figure 12 shows an exaggerated deformed shape of the slope. This view, combined with the failure locations across the slope in Figure 11, shows how the deep-seated flexural toppling mechanism is developing in the upper right section of the slope.

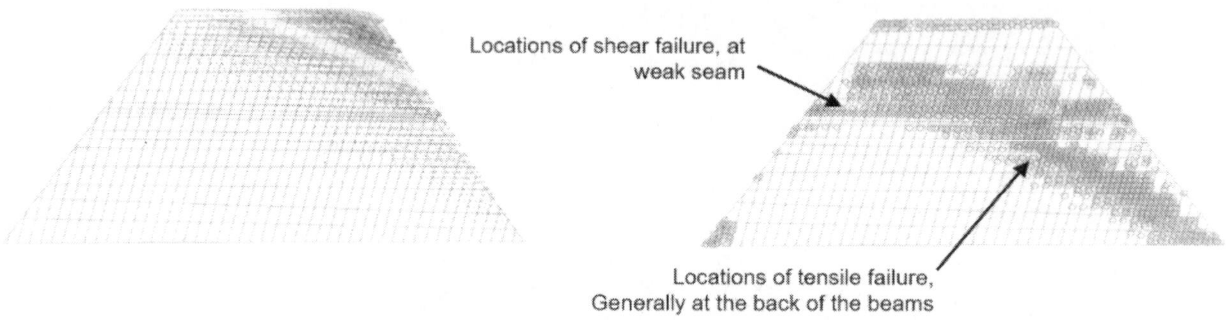

Figure 11 – Flexural toppling displacement vectors and failure locations (Station 194+00, deterministic analysis, 1.2 strength reduction factor).

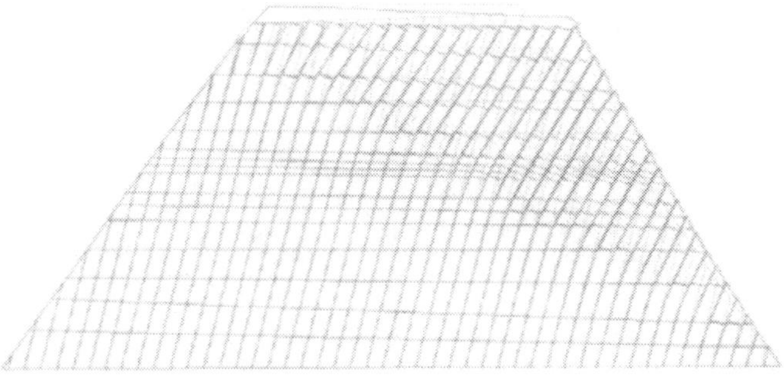

Figure 12 – Exaggerated (100x) deformed shape (in green) compared to the original shape (in grey), illustrating the nature of the flexural toppling mechanism.

Eighteen stochastic models were run for the 75-degree model. Joint and bedding properties were sampled from the distributions in Figure 10 and assigned to the model. The factor of safety results, shown in Table 2, decreases closer to the bridge and increases further from the bridge, where the height of the slopes also decreases. Additionally, the averaged factor of safety for each stochastic model compared to the corresponding deterministic model was lower. This was likely due to the assignment of below-average strength properties sampled from the distribution.

Table 2 – *UDEC* Factor of Safety Results for Deterministic and Stochastic Analyses							
Case		198+00	194+00	189+00	183+00	177+00	173+50
Deterministic		1.08	1.12	1.17	1.80	1.71	2.0
Stochastic	Run1	0.94	0.94	0.95	1.68	1.92	2.02
	Run2	1.05	1.06	1.19	1.89	1.78	1.62
	Run3	1.06	1.07	1.21	1.37	1.68	2.12
Average of stochastic analyses		1.02	1.02	1.12	1.65	1.79	1.92

THREE-DIMENSIONAL STABILITY ASSESSMENT

Motivation for three-dimensional model

Initial *UDEC* models showed an unacceptable factor of safety, which motivated the creation of a three-dimensional model. In *UDEC*, a single 2D section cannot be cut perpendicular to strike on all joint sets, which means a 2D model is capturing apparent dip rather than true dip. Thus, a two-dimensional analysis cannot fully capture the three-dimensional interaction of the joints and the isthmus. Figure 13 illustrates the distribution of dips for Joint Set 2, the cumulative distribution, and the values used in the *UDEC* and *3DEC* models. The *UDEC* dip angle was chosen based on apparent dip. As Figure 13 illustrates, the *UDEC* apparent dip angle is at about the 30th percentile of the actual dips, while the *3DEC* dip angle is at about the 60th percentile.

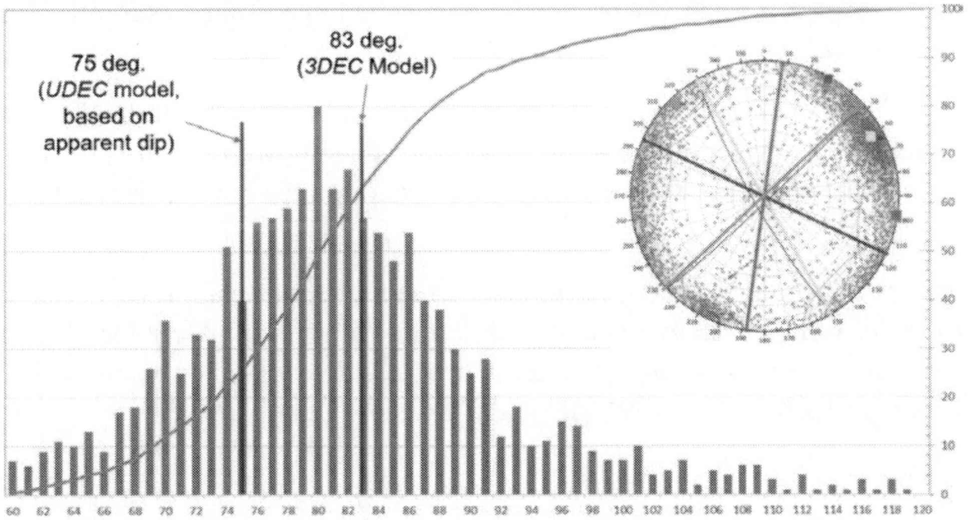

Figure 13 – Dip histogram for Joint Set 2.

To understand the effect of dip angle on the slope safety factor, the *UDEC* model for Station 194+00 was rerun by varying the dip angle for Joint Set 2. Table 2 illustrates the results. The safety factor ranges from 1.06-1.25 over a dip angle change of 10 degrees. These results confirmed the need for a 3D model to compare to the *UDEC* results.

Table 2 – Safety Factors for Station 194+00 and Varying Dip Angles

Dip	Safety Factor
Base Case 75	1.12
77	1.08
79	1.09
81	1.06
83	1.17
85	1.25

3DEC Model Creation

Joints created in the *3DEC* model were based on the vertical-to-subvertical joint sets and subhorizontal bedding planes collected on site. In *UDEC*, all joints were discretely modeled, whereas in *3DEC*, the subhorizontal bedding planes were not discretely modeled (to increase runtime). Bedding planes in *3DEC* were modeled using the ubiquitous joint constitutive model, defined by weak planes of a specific orientation embedded in a Mohr-Coulomb solid (Itasca Consulting Group Inc., 2016).

The *3DEC* model used the same material and joint properties as the *UDEC* model. The isthmus and nose were divided into five material groups, as shown in Figure 14. The four joint sets were cut into the nose and isthmus, but were not cut into the boundary materials or the base material (PQ).

The model was brought to equilibrium using the initial material properties. Then the material properties were gradually reduced to determine a factor of safety for the slope. Stability was determined based on displacement magnitude across the slope. For example, if displacements continuously increased, the model was considered unstable.

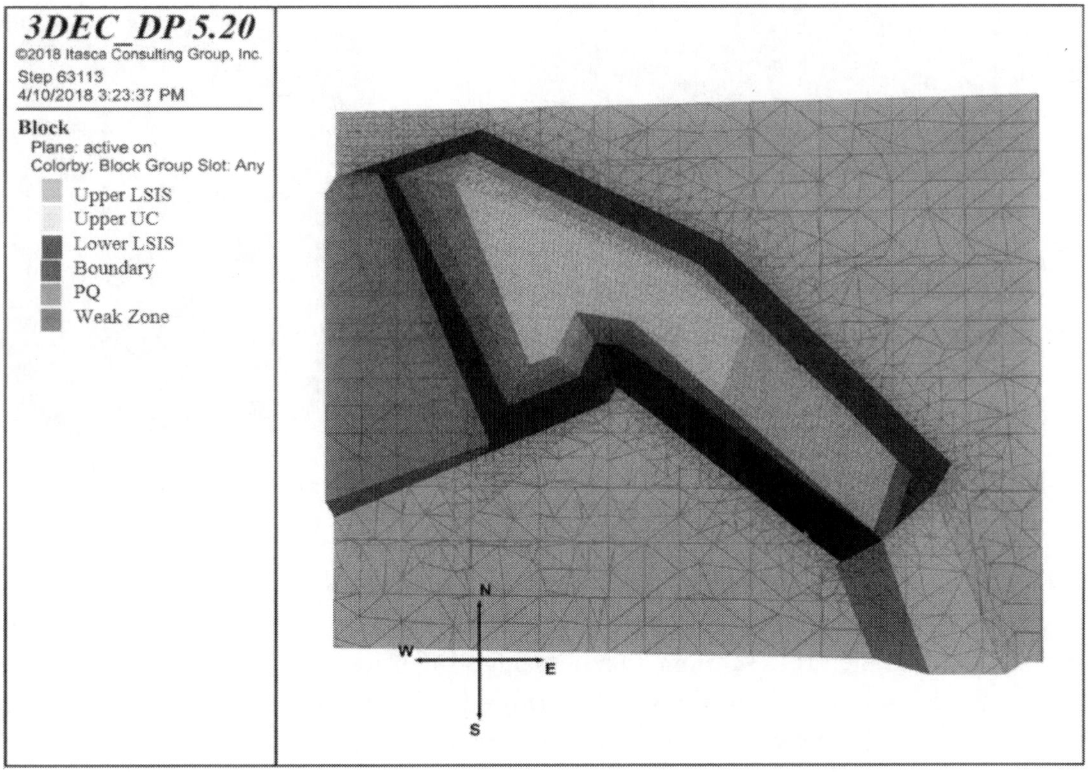

Figure 14 – Geometry of the comprehensive *3DEC* model.

The final *3DEC* model was stable up to a strength reduction factor of 1.5, but unstable when the strength reduction factor was increased to 1.8. The left image in Figure 15 shows the surface displacements across the isthmus for a factor of safety of 1.5, where the maximum displacement is approximately 0.16 ft. When the strength reduction factor was increased to 1.8, the displacements continued to increase across the slope, so the slope was determined to be unstable. The right image in the same figure shows the displacements extending across the isthmus with a factor of safety of 1.8. Note: for a factor of safety of 1.8, the scale is truncated at 0.2 ft, so all displacements above 0.2 ft plot as red.

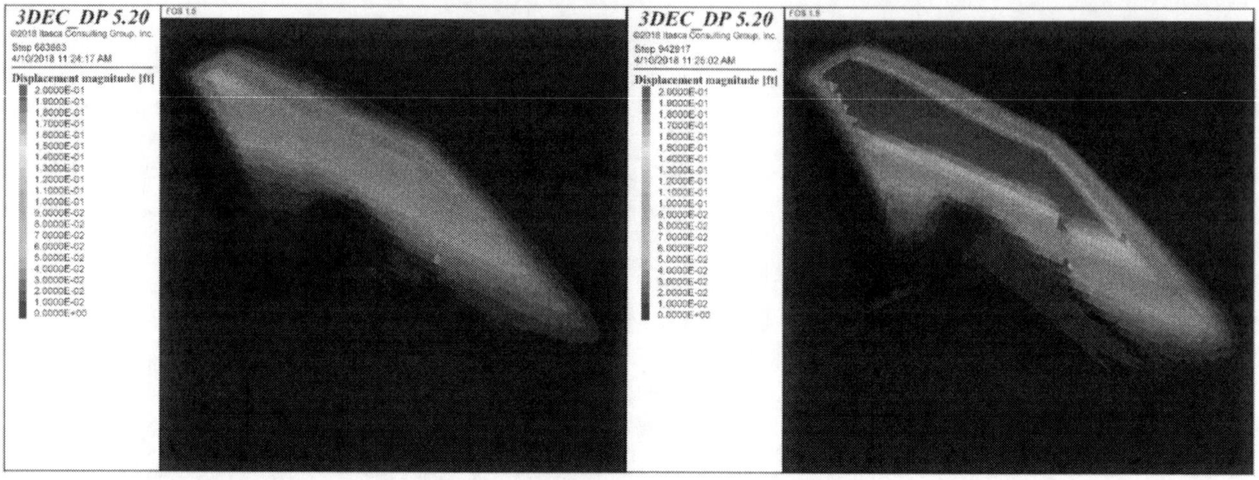

Figure 15 – Predicted surface displacements at equilibrium for a safety factor of 1.5 (left) and 1.8 (right).

To further illustrate the failure mechanism, cross sections were made at multiple locations along the isthmus to determine stability and the mechanism of failure. Figure 16 shows the location of one of the cross sections taken through the bridge abutment. Figure 17 compares the displacements through this section for strength reduction factors of 1.5 and 1.8, where it is possible to see the deep-seated movement in the slope associated with flexural toppling. Figure 18 compares the zone failure in the bridge abutment cross sections for strength reduction factors of 1.5 and 1.8. When the strength reduction factor increases to 1.8, there is a significant increase in tensile and shear failure deep in the slope, indicating flexural toppling.

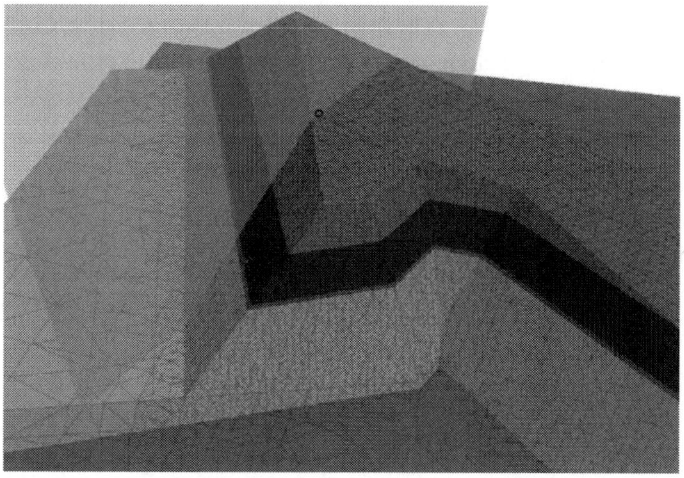

Figure 16 – Location of cross-section through the approximate location of the bridge abutment.

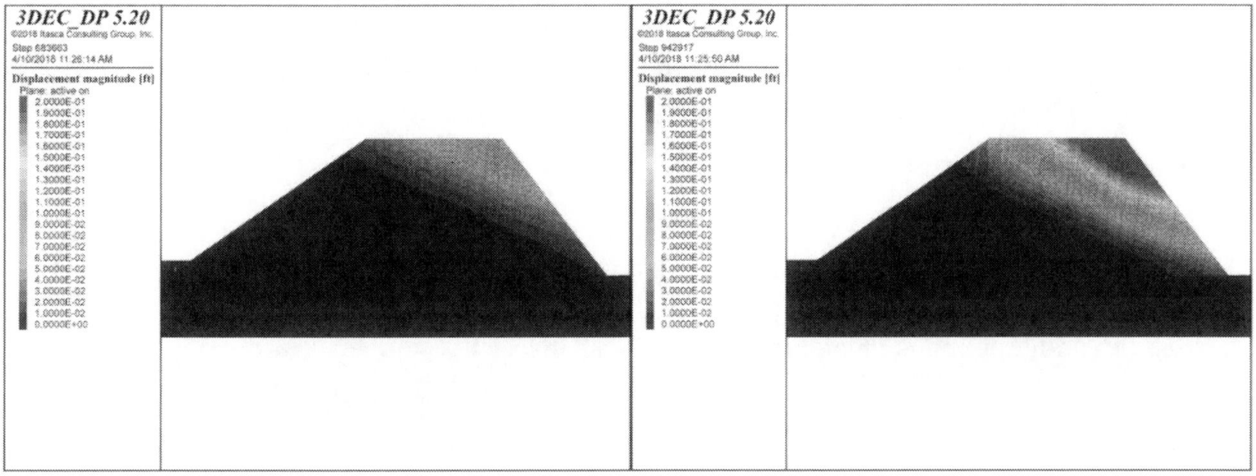

Figure 17 – Model displacements for the cross section at the bridge abutment for a factor of safety of 1.5 (left) and 1.8 (right)

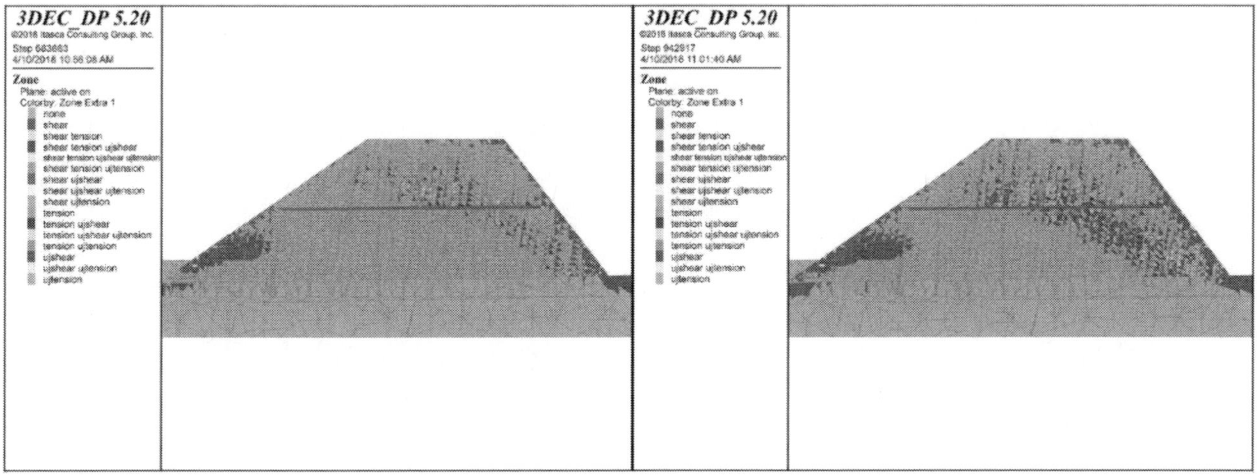

Figure 18 – Zones with various material failure behavior for the cross section at the bridge abutment for a factor of safety of 1.5 (left) and 1.8 (right).

SUMMARY AND CONCLUSIONS

Summary

A series of kinematic and numerical models were developed to assess the stability of T.H. 53 isthmus rock slopes. The rock mass and discontinuity properties input into the models were based on an extensive field campaign which included core logging, down-the-hole-logging, laboratory tests and photogrammetry. Using the collected discontinuity data, a kinematic analysis was performed and the resulting factor of safety was found acceptable. Then, two-dimensional *UDEC* analyses provided insight into the failure mechanism, but due to the nature of 2D models, did not fully capture the three-dimensional geometry of the subvertical joints and the isthmus slope. Thus, the safety factors determined from *UDEC* were not fully reliable. This was confirmed by varying the joint dip angle and comparing the safety factors in *UDEC*. A three-dimensional *3DEC* model was created to capture the interaction of the joints with the isthmus slopes. The *3DEC* model determined that the factor of safety of the slope is at least 1.5, which met MNDOT's minimum safety requirements.

Conclusions

The *UDEC* and *3DEC* models both confirmed that flexural toppling is a mechanism of failure for the isthmus slope. By analyzing the displacement plots, flexural toppling-induced displacements extend across the isthmus. The *3DEC* model predicted that the factor of safety against flexural toppling is greater than 1.5 and less than 1.8.

REFERENCES:

- Report
  1. Hittinger, M., and Goodman, R. E. (1978). JTROCK, A Computer Program for Stress Analysis of Two Dimensional, Discontinuous Rock Masses. Report No. UCB/GT/78-04, University of California, Berkeley.
- Journal
  1. Nichol, Susan L., Oldrich Hungr, and S. G Evans. Large-scale Brittle and Ductile Toppling of Rock Slopes. *Canadian Geotechnical Journal* vol. 39, 2002, pp. 773-788.
- Software Manual
  1. Itasca Consulting Group, Inc. (2016) 3DEC — Three-Dimensional Distinct Element Code (Version 5.2). Minneapolis: Itasca.

Innovative Socketed Pile for Accelerated Bridge Construction in Naples, Maine

Blaine M. Cardali, E.I.T
GZA GeoEnvironmental, Inc.
477 Congress Street
Suite 700
Portland, Maine 04101
(207) 358-5131
Blaine.Cardali@gza.com

Andrew R. Blaisdell, P.E.
GZA GeoEnvironmental, Inc.
477 Congress Street
Suite 700
Portland, Maine 04101
(207) 358-5117
Andrew.Blaisdell@gza.com

Christopher L. Snow, P.E.
GZA GeoEnvironmental, Inc.
477 Congress Street
Suite 700
Portland, Maine 04101
(207) 358-5118
Christopher.Snow@gza.com

Laura Krusinski, P.E.
Maine Department of Transportation Bridge Program
16 State House Station
Augusta, Maine 04333
(207) 624-3441
Laura.Krusinski@maine.gov

Garrett Gustafson, P.E.
Maine Department of Transportation Bridge Program
16 State House Station
Augusta, Maine 04333
(207) 624-3419
Garrett.A.Gustafson@maine.gov

Prepared for the 69th Highway Geology Symposium, September 2018

Acknowledgements

The authors would like to thank the following individuals/entities for their contributions to this project:

Erik Friede – GZA GeoEnvironmental, Inc.
Nicholas Williams – GZA GeoEnvironmental, Inc.
Peter Marcotte – Maine Drilling & Blasting

Disclaimer

Statements and views presented in this paper are strictly those of the authors(s), and do not necessarily reflect positions held by their affiliations, the Highway Geology Symposium, or others acknowledged above. The mention of trade names for commercial products does not imply the approval or endorsement by HGS.

ABSTRACT

In fall 2016, the Maine Department of Transportation (MaineDOT) replaced a bridge carrying State Route 11/114 over Muddy River in Naples, Maine. Route 11/114 is a rural road, but it is the primary route for travel around the west side of Sebago Lake, being used heavily by residents, trucks and tourists. Based on the road usage and lengthy detour, Accelerated Bridge Construction was selected for the project.

The geologic conditions included shallow, irregular, sloping bedrock at the abutments. Bedrock consisted of hard granite, with very hard, intrusive, trachyte dikes. The preferred configuration for this type of bridge is an Integral Abutment Bridge supported by driven H-piles. However, considering the shallow and sloping rock, driven piles were not feasible at one of the abutments. GZA GeoEnvironmental, Inc. (GZA) and MaineDOT collaborated to develop and design an innovative drilled pile foundation, the spun pipe pile, which consisted of a conventional micropile casing within an uncased socket in bedrock. The casing is installed into bedrock to gain lateral fixity and to gain axial resistance through end-bearing, eliminating the typical drilled socket with reinforcing bar and saving construction time.

Thanks in part to this innovative foundation system, the contractor completed the work in 18 days, shaving eight days off the 26-day closure allowed in the contract. Foundation installation was completed in 3 days, with 24-hour per day oversight by GZA throughout construction.

INTRODUCTION

A bridge replacement project was undertaken by the Maine Department of Transportation (MaineDOT) on Maine State Route 11/114 (Sebago Road) over the Muddy River in Naples, Maine. GZA GeoEnvironmental, Inc. (GZA) was retained by MaineDOT, the bridge designer for the project, to serve as the geotechnical consultant for the project. The scope of services included: review of initial field investigations, designing and overseeing a supplemental field investigation program, developing preliminary recommendations and feasible foundation types, and developing engineering solutions and recommendations for abutment foundations capable of being installed during the construction window, developing foundation specifications, and providing on-site observation during foundation installation.

The project plans called for the bridge to be constructed in the fall of 2016 using Accelerated Bridge Construction (ABC) inside of a 26-day road closure, during which traffic would be detoured. All of the bridge demolition and construction activities were required to be completed within this window, and at least one lane was required to be re-opened to traffic at completion of the closure.

PROJECT AREA

Bridge Project

Crockett Bridge No. 2199 carries Maine State Route 11/114 (Sebago Road) over the Muddy River between Sebago Cove to the west and Sebago Lake to the east in Naples, Maine as shown on the annotated aerial photograph, **Figure 1**.

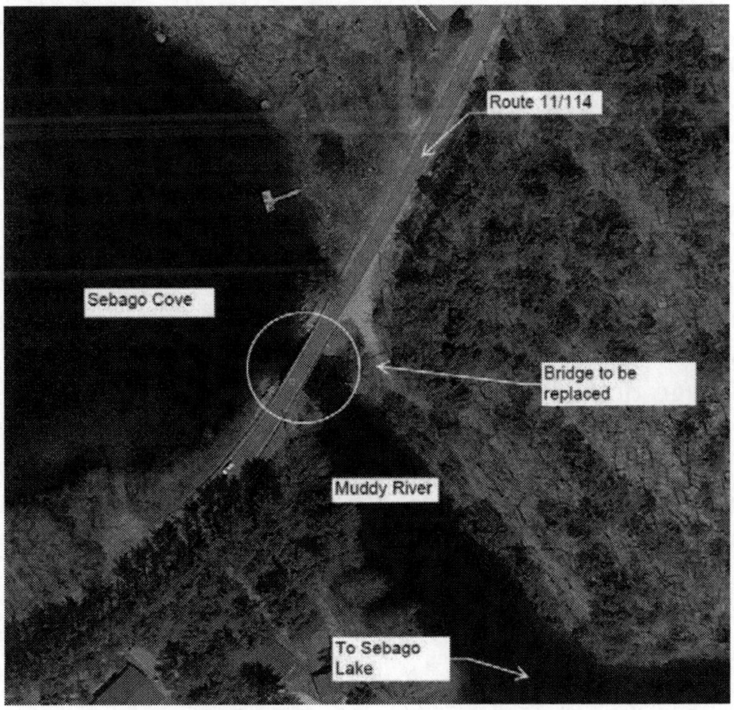

Figure 1 – Project Site

The original bridge was constructed in 1930 and consisted of a single-span, cast-in-place concrete rigid frame with a clear span of 20 feet. The original southwest abutment was supported by a spread footing bearing on bedrock and the original northeast abutment was supported on timber piles. Depths to the bottom of both of the abutments ranged from 15 to 20 feet below the existing roadway.

During preliminary design, MaineDOT proposed a replacement bridge consisting of an 80-foot-long, single-span bridge following the same alignment as the existing bridge. The longer bridge length was selected to provide a wider opening for the Muddy River, and the new abutments were located behind the existing abutments. This new bridge was proposed to include a superstructure consisting of four precast NEXT 36 F beams with integral abutment substructures. In Maine, integral abutments are typically supported by a single row of driven H-piles.

OBJECTIVES AND PROJECT APPROACH

The primary objective was to provide geotechnical engineering recommendations for design of the Crockett Bridge foundations. Based on GZA's initial review of the project and discussions with MaineDOT, it was apparent that two primary factors would drive design considerations for the project:

1. Variability in bedrock elevation, inclination and type; and

2. Ability to construct deep foundations rapidly during the scheduled closure.

GZA proceeded with a subsurface characterization approach that would identify bedrock conditions across each abutment by way of three cored test borings, including two at the abutment with shallow bedrock, to allow a reasonable estimate of variation in pile lengths and installation considerations. The design approach was focused on identifying, designing and detailing deep foundation types for integral abutment support that were cost effective and could be constructed within a time window of several days as would be required by ABC.

GEOLOGIC SETTING

Based on the Maine Geologic Survey Map of the Naples Quadrangle (*1*), surficial geologic units in the site vicinity are mapped as Glacial Lake Sebago Bottom Deposit and Till. The Glacial Lake Sebago Bottom Deposit is described as massive to stratified and cross-stratified sand, and massive to laminated silt and silty clay, sometimes containing boulders and gravel, varying in thickness from 1 to 60 feet. Till is described as light to dark gray, poorly sorted mixture of clay, silt, sand, pebbles, cobbles, and boulders. The bridge approach embankments are mapped as Artificial Fill. Bedrock outcrops are identified on the northeast side of the bridge and are shown as a horizontal hatch pattern in **Figure 2**.

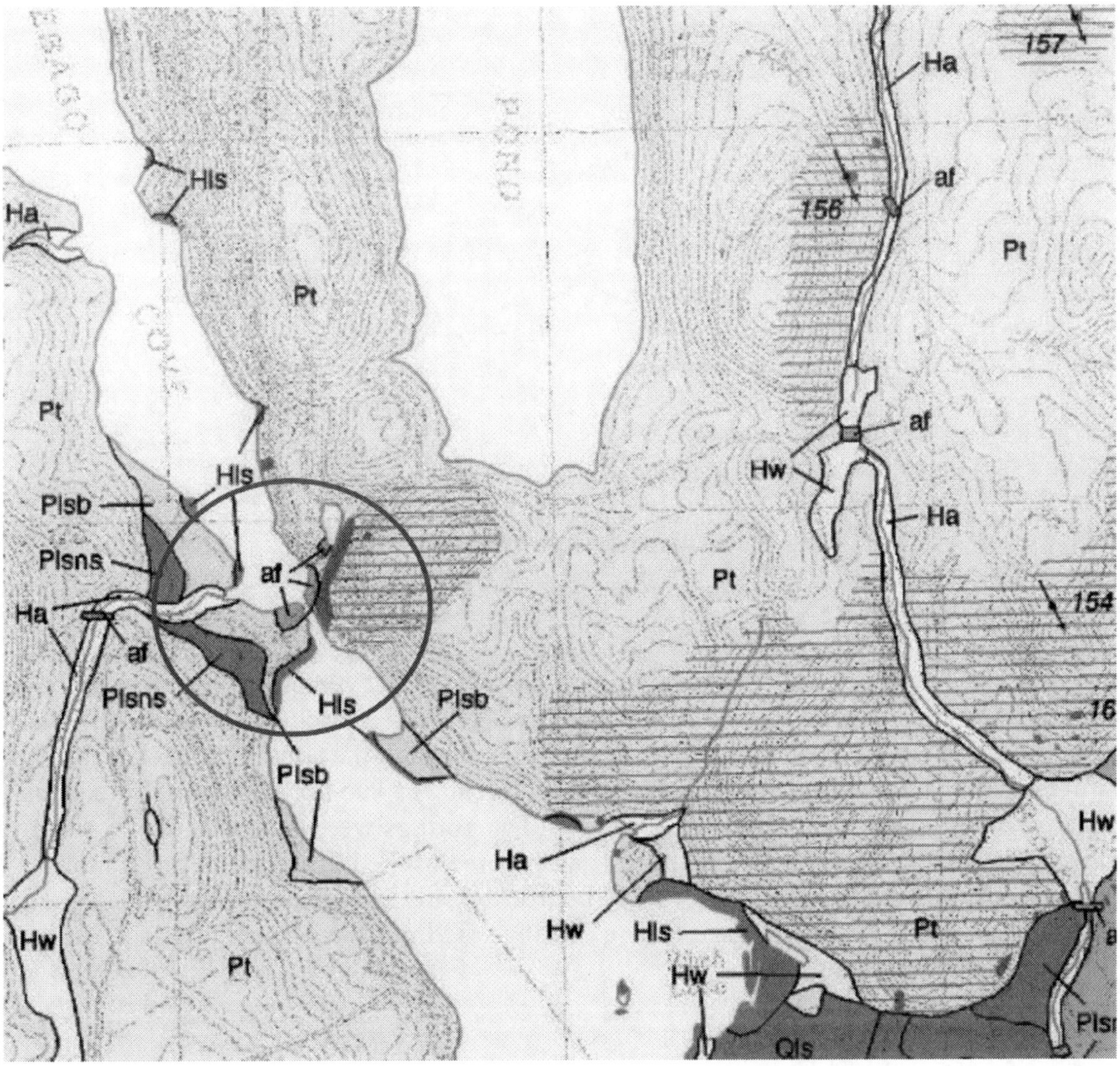

Figure 2 – Surficial Geology Map

Bedrock at the site is mapped as the Sebago pluton (*2*), shown as the pink shading on **Figure 3**. The Sebago pluton in the site vicinity is described as medium grained equigranular, biotitic-muscovite Granite (CG), white to pale pink, locally pegmatitic. Two intrusive dikes are also mapped in the immediate site vicinity, including a mafic dike (red line) described as reddish-brown weathering, black basaltic dikes and a trachyte dike (blue line) described as dark gray weathering, chocolate-brown feldspar-bearing dikes.

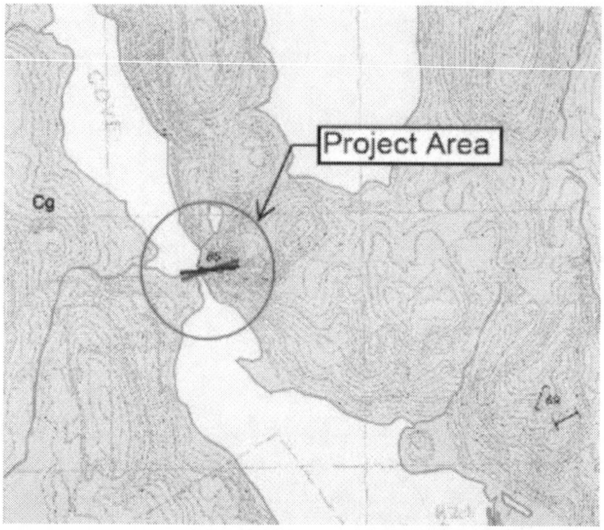

Figure 3 – Bedrock Geology Map

FIELD INVESTIGATIONS

The subsurface exploration program was completed between May and December 2015 and included five test borings. The borings were drilled using a truck-mounted drill rig through the existing bridge approaches, including two borings through the south approach (Abutment 1) and three borings through the north approach (Abutment 2). The borings were drilled using cased rotary wash methods to allow rock coring. The borings were drilled to depths ranging from 29 to 50 feet below existing ground surface. Three of the borings were drilled to split-spoon refusal followed by 7 to 9 feet of bedrock coring, and two borings were terminated in the overburden due to casing damage during drilling. Bedrock cores were obtained using NQ2 wire-line coring equipment.

SITE SUBSURFACE CONDITIONS

Upon completion of the supplementary borings, four soil units were identified as follows: Fill, Gravelly Sand, Silt, and Gravel, which were encountered below pavement and above bedrock in the test borings. The profile in **Figure 4** below shows the approximate strata thicknesses, generalized soil and rock descriptions and interpolated bedrock surface along the centerline of the bridge.

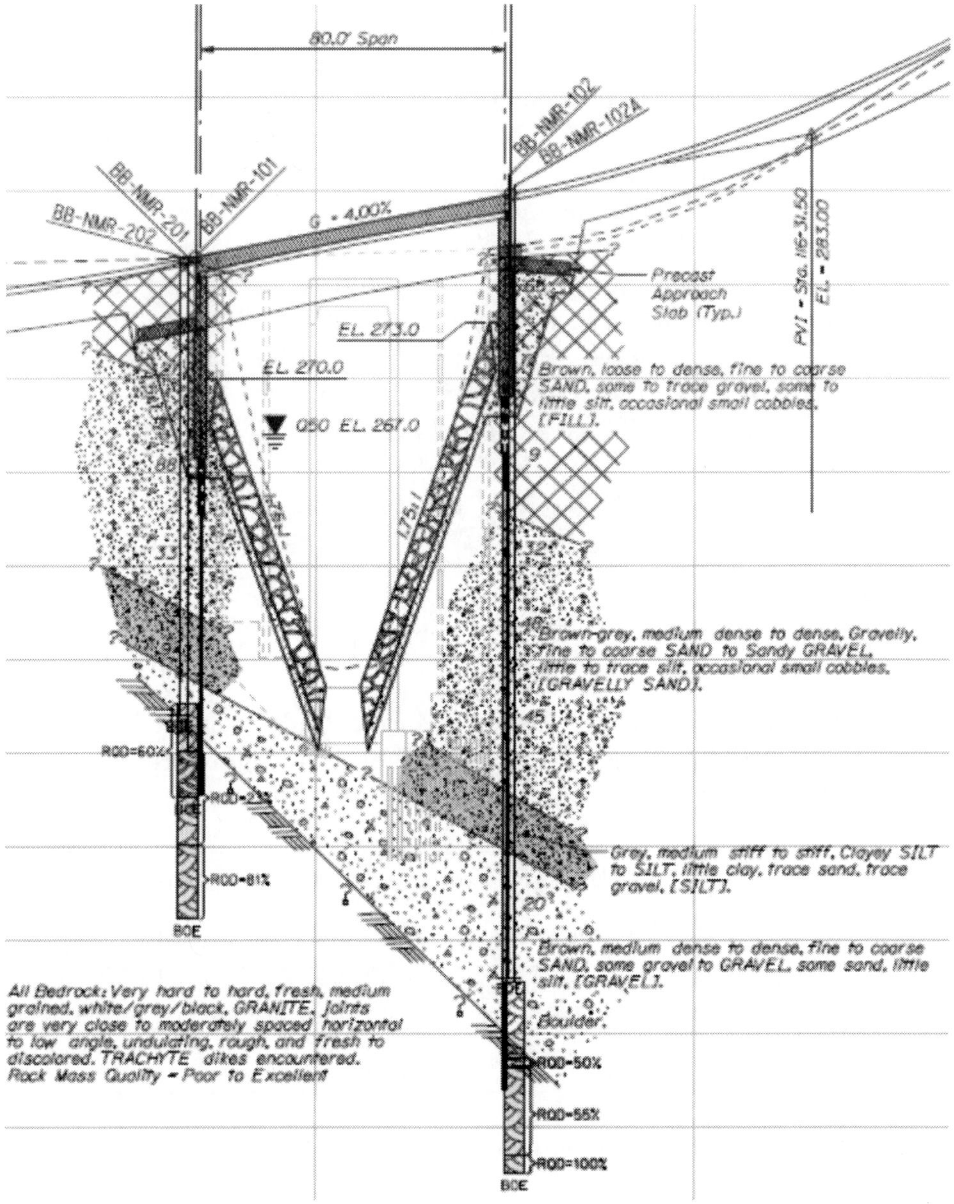

Figure 4 – Interpretive Subsurface Profile

Bedrock encountered in the borings consisted of Granite with Trachyte dikes. The depth to the top of bedrock encountered in the borings varied from 22 to 26 feet bgs at Abutment 1 and was approximately 40 feet at Abutment 2. Two rock types were encountered in the borings; Granite of the Sebago Pluton, and Trachyte associated with one of the mapped dikes in the site vicinity.

Granite was encountered in all three of the cored borings and was generally described using the Modified ISRM Rock Classification system as very hard to hard, fresh, medium grained, and white/gray/black. Joints were very close to moderately spaced, horizontal to moderately dipping, undulating, rough, fresh to discolored, and tight to open. Trachyte dikes were encountered in one boring at Abutment 1 from the top of rock (26 feet bgs) to 29.3 feet bgs and from 33.5 to 34.8 feet bgs, with Granite encountered between and below the Trachyte dike layers. Trachyte was generally described as very hard, fresh, aphanitic and red-brown. Joints were very close to close, low angle to moderately dipping, undulating, rough, discolored, and tight to open. The Rock Quality Designation (RQD) was 23 percent in the Trachyte and ranged from 55 to 100 percent in the Granite. The bedrock profile was identified to be sloping down toward the north and west at average inclinations ranging from 2H:1V to 4H:1V based on the encountered top of rock elevations in the borings.

Two laboratory unconfined compressive strength tests with strain measurements were conducted on bedrock core samples, one on Trachyte and one on Granite. The Trachyte had an unconfined compressive strength of 34.3 kips per square inch (ksi), a Young's modulus of 4,580 ksi and a Poisson's ratio of 1.38. The Granite had an unconfined compressive strength of 14.9 ksi, a Young's modulus of 3,230 ksi and a Poisson's ratio of 0.94.

In general, the results of the rock coring and testing indicated the Trachyte dikes were stronger and harder but more fractured than the Granite Pluton, which was an important consideration for foundation design.

ENGINEERING CHALLENGES AND SOLUTIONS

GZA identified two primary considerations for the bridge replacement project that would have a significant impact on the success of the overall project. These project elements and the associated engineering challenges are summarized in **Table 1**.

Table 1 – Primary Engineering Challenges

Geotechnical/Construction Consideration	Associated Design and/or Construction Challenge
Shallow and Sloping Bedrock Across the Site	Integral abutment deep foundations must achieve fixity through embedment in soil or bedrock and remain plumb and on-location during installation.
Accelerated Bridge Construction	A foundation type must be selected that can be installed within several days to allow rapid construction of the abutments.

GZA's evaluations and recommendations to address these conditions are described in the following sections.

Based on constructability and cost considerations, MaineDOT had selected an integral abutment bridge (IAB) for the project. IAB bridges are preferred by the MaineDOT and by many other states for single-span bridge replacements. An IAB is defined as a bridge with no expansion joints, instead fixing the substructure elements to the superstructure and allowing them to move with the thermal expansion and contraction of the bridge, shown in **Figure 5**.

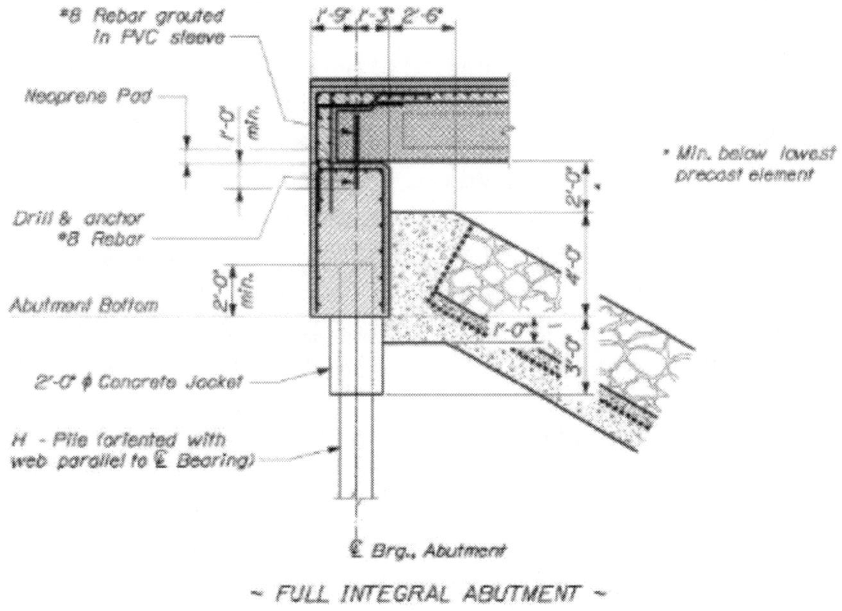

Figure 5 – Integral Abutment Detail

Due to the rigid connection between the bridge superstructure and the abutment, the lateral load from thermal expansion is resolved in the abutment and foundations. The University of Maine published a paper entitled "Behavior of pile-supported integral abutments at bridge sites with shallow bedrock" (*3*). This paper describes the interactions of short piles that support integral abutment bridges and concluded that piles less than 4 meters (13 feet) in length performed differently under the combined lateral and axial loading than a longer pile, resulting from a lack of fixed or pinned toe condition. Therefore, the shallow, steep bedrock surface was identified to be one of the primary challenges to resolve the lateral loading expected from thermal expansion. In addition, driven H-piles have encountered constructability issues on sloping rock on similar projects where the driven pile moved off location and/or out-of-plumb and "walked" as it moved diagonally along the rock surface.

Three foundation types were considered during design development for support of the proposed IAB: driven H-piles, conventional micropiles (with rock sockets and central steel threadbars), and spun pipe piles (without sockets or threadbars).

At this site, driven piles would likely have been driven to refusal on rock due to the relatively shallow soil profile, especially at Abutment 1. Integral abutment support with driven H-piles relies on a soil profile that is deep enough to develop fixity, or at least a pinned end condition. The subsurface data at Abutment 1 indicates that the depth from bottom of integral abutment to top of rock was as little as 10 feet, and the rock surface was sloping. Preliminary lateral pile evaluations were conducted for H-piles assuming a 10-foot-deep soil profile, and the results indicated the piles would not achieve a pinned condition under the imposed thermal deflection. In addition, piles could potentially "walk" when driven to sloping rock, which would induce additional stress, and is a concern given the tight tolerance for location and inclination of integral abutment piles. H-piles would be feasible at Abutment 2, but considering the planned ABC, it was not desirable to mobilize two different foundation operations. Therefore, driven piles were not considered further.

Conventional micropiles were considered to be a feasible foundation type. Micropile casing is typically advanced through the overburden and into bedrock using an air percussive hammer. An air hammer with a smaller bit is conventionally used to drill a rock socket below the bottom of the casing. For a conventional micropile, a threadbar or inner hollow casing is used to transmit vertical loads to the socket, and the micropile gains axial compression resistance primarily through friction along the grout-rock interface. The outer casing is typically advanced a moderate distance into bedrock to promote fixity under lateral loading, thereby eliminating the "walking pile" effect associated with driven pile. Preliminary lateral pile evaluations indicated that micropiles could achieve adequate fixity with a casing embedment depth into rock of 3 feet.

A third option of spun pipe piles was a concept that was developed by the MaineDOT/GZA design team specifically for this project. This concept was identified to be feasible when the analyses of the conventional micropile achieved fixity with only casing advancement into bedrock. A spun pipe pile is essentially a micropile with no central reinforcement and where the bottom of the casing sits on the bottom of the rock socket. The spun pipe pile gains axial compressive resistance through end bearing on the rock surface at the bottom of the casing, which requires that the casing be filled with grout to provide end bearing resistance over the entire tip area, similar to a rock-socketed drilled shaft. The primary advantage of the spun pipe pile over conventional micropile is reduced construction time since a deeper, second stage of drilling and internal reinforcement installation is not required. Spun piles can be grouted and completed immediately after completion of drilling. In addition, the spun pipe pile could be designed using AASHTO LRFD resistance factors appropriate for end-bearing drilled shafts, which eliminates the requirement for testing and saves more time in the schedule.

Based on schedule and cost considerations, the project team selected spun pipe piles over conventional micropiles as the preferred pile type for this project. Preliminary lateral pile evaluations conducted for readily-available 9.625-inch-outside-diameter pipe sections indicated that a wall thickness of 0.545 inches would be required to limit the combined axial and bending stress in the spun casing to less than 90 percent of the yield stress. Therefore, a 9.625x0.545 pile section consisting of American Petroleum Institute (API) 5CT N80 steel pipe with a minimum yield strength (fy) of 80 ksi was selected for spun pipe piles for the project.

DESIGN OF SPUN PIPE PILES

Design evaluations were conducted for axial compressive geotechnical resistance and lateral load resistance of the piles. The geotechnical static resistance of spun pipe piles was calculated using the drilled shaft tip resistance on rock methodology in accordance with AASHTO LRFD Article 10.8 (*4*). Side friction was not assumed to provide any resistance to axial compressive loads.

Axial Pile Resistance

Each abutment included a single row of five, 9.625x0.545 spun piles. The maximum factored axial load for the strength condition provided by MaineDOT was 365 kips per pile. The piles were designed at the strength limit state considering geotechnical resistance of the piles using a resistance factor of 0.50, for tip resistance on rock, per AASHTO Table 10.5.5.2.5-1. Therefore, the required nominal axial compressive resistance was 730 kips per pile.

Spun pipe piles were designed to gain axial compressive resistance through end bearing in bedrock. The nominal tip resistance was estimated using procedures described in AASHTO Article 10.9.3.5.3, which references Article 10.8.3.5.4c for tip resistance on rock. The tip resistance on rock is based on the strength of jointed rock masses evaluated using the Hoek-Brown failure criterion (*5*). The primary input parameters used to calculate rock mass strength and tip resistance on rock include the Geologic Strength Index (GSI), unconfined compressive strength (q_u) and the rock group constant (m_i). Based on the results of the borings, it was concluded that the spun pipe piles could bear in either the encountered Trachyte or Granite, and the subsurface characterization was not adequate to predict bearing materials for each pile. Therefore, we evaluated tip resistance for both rock types with the intent of utilizing the controlling geotechnical resistance for design. The bedrock input parameters selected for our evaluation are summarized in **Table 2**, below.

Table 2 – Bedrock Properties and Spun Pile Tip Resistance

Parameter Description	Parameter Symbol (units)	Value for Granite	Value for Trachyte	Reference
Unconfined Compressive Strength, Intact Rock	q_u (psi)	14,930	34,300	Laboratory test data
Geologic Strength Index	GSI	60	60	AASHTO Figure 10.4.6.4-1
Rock Group Constant	m_i	32	25	AASHTO Table 10.4.6.4-1
Nominal Unit Tip Resistance, Jointed Rock Mass	$q_{p,jointed}$ (ksf)	2,553	4,806	AASHTO Eq. 10.8.3.5.4C-3
Nominal Geotechnical Tip Resistance, Jointed Rock Mass	$R_{p,j}$ (kips)	1,290	2,428	AASHTO Eq. 10.8.3.5-2

The granite was selected as the controlling tip resistance for the spun piles. A factored geotechnical resistance of 645 kips per pile was calculated by applying the 0.5 resistance factor to the controlling nominal resistance of 1,290 kips. Since the factored resistance significantly exceeded the factored load, we concluded end bearing resistance on rock is suitable to support the axial design loads.

Lateral Pile Analysis

GZA conducted lateral pile analyses using L-PILE 2015® using two end conditions provided by the bridge designer: horizontal thermal displacement of the pile top of 0.44 inches, and pile top slope of 0.00245 in/in, induced by the live loads. The assumed axial load was the maximum factored axial load, 365 kips.

The spun pile section was analyzed assuming: 1) empty casing and 2) casing with grout infill with a compressive strength of 6 ksi. This grout compressive strength was recommended by the MaineDOT bridge designer to model grout that achieves a higher unconfined compressive strength than the design value, which is intended to model the upper-bound bending stiffness. The graphs shown in **Figure 6**, display the results for a pile from Abutment 1, which was expected to be the shorter pile due to shallower bedrock. This model represents the casing without grout filling to analyze the worst case for bending stresses in the steel casing. This case shows that the moment and deflections applied at the pile head are resolved within the first three feet into bedrock, and fixity is developed.

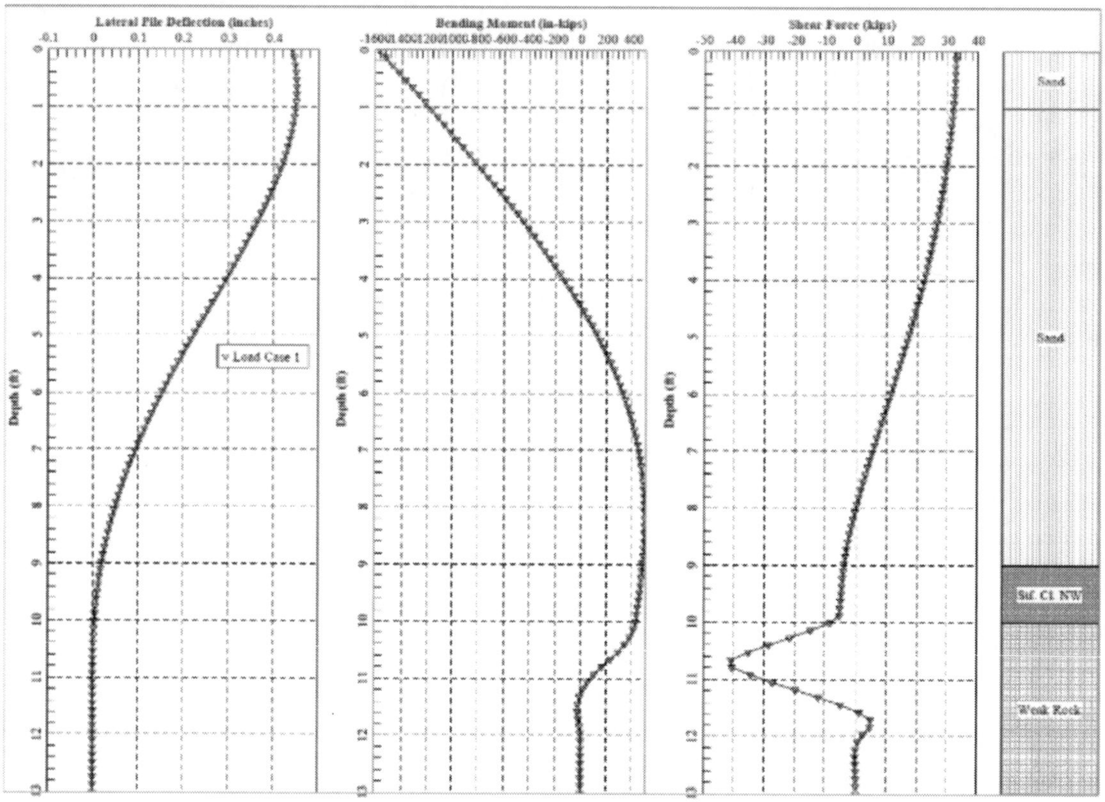

Figure 6 – L-Pile Results

L-PILE 2015 models the combined steel and grout section using a "cracked section" when the bending stress exceeds 75 percent of the unconfined compressive strength of the grout, resulting in a reduced bending stiffness. This condition occurred in approximately the upper 3 feet of the pile based on our evaluation, and it would occur over a longer distance for lower strength grout. The analyses show that the piles were able to achieve fixity within the top three feet of bedrock while limiting combined axial and bending stress in the casing to less than 90 percent of the yield strength, indicating the 9.625x0.545 pile section met the design criteria, as summarized in **Table 3**.

Table 3 - L-PILE® RESULTS							
Abutment	Pile Type and Size	Factored Axial Load (kips)	Shear Force for Lateral deflection of 0.44 in. (kips)	Moment at Pile Head (ft-kips)	Total Stress at Pile Head (ksi)	Bending Stress at Pile Head (ksi)	Axial Stress at Pile Head (ksi)
1	9-5/8x0.545 (Empty Casing)	365	32.6	-1569.8	70.5	47.0	23.5
1	9-5/8x0.545 (6 ksi Grout Infill)	365	36.0	-1787.1	60.1 / 5.94	--	--
2	9-5/8x0.545 (Empty Casing)	365	34.1	-1618.3	71.9	48.4	23.5
2	9-5/8x0.545 (6 ksi Grout Infill)	365	37.4	-1823.8	61.5 / 5.96	--	--

Spun Pipe Pile Recommendations and Specifications

The spun pipe piles were specified as 9.625x0.545 spun pipe piles (80 ksi yield stress) infilled with grout with a 28-day compressive strength of 4 ksi. A minimum spun pile embedment of 5 feet below the top of rock elevation was judged deep enough to encounter sound bedrock suitable to provide the desired axial and lateral resistance. Pile testing was waived in consideration of the resistance factors used for geotechnical design.

The N80 casing was anticipated to consist of flush-joint casing with threaded connections. In order to avoid casing joints in the high moment zone, the specifications called for the uppermost joint to be at least 4 feet below the bottom of the abutment, corresponding to 6 feet below the top of the pile.

Construction recommendations were critical to the performance of the spun piles. Because the piles relied solely on tip resistance for axial support, the condition of the rock beneath the pile tip required careful preparation and verification. The following construction recommendations were provided for spun piles:

1. Thoroughly clean spun pile holes at the completion of drilling using high-pressure air or water to provide a clean end bearing surface.

2. The depth and soundness of the hole should be assessed using a weighted tape prior to grouting.

3. If soil was detected in the casing following drilling and cleaning and was suspected to be washing in, additional measures would be required to achieve a seal before grouting. This could include advancing the casing further into rock, and/or retracting the casing, grouting the area just above and within the socket, and re-drilling to rock, below the original socket depth.

4. The drill holes should be tremie-grouted from the bottom, up. A plug should be placed in the tremie pipe prior to insertion into the pile to prevent water entry into the pipe. The tremie pipe should remain at least 5 feet below the top of grout level throughout the grout placement, if it is pulled during grouting.

Because load testing was not planned, the presence of a Geotechnical Engineer was specified throughout advancement of steel pipes, final cleaning, and grout placement to ensure that the intent of the design and special provisions are met. The Geotechnical Engineer was also specified to observe and assess the depth to top of rock, embedment in the rock, bottom cleanliness, depth of hole, length of casing installed, and theoretical versus actual grout volumes.

CONSTRUCTION

During the construction phase, GZA was onsite 24 hours per day on behalf of MaineDOT to provide quality control for the spun pipe pile installation, assessing the work for compliance with the project plans and specifications. The following sections describe the installation equipment and process that was observed.

Pile Installation Equipment

The spun pipe piles were installed using a non-displacement drilling method, which utilized a Robit air-hammer system. The drill rig that was used for this project was a Hutte 504, equipped with the Robit air hammer shown in **Figure 7**. This procedure involved the use of an under-reamer bit that was used to socket the pipe into bedrock (**Figure 8 and 9**).

Figure 7 – Drill Rig and Robit Hammer

Figure 8 – Robit Air Hammer Tip (side)

The first piece of casing for each pile was equipped with a ring bit (**Figure 10**), which allowed the Robit drill head to lock into the casing. The groove inside the ring bit which the hammer locks into is shown in **Figure 11**. This system uses an air percussion hammer to advance through overburden materials and bedrock. Since the casing is locked into the hammer, the casing advances along with the hammer, and additional casing and rod sections are added to continue advancement. As the hammer breaks through material, the cuttings are air-lifted out of the drill head.

Figure 9 – Robit Hammer (Bottom)

Figure 10 – Ring Bit

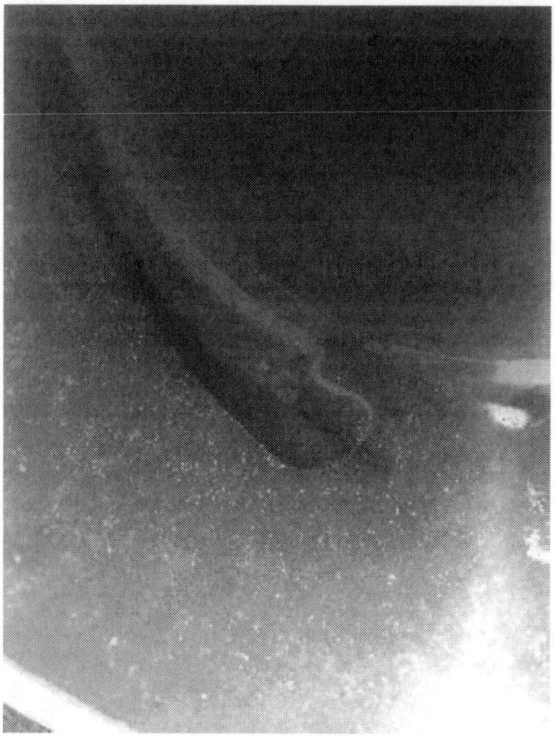

Figure 11 – Ring Bit Groove

Figure 12 – 10-foot Casing Section

This method allowed the central drill hammer to be removed and reinstalled if the pile was required to be advanced further into rock, if rock surface was observed to be inadequate by the field engineer, or if the rock seal was poor and allowed fine grained materials to accumulate on the pile base.

Installation Sequence

The first section of each pile was 6 feet in length, which included a 5-foot section of casing equipped with a 1-foot long Ring bit. The casing was loaded onto the rig followed by the hammer and drill rod as shown in **Figure 12**. Once the drill head was locked into the Ring bit, the location and plumbness were checked. The drill then began advancing the casing. 10-foot casing sections were added until the casing had advanced a minimum of 5 feet into the bedrock. To meet the specified uppermost pipe section length of 6 feet, some piles were extended further into bedrock. For example, if the pile was only 5 feet into bedrock at completion of a pipe section, an additional 9 feet of the next section would be required to provide the required distance to the first joint, resulting in a 14-foot deep rock socket.

At the completion of drilling, the holes were cleaned thoroughly under air and water to provide a clean end-bearing surface. The surface was confirmed to be clean by the geotechnical engineer by sounding, using a weighted tape until a hard surface became evident throughout the base of the drilled pile.

Pile bedrock embedment material was judged based on the color of the cuttings and variations in drill characteristics. Two out of the ten piles installed were judged to be bearing in the Trachyte dike. The more fractured Trachyte was believed to be creating a hydraulic connection between the drilled hole and the river, based on repeated detection of fine-grained materials in the base, and the nearly immediate equalization of water level in the drill hole to the river level. As a result, additional flushing was required at these two locations before grouting to maintain a clean bedrock interface with the grout. This soil infilling was judged to be from joints in the bedrock rather than a poor seal. Therefore, additional casing advancement was not required. For the eight piles installed into granite, the bottom of the socket remained clean and dry.

The onsite engineer also recorded the plumbness of the spun piles prior to grouting utilizing a 4-foot level. The tolerance for plumbness was 1 inch of lateral deviation over 4 feet. Since the drill head of the system was fixed in place and the pile was locked into the drill rod at the tip and at the drill head, plumbness checks were readily made by the drillers and geotechnical engineer during installation.

After the spun pipe piles were installed at each abutment, the drill holes were tremie-grouted from the bottom, up. A plug was placed in the tremie pipe prior to insertion into the pile to prevent water entry into the pipe. The ½-inch PVC tremie pipe remained at least 5 feet below the top of grout level throughout the grout placement to prevent contamination of the grout.

Figure 13 – Grouting

Figure 14 – Grouting

Schedule

The entire spun pile installation occurred over a three-day period, September 27, 2016 through September 29, 2016, taking less than 60 hours from start to finish. **Table 5** describes the major activities that were completed during that time and the duration of each task.

Table 5 – Foundation Installation Schedule

Task	Start Date and Time	End Date and Time	Approximate Duration (Hours)
Abutment 2 Spun Pile Installation	9/27 – 11:00 AM	9/28 – 10:00 PM	35
Mobilize Equipment to Abutment 1	9/28 – 10:00 PM	9/29 – 2:00 AM	4
Abutment 1 Spun Pile Installation	9/29 – 2:00 AM	9/29 – 5:30 PM	16
Abutment 2 Spun Pile Grouting	9/29 – 1:30 PM	9/29 – 6:30 PM	5
Abutment 1 Spun Pile Grouting	9/29 – 6:30 PM	9/29 – 9:30 PM	3

As shown in the table, the Abutment 2 piles were installed first and took approximately 35 hours and were followed by the Abutment 1 piles. The contractor began grouting the Abutment 2 piles while installing the final two piles at Abutment 1, with both activities shown in **Figure 14.**

Figure 14 – Construction Sequence

This allowed for continuous operation of the grout plant. Using an extended grout line, the Abutment 1 piles were grouted directly after the completion of the Abutment 2 grouting, without relocating the grout plant across the river.

Thanks in part to this innovative foundation system, the contractor completed the foundation construction in three days, and the entire closure in 18 days, shaving eight days off the 26-day closure allowed in the contract.

CONCLUSIONS

The geotechnical challenge of the Crockett bridge replacement project was to develop and design a deep foundation system which could be installed within the proposed closure window for the project. The selected spun-pile design was able to overcome the geologic conditions which included shallow, irregular, sloping bedrock at the abutments and was able to be installed within the allotted time. The design of the spun pile included a 9.625-inch diameter casing installed into the bedrock that relied on end bearing for resistance. This foundation type is installed like a micropile but without central reinforcement, eliminating the construction steps of drilling a socket beneath the casing and placement of the reinforcing bar. The design methodology for the spun piles allowed elimination of the requirement for pile testing during installation, removing a significant schedule impediment from the ABC project. Foundation installation was completed in three days, with 24-hours-per-day oversight by GZA throughout construction to ensure that the completed foundations met the intent of the geotechnical design and were in accordance with the project plans and specifications.

The spun piles were found to provide the following benefits:

- By using a design methodology for a drilled shaft tip resistance in rock, the requirement for load testing can be eliminated, resulting in reduced construction duration;
- Delays such as obstructions that are associated with driven piles are mitigated with the air-hammer drilling technique;
- Pile “walking” and lack of fixity development that can be associated with driven piles is prevented; and
- By removing the bond zone length of the typical micropile, the drilling time is reduced and it removes the need for a central bar.

Shallow bedrock can dissuade agencies from ABC but spun piles provide an alternative. MaineDOT is continuing to utilize the spun pipe pile foundation system for similar IAB projects with shallow bedrock of sufficient strength and quality to be able to resist design loads in end bearing.

REFERENCES

1. Hildreth, Carol T., 1997, Surficial geology of the Naples quadrangle, Maine: Maine Geological Survey, Open-File Map 97-50, map, scale 1:24,000. *Maine Geological Survey Maps*. 762. http://digitalmaine.com/mgs_maps/762

2. Creasy, John W., 1996, Preliminary report: Bedrock geology of the Naples and Raymond quadrangles: Maine Geological Survey, Open-File Report 96-4, 9 p. report, 6 figs., 2 maps, scale 1:24,000. *Maine Geological Survey Maps*. 223. http://digitalmaine.com/mgs_maps/223

3. Delano, John G., 2004. Behavior of Pile-Supported Integral Abutments at Bridge Sites with Shallow Bedrock: The University of Maine.

4. American Association of State Highway and Transportation Officials. 2014 and Interims. "AASHTO Load Resistance Factor Design (LRFD) Bridge Design Specifications", Washington, DC

5. Hoek,E., and E. T. Brown. 1997. "Practical Estimates of Rock Mass Strength," *International Journal of Rock Mechanics and Mining Sciences.* http://www.rocscience.com

6. Maine Department of Transportation. 2003 with updates through 2014. "Maine Bridge Design Guide"

Design and Construction Considerations for Innovative Rockfall Protection Systems

Robert Huber, GIT, Field Engineer
GeoStabilization International, 543 31 Road, Grand Junction, CO 81504
724-316-2088, robert.huber@gsi.us

Martin J. Woodard, PhD, PG, PE, Chief Rockfall Engineer
GeoStabilization International, 1419 Madison St, Radford, VA 24141-3711
540-315-0270, marty@gsi.us

Abstract

As professional engineers and geologists we are tasked with providing solutions to protect people and infrastructure from hazards such as rockfalls both during and after construction or remediation of rock slopes. Rockfall hazards can be found in a variety of environments including mining operations, highway transportation, and railroad corridors, just to name a few. Each of these environments necessitate unique project requirements from available work times, to maintaining proximity access to the slope (travel lanes open to public), to environmental and aesthetic constraints. This paper discusses standard rockfall protection systems in both permanent and temporary configurations, as well as a brief discussion on how to determine the design criteria that must be included in the design. These criteria include the geometry of the slope, the approximate size/shape of falling material, coefficients of restitution, and slope roughness. Also in this paper are considerations for construction such as the constructability (cost and project duration) of the protection systems and the safety of crew members who are performing the work. Case studies include a temporary rockfall barrier with a road closure, a temporary barrier with the road fully open, a semi-permanent barrier using alternative materials, and a permanent shotcrete facing using alternative materials. These case studies demonstrate unique strategies and materials that have been used recently to best serve the public by reducing the impact to normal operation in both mining and transportation environments.

Introduction

Rockfall is a constant threat to motorists in mountainous terrain and it is the responsibility of engineers and geologists to mitigate the risk of this hazard through design and implementation of engineered controls. Part of this responsibility is knowing what systems are available for various conditions and being able to select the appropriate system for each project based on effectiveness, constructability, and cost. Occasionally, due to project constraints, the use of standard/typical methods is not practical and unconventional approaches in dealing with rockfall hazards is necessary. Implementation of unconventional materials and methods of installation of these systems is also within this responsibility of these professionals to minimize the impact that rockfall hazards have on the public's lives as well as fit within project constraints. Within this paper, an emphasis is placed on design and construction considerations for solutions to rockfall problems. In addition, case studies where temporary barriers and strategies to reduce time for project completion were used to reduce impact to normal operation during rockfall remediation will be discussed.

Design Considerations

Standard Rockfall Protection

Rockfall protection systems such as barriers, drapes, and attenuators are utilized when stabilization of a slope is not possible due to physical or economical restraints. These systems come in many forms and can be installed on the slope itself or at the base of the slope. Examples of on-slope rockfall protection include but are not limited to draped mesh, pinned mesh, and shotcrete facing. Examples of protection at the base of the slope includes but is not limited to catchment areas, rigid barriers, and flexible fences. Determining which system should be selected for each individual site depends on a number of factors such as anticipated size and trajectory of rock fall, site constraints, cost, and acceptable levels of maintenance. Table 1 is a condensed reference of the suitability of each protection system. A more substantial reference can be found in the Rockfall Characterization and Control manual from the Transportation Research Board (Turner, Schuster, 2012).

Table 1. Suitability of standard rockfall protection methods based on material size, trajectory of rockfall, construction costs, and maintenance required for the system.

Protection Type	Material Size		Trajectory		Cost		Maintenance	
	Small	Large	Low	High	Low	High	Yes	No
Drape Mesh	☑				☑		☑	
Pin Mesh	☑	☑				☑		☑
Small Rigid Barrier	☑		☑		☑		☑	
Barrier Fence	☑		☑	☑		☑	☑	
Flexible Fence	☑	☑	☑	☑		☑		☑

Rockfall Modeling

Predicting rockfall is not an exact science but tools can be used to estimate potential trajectories and magnitudes of energy capacity requirements for a particular rock slope and project constraints.

The most common method of doing this estimation is with computer modelling. Two of the most common programs include Colorado Rockfall Simulation Program (CRSP) and RocFall which was developed by RocScience. The programs are capable of calculating energies and trajectory of falling rock using Newtonian mechanics and can assist geologists and engineers in selecting the best means and methods of rockfall protection. The caveat with these programs is the same in any modeling software, which is that the model is only as good as the design parameters entered into the program. Care should be taken while utilizing such programs to ensure that the output is appropriate for the project. This can be accomplished in several ways such as confirmation with actual site testing, comparison with historic rockfalls, or comparison with design guides (Pierson, Gullixson, Chassie, 2001)

The primary design parameters input into these software programs include slope geometry, approximate size/shape of falling material, coefficients of restitution (normal and tangential) and roughness of the slope. Slope geometry can be acquired using various methods including the use of a laser range finder, or more recently, point clouds derived from Lidar or drone surveys. Each of these methods are suitable and the decision on what to use should come down to available equipment and budget.

The approximate size/shape of falling material can be determined through visual inspection of the slope and material at the base of the slope. The design coefficients are a little more challenging to assign values for without testing that is not typically feasible due to time and budget restraints. Alternatively, field observations of previous rockfall events can be used to fine tune values published in literature to perform a back analysis. This approach is taken by observing rock at the base of the slope that has previously fallen or evidence of fallen rock such as gouges in pavement.

Slope roughness is another key component in rockfall modeling. Surface roughness is incorporated into models to account for irregularities in the slope such as material that has already fallen or weathered on the slope. These irregularities can increase the impact angle (Wu, 1985), which has the effect of increasing the bounce height, while also decreasing velocity and energy of falling rock (FHWA, 1993). Values for slope roughness should be determined from conditions on the slope, but must also be calibrated in the same manner as the coefficients of restitution.

Site Constraints

Site constraints for a rockfall project can include maintaining proximity access, environmental concerns, and aesthetic constraints. Maintaining proximity access is often required where a reasonable detour cannot be created and a road closure is too disruptive to the public. A solution for this is the use of temporary fences and barriers to capture material while maintaining at least one lane of traffic open. When implementing these systems, the main consideration must be the size and amount of material that will fall from the slope. Most small barriers (e.g. Jersey Barriers) can offer good containment for small low-energy events, while flexible fences will provide the most energy dissipation. Larger rigid barriers can be constructed that will have the highest capacity, but will also have the highest cost. One consideration for using flexible fences for proximity access is that large rockfall events will cause the fence to elongate and allow contained material to reach beyond the limits of the fence posts. This elongation has the potential to extend into the area that is being protected if the system is not appropriately designed with this in mind.

Construction Considerations

Constructability

Labor intensive installations prolong construction schedules and increases the overall costs of projects. Therefore, it is critical to a project's success to improve the constructability of the rockfall protection system to minimize the duration of the project. One approach to reduce the duration of a project is to minimize time that technicians must be on rope. This can be accomplished using equipment such as man lifts, cranes, or long reach excavators with a drill mount to reduce the amount of drilling that needs to be performed on rope. Most of this equipment can also be used for installing any mesh required for the project, which can also be installed using a heavy lift helicopter. A cost analysis should be performed to determine if the cost of this equipment is worth the time savings that they can provide.

Safety

Rockfall projects are inherently dangerous and require that precautionary measures be taken to avoid injury to the crews working on site. The most common hazards to crew members are falling rock/debris and falling from heights that could cause injury/death. Safety may be thought to be the responsibility of the individuals performing the work, however, a successful safety program should start during the conceptualization of the project.

For constructability, it is important for a designer to be aware of the risks and hazards of performing the work. Tasks that should be avoided are drilling on or below unstable material to avoid having a worker present if the material begins to move. Additionally, prolonged activity below overhangs and in chutes where rock fall may be more naturally prone to occurring should be avoided or minimized. Alternative solutions that do not require crew members to put themselves in harm's way should be utilized for these conditions.

Project owners may be able to play a role in project safety by including safety protocols in the contract. This could prohibit inexperienced crew members from performing tasks that may present opportunity for injury. These provisions could also include that each crew member has a designated number of hours of experience for the task at hand and requiring third party rope access training. Examples of these programs are offered by the Professional Climbing Instructors Association (PCIA) and Society of Professional Rope Access Technicians (SPRAT), which teach and/or test the abilities necessary for working safely on rope.

In addition to these programs, two-rope safety systems are becoming more common in North America. The Association of Geohazard Professionals (AGHP) has developed a Rope Access Assessment (RAA) form which looks at site conditions and the tasks to be completed and aids in determining what protocols should be followed to create the safest working conditions as possible (Duffy, Fish, Barrett, 2018). The form can be filled by the project owner or representative if they have enough experience. However, simply including a provision that the AGHP RAA form be completed and adhered to prior to the start of the project could help to reduce the risk of injury/death in rockfall projects.

Case Studies

Temporary Rockfall Fence with Road Closure

Recently, GSI utilized a temporary flexible fence to conform to environmental constraints. The project was along the Stillwater River near Absarokee, MT and specifications in the contract prohibited material from entering the river. The original plans called for bolting several large blocks (100 - 200 CY) into the slope and installing mesh for weak zones of rock (ash layers). These plans were developed assuming that the majority of material had been previously scaled and the temporary flexible fence was installed to catch material from safety scaling. Once crews were on site, it became obvious that the scaling efforts were beyond what would be expected of a slope that was previously scaled. Further inspection of the repair areas that were initially designed to be bolted were identified as being hazardous to the rockfall technicians on the slope as the blocks either weren't intact or were not sufficiently supported to permit drilling (Figure 1). It was determined that these blocks needed to come down in the best interest of the crews safety and the long term stability of the slope.

Although the fence was robust (2000 kJ), it was not designed for the magnitude of material that would need to be contained. An earthen berm was built in front of the fence to help dissipate the energy and lessen the impact on the fence. During removal, a seventy-ton pneumatic lifting bag was able to topple the largest of blocks from the slope and the blocks came apart on their way down. For the most part, the largest portions of the blocks were captured by the berm and did not engage the fence. Large blocks that had separated during the topple (8+ CY) made a direct impact on one of the fence posts which collapsed upon impact (Figure 2). Despite this, essentially all material dropped from the slope was contained (Figure 3). These efforts saved the client nearly 16% of their budget and enabled GSI to address other areas on the slope that were initially out of the client's budget.

Temporary Barrier with Road Open

Another project where GSI utilized a temporary barrier was on I-77, a 4-lane divided highway near Fancy Gap, VA. This project included operations in scaling, drilling, and hanging mesh. This required temporary protection as the contract stipulated that both lanes of I-77 remain open during the duration of the project. This was accomplished by a hanging mesh that was secured to two cranes at each end of the active work zone during scaling activities (Figure 4). Since no material was able to pass to maintain a safe traffic environment, mesh of various sizes was included in the system so that no material was able to pass. Limitations of this system were typical of crane operations which included stand down time during storms/high winds and the cost of crane operations, in addition to limited containment capacity. Benefits of this system were a rapid installation and leaving no parts of the system, such as anchors, after the project was completed.

Semi-Permanent Barrier with Alternative Materials

A semi-permanent barrier was installed by GSI at a mine site in northeast Canada which was installed to protect temporary infrastructure. Knowing that the protection was only needed for a limited time, the client requested that the rockfall protection be constructed with the ability to be relocated to another area in the mine at a later date. To accommodate this, above-grade concrete footings for the fence posts were constructed using retired tires of a haul truck for the form work (Figure 5). The weight of these footings are in excess of 20,000 lbs. This mass provides the

Figure 1. Photograph of a rock block (30'x20'x10) resting on weak material less than 2 feet wide. Drilling this block may have initiated movement of the block which may have resulted in injury.

Figure 2. Photograph of from across the river showing the containment of essentially all material that was to be pinned to the slope by the temporary flexible fence.

Figure 3. Photograph of from across the river showing the containment of essentially all material that was to be pinned to the slope by the temporary flexible fence.

Figure 4. A temporary barrier hung from cranes on each end of scaling extents. System consists of various sizes of mesh to contain 100% of material.

Figure 5. Photograph showing the above-grade footings for a semi-permanent rockfall which were constructed using recycled tires as form work for the concrete.

frictional resistance to sliding along the surface and the resistance to overturning from impact. This fence has been in service for nearly a year at the time of this publication and has performed adequately. The posts secured to the above-grade footings have performed as though they were anchored into the ground, and could likely serve as a permanent post installation if drilling was not possible due to utilities or environmental constraints.

Permanent Solution with Shotcrete Facing

GeoStabilization International (GSI) has recently completed a project in which synthetic fiber shotcrete facing was applied to a slope nearly 500 feet long and up to 60 feet tall at its highest point (Figure 6). The site is along a rail line in Hudson, NY which was experiencing raveling of the heavily weathered surface, although the dip of the most prominent joint surface was into the slope. The project consisted of installing rock bolts, improving drainage on the slope, and applying a sculpted shotcrete facing to preserve the natural appearance of the site. Fiber dosage and mix design was recommended by the fiber manufacturer and had to be relied upon as no design manual is thought to exist for designing fiber reinforced shotcrete facing at this time. The estimated time savings on this project is thought to be over a week, which reduced the project cost and allowed the rail line to open in full capacity sooner than it would have been if welded wire reinforcement was installed.

Figure 6. Photograph of sculpted shotcrete facing on a slope along a rail line in Hudson, NY. The project included over 300 cubic yards of shotcrete and rock bolts which would be a permanent solution to the issue of rock raveling into the track.

Conclusions

Protecting the public from geohazards includes not only physical protection, but also from disruption of their day to day activities that can be caused by rockfall events. By thoughtfully considering the design and construction aspects of these projects, engineers and geologists can help to reduce the impact of rockfall events on the public. By including the design and construction considerations before and during the construction of rockfall protection systems, engineers and geologists can effectively reduce the time that the public is affected by rockfall events. Additionally, calculated risks can be very beneficial to the overall project budget and schedule which provides the best service possible to both clients and the public, as was demonstrated by the case studies.

References

Duffy, D., Fish, M., Barrett, C., 2018, Safe work on dangerous slopes: Geotechnical News, v. 36, p. 55 – 57

FHWA, 1993, Rockfall hazard mitigation methods: Publication FHWA SA-93-085. U.S. Department of Transportation

Pierson, L., Gullixson, C., and Chassie, R., 2001, Rockfall catchment area design guide: Oregon Department of Transportation, 91 p.

Turner, A., and Schuster, R., 2012, Rockfall characterization and control: Transportation Research Board of the National Academics, 658 p.

Wu, S., 1985, Rockfall evaluation by computer simulation: Transportation Research Record 1031, Transportation Research Board, National Research Council, Washington, D.C. pp. 1-5

Laboratory Investigations of the Oldest Concrete Pavement in America – Applied Geology in Civil Engineering

Blake Lemcke, PG
Senior Petrographer/Geologist
American Engineering Testing, Inc.
550 Cleveland Ave. North
St. Paul, MN 55114
651-659-9001
blemcke@amengtest.com

Prepared for the 69th Highway Geology Symposium, September 2018

Acknowledgements

I would like to thank several of my colleagues, mentors, and co-workers who have exhibited patience, guidance, and a sense of humor not only in the writing of this paper but in my adventure through the tiny world of concrete petrography and concrete technology. My accomplishments were possible because of the time they've invested in me:

Gerard Moulzolf, PG – American Engineering Testing, Inc.
Lawrence Sutter, PhD, PE, FACI – Michigan Technological University
Terrance Swor, PG – American Engineering Testing, Inc.
Scott Wolter, PG – American Petrographic Services, Inc.
Richard Stehly, PE – American Engineering Testing, Inc.
Karl Peterson, PhD – University of Toronto
Dan Frentress, PE – Frentress Enterprises LLC

...and finally, all my fellow lab-workers

Disclaimer

Statements and views presented in this paper are strictly those of the author(s), and do not necessarily reflect positions held by their affiliations, the Highway Geology Symposium (HGS), or others acknowledged above. The mention of trade names for commercial products does not imply the approval or endorsement by HGS.

Copyright

ABSTRACT

In summer of 2016, American Engineering Testing (AET) was contacted by the American Concrete Paving Association (ACPA) to investigate concrete from Court Avenue in Bellefontaine, Ohio – the oldest concrete pavement in the United States (ACPA, 2016). The study involved several collaborating parties in both the private and academic sectors to assess the physical, chemical, and geologic properties of the historic pavement, which is still in service. The goals of the investigation were to understand how a concrete pavement placed in 1893 is still performing and what implications can be drawn to modern portland cement concrete pavements (PCCP's) used in highway construction.

AET received two pavement sections from the ACPA for the laboratory study. Representative sub-samples were procured by AET and sent to four separate laboratories to perform analysis of their choosing. Techniques utilized in the study included: petrography (optical microscopy), air void system analysis, scanning electron microscopy, electrical resistivity, neutron imaging, thermogravimetric analysis, and low-temperature differential scanning calorimetry.

These combined studies revealed the nature of the pavement's construction, the properties and attributes of the raw materials utilized, and led to an understanding of the pavement's durability and longevity. Specific material properties obtained by the study were as follows: aggregate characteristics which included lithology/type/size/grading, hardened paste properties which included air void system parameters/water-cement ratio/cement clinker chemistry and morphology. Physical concrete characteristics obtained included permeability, volume of permeable pores, and sorptivity. Petrography in-particular has proven a beneficial tool in the assessment of concrete (both young and old) and is a testament to the power of applied geology in highway engineering and construction.

INTRODUCTION

This paper primarily presents data and results of petrographic analysis performed by AET on historic concrete pavement from Bellefontaine, Ohio. The pavement from Court Avenue was placed in 1893 and is the oldest known example of portland cement concrete pavement (PCCP) in the United States (Figure 1, a). The laboratory study was envisaged by the ACPA and involved three other entities, including: Braun Intertec Corp., Oregon State University, and the University of Toronto. Bellefontaine, located in west-central Ohio, lies directly in a severe freeze-thaw environment according to the American Concrete Institute (ACI). The longevity of the pavement in such an environment is a testament to the material's durability and the ACPA felt it warranted an in-depth study. The obvious question to be answered: what characteristics allowed the concrete to last 125 years? The paper also draws comparisons between the historic pavement and modern PCCP's, which are widely used in highway construction.

BACKGROUND

George W. Bartholomew (Figure 1, b) has been attributed with the vision, mix-design, and financial backing of this pavement – including the production of the cement used in the mixture (Sutter, 2017). During this time in the United States portland cement was in its infancy, and many preferred the more reliable German-produced portland cements or the American 'Rock/Natural Cements'. Bartholomew learned of the cement making process while visiting Germany and later at the Alamo Cement Company in San Antonio (Snell, 2002). He began experimenting with local marl deposits in an area north of the city upon returning to the area in 1886. Upon developing the correct materials for cement production, he founded the Buckeye Cement Company in 1889. They first used vertical-style kilns which utilized a continuous feed which reportedly improved production rates. Ground marl and 'blue clay' (raw-feed) was fed into the top of the kilns and raked into the underlying combustion chamber (Sutter, 2017). Clinker was eventually removed from the base of the kiln. Information regarding the clinker grinding process is sparse, however; the clinker was apparently milled with rounded imported 'flint stones' from Iceland (Pardi, 2017). The fuel utilized in the kilns at that time was petroleum crude.

(a) (b)

Figure 1 – recent photograph of Court Avenue (a) and photograph of George Bartholomew, date unknown (b) (photos accessed from http://explorer.acpa.org/explorer/places/united-states/ohio/bellefontaine/street/old-us-30-lincoln-highway/)

Historical accounts say that the streets of Bellefontaine were a sea of deep mud in the spring and fall of each year (Pardi, 2017). During drier months the streets were hard and very dusty. George Bartholomew saw this as an opportunity to increase commerce in the city, as horses and buggies experienced great difficulties on the earthen thoroughfares. Having previously placed a driveway for a local lumberyard, the City Council was impressed and commissioned a 220-foot-long test strip in 1891. W.T.G. Snyder, a local cement contractor, was hired to place the test strip.

The installation followed a similar technique to sidewalk construction, the slabs were formed in 5-foot square sections (Snell, 2002). Tar paper was placed between adjacent slabs and a 'two-lift' system was utilized. The base coarse contained coarser rock and a higher water to cement ratio (w/c) while the upper lift or wear-course contained smaller aggregate and a lower w/c. Concrete mixing was done without heavy equipment; the sand, stone, and cement were unloaded into a pile and mixed by hand and tamped into the forms. The pavement was cured by placing a few inches of wetted sand over its surface for one week. The original surface finish of the pavement was tined to aid in horseshoe traction. In 1893, portions of the pavement were sent to the International Exposition in Chicago (World's Fair) where Bartholomew won first prize for Engineering Technology Advancement in Paving Materials (Snell, 2002).

Sampling

Two large segments of full-depth pavement were received by AET in summer of 2016. One of the segments was unadulterated (Figure 2, a) and one had been previously core sampled and exhibited several cylindrical core hollows (Figure 2, b). Both pavement segments were fairly large; the intact sample measured approximately 12" x 12" and the previously-cored segment slightly smaller. Both segments represented a nearly 'full-depth' cross-section of the pavement and were about 6" to 8" thick. The larger non-cored segment was wet saw-cut into several 'slabs' for the laboratory analysis and as display pieces. The slab sections were flattened and polished with loose grit abrasives on a lapidary wheel using water as a lubricant. The slabs were worked from a coarse 80 grit up through progressively finer grits and eventually finished on 600 to produce a matte finish suitable for microscopic observations. The smaller pavement segment was treated in a similar fashion, with smaller slabs and sections sub-sampled.

(a)

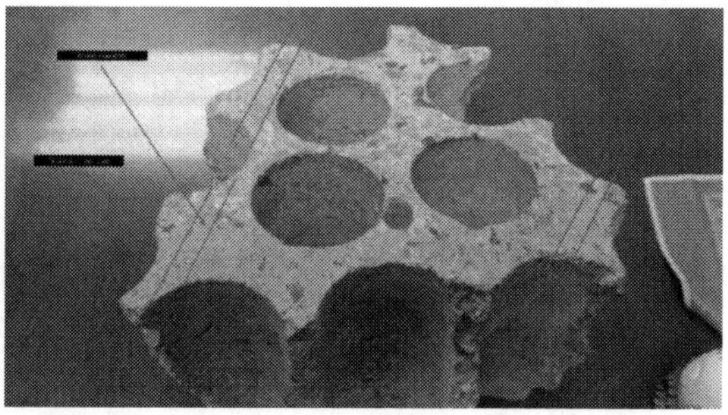

(b)

Figure 2 – pre-submittal photos of intact concrete pavement segment (a) and previously-cored segment (b) (photos supplied by the ACPA)

Sub-samples and polished slabs were distributed to the University of Toronto, Oregon State University, and Braun Intertec for analyses of their choosing. Upon slab preparation, the distinct layers or 'two-lift' construction of the pavement was obvious (Figure 3). Thin sections of the pavement were produced by AET from the denser wear-course, porous base-course, and the interface between the two concrete placements.

Figure 3 – saw cut and lapped cross-sectional profile of Bellefontaine pavement sample. Note the two-layer construction is evident by both paste coloration and aggregate sizing

PETROGRAPHIC FINDINGS

The base concrete of the examined pavement section ranged from 44 mm (1-3/4") to 102 mm (4") in thickness, contained a 38 mm (1-1/2") nominal-sized carbonate-rich gravel coarse aggregate, was placed at a moderately high w/c, and was fully carbonated throughout its thickness (Figure 4). The concrete topping ranged from 32 mm (1-1/4") to 70 mm (2-3/4") in thickness, contained a mixture of natural 'pea gravel' and crushed granite coarse aggregate, was placed at a moderately low w/c, and exhibited an overall negligible carbonation profile measured from the top surface of the pavement. The two concretes appeared to be very well bonded to each other. The total thickness of the base layer in the examined section was likely not full-depth, as much of its bottom surface exhibited a fractured rather than formed appearance. The base concrete of the pavement contained a measured 7.9% total air void content and the top or wear-course contained 7.5% total air.

Figure 4 – saw cut and lapped cross section of pavement after application of pH indicator. Magenta stain represents pH levels over 8.3. Note the base concrete exhibited lower pH associated with paste carbonation.

Air Void Analysis

Air void system analysis (ASTM C457, Procedure A) were performed individually on both layers of the concrete pavement section. This testing involves a 'linear traverse' of the lapped section under high magnification in which individual void spaces are measured (chord lengths) and tallied. This test method was developed by ASTM for modern concrete mixes in which air entraining admixtures are utilized to create a system of microscopic bubbles (voids) which protect the paste from frost damage. Air entrainment has been widely utilized since the middle of the 20th century in high performance concretes. The American Concrete Institute (ACI) has developed a series of air void system parameters which they consider necessary for freeze/thaw durability. These parameters are the total volume of entrained air, spacing factor (average distance of the air voids), and specific surface value (essentially a ratio of void diameters to void volumes).

The base concrete layer contained a measured air void content of 7.9% with a spacing factor of 0.012" and specific surface value of 240. Approximately 4.3% of the measured air was 'entrained-sized' or less than 1 mm (1/32") chord length and 3.6% of the air was considered 'entrapped-sized' or greater than 1 mm chord length. The vast majority of the air in the base layer consisted of coarse, irregular-shaped, consolidation-like voids (Figure 5, a) which resulted from incomplete tamping or packing of the mixture. Some areas of the base layer even exhibited a 'honeycomb' appearance from the under consolidation of the mixture (Figure 5, b). This type of 'coarse' air is generally not beneficial to protecting the paste in a freeze/thaw environment, but is however ideal for drainage.

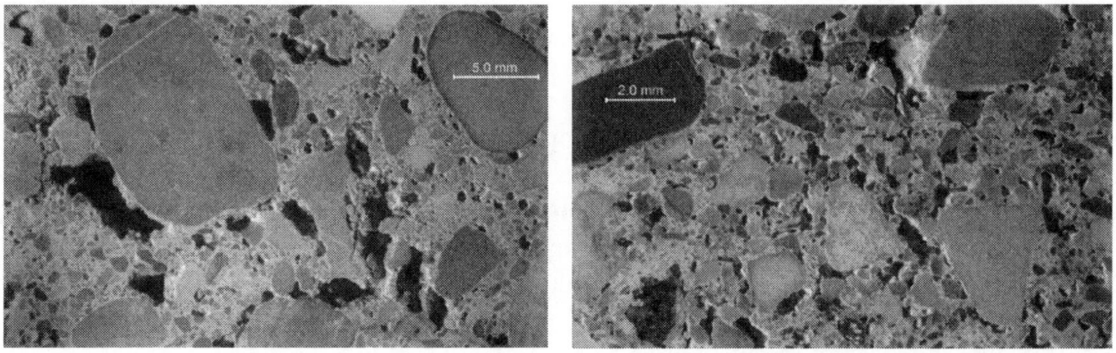

(a) (b)

Figure 5 – saw cut and lapped surfaces of the base concrete layer showing coarse irregular-shaped voids under 5x mag (a) and abundant voids producing a honeycomb' texture under 10x mag (b)

The concrete topping layer or wear-course contained a measured air void content of 7.5% with a spacing factor of 0.008" and specific surface value of 660. Remarkably, these air void system parameters were consistent with the modern recommendations for freeze/thaw durability outlined in ACI 212.3R: "The cement paste in concrete normally is protected against the effects of freezing and thawing if the spacing factor does not exceed 0.008", as determined in accordance with ASTM C457. Additional requirements are that the surface area of the air voids should be greater than 600 in^2/in^3..." The air void system of the wear-course closely resembled those observed in modern concretes produced with intentional air entraining admixtures (Figure 6, a & b). The top layer or wear-course concrete would be particularly susceptible to saturation from meteoric water and freeze/thaw cycling; the air void system (and low w/c, discussed later) certainly played an important role in the topping's durability/longevity. The possible origins of such an air void system in this historic concrete which was produced prior to the discovery of air entrainment is discussed later.

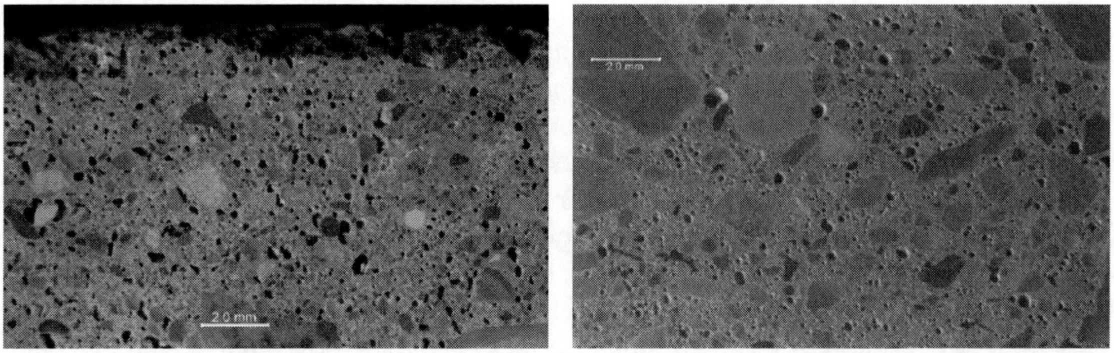

(a) (b)

Figure 6 – lapped section of top layer or wear-course concrete showing abundant (sub) spherical air voids within its paste under 10x mag (a) and comparison to a modern concrete with purposeful air entrainment under 10x mag (b)

Paste Characteristics

A controllable parameter in designing concrete for durability is the water to cement ratio (or w/c). Concrete is a porous material often likened to a 'hard sponge' and contains three different types of voids or pores. The smallest of these are the pores present within the gel of the amorphous cement hydration products (calcium-silicate hydrates, CSH). These pores are present on a nano-scale (0.5 to 10 nm) and their role regarding durability is relatively insignificant. In contrast, the capillary or interstitial pores of the paste are of great importance regarding the overall strength and freeze/thaw durability of any concrete. This porosity is typically on the range of 10 nm to 10 μm and results from the residual spaces between cement hydration products, residual cement grains, and aggregates. The nature of this porosity is a direct result of the concrete's w/c as it is placed, representing spaces which were originally filled with mix water. The largest of the pores in concrete are the coarse voids which become incorporated during mixing or consolidation, whose importance was covered in the previous Air Void Analysis section.

Water to Cement Ratio (w/c)

Just as variation was noted in the total air void content of the two pavement 'lifts', a variance in w/c between the two layers was equally evident from the analysis. Simple physical characteristics such as paste color and paste hardness can be used to qualitatively assess the w/c of any concrete mixture. The paste coloration of the upper concrete placement was light to medium gray (Munsell® Rock Colors N7 to N5). In contrast, the base concrete placement exhibited a much lighter paste coloration which ranged from yellowish gray to very pale orange (Munsell® Rock Colors 5Y 8/1 to 10YR 8/2). These coloration differences are quite clear in Figure 3. The gray or general dark coloration of the concrete topping layer was consistent with placement at a moderately low w/c and the lighter overall color of the base with placement at a moderately high c/m. Although paste carbonation (discussed later) can also influence coloration. Paste hardness, as one might guess, is also directly related to w/c. In general, harder pastes are indicative of lower w/c (being less porous) while softer paste indicate higher w/c. The paste of the wear-course exhibited variable hardness, but was generally considered to be moderately hard overall (Mohs 3.5 – 4). The paste of the base concrete was judged to be moderately soft (Mohs 2.5 – 3).

A more detailed and quantitative estimate of w/c can be drawn from thin section analysis. Aged or historic concretes add an extra challenge to the petrographer, as portland cement manufacturing and grinding technology has changed drastically in the last 60 years. Both layers of the concrete pavement exhibited abundant residual cement clinker grains for observation, many of which were coarsely-ground (up to 3 mm) and even visible in hand sample (Figure 7, a). Additionally, the clinker morphology was somewhat inconsistent, with many grains being of unique composition (Figure 7, b). For example, some unhydrated particles within the wear-course contained very fine and well-rounded belite particles with interstitial alite, and many particles were free of the ferrite and aluminate cement phases. Additionally, several of the residual calcium silicate grains (alite and belite) were a moderately dark tan to brown coloration – not a common feature of modern cements. The abundance of residual cement clinker was not surprising given the very coarse grind of the cement. The topping or wear-course contained a visually estimated 8 to 10% residual cement and an approximate w/c of 0.30 to 0.45, depending on exact location. The base concrete was estimated to contain between 4 and 6% residual cement particles and w/c between 0.55 and 0.65. Interestingly, a historical marker placed at the site in Bellefontaine claims the bottom course of the concrete pavement *had 18 sacks of cement, 104 cubic feet of aggregate, and water; making a 1:2:4 ratio of materials. Then a 2 inch 1:2 mortar top was spread on the base and tamped.* (Sutter, 2017). Petrographic estimates were generally in-line with these figures from the historical marker.

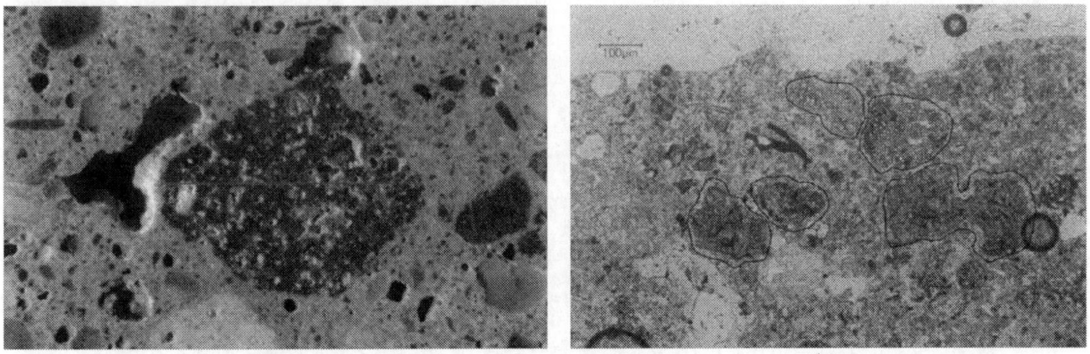

(a) (b)

Figure 7 – a very coarse dark-colored remnant clinker particle with a max. dimension of 3 mm in the wear-course concrete under 25x mag (a) and variable clinker morphology (blue outlines) as viewed in thin section of the wear-course under plane polarized light at 100x mag (b)

Residual Cement Characteristics

Scanning electron microscopy (SEM) and energy dispersive spectroscopy (EDS) were performed at the University of Toronto. The work revealed many of the residual clinker particles to be of compositions quite like modern portland cements. Figure 8 shows one such residual particle and details its elemental composition based on emitted x-ray energy upon electron-beam stimulation. The phases documented in the particle include the four main phases of modern portland cements, which include: alite (tricalcium silicate), belite (dicalcium silicate), aluminate (calcium-aluminate), and ferrite (calcium-iron-aluminate). Euhedral grains of tabular alite are arranged with sub to anhedral grains of belite within the particle. Filling in the interstices are the aluminate and ferrite phases. Of interest, the presence of a Mg-rich phase was also noted in the interstitial material and was attributed to the presence of periclase (MgO) (Avdyllari, 2017). Periclase is not commonly found in modern cements as Mg contents of raw-feed are now kept to a minimum as periclase hydration can lead to soundness issues. Its presence in this historic material is consistent with the use of relatively 'impure' raw ingredients. Although the manufacture of this historic cement in a vertical-style kiln would now be considered crude; the resulting material was very similar to modern cements produced in rotary-style kilns.

Also of interest regarding the cement and cement hydration is the lack of gypsum utilized in the ground clinker. Gypsum is currently inter-ground with cement in order to control its setting time. The discovery of gypsum addition was not realized until several years after the Court Avenue pavement had been placed. Portland cements which lack a source of sulfate will 'flash' set, reducing workability and disallowing longer placement times. Further, long-term storage of ground cement is problematic without gypsum as it will prematurely hydrate from atmospheric humidity, leading to clumping of the gray powder.

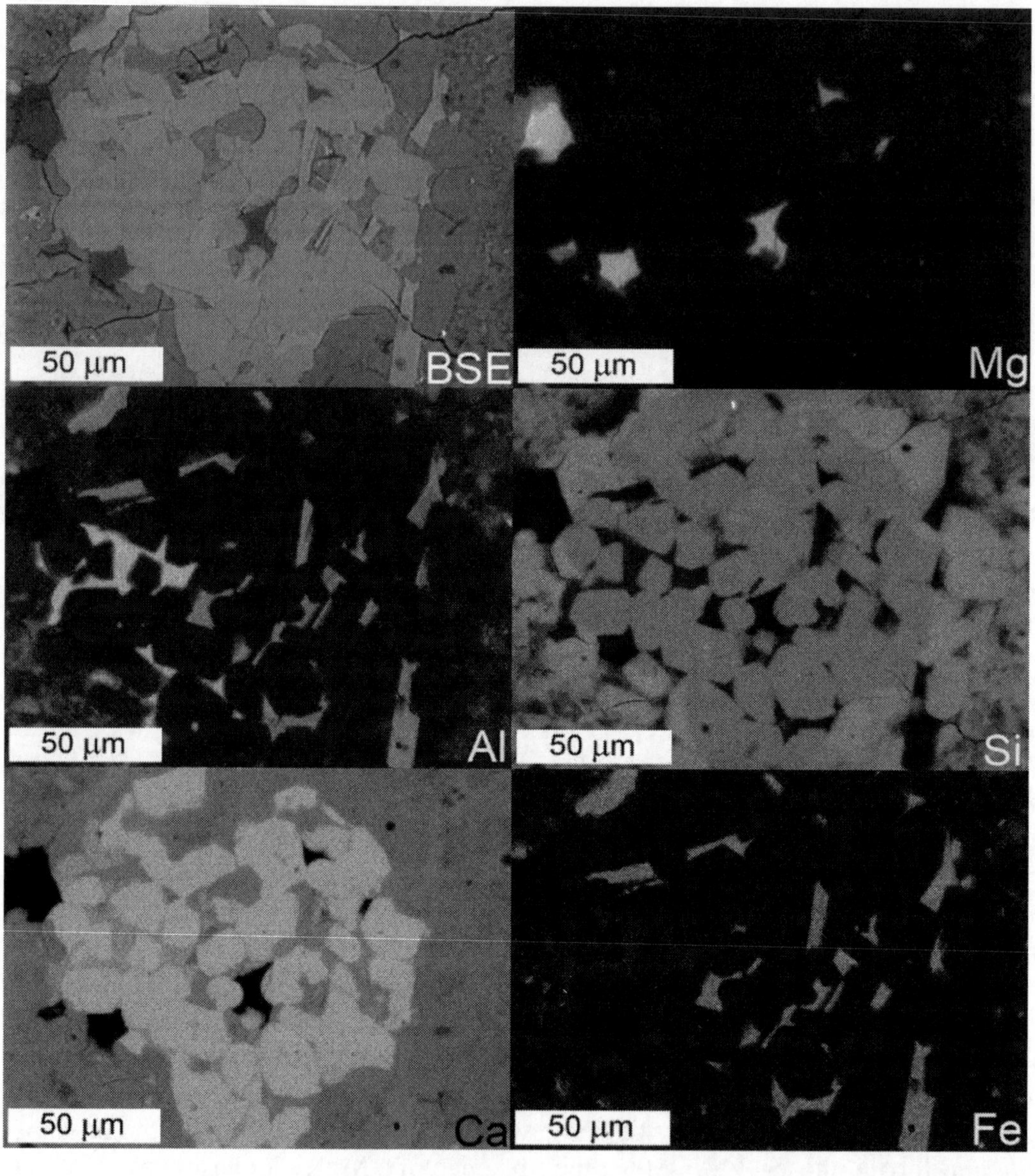

Figure 8 – series of images consisting of a back-scattered electron (BSE) image (upper left) and subsequent elemental maps obtained from the same region. The element listed in the bottom right corner of the images denotes its presence as brightly-lit areas. Note the few brightly-lit areas in the Mg map which are attributed to the presence of magnesium oxide or periclase (MgO). Images from (Avdyllari, 2017).

Carbonation

The carbonation of portland cement paste involves the reaction of carbonic gases (carbon di/monoxide) in the air or dissolved in moisture and the cement hydration products. The hydration products altered from this reaction include the crystalline portlandite phase ($CaOH_2$)

and the amorphous CSH gel. As concrete is exposed to the atmosphere, the reaction slowly converts the portlandite and CSH into the more stable calcium carbonate phases of calcite and/or vaterite. The rate of this reaction is dependent upon several factors, with the most influential being paste porosity and permeability (a function of w/c) and the exposure conditions of the concrete. With the carbonation reaction comes a drop in the pH level due to the consumption of alkalis. Concrete is a very alkaline or basic material and when freshly mixed typically exhibits pH levels in the 12 to 13 range. Phenolphthalein is an ideal indicator for the drop in pH that occurs from the carbonation reaction, being a bright magenta at pH levels between approximately 8.2 and 13.0. The indicator is colorless below 8.2, and as can be readily seen in Figure 3, which shows the base concrete of the Bellefontaine pavement to exhibit nearly complete carbonation. This is a direct result of the base concrete's porosity/permeability which was the product of both its higher w/c and abundant coarse consolidation voids, readily allowing the passage of moisture. The high w/c and permeance of the base layer may at first seem like a negative attribute; however, this design produced relatively strong base material which also allowed for adequate drainage to ensure the passage of moisture/meteoric water. The negligible level of carbonation of the wear-course after over 100 years of service is a direct result of this layer's density or low porosity/permeability. This strong surface, the result of low w/c, has proven very durable to overhead traffic (both horses and automobiles) and general exposure.

It is important to note that once the carbonation reaction begins, the hydration reaction is halted. This is paramount in the curing of modern portland cement concretes as adequate cement hydration (or curing) is needed to realize the full-strength potential of the material, which establishes its long-term durability. Tying this into the residual cement observed in the wear-course, one could assume that the concrete of the wear-course is still undergoing the hydration reaction and is slowly gaining strength after over 100 years of service!

Aggregate Characteristics

It has been said that you can make poor concrete from good materials, but never make good concrete from poor materials. As demonstrated, the paste/cement portions of the Bellefontaine pavement were of great quality for their time, and so, the aggregates also deserve some credit for the longevity of Court Avenue.

The aggregates of the base concrete layer were of a coarser gradation and of slightly different composition than those of the wear-course topping. Aggregate in the base layer consisted of 38 mm (1-1/2") to 51 mm (2") nominal-sized natural carbonate-rich gravel. Lithologies documented primarily included: micritic, argillaceous, and sandy dolostones. The base layer also contained many pea-gravel sized particles of similar composition. The coarse aggregates were mostly rounded to sub-rounded with only a few sub-angular particles present. This aggregate property likely made for easier workability when hand mixing the plastic concrete mixture in its formwork. Additionally, the predominant carbonate lithology of the aggregates lent itself to a very good bond with the surrounding paste, aiding strength. It has been well-documented in concrete literature that carbonate lithologies exhibit exceptional bonding properties with portland cement pastes. Overall, the coarse aggregates of the base concrete were considered hard and durable; however, some aggregate deterioration was noted in the examined pavement and are discussed later. The finer, sand-sized particles of the mixture comprised quartz, feldspar, carbonate, and other lithic particles (including chert).

The wear-course topping concrete visually contained a lesser amount and finer coarse aggregate relative to the base concrete. This was consistent with the general paste-rich appearance of the topping compared to the more paste-lean base layer. The topping layer contained a 12 mm (1/2") nominal sized coarse aggregate that appeared to be a mixture of natural carbonate-rich pea gravel and crushed (?) or angular igneous rock. The pea gravel was of similar lithologic composition to that documented in the base concrete. The abundant angular igneous particles consisted primarily of a dark-colored amphibole-bearing gabbro and lighter-colored granitic lithology. It is not known if the igneous material was a natural feature of local gravel deposits at that time or was intentionally added to the topping mixture to add durability. As can be seen in Figure 9, the dark igneous particles were primarily present within the topping concrete, and a few also residing within the base layer. Like the base concrete aggregates, the coarse aggregate in the wear-course was considered very hard and durable. Without question, the harder siliceous aggregates in the topping would provide an abrasion resistant surface that could withstand the impacts of overhead traffic.

Figure 9 – saw cut and lapped pavement section showing many dark and angular igneous rock particles within the wear-course topping, a few were also noted in the base layer.

Deterioration Mechanisms and Secondary Features

Some wear-and-tear would be expected from any material exposed to the forces of nature for over 100 years. And although still in service, the Bellefontaine pavement exhibits some evidence of deterioration driven by the universal solvent – water. As in with modern concretes, the primary culprit in most deterioration is water or moisture. Perhaps the most mundane deterioration documented in the pavement is the wearing of its top surface. While some of this surface erosion was likely derived from physical wear or impacts from traffic, much of it was likely due to paste denudation from slightly acidic meteoric water. Similar to the acid erosion of ancient marble statues, acid rain will slowly dissolve both the carbonate aggregate and surrounding cement paste binder. While the degree of this weathering was not directly measurable in the examined specimens, much of the paste on the exposed top surface of the material was recessed and surrounding siliceous aggregates 'stood proud' from the surface – good evidence of chemical weathering.

While not a direct deterioration mechanism, the topping material exhibited several deep drying-shrinkage cracks. These cracks were apparent on the top surface of the pavement and reflected sub-vertically through the full-depth of the wear-course. The shrinkage results from both the long-term drying of the paste and from the progression of cement hydration. This cracking, while not necessarily detrimental itself, act as conduits for water/moisture to infiltrate the pavement system. Shrinkage in modern concretes is expected, and engineers typically plan for this in the design of pavement and floor slabs.

Freeze-Thaw Deterioration

Some evidence of freeze-thaw deterioration was documented in both concrete layers, within both the paste and aggregates. Freeze-thaw deterioration is essentially the overcoming of the tensile strength of the concrete by the expansive force of freezing (expanding) water. Of course, the water in concrete would be present within the voids and capillary porosity of the paste. A good entrained air void system typically alleviates these internal pressures by allowing the freezing water to expand into the small spherical voids of the air void system. As previously discussed, the wear-course topping contained an air void system which meets the current recommendations to resist frost damage. However, the paste of the wear-course contained several anomalies which consisted of pebble-sized paste nodules or 'agglomerations' which did not contain any of the observed sub-spherical air voids (Figure 10, a). This is where deterioration was noted, most commonly occurring within these anomalous zones present near the top surface of the pavement. These areas of paste were noted to contain abundant sub-horizontal microcracks – consistent with damage from cyclic freezing and thawing (Figure 10, b).

The origin of these anomalous 'void-free nodules' was not entirely clear, as was the origins of the sub-spherical voids themselves. One hypothesis is that the nodules represent cement which had prematurely hydrated due to the lack of gypsum or sulfate in the cement, as previously discussed. These pre-hydrated paste clumps were not broken up during the mixing/placement of the pavement when the air voids were apparently incorporated or 'entrained' into the mixture. It is plausible though somewhat unlikely, that the mixing alone led to the formation of the entrained-like air voids. Some have speculated that the residues of the crude petroleum used to fire the vertical cement kilns was present in the final ground cement product. Upon mixing with water, the residue acted as a surfactant similar to modern air entraining admixtures, and produced the abundance of microscopic 'bubbles' within the paste. The pre-hydrated 'clumps' would be protected from this mixing and thus contain no bubbles or voids. Despite the origins of the anomalous paste nodules and the voids themselves, the air void system of the wear-course has clearly been essential in protecting the paste from freeze-thaw damage.

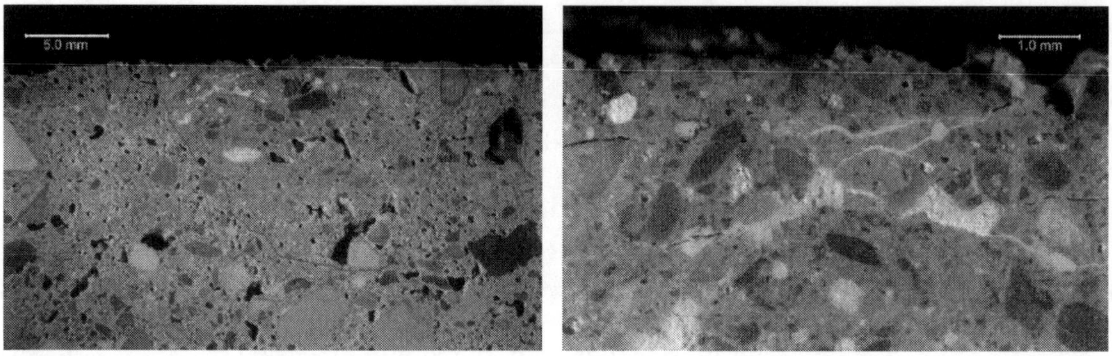

(a) (b)

Figure 10 – anomalous paste 'nodule' (red outline) near the pavement surface which lacks fine air voids under 5x mag (a) fine sub-horizontal microcracking within the nodule highlighted by white secondary deposits under 25x mag (b)

It can be seen in Figure 10b that much of the horizontal microcracking (and some surrounding pores) are filled with white-colored secondary deposits. The deposits primarily consisted of portlandite and calcite with some minor ettringite. This 'self-healing' was possible from paste leaching and transport of hydration products within the concrete system. Overall, this freeze-thaw damage within the wear-course was very minor and had little effect on the bulk condition of the examined pavement samples.

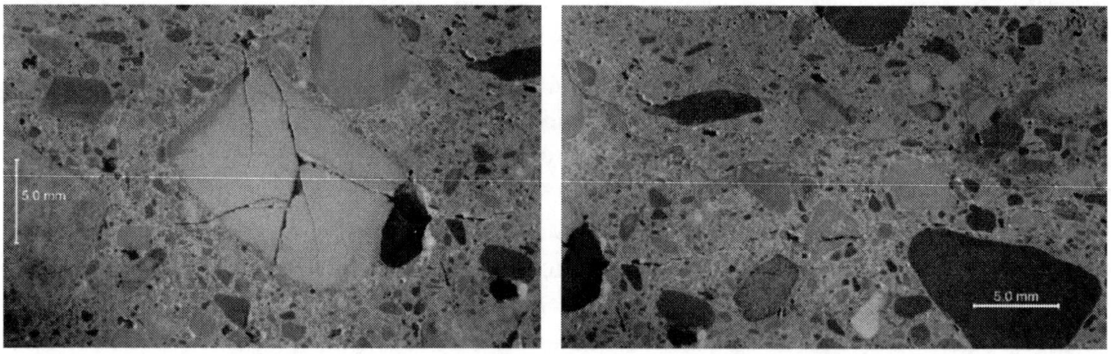

(a) (b)

Figure 11 – cracking within soft dolostone coarse aggregate particle within base layer concrete under 5x mag (a) and sub-horizontal cracking within base concrete paste near the contact with the wear-course (red line) under 5x mag (b)

Freeze-thaw damage within the base concrete was more extensive than that observed in the wear-course topping. Although the base layer lacked entrained air and was more porous, most of the frost damage was documented within softer dolostone coarse aggregate particles (Figure 11, a). Currently, this type of aggregate deterioration in pavements is termed 'D-cracking' as it usually manifests near control or construction joints where the infiltration of water leads to saturation. The subsequent damage results in cracking that reflects towards the pavement surface and creates a 'D' shaped crack along the pavement's edge. Relatively few of the carbonate gravel particles exhibited this cracking, and of those that did, most was relatively minor. The cracking was observed propagating into the surrounding paste, though typically not far. Some minor freeze-thaw damage was also observed within the paste itself and consisted of sub-horizontal microcracking, mostly present near the interface between the base layer and overlying wear-course (Figure 11, b).

Alkali-Silica Reactivity (ASR)

Alkali-silica reactivity was documented within the wear-course topping concrete of the examined samples. ASR can be a destructive force which is caused by the swelling of an expansive gel byproduct. The reaction occurs between unstable forms of silica within aggregate particles and alkalis present within the paste (typically Na and K). Unstable silica is essentially 'attacked' by alkaline-rich pore solutions and dissolved to form a gel. The gel is extremely hygroscopic and at relative humidity greater than 80-90% it will absorb water and expand – initiating cracking and destroying concrete from the inside-out. Another water-driven deterioration mechanism, ASR was discovered in the 1930's and first written about in 1940 by Thomas E. Stanton while studying concrete expansion in California. ASR has affected infrastructure throughout the world, in worst-case-scenarios leading to structural deficiencies and demolition. It is important to note that ASR typically takes many years to fully manifest and lead to this destruction.

The level of reactivity in the Bellefontaine pavement was considered innocuous and had not induced any bulk deterioration. Several chert particles within the sand and pea gravel of the wear-course exhibited proximal deposits of ASR gel (Figure 12, a & b). Only a few of the particles exhibited the associated expansive cracking, which was very minor and did not extend far into the surrounding paste.

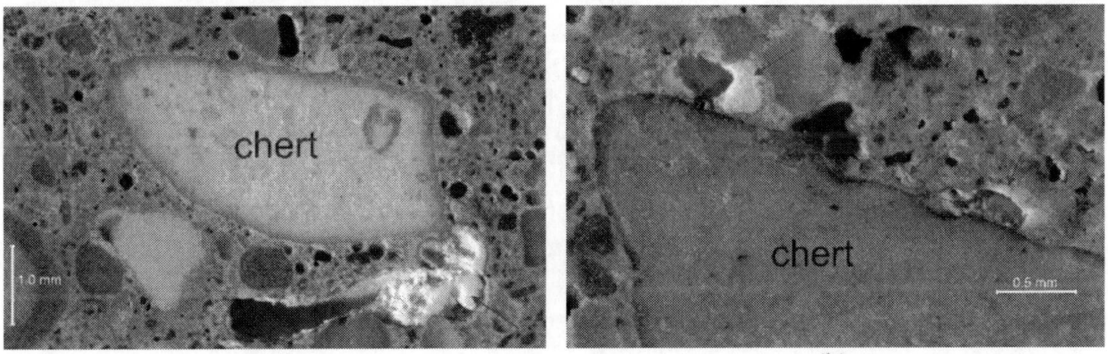

(a) (b)

Figure 12 – chert fine aggregate particles undergoing minor ASR within the wear-course concrete, red arrows indicate proximal deposits of secondary bright white ASR gel under 25x mag (a) and 50x mag (b)

ANCILLARY LABORATORY TECHNIQUES

As mentioned, several sub-samples of the pavement were distributed to various laboratories/Universities for study and testing. The primary tests applied were physical in nature and involved those associated with fluid transport which have been developed for modern concretes. The most striking results from these analyses were the data obtained from the electrical resistivity testing. This testing relates to the connectivity and tortuosity of the pore structure of the paste. The resistivity testing was performed at Oregon State University and relates a number, called the formation factor, to the microstructure of the paste. The base layer had a measured formation factor of 137 and the wear-course a measured value of 987. For example, modern concretes which are low in permeability (as determined by ASTM C1202) have a typical formation factor between 140 and 150. The measured value for the wear-course indicates a permeability much lower than modern high-performance concretes at 28 or 56 days of age.

The results for other physical testing (porosity, sorptivity, calcium hydroxide content) produced variable results from which not much further information could be drawn. The direct measurement and modern test procedures applied to small fragments of historic material is a likely cause for this. Several of these tests require large specimens, for example 6" diameter x 12" long cast cylinders, and their application was not ideally-suited for the pavement sections. Further, paste alterations (carbonation, secondary deposits, etc.) likely influenced the outcome of physical testing results.

CONCLUSIONS

The concrete pavement from Court Avenue in Bellefontaine, Ohio is the oldest known concrete pavement in the United States and is still in service today. The longevity of the pavement is a product of the raw materials and processes used in cement production and in the manufacture of the concrete itself. The pavement also owes its longevity to its 'two-lift' design, now commonly referred to as granitoid-type construction (Lemcke, 2017). The upper wear-course was very dense and impermeable due to placement with a low w/c, keeping moisture out and providing a hard and solid wearing surface. The base layer was less dense and more permeable, but still somewhat hard and durable, providing a solid yet permeable sub-base for the protective wear-course topping. Apparently incidental microscopic air bubbles were incorporated into the paste of the wear-course and have kept keep freeze-thaw deterioration to a minimum. The 'air void system' of the wear-course closely resembles those produced in modern concretes with specialized air entraining admixtures. While trying to understand the pavement's longevity through modern physical testing; petrography proved the most powerful and beneficial tool in determining its general properties and overall success as a pavement. Petrography is also applied to modern concretes of any construction type to determine the cause(s) of performance issues or to aid in condition assessments, for quality control and material screening purposes.

REFERENCES:

American Concrete Pavement Association. Historical Concrete Pavement Explorer. *First and Oldest Concrete Pavements in U.S. – Main Street and Court Avenue, City Square in Bellefontaine, OH.* ACPA. http://explorer.acpa.org/explorer/places/united-states/ohio/bellefontaine/street/old-us-30-lincoln-highway/#post_profile. Accessed May 4, 2018

Avdyllari, K. Avdyllari, S. Chan, C. Rahman, T. *Report on 1891 Concrete from Bellefontaine Ohio*. Department of Civil Engineering, University of Toronto, 2017

Lemcke, B. Moulzolf, G. *Report of Concrete Analysis.* American Engineering Testing, Inc. St. Paul, MN, 2017.

Pardi, M. What Does Bellefontaine's 125-Year Old Concrete Tell Us About Pavement Performance. *American Concrete Pavement Association,* ACPA, San Francisco, CA, 2017.

Snell, L. Snell, B. Oldest Concrete Street in the United States. *Concrete International* March 2002, pp. 72-74.

Suraneni, P. Qiao, C. Khanzadeh Moradllo, M. Weiss, J. *Progress Report of the Tests on Concrete from the Bellefontaine Pavement*. School of Civil and Construction Engineering, Oregon State University, Corvallis, 2017.

Sutter, L. *A Laboratory Examination of Court Avenue, Bellefontaine Ohio: The Oldest Concrete Pavement in the United States.* Department of Materials Science and Engineering, Michigan Technological University, Houghton, MI, 2017.

From Field Data Collection to Soils Analysis in A Few Mouse Clicks:

Going (Even More) Digital at North Dakota DOT

Jesse Greenwald
Bentley Systems, Inc.
2019 E 4th Ave.
Port Angeles, WA 98362
(858)-410-6877
jesse.greenwald@bentley.com

Colter Schwagler
North Dakota Department of Transportation
608 E. Boulevard Ave.
Bismarck, ND 58505
(701)-328-6975
cschwagler@nd.gov

Prepared for the 69th Highway Geology Symposium, September, 2018

Acknowledgements

The authors would like to thank the following individuals for their contributions in the work described:

Dave Kyllonen – GeoDatabase Solutions
Sharon Taylor – NDDOT

Disclaimer

Statements and views presented in this paper are strictly those of the author(s), and do not necessarily reflect positions held by their affiliations, the Highway Geology Symposium (HGS), or others acknowledged above. The mention of trade names for commercial products does not imply the approval or endorsement by HGS.

ABSTRACT

Most organizations recognize and acknowledge inefficiencies in the process of collecting, processing, and reporting geotechnical data. However, many stick to outdated workflows as they perform business as usual. North Dakota DOT recently partnered with Bentley Systems, Inc. on a project to migrate legacy systems to gINT Software that provides a model to other organizations looking to optimize the management of geotechnical data.

NDDOT records borehole data electronically in an Excel spreadsheet in the field. Automated routines were created to import the Excel field data into a gINT database, eliminating the tedious and error-prone transcription effort required with paper-based data collection. The database was also customized so that lab personnel could easily enter index test data, as well as import advanced test results. Thus, data entry requires very little effort from engineers, who can then automatically create custom reports that were previously drafted by hand. Additional features were developed to significantly streamline workflows for linear soil surveys. With a few mouse clicks, engineers can create to-scale, color-coded profiles that summarize soil conditions across miles of highway and any number of boreholes. Further, data from over 1,200 legacy projects was migrated to a gINT SQL Server database for easy access to historical data.

The improvements at NDDOT demonstrate that geotechnical data management requires continuous questioning and elimination of inefficiencies. The result is systems that allow engineers to focus on analysis and recommendations rather than compiling information.

INTRODUCTION

Many organizations struggle with efficiently managing geotechnical data. They tolerate repetitive data entry, laborious manual data validation, time-consuming report drafting, and data that is difficult to access. There are many ways in which these workflows can be improved, and the benefits of doing so are clear, such as time and cost savings, error reduction, improved data quality, and higher morale through the elimination of repetitive tasks. Yet, these inefficiencies persist for many reasons. One of the primary obstacles is uncertainty of how and where to start optimizing these routine workflows.

North Dakota Department of Transportation (NDDOT) recently partnered with Bentley Systems, Inc. (Bentley) to improve its management of geotechnical data by implementing gINT Software (gINT). gINT provides centralized data management and reporting for subsurface geotechnical projects. In this paper, we'll trace the flow of data from the field to final analysis for a typical project at NDDOT. We'll look at the work done by Bentley and NDDOT to streamline that process to just a few figurative mouse "clicks". For each click, we'll examine improvements that were made, problems that arose, and key takeaways for other organizations looking to optimize the collection, processing, and reporting of geotechnical data.

HISTORY OF GEOTECHNICAL DATA MANAGEMENT AT NDDOT

The NDDOT Geotechnical section performs in-house drilling and laboratory testing for roadway improvement, structure improvement/replacement, borrow areas, landslide areas, miscellaneous roadway issues and distresses, and forensic studies.

NDDOT collects borehole information in the field using Esri ArcPad installed on a rugged laptop. ArcPad is a software package for mobile field mapping and data collection. NDDOT used ArcPad prior to implementing gINT, and still uses it for field data collection. Once all data is populated, NDDOT exports the data from ArcPad into an Excel spreadsheet. These files are referred to as "Driller's Data" files.

Previously, the Driller's Data Excel files were imported into a Microsoft Access application that would perform various calculations and store the project information in a database. This database contained over 1,200 projects dating back to 1991. Laboratory personnel would then perform index tests and enter the data into the database. Other data, such as Proctor, unconfined compression, and triaxial tests, was saved in paper format, but not entered into the database. Borehole logs and other reports were created manually in Excel, CAD, and other formats.

GOING (EVEN MORE) DIGITAL

The concept of "going digital" can have various meanings or connotations in different contexts. In this paper, going digital refers to utilizing modern technology to automate and streamline routine processes. An essential component to going digital is the implementation of appropriate software and IT infrastructure, but equally critical is an organizational mind-set that is committed to continuous improvement as technology, business requirements, and other factors change. An important implication of this concept is that going digital is not a destination, but rather an ongoing process (*1*), as the "Even More" qualification in the subtitle of this paper suggests.

The digital workflows that NDDOT already had in place, such as electronic field data collection and automated data import and calculations, are the sort of processes that many organizations strive to implement. And while these workflows are certainly digital, NDDOT realized that some procedures had become obsolete and could be improved. Limitations and problems that NDDOT was experiencing included:

- No easy way to search and locate historical data
- Redundant data entry, resulting in issues with data accuracy and quality
- Correcting errant data required changes in multiple locations in the database
- Engineers spent a significant amount of time re-entering data to create reports and boring logs
- Manual, time-consuming report drafting
- Lack of standardization in reporting
- Limited to creating tabular reports
- Difficulty sharing data with CAD and GIS software
- Decentralized data storage in paper, Excel, Access, CAD, and other formats
- Compatibility issues with newer versions of Microsoft Access

Many of these issues are solved by fundamental gINT capabilities, such as centralized data management, automated reporting, and interoperability with other software. However, in this paper, we'll review specific pain points in the data management process at NDDOT that were addressed with optimization and automation beyond what is possible with a simple off-the-shelf implementation. These improvements are the most instructive to other organizations looking to go even more digital.

Click 1 – Locating and Accessing Historical Data

The first step in the data management cycle for a new project at NDDOT is locating relevant historical data. NDDOT's legacy Access database stored data from over 1,200 past projects, but as described above, the database had a number of limitations. There was no way to search the database to find what previous projects had been done along a given section of highway. The only way to locate data was by happening to know the project number. This meant that it was often impossible to find data, and the only option was to attempt to scroll through hundreds of records. Thus, the database had become primarily a reporting tool for current projects rather than a data management tool. This issue was compounded by the fact that much of the historical data was only available in paper format.

Accordingly, custom routines were created to convert each project in the Access database to gINT format and then migrate each project to a single Microsoft SQL Server database. A SQL Server database allows data to be managed at the enterprise level, and can contain data for all an organization's projects in a central location. At many organizations, geotechnical data is managed and stored on a project-by-project basis using individual gINT project files and other formats. However, this practice has limitations for accessing, reporting, and sharing data from historical projects. Archival in an enterprise database is the recommended best-practice for organizations looking to most effectively utilize historical data.

Furthermore, NDDOT had legacy data stored in PDF, Excel, and other formats that could not be converted to gINT format as efficiently as the Access data. A table was created in the SQL Server database to link to external files associated with each project that were stored in other locations. This ensures that these files are easy to locate and access.

Archiving data in an enterprise SQL Server database significantly streamlined NDDOT's historical data review phase at the start of a typical project. By utilizing an enterprise database, NDDOT can now search across all legacy projects in seconds using stored queries and filters based off project attributes. Likewise, GIS software can connect to the SQL Server database, allowing for spatial querying of historical projects in the work area. These improvements streamlined what was previously a very manual and often unsuccessful search through unstructured data, to what is now the relative "first click" in the data management cycle for a project.

Click 1 Takeaways

Historical data has clear value at any organization. It provides additional information to improve site characterization and design work. It can lower costs by reducing or better planning the scope of site investigation work. And for a consulting firm, an archive of relevant historical data can be a competitive advantage by demonstrating experience with a specific area, client, or site conditions.

Unfortunately, historical data is commonly stored in unstructured and unsearchable formats. At worst, it may be archived as uncatalogued paper reports in storage boxes. Consequently, it is often frustratingly time-consuming or, in some cases, impossible to locate and access historical data.

There is no one-size-fits-all solution to better manage historical data as the approach will vary depending on the state and format of legacy data. If the bulk of an organization's data is in a consistent, structured format, it may be appropriate to fully migrate this data to new systems. This was the case with NDDOT's Access database where automated routines were used to convert and migrate the data to gINT format.

However, a large upfront data conversion effort is not always appropriate. Data conversion can be messy and costly if there are many data formats, and the quality of the converted data can be poor. Also, much of the conversion process may be a waste, as a certain percentage of legacy data will never have future value.

When it is not practical to convert legacy data upfront, it is still important to make that data searchable and accessible from new systems. For geotechnical applications, this typically means making past projects searchable by geographic location. Ideally, this can be achieved by migrating borehole or project coordinates to new systems where they can be represented as points and polygons on a map that can be spatially queried. Where coordinate information is not available, there are often project attributes such as address, city, county, or other location information that can be used to locate a project. Project name, number, or client information can also provide useful search parameters. As such, initial data conversion efforts should focus on compiling only this information in new systems, and linking to associated data archived in secure, external locations. When a relevant past project is identified, it can then be determined on a case-by-case basis if there is value to convert the externally-stored data. At NDDOT, this approach was exemplified by adding a table to the database to link to PDF, Excel, and other formats rather than attempting to convert these file types to gINT format upfront.

Click 2 – Digital Field Data Collection

The next step in the data management cycle at NDDOT is the collection of field data. NDDOT performs site investigations with in-house operated drilling rigs. As described in *History of Geotechnical Data Management at NDDOT* above, borehole data is digitally recorded in the field using Esri ArcPad installed on a rugged laptop. The ArcPad application outputs data in "Driller's Data" Excel files. The format of the Driller's Data files varies depending on the investigation, which includes borrow area, linear soil survey, and deep foundation project types.

Bentley configured automated routines to import the Driller's Data files to gINT format for each of the project types. This eliminates the need for the tedious and error-prone data transcription that is required with the pen and paper approach used at many organizations. All data is imported and preliminary field logs can be output from gINT with essentially no manual data entry, a workflow that represents *Click 2* in the data management cycle at NDDOT.

Click 2 Takeaways

Pen and paper is still the standard practice for recording field geotechnical data. The primary hurdle preventing more organizations from going digital is the availability of an efficient solution that meets the diverse range of industry requirements. As mobile hardware and software offerings mature, more flexible and capable options will be available, and more organizations will be able to digitally collect field data without having to develop custom applications.

The primary and obvious advantage of digital field data collection is the elimination of redundant data entry and quicker turnaround of preliminary borehole logs, but the secondary benefits are important to acknowledge. Digital collection allows data to be transferred more frequently and efficiently between field and office, improving the pace and quality of decision-making as the site investigation progresses. Additionally, with pen and paper there is no way to enforce consistency with organizational standards other than an iron fist. Custom digital forms ensure that boreholes are logged completely, consistently, and legibly. For many organizations, this has the potential to shave hours off the log revision process and boost morale for staff who would rather be engaged in more rewarding endeavors.

Click 3 – Efficient Lab Data Entry

After field data is collected, the next step in the data management cycle is to perform laboratory testing and input the results into the project database. This step is another potential opportunity for tedious and error-prone data entry, but work was done to automate and optimize lab workflows at NDDOT.

Bentley developed automated import routines for triaxial test data. Triaxial data is output from NDDOT's testing equipment as an ASCII (text) file. All raw data is imported into the project database, and processing is automatically performed to calculate final test parameters. Test reports can then be created automatically, and results can also be included on borehole logs and other reports.

Another component of NDDOT's laboratory workflows is that all testing container tare weights are pre-measured. Tables were set up in NDDOT's gINT files to store the tare weights so that lab technicians need only enter the container number, and then automated lookup routines retrieve the tare weight. This eliminates the need to repeatedly weigh or manually lookup the weight for each test.

Additionally, the gINT interface was customized to be more user-friendly for the laboratory technicians, who enter raw lab data directly into the database. This included color-coding fields or renaming fields to familiar names from lab testing worksheets. Built-in data validations help to ensure the reasonableness of data entered and automatically detect many data entry errors. Pre-programmed calculations automatically compute final test parameters. Consequently, gINT is used to both perform lab testing calculations and to create reports, significantly streamlining the data entry and reporting process.

Thus, lab data entry is automated or delegated to lab technicians, reducing engineer involvement in the process to what is a mere *Click 3* in NDDOT's data management cycle.

Click 3 Takeaways

Many lab workflows are ideally suited for automation. Data is often collected in electronic formats that can be automatically processed, as is done with NDDOT's triaxial test data. And by nature, many tests involve standard, repeatable procedures that can be automated, such as the container tare weight lookup.

However, laboratory testing also requires human involvement to mechanically perform tests, as well as record and enter data. While it may not be possible to automate these tasks, efficiency can be gained by utilizing the appropriate set of software tools and division of labor.

At many organizations, engineers are involved with some aspect of the data entry process, whether that be entering raw data or compiling test results into figures and reports. There are a number of reasons why this is done – to reduce license costs for software packages, perceived quality control benefits, lab technicians that are not comfortable with the software, or because "it's always been done that way."

But, if one cannot automate, the next best option is to delegate, and there is a lot that can be done to efficiently delegate data entry and preliminary reporting tasks to lab technicians. Proper training and guidance on the use of that software is critical. Also, software interfaces should be configured for ease of use, such as by removing extraneous fields, color-coding fields, and setting up the interface to mirror familiar laboratory worksheets as much as possible.

By having technicians enter lab test data directly into gINT, NDDOT has recognized improvements in the quality and accuracy of laboratory data. Previously, errors were often identified by engineers when compiling boring logs or using lab data for analysis. Now, gINT's built-in data validation capabilities bring many errors to the technician's attention during data entry. Additionally, lab technicians can preview lab test reports directly from gINT, allowing them to visually spot errors and make corrections before handing over data to engineers.

Click 4 – Automated Data Processing and Reporting

The next step in the data management cycle at NDDOT is generating reports that assist with data analysis and preparing design recommendations. One of the core capabilities within gINT is the ability to automatically create any number of custom reports. Accordingly, Bentley developed several boring log and other report templates for NDDOT. Particular attention was devoted to data processing and reporting tools for linear soil survey analyses.

A linear soil survey is conducted along a stretch of roadway that is planned for improvement, reconstruction, or realignment to characterize soil, groundwater, and other subsurface conditions. The survey involves drilling boreholes at regular intervals along the length of the roadway (*2*). Bentley worked with NDDOT to develop reporting tools that automatically create to-scale, color-coded profiles that summarize soil conditions across miles of highway and any number of boreholes.

Previously, NDDOT compiled linear soil survey data into an Excel spreadsheet, referred to as a "Color Sheet", that used color-coding to summarize soil parameters for each borehole along the roadway. The Color Sheet is used to characterize and better understand conditions when preparing design recommendations for linear soil survey reports. An example Color Sheet is shown in Figure 1, and the Color Key is shown in Figure 2.

The old Excel Color Sheet had a number of inefficiencies and weaknesses. Depth-related results were depicted horizontally instead of vertically. There was also no easy way to tell the spacing between boreholes without referencing a separate location map, and there was little flexibility to pick and choose which borings were shown on the sheet. This made it difficult to visualize soil conditions and how they changed along the highway alignment.

Project No.: SS-3-030(029)105
PCN: 19740

Color Key
Swell Potential: Low Marginal High
Moisture Content: Below PL 0-5% Over PL >5% Over PL
Group Index: GI≤20 GI>20
Ave. In-Place Moisture Content: MC < Opt. 0 ≤ MC < 6% Over opt. 6 ≤ MC < 10% Over opt. 10 ≤ MC < 16% Over opt. MC ≥16% Over opt.

Lab No.	STA	Offset	Depth of Sample (ft)	Plastic Limit (PL)	AASHTO Class	Group Index	Optimum Moisture	Swell Potential (PI)	Ave. In-Place Moisture	Moisture Content @ Depth (ft) 2	3	4	5	6	7	8	9	10	11	12	13	14	15	16	17	18	19	20	21	22	23	24	25	Frost Class
356	107+2731	Lt 7 SB	0.7-5	15.0	A-6(8)	8	9.0	17	14.2	13.7	13.4	14.9	15.0																					F3
355	107+2760	Lt 7 SB	0.6-5	15.3	A-6(7)	7	6.9	16	15.7	15.2	15.7	15.3	18.7																					F3
354	107+2787	Lt 7 SB	1.9-5	18.7	A-1-b(0)	0	6.9	8	6.3	4.9	6.9	6.0	7.3																					F2
357	109+2012	Rt 10 NB	0.6-5	19.0	A-6(8)	8	10.3	12	21.8	21.7	21.6	21.0	22.9																					F3
358	109+2091	Rt 10 NB	0.7-5	16.3	A-6(3)	3	8.9	11	16.6	19.8	15.2	15.2	16.3																					F4
359	109+2197	Rt 10 NB	0.4-5	17.3	A-4(4)	4	9.8	9	20.1	22.0	21.9	18.2	18.3																					F4
360	111+3921	Rt 10 NB	1.1-5	15.7	A-6(8)	8	9.4	14	18.0	21.2	18.9	17.1	14.8																					F3
361	111+4017	Rt 10 NB	0.8-5	15.6	A-6(10)	10	9.8	17	11.8	11.9	11.6	13.4	10.2																					F3
362	111+4068	Rt 10 NB	0.9-5	14.8	A-6(6)	6	8.3	14	12.3	11.9	13.8	11.5	11.9																					F3
363	116+2230	Rt 8 NB	0.8-5	17.9	A-6(10)	10	12.2	19	19.7	15.6	17.6	23.6	22.1																					F3
364	116+2282	Rt 8 NB	0.9-5	19.7	A-6(10)	10	12.7	18	20.0	12.8	19.0	25.8	22.3																					F3
365	116+2330	Rt 8 NB	0.7-5	17.2	A-6(9)	9	12.7	18	22.3	15.6	23.9	20.3	23.4																					F3
366	116+2380	Rt 8 NB	0.5-5	17.2	A-6(12)	12	11.7	20	20.2	20.9	26.0	18.2	15.9																					F3

Figure 1 – Example of NDDOT's old Color Sheet created in Excel.

Accordingly, Bentley created a version of the Color Sheet using gINT's Fence report format. A Fence report provides a cross-section view of subsurface conditions along a baseline. Boreholes are projected perpendicularly onto the baseline and depicted as "fence post" stick logs. Data can be represented graphically and as text posted at its corresponding depth on the stick log.

The gINT Fence Color Sheet has the advantage over the Excel version that conditions can be plotted to-scale, allowing for a better understanding of spatial variation of soil conditions, both horizontally and vertically. A basic site map further assists with this, and allows the Color Sheet to be used without referencing other documents. The parameters are depicted on the Fence as columns using the same color-coding scheme as the Excel sheet, as shown in Figure 2 and Figure 3. gINT's automatic soil classification tools are used to plot the USCS and AASHTO classification for each soil layer. The user also has the option to remove any of the columns. This is all accomplished without any redundant data entry or user input. The Fence report automatically queries the data from the database, performs any necessary calculations, and applies pre-programmed logic to determine the color-coding.

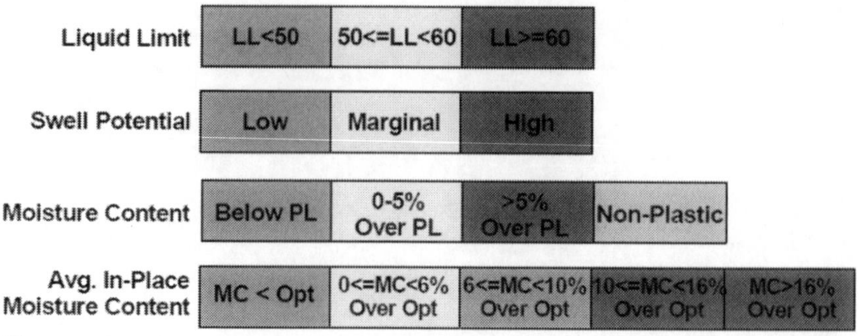

Figure 2 – Color Key for NDDOT's gINT Fence Color Sheet.

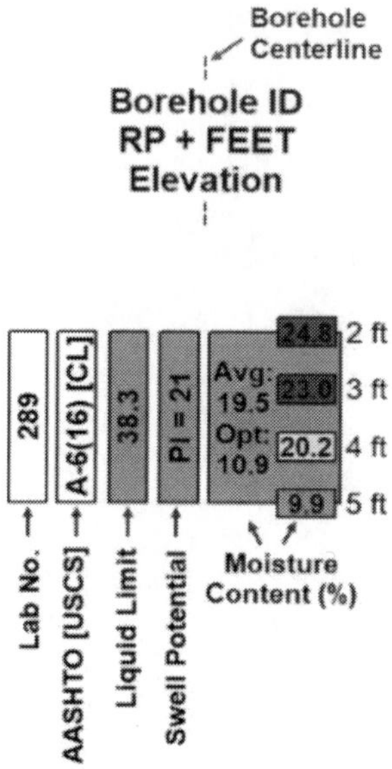

Figure 3 – Legend depicting a typical borehole on the NDDOT gINT Fence Color Sheet.

As NDDOT started using the Fence report, it noticed ways the report could be improved. The first issue was that the strata breaks for each soil layer had to be manually entered. However, NDDOT assigns a "Grouped Sample Number" to all samples in a layer, which is imported into the database during the Driller's Data import process. Thus, it was possible to write gINT Rules (VBA-like code) to automatically determine the top and bottom depth for each layer based on the depth range for the Grouped Sample Numbers, eliminating a time-consuming manual process.

Second, the standard functionality for gINT Fences allows for the output of a single Fence report (one page) at a time. However, a typical survey will have too many boreholes to fit on a single page. A user can manually select a subset of the project boreholes for each Fence, but this is inefficient for linear soil survey workflows where there might be numerous pages over miles of roadway. To address this, Bentley developed a gINT Rules Add-in that leverages gINT's Alignment module. The module stores alignments for a project, which in the case of a linear soil survey, would be a roadway centerline. The Add-in divides the alignment up into specified intervals and generates a separate page of the Fence for each section of the alignment. The only parameter the user must enter is the footage along the alignment to include on each page. All other parameters are optional or pre-populated. This allows the user to quickly and easily adjust Fence options to determine the optimal settings for a given project.

The interface for the Add-in is shown in Figure 4. The optional parameters further enhance the Fence report functionality. For example, when there is a large vertical range for a Fence report, thin soil layers will be shrunk to a small scale and become illegible. This can happen when there are large elevation changes or a single deep borehole along that section of the roadway. To improve legibility, the user has the option to filter out deep boreholes, plot the data versus depth (instead of elevation), or manually override the default vertical range. If boreholes overlap horizontally, the user can choose to equally space all boreholes, or offset the borehole plot positions individually. Figure 5 provides an example of the Fence using the default settings with all boreholes plotted to horizontal and vertical scale. In Figure 6, boreholes are plotted equally-spaced and with a depth scale for improved legibility.

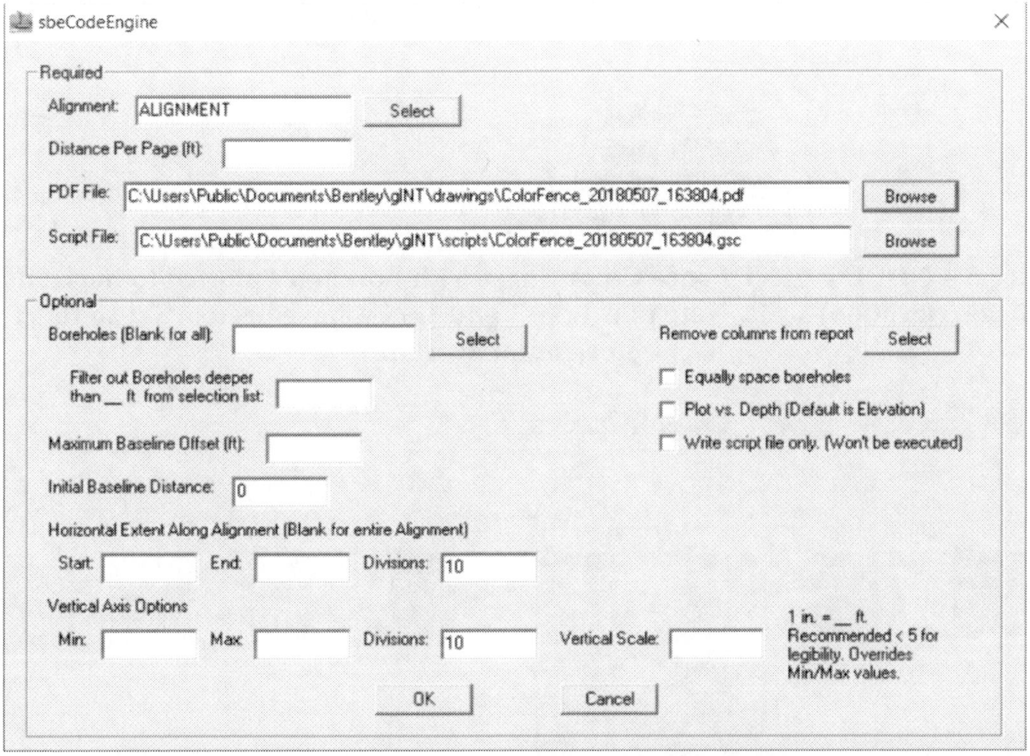

Figure 4 – Interface for the Add-In to create the NDDOT gINT Fence Color Sheet.

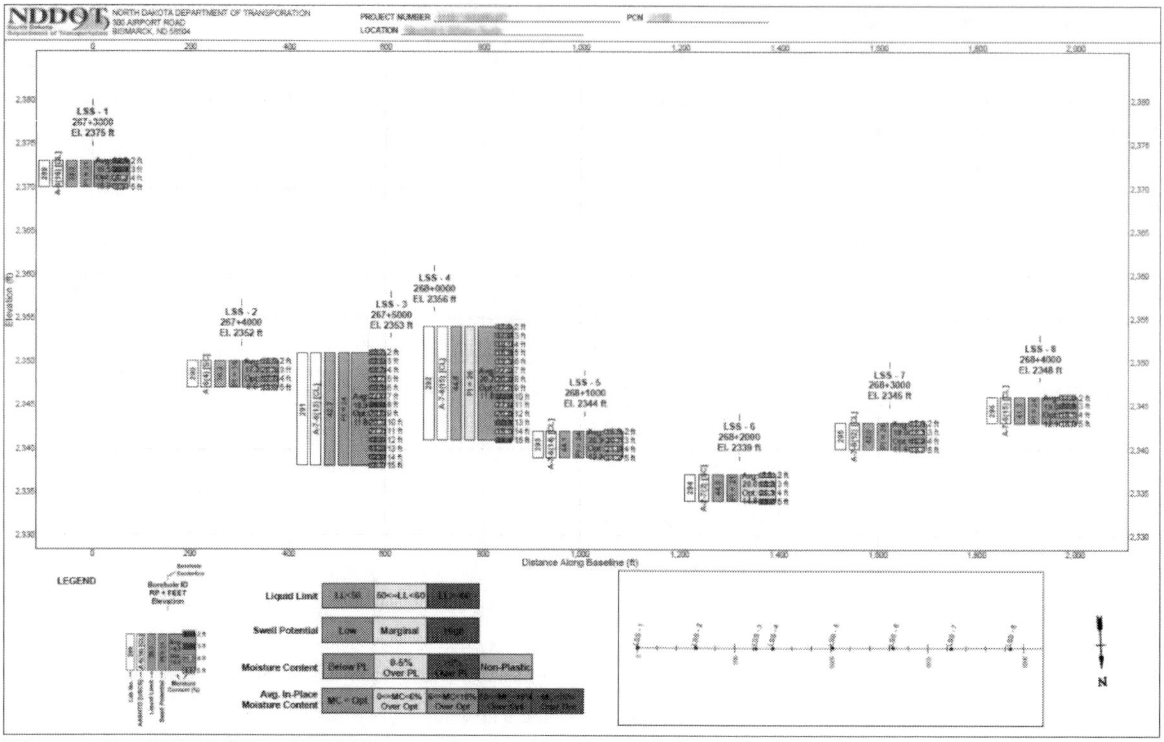

Figure 5 – NDDOT gINT Fence Color Sheet with boreholes plotted to horizontal and vertical (elevation) scale. Note that LSS-3 and LSS-4 have been offset so there is not horizontal overlap.

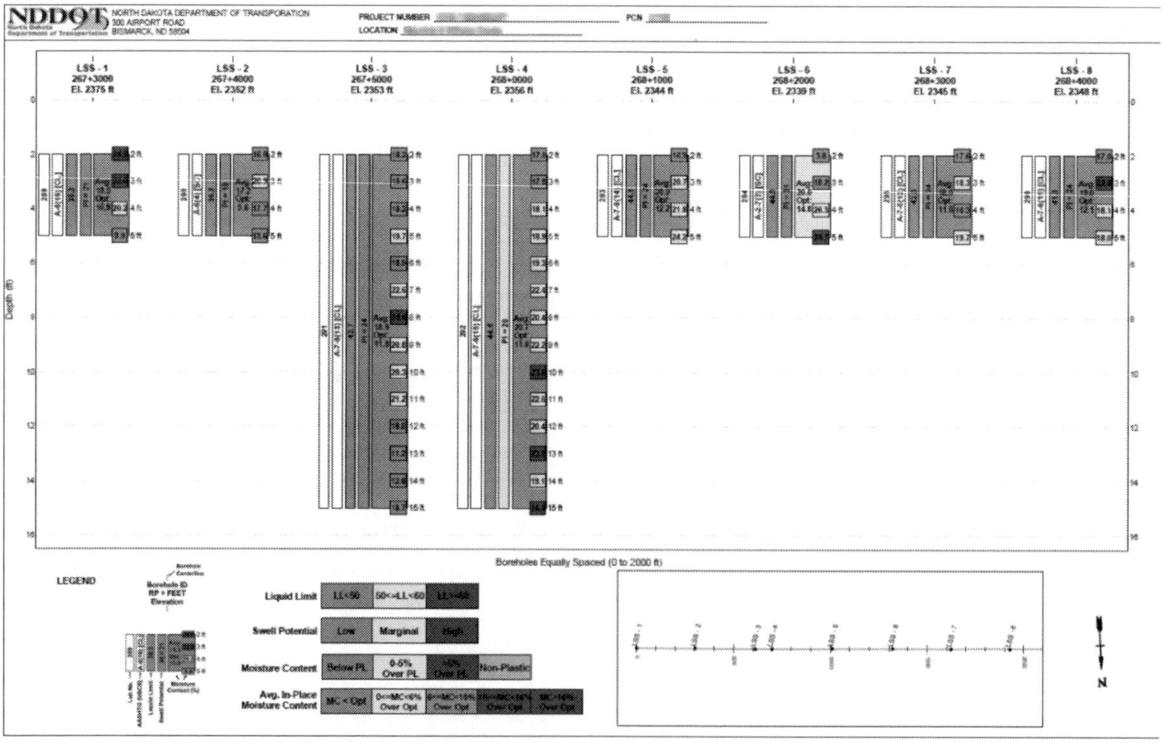

Figure 6 – NDDOT gINT Fence Color Sheet with boreholes plotted equally-spaced and with a depth scale.

The Add-In generates two outputs. First, it creates a PDF version of the Fence report. It also generates a gINT Script file. A Script file remembers the settings for each page of the Fence PDF. If data is updated for the project, NDDOT can simply re-run the Script file to create updated versions of the Fence without having to remember or re-enter the optimal settings into the Add-In interface. Also, the Script file can be edited either manually or using the Add-In, allowing NDDOT to vary the Fence report settings for different sections of the Alignment.

In summary, linear soil survey workflows were improved by incrementally identifying and addressing NDDOT's specific pain points. Tools were developed that automate and improve the quality of data processing and reporting, yet still offer flexibility. Tedious and repetitive steps were eliminated, streamlining the process to *Click 4* in NDDOT's data management cycle.

Click 4 Takeaways

When many organizations make an investment in new technology, they want the maximum return on investment so they immediately aim for a "gold-star" implementation thinking that their ambition will be rewarded. However, this approach is frequently less successful as it often involves much wasted time trying to replicate existing workflows that are not appropriate for the new platform. The new platform may have alternative capabilities that the organization doesn't fully understand. It also takes time to train staff and integrate other systems with the new platform. As with any endeavor, trying to do too much at once can be counterproductive. A phased approach is preferable because it allows the organization to determine if it will be better served by adjusting workflows to the new technology, or modifying the technology to meet its workflows.

Improvements to the linear soil survey workflows illustrate this ongoing and iterative process. NDDOT identified the need to automatically create an improved version of the Color Sheet, and after some time using it, realized ways to improve that process. This continuous questioning of inefficiencies led to a successful outcome and should be an example to other organizations. Many organizations stick with inefficient workflows because "it's the way things have always been done." But, all organizations would be better served by a mindset that actively seeks to improve the way routine tasks are performed. It's nearly impossible to predict every requirement from the outset, so an iterative approach that continuously addresses the most critical pain points is a necessity. It is a process that is never complete, and certainly there are further opportunities for improvement at NDDOT.

The automation of the Color Sheet also demonstrates a potential pitfall with automation. Automation does not excuse robust quality control, and in fact makes it more important. Consider that a new factory that can output 10 times the number of widgets in a day can just as easily output 10 times as many defective widgets that day. One of the early versions of the gINT Fence Color Sheet had a bug that depicted conditions as more-favorable than they actually were for one of the parameters, which was not immediately identified. Ultimately, automation promises higher-quality data since calculations are more reliable than those performed manually. However, quality control is critical to ensure that any automated capabilities are providing the expected results.

The linear soil survey workflow improvements also demonstrate that certain tasks are more suitable to automation than others. Linear soil surveys are conducted systematically with a routine and standardized approach that is appropriate for automation and optimization. Other tasks or project types may require more complex analysis and a standard approach may not be feasible. Before any attempt is made to automate a task, it is critical that standardized procedures are in place. And then it is the automation of these routine tasks that frees the engineer to direct time and focus towards more complex endeavors.

Click 5 – Storage and Archival of Project Data

The final step in the data management cycle at NDDOT is the storage and archival of project data. At NDDOT, geotechnical data is managed in a gINT project file during the project execution stage. As mentioned in *Click 1 – Locating and Accessing Historical Data*, storing data in individual project files has limitations for searching, accessing and re-using that data on future projects. Accordingly, once a project is completed, NDDOT archives project data in an enterprise SQL Server database. gINT contains built-in tools to easily migrate data between project files and enterprise databases making this final step little more than *Click 5* in NDDOT's data management cycle.

Click 5 Takeaways

Proper archival extends the value of geotechnical data beyond the original project. This ensures that historical data can be easily accessed in *Click 1* for re-use on future projects. Thus, the data management cycle does not begin and end on a single project, but is an ever-ongoing endeavor.

SUMMARY & CONCLUSION

In summary, going (even more) digital at NDDOT streamlined core data management and reporting workflows to a few figurative clicks:

- *Click 1* – Easily locate historical data archived in an enterprise database that can be searched or spatially queried.
- *Click 2* – Automatically import digitally-collected field data into the project database with little to no manual data entry.
- *Click 3* – Compile laboratory data with little engineer involvement through automated import routines and delegation of data entry to lab personnel.
- *Click 4* – Automatically generate reports and process data for Linear Soil Survey workflows.
- *Click 5* – Archive project data for potential re-use on future projects.

The work performed at NDDOT provided a number of lessons learned and key takeaways for organizations also looking to streamline their data management workflows:

- There is no one-size-fits-all approach to managing legacy data. A large upfront data conversion effort will often be messy and cumbersome, so effort should focus on making

historical data easy to search (spatially or otherwise) and only converting data on an as-needed basis.
- Digital field data acquisition has clear efficiency, quality, and morale benefits. Organizations should keep their eyes open for new applications that provide improved functionality.
- If you can't automate a task, delegate it.
- If you delegate a task, set others up for success by providing the proper training and accommodations based on their comfort-level and existing workflows.
- Don't try to implement too much all at once. Going digital is an ongoing process, often best-accomplished through a phased and iterative approach.
- An organizational mindset that questions inefficiencies and actively seeks to optimize workflows is critical.
- Automation does not replace quality control.
- Tasks and workflows with standardized procedures are more suitable for automation than those that require complex or non-routine analysis.
- Geotechnical data has value beyond the original project. Proper storage and archival ensures data can be efficiently accessed and re-used on future projects.

The ultimate goal of going digital is to improve the efficiency and quality of project delivery. Removing engineer involvement from manual and routine tasks to focus on higher value work is a key component in achieving this. The successful outcomes at NDDOT provides a model to other organizations looking to optimize their management and reporting of geotechnical data.

REFERENCES:

1. *Digital Workflows, Digital Components, Digital Context, Going Digital: Greg Bentley, Bentley Systems*. ARC Advisory Group, USA. www.bentley.com/en/about-us/news/2018/april/23/digital-workflows-digital-components-digital-context. Accessed May 7, 2018.

2. *Design Manual – Chaper VII: Geotechnical Studies and Design*. North Dakota Department of Transportation. www.dot.nd.gov/manuals/design/designmanual/chapter7/DM-07_tag.pdf. Accessed May 7, 2018.

RATTLESNAKE HILLS LANDSLIDE:

OVERVIEW AND MONITORING

George Machan, PE
Charlie Hammond, CEG
Thomas Westover, PE
Landslide Technology (A Division of Cornforth Consultants, Inc.)
10250 SW Greenburg Road
Portland, OR 97223
(503)-452-1200
georgem@landslidetechnology.com

Prepared for the 69th Highway Geology Symposium, September, 2018

Acknowledgements

The authors would like to thank the individuals/entities for their contributions in the work described:

James Struthers – Washington Department of Transportation
Noah Kimmes – Landslide Technology

Disclaimer

Statements and views presented in this paper are strictly those of the author(s), and do not necessarily reflect positions held by their affiliations, the Highway Geology Symposium (HGS), or others acknowledged above. The mention of trade names for commercial products does not imply the approval or endorsement by HGS.

ABSTRACT

The Rattlesnake Hills Slide is located south of Yakima, WA, on the southeast side of Union Gap, where the Yakima River cuts through an east-west-trending ridge. The ridge is a tectonic anticline that rises 2,000 feet above Yakima Valley. The slide is on the south flank of the asymmetric anticline, which dips 10 to 20 degrees. The landslide consists of a translating block approximately 4 million cubic yards in volume, 1,700 feet long (north-south), up to 850 feet wide (east-west), and approximately 200 feet thick. The landslide block is comprised mainly of basalt which is moving downdip on an interbed. An open pit quarry is located at the toe of the landslide.

Landslide movement was visually detected in early October 2017 when scarp cracks were observed. The quarry operator retained a geotechnical firm (Cornforth Consultants) and implemented a monitoring program. Landslide movement slowly increased through late-December, until the movements reached constant velocity (approximately 2 to 3 inches/day). Other stakeholders became involved due to the proximity of a county road, an interstate highway, a cluster of residences, an irrigation pipeline, utilities, and the Yakima River.

Landslide geometry and mechanisms have been evaluated, based on monitoring measurements and geologic studies, allowing for preliminary assessments of landslide impacts on nearby properties and facilities. Precautionary measures were implemented to protect facilities and to minimize impacts to people. Landslide stability and rates of movement were analyzed to predict long-term landslide consequences.

INTRODUCTION

The Rattlesnake Hills Landslide is located less than 3 miles south of Yakima, WA, on the southeast side of Union Gap. The vicinity map is shown on Figure 1 *(1)*. Landslide cracks were first identified in early October 2017 by a neighbor flying over the property. This prompted a geologic reconnaissance of the hillside, which verified active landslide features and conditions.

The toe of the 20-acre landslide daylights into a quarry, approximately 30 feet above the quarry floor, and the headscarp extends into Yakama Nation land. The west flank of the landslide daylights in the Union Gap hillside, and the east flank follows a steep fracture zone. An irrigation pipeline conduit and a county road are located downslope to the south and the west flank of the landslide. Interstate I-82 is located adjacent to the county road, and the highway is bounded by the Yakima River on the west. A small residential community is located between the county road and the highway, to the south of the landslide.

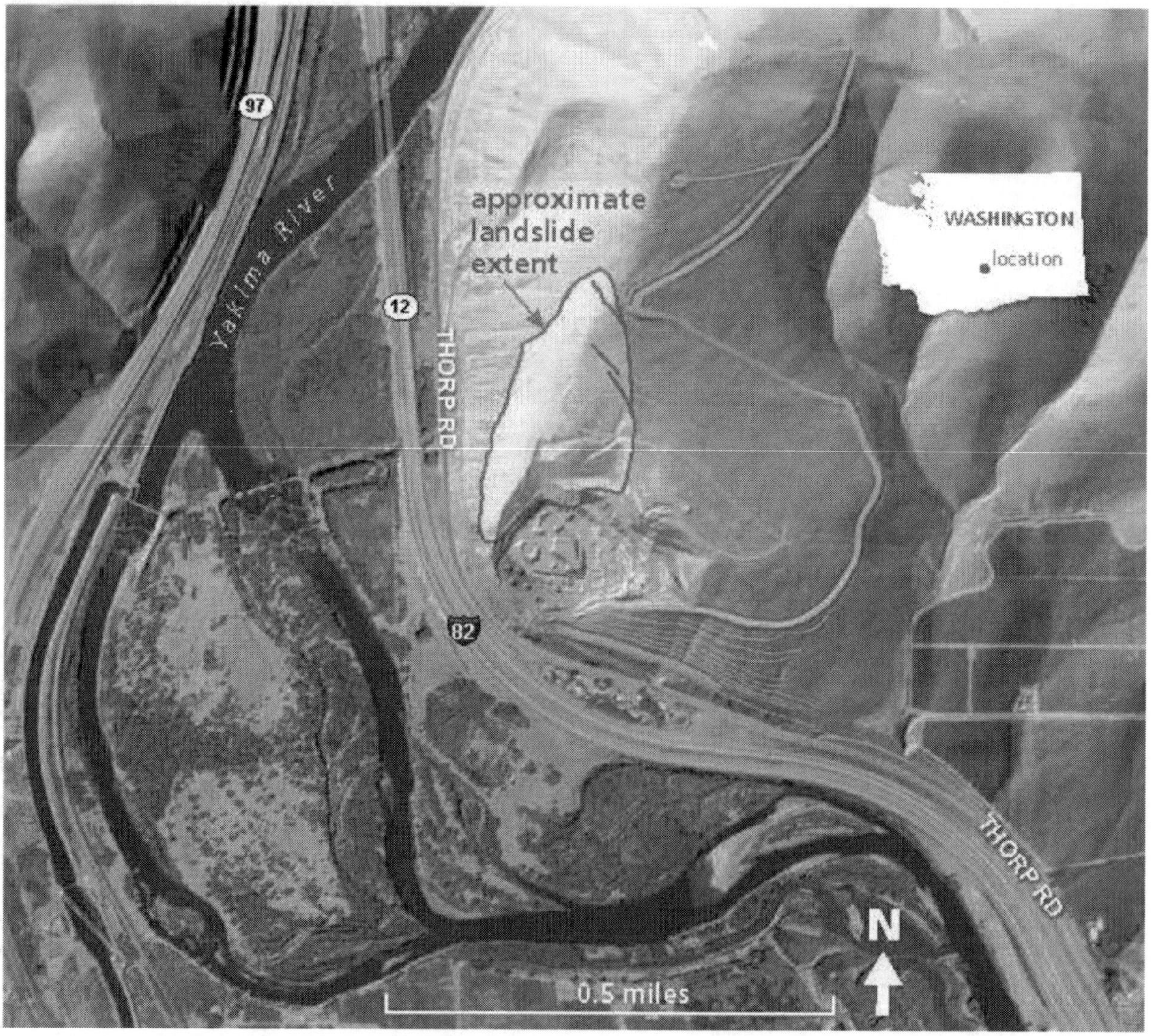

Figure 1 – Vicinity Map (Washington Geological Survey *(1)*)

Stakeholders were appraised of landslide activity and potential consequences and risks. The County and the State Department of Natural Resources initiated emergency response discussions and planning efforts. Possible hazards from the landslide included rockfall and slide debris moving toward the county road and the quarry pit. Figure 2 is an oblique image of the landslide, created using point cloud data from drone images (courtesy of Washington Department of Transportation).

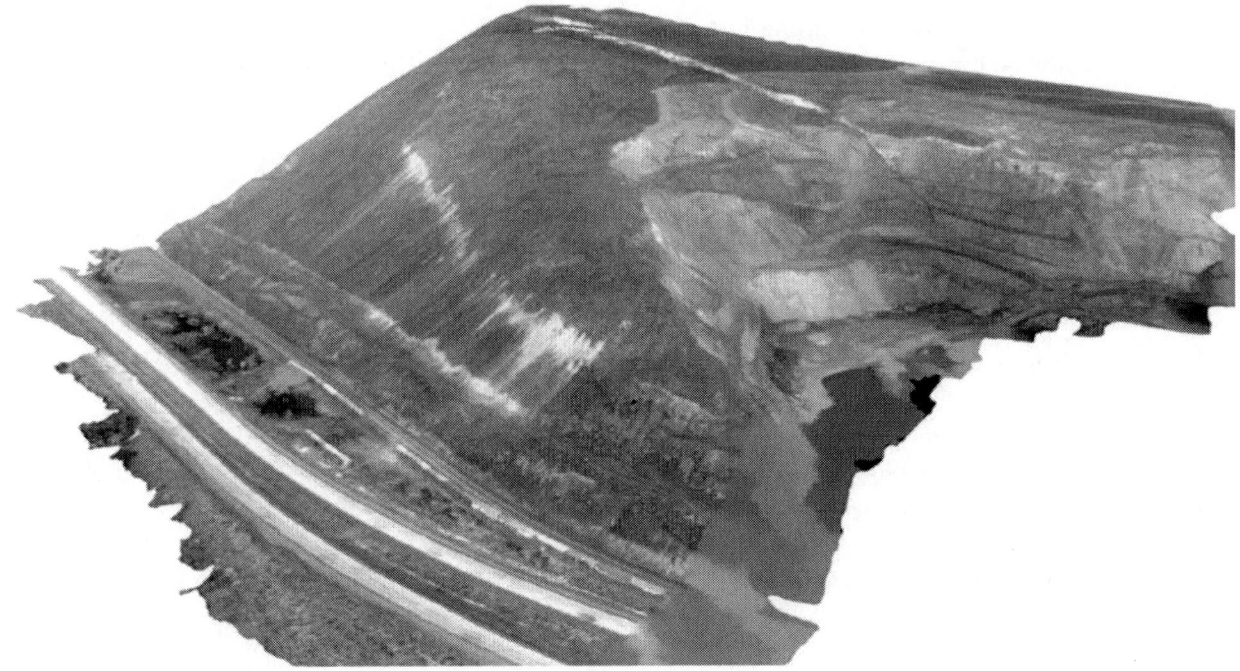

Figure 2 – Oblique Image of Landslide, looking northeast

Preliminary mitigation measures included restricted use of the quarry, closure of the county road next to the landslide to prevent impacts to local traffic, placement of barriers made with shipping containers weighted with concrete blocks to protect interstate highway traffic from rockfall, warning signs on the highway, planning potential detour routes, rockfall patrols, and public notifications. Residents were evacuated in January 2018 while monitoring and independent assessments were performed, and they were allowed to return to their homes when it was concluded that rapid slide movement would be unlikely and that slide debris runout would have limited travel.

GEOLOGIC CONDITIONS

The landslide is located on the southeast side of Union Gap, a water gap where the Yakima River flows through an east-west-trending ridge that rises approximately 2,000 feet above Yakima Valley. The ridge is a tectonic anticline and is known as Rattlesnake Hills (to the east of the gap), and Ahtanum Ridge (to the west). The quarry is on the south flank of the asymmetric anticline in the Columbia River Basalt Formation (CRB), the flanks of which dip more steeply on the north and gentler to the south. The rock formations are the Saddle Mountains Basalt and Wanapum Basalt of the CRB, and an interbed of the Ellensburg Formation *(2, 3)*.

Geologic structures include tilted bedding at 10 to 20 degrees in an Azimuth direction approximately 190 degrees, and high angle fracture zones that trend generally north, north-northwest, and east-northeast.

The Saddle Mountains Basalt, previously mapped as the Pomona and Umatilla Members, is approximately 200 feet thick and overlies an Ellensburg Formation interbed, which is approximately 5 feet thick and contains clay, silt, sand and fine gravel, including coal and apparent lahar seams. The basalt is moderately to highly jointed and slightly fractured with through-going high angle fracture zones. The Ellensburg layer exhibits shear textures, interpreted to be associated with fault movement within the bedding from flexural slip that occurred during the tectonic folding.

Tectonic faults mapped locally include generally east-west-trending normal faults near the crest of the anticlinal fold (the overall east-west ridgeline), and a south-verging thrust fault located low on the slopes of the anticline's south flank. The normal faults appear to represent the north and south sides of a tensional zone within the crest of the anticlinal fold, and with fault displacement that terminates with depth in the layers of CRB (exposed on the west-facing slope of Union Gap).

Landslide deposits are mapped in available geologic reports to the north and west of the active landslide, but not within the subject landslide *(2)*. However, re-assessment of geologic maps and slope shading imagery from recent LiDAR data *(4)* indicates that ancient landslide features may also occur within the area of and surrounding the quarry property. Normal faults a few hundred feet upslope and north of the quarry property and the thrust fault to the south are suspiciously coincident with the head graben of a translated landslide (down dropped area between parallel normal faults) and the toe of a paleo-landslide (geologically ancient). A preliminary assumption could be that a massive slide or series of slides on the south flank of the Rattlesnake Hills anticline may have occurred during a prior geologic environment.

LANDSLIDE CONDITIONS

The landslide mass is approximately 1,400 to 1,700 feet long (north-south), 600 to 850 feet wide (east-west) and up to 200 feet thick, as shown on the Site Plan, Figure 3. The natural ground slope of the landslide is generally 10 to 20° to the south. The west flank of the landslide daylights on the Union Gap hillside, which is generally sloped approximately 38°, and becomes gentler at the base of the slope. The cut slope for the north quarry high wall ranges from 35 to 45° and includes benches and access ramps. The landslide mass is approximately 4 million cubic yards in volume.

The exposed slide mass is comprised of hard basalt that is highly to moderately jointed (spaced 3 to 18 inches) forming slender columns in near vertical, wavy and fanning patterns. Fractures also occur in the rock formation, which appear continuous through the formation, as singular fractures or multiple near-parallel fractures, and with spacing that varies from moderate to wide (<1 foot to 10s of feet). Fracture trends are generally north-northwest to north, and with dips that are vertical to steep (>70°). The rock has high shear strength, even when fractured, due to the angularity and interlocking of the jointed rock fragments. Interbeds are not observed within the slide mass, which appears to consist of one CRB flow unit. The active wedge (head scarp/graben area) crosscuts through the basalt flow.

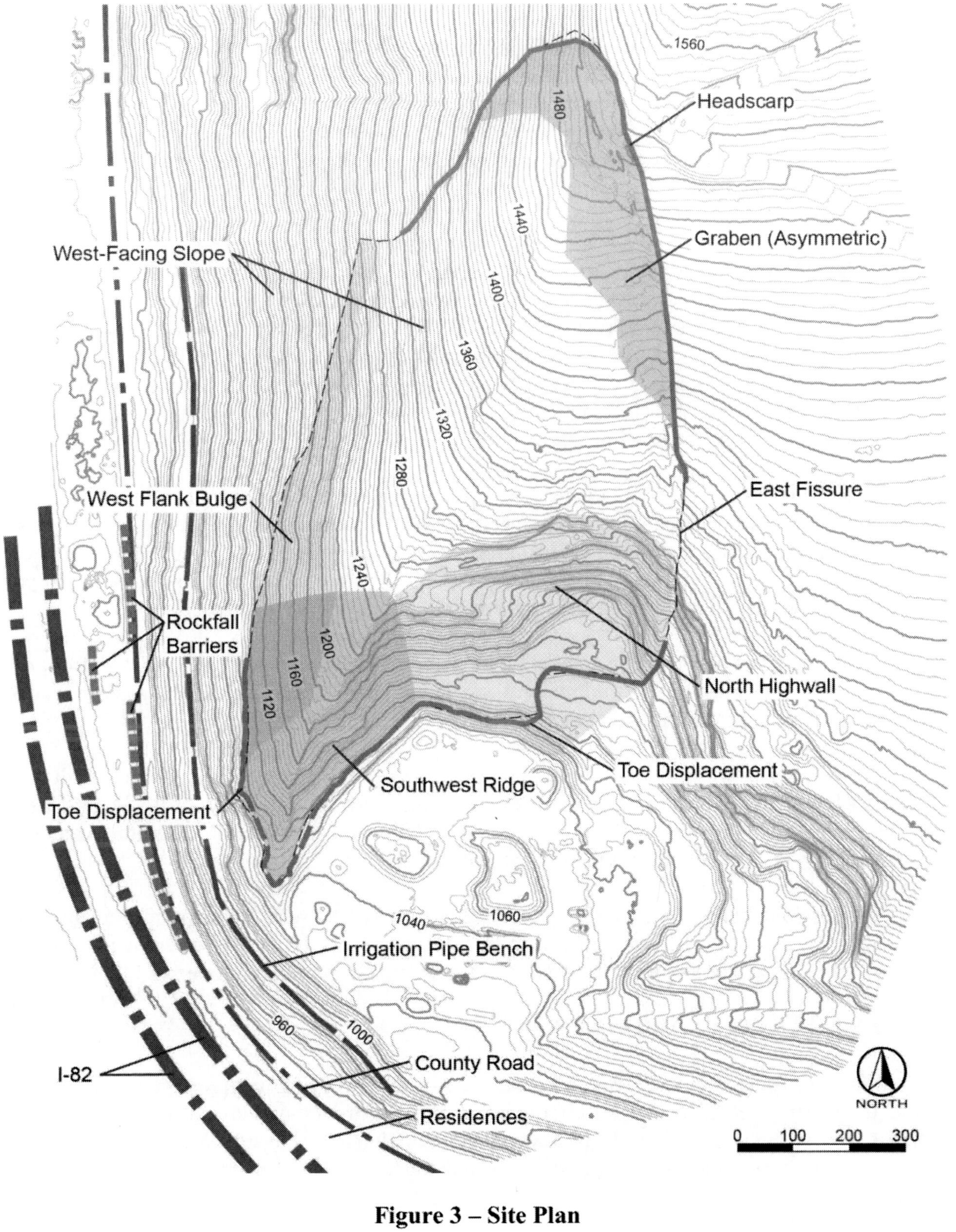

Figure 3 – Site Plan

The landslide developed on the south flank of a tectonic anticlinal fold, occurring on a basal shear that likely originally developed as a flexural slip displacement within a sedimentary interbed between Columbia River Basalt flows. The landslide consists of translational rock blocks separated by evolving tension cracks. The lower slide mass is moving slightly faster than the middle area, indicating tensional spreading of the slide mass. Examination of scarps and displaced blocks indicated the general southward direction of slide movement, with lower blocks initiating movement and causing tensional extension of the slide mass, thus forming many intermediate tension cracks (scarps). As the slide blocks moved to the south and caused fissures and voids upslope, grabens developed as upslope blocks collapsed into the voids. The east slide margin formed along a steeply dipping fracture zone. Landslide features are shown on the foregoing Site Plan, Figure 3.

The landslide toe daylights in the north quarry cut slope/highwall, approximately 30 feet above the quarry floor. The landslide toe is moving on a thin interbed that was visible between the southwest ridge and the middle of the slide toe. The sedimentary interbed near the west side of the slide toe is dipping less than 10 degrees toward Azimuth direction 190 degrees. The east portion of the slide toe appears to be partially buttressed by a ramp within the quarry excavation (less material removed at the southeast corner of the landslide toe), thus forcing a passive wedge to form at the southeast corner of the slide.

The southwest ridge at the toe of the landslide appears to be slightly resistant and is deforming in response compared to the movement of the slide area upslope. It appears the west portion of the slide mass is pushing against the southwest ridge and is shearing as it moves toward the ridge, causing inflation on the west flank of the slide. It appears that the weakest path for the landslide is to shear through the southwest ridge rather than underneath it.

The southeast portion of the landslide toe is buttressed by unexcavated mass in front of the east side of the landslide toe, and the weakest shear path is occurring along a passive wedge that is causing bulging and thrusting of material within a bench in the quarry cut slope.

Groundwater was not observed within the quarry and adjacent hillside bordering the Yakima River valley. There are no springs or seeps and stormwater readily infiltrates into the ground, even within the floor of the quarry. Nearby wells indicate the regional groundwater table is substantially underneath the basal shear zone of the landslide.

A possible interpretation is that the active slide may be a reactivation of a remnant of an ancient paleo-landslide. The west flank of the landslide appears to have been removed in past geologic time, exposing the basal shear zone in the west-facing hillside slope.

INVESTIGATIONS AND MONITORING

Investigations included geologic reconnaissances, surveying/mapping, geomorphological evaluations using LiDAR and drone photogrammetry, and ground movement monitoring. Deep borings were planned; however, drilling was postponed due to increased slide movements which could possibly have caused binding of drill rods/casing, as well as concerns for safety.

Stakeholders and academia volunteered additional monitoring methods, including video cameras, total station prism monitoring, terrestrial LiDAR, seismometers, InSAR, and ground-based radar (GPRI).

Ground features were evaluated using orthomosaic images, hillshaded oblique images, and a site topographic map that were generated from drone photogrammetry point cloud data and GPS Surveys. In addition, the evaluations utilized available LiDAR imagery *(4)*.

Unmanned Aerial Vehicle (UAV/drone) photogrammetry with post processing kinematic (PPK) technology was used to document landslide conditions on various dates. The photography was processed to create orthorectified point clouds. The point clouds were aligned (shifted and rotated) to match "stable" reference points outside the landslide area and were resampled to create a consistent point cloud density between scans. Resampled point clouds were "hillshaded" using a consistent sun zenith and azimuth for uniform lighting and visualization of relief features. Hillshade images were used to develop animations of the progression of landslide displacement (time lapse images). Example hillshade images are presented in Figures 4A and 4B. Visible landslide features include scarps, grabens, raveling, compression in SW ridge, overall movement to the south, slight displacement to the west and bulging of the west flank.

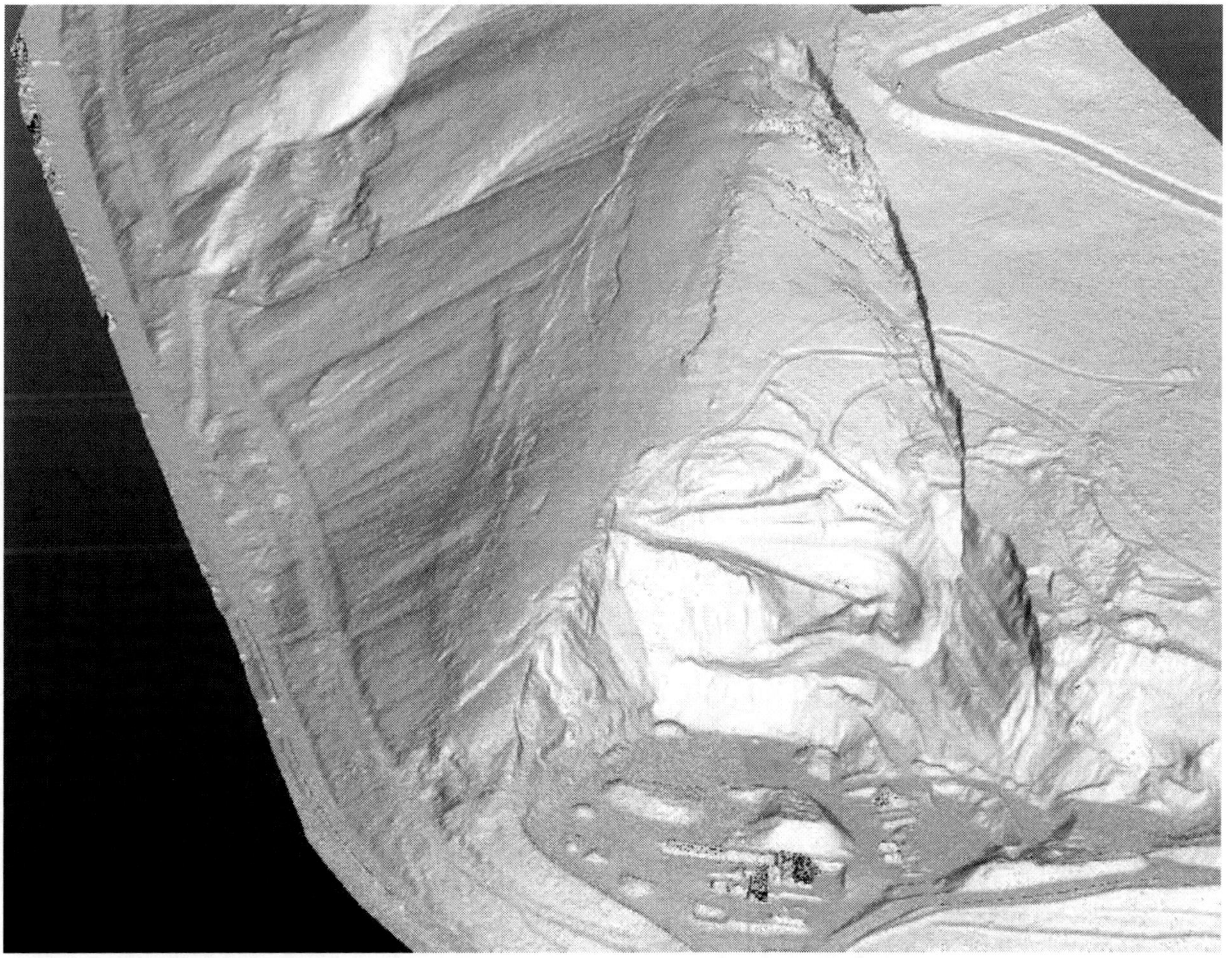

Figure 4A – Hillshade Image, looking to north (May 9, 2018)

Figure 4B – Hillshade Image, looking to northeast (May 9, 2018)

GPS surveys and monitoring were initiated in early October 2017 to map and quantify slide movement and direction (using real time kinematic technology, RTK). In January 2018, the RTK manual survey monitoring was replaced with an automated robotic total station laser system with 21 prism targets, along with 3 telemetered GPS units positioned on the landslide mass. The robotic total station instrument is located approximately 3,000 feet south of the landslide on the opposite side of the Yakima River. Control points are included to check for data deviations (2 control prisms are located close to the instrument to check its stability; and the 3rd control prism is located across the river next to the landslide to record environmental effects). Data collected by the instrument was sent through the internet to develop trend plots and update reports at the manufacturer's website.

The coordinates for each survey point were measured for each monitoring cycle, and displacements were calculated by comparing datasets over time. Measurement made in direct line of sight (Northings) had less data scatter than Eastings and Elevations, due to the methodology and limitations of laser scanning. Environmental effects caused errors, particularly in the transverse (Eastings) and vertical directions. Landslide movement trends were calculated as horizontal vectors (combining Northing and Easting displacements).

The directions of surficial slide movements were based on the calculated displacement vectors, as shown on Figure 5. The length of each vector arrow is scaled relative to the ground movement velocity at each location.

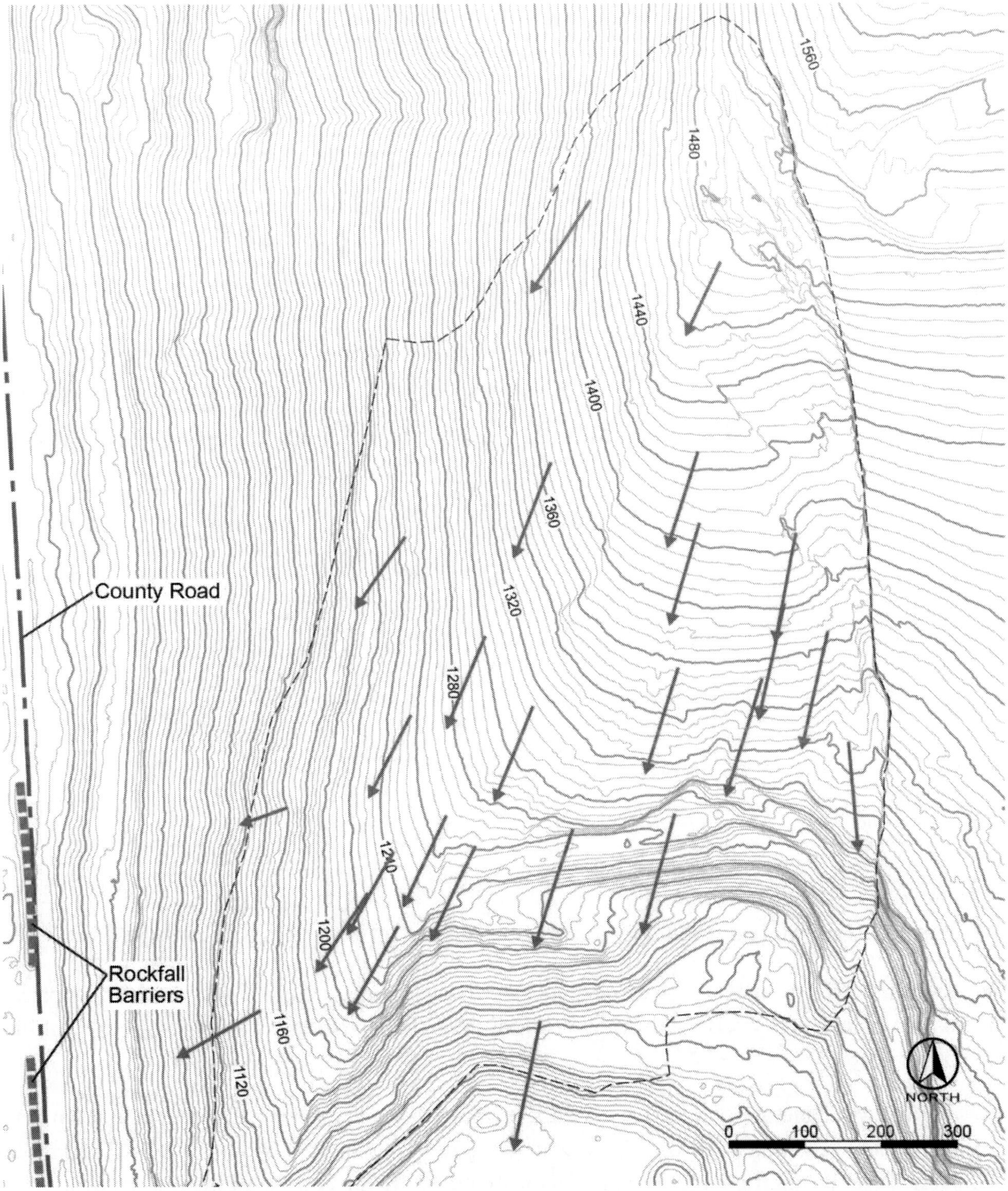

Figure 5 – Landslide Horizontal Movement Vector Directions

Example plots of horizontal movement vector for the RTK (October 2017 to mid-January 2018) and robotic total station (mid-January to May 2018) measurements for representative survey prisms in the middle of the landslide are shown on Figures 6 and 7, respectively.

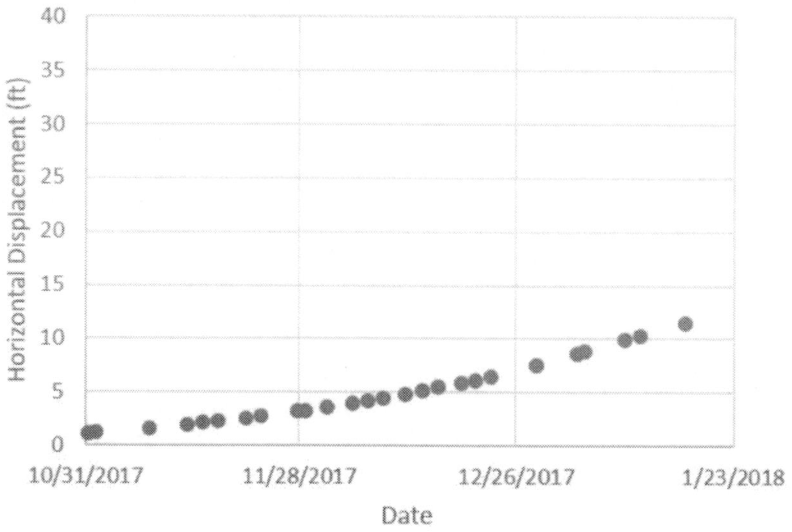

Figure 6 – Landslide Horizontal Movement (RTK, October 2017 to mid-January 2018)

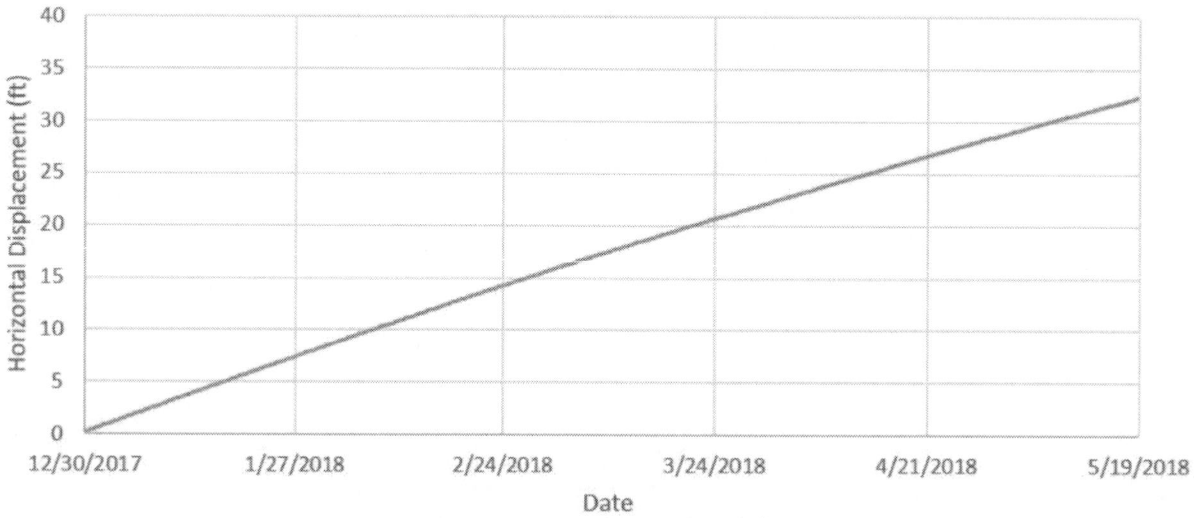

Figure 7 – Landslide Horizontal Movement (robotic total station, January to May 2018)

Figure 7 indicates the mean velocity of the landslide is approximately 1.5 feet/week. Comparing the data from all survey prisms, the movement of the main landslide mass ranges +/- 10% of the mean velocity.

The Washington DOT retained Wyllie & Norrish to provide an independent assessment of the landslide and potential risks the interstate highway and other nearby facilities. Their report concluded that the multiple monitoring methods have verified a consistent trend in landslide areal extent, bounding features, and movement rates, direction and inclination (Norrish, 2018).

The Pacific Northwest Seismic Network used a series of broad-band and short-period seismometers at the landslide to identify microseismicity due to slide movement and rockfall activity. Frequent very short broad-band signals representing local motions were detected.

The progression of rockfall and the buildup of a talus fan was monitored using video cameras and periodic patrols. A grid was painted on the quarry floor with 20-foot spacing to allow visual estimation of the advance of the slide debris and talus fan southward into the quarry, The monitoring indicates that the breakup of slide mass at the toe of the landslide results in rockfall and development of a talus fan. The movement is relatively slow and the rockfall energy is generally low, resulting in limited runout. An example view of the advancing talus fan and rockfall is shown on Figure 8.

Figure 8 – Landslide Debris, Talus Fan, and Rockfall in Quarry Floor (March 22, 2018)

Rockfall is also occurring on the west-facing hillside, where the flank of the landslide is bulging and raveling (Figure 9). Periodic patrols on the county road identify rockfall events and mark the locations of rocks that land on the pavement. Approximate dates of rockfall events reaching the road were documented. Most of the raveled material from the edge of the landslide moved short distances and stopped at various locations on the west hillside slope, and a smaller percentage of the rocks rolled all the way downslope and reached the county road. Talus fans are evident on the slope and occasional rocks exist in the ditch and road. An example of the rockfall on the county road is shown on Figure 10.

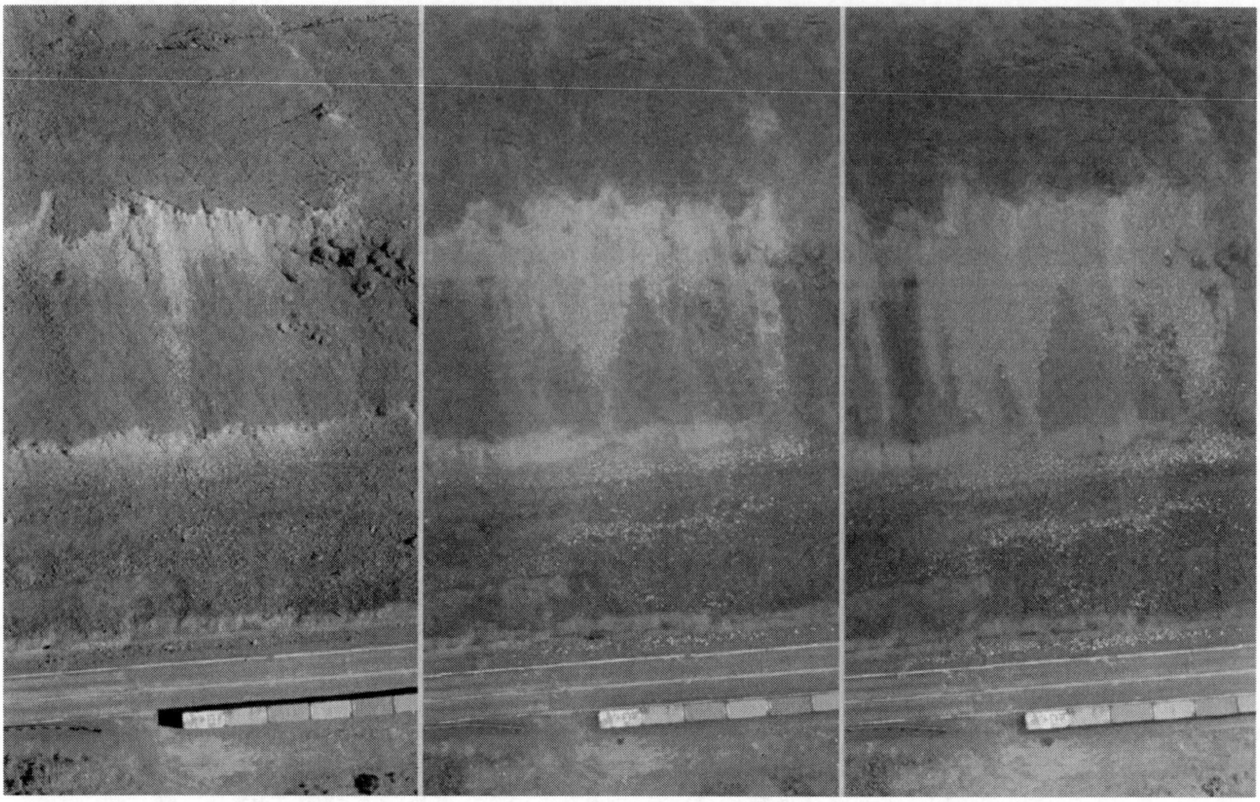

Figure 9 – Raveling Along West Flank of Landslide, Producing Rockfall (February 26, April 11 and May 9, 2018)

Figure 10 – Landslide Rockfall on County Road (March 22, 2018)

EVALUATIONS

Landslide Shear Zone Geometry

Evaluation of the exposed basal shear zone in the west flank of the landslide suggests the shear zone is roughly parallel to the natural ground surface. The survey monitoring data was used to analyze the apparent slope of the underlying shear zone. Assuming the shear zone is parallel to movement vectors at the ground surface, the basal shear zone under the main slide block is interpreted to be inclined 10 to 14° to the south (approximately Azimuth 190°). 3-D studies were performed to develop interpretations of the geometry of the basal shear zone, resulting in the development of interpreted contours of the shear zone and headscarp superimposed on the topographic site map (Figure 11).

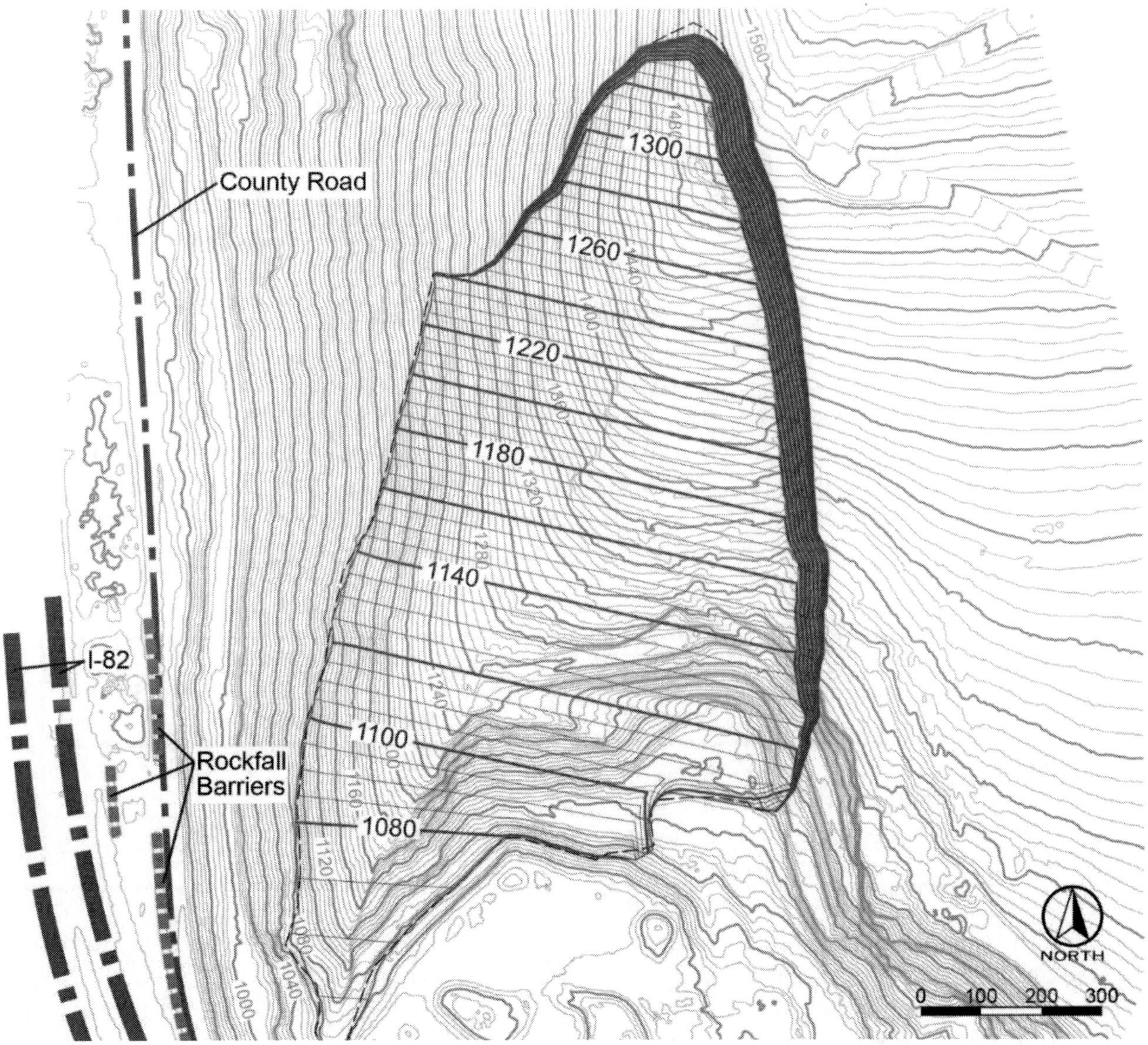

Figure 11 – Interpreted Geometry of Landslide Shear Zone

Preliminary Stability Analyses

Parametric stability analyses were performed to estimate the natural buttressing effect as the landslide continues to move toward the quarry and sheds material down its toe while the graben drops. Slope stability analyses were performed using industry-accepted 2D limit equilibrium computer modeling software. Interpreted geologic cross sections near the middle of the landslide were used to develop models for analysis. Figure 12 presents a representative cross section. Interpretations were necessary for the locations and orientations of the basal shear zone and active and passive wedges.

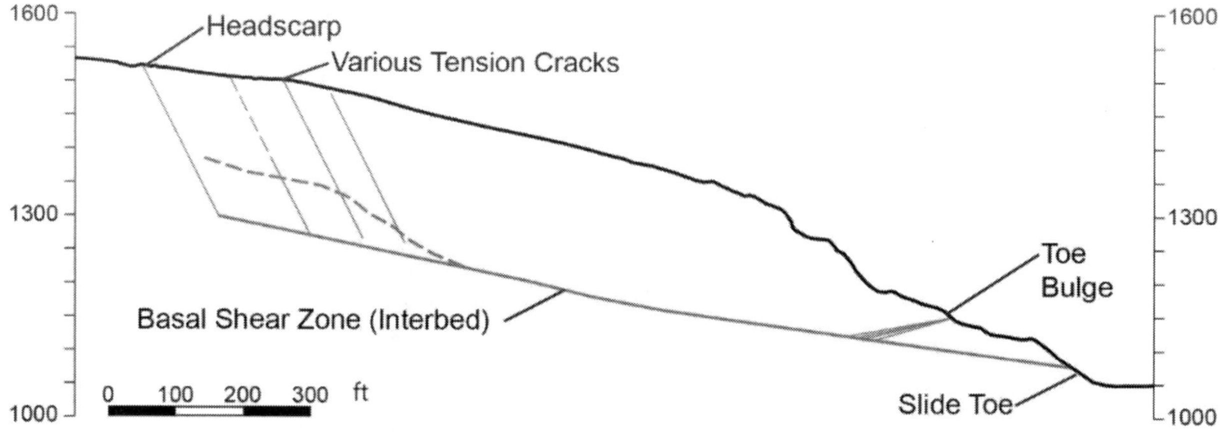

Figure 12: Landslide Cross Section

The back-analysis method was used to estimate values of the average residual shear strength (ϕ'_r) of the shear zone materials, assuming a Factor of Safety FS of about 1.0. The analysis results were expressed as percent FS increase since the actual FS while the landslide continues to move is less than 1.0. The back-analyzed residual shear strength ϕ'_r along the basal shear zone was approximately 10 to 12°, which is reasonable considering test results on similar basal shear material at other landslides. Table 1 presents the results of the preliminary stability analyses.

Table 1 – Preliminary Parametric Stability Analyses	
Width of Buttress (in direction of slide movement)	Relative increase in Factor of Safety, FS
0 (initiation of slide condition, October 2017)	0
100 feet	6 %
200 feet	20 %

The time for the landslide to naturally displace 100 and 200 feet was estimated to be 15 and 30 months, respectively, assuming the landslide moves an average 1½ feet per week. The estimated time could be longer if the active wedge (graben) downdrops and decreases with time and if some debris sheds westward toward the county road instead of all toward the quarry. Complicating predictions of buttressing is that the movement rate of the landslide could vary over time. Another complexity with predicting stability is the uncertainty of the current Factor of Safety (not knowing how much the existing FS is less than 1.0). If the correlation between FS and slide velocity presented by Cornforth *(5)* is used, the apparent existing FS could be as low as 0.5. The primary takeaway from the parametric analyses is that the natural evolution of a large buttress could take years to become effective.

The instability of the west flank of the landslide was also evaluated. As the landslide flank bulges, it becomes locally loosened and oversteepened, causing raveling of material, primarily as rockfall and talus. The slide mass is primarily comprised of basalt and the apparent dip of the shear zone appears to be inclined horizontally in the west direction (transverse to the south dip of the interbed and shear zone). The slope stability of a theoretical slump or slope failure would benefit from the high shear strength of the hard rock and its drained condition, resulting in Factor of Safety much greater than 1.0. The independent study confirmed this evaluation *(6)*.

Landslide Displacement

The movement of the main body of the landslide gradually increased between early October and mid-December 2017, when it reached a peak velocity of approximately 2 to 3 inches per day. The peak velocity was relatively constant in January through March 2018. By mid-April 2018, the landslide toe had advanced approximately 30 feet. The movement and protrusion of the toe of the landslide has caused internal stresses and strains that resulted in loosening, fracturing and inflation of the columnar basalt flow slide mass, resulting in raveling of rock fragments (typically 6 to 24-inch size) along the south-facing landslide toe and the west-facing flank of the slide. The majority of rockfall debris within the quarry accumulates as talus, essentially forming its own buttress. The head of the landslide has subsided, as would be expected for a translational slide, due to voids created by extension of the slide mass (tension cracks and down-dropping graben).

The evolving subsidence and buttress should be slowly increasing the stability Factor of Safety (FS), by increasing resisting forces and reducing driving forces. As the FS increases towards 1.0, the velocity of slide movement would be expected to decrease. In April 2018, some portions of the landslide have experienced slight decreases in velocity (less than 10% decrease).

The majority of rockfall debris within the quarry accumulates as talus, with small runout on the quarry floor up to approximately 25 feet. To date, the rockfall events have been relatively small, typically less than 1 to 5 cubic yards and rarely in the range of 100 cubic yards. Based on site observations it appears approximately 90% of rockfall is falling within 15 to 25 feet of the slide toe, while one individual rockfall stopped rolling about 50 feet from the slide toe.

The source areas on the west-facing flank of the landslide are the bulging ground above the slide shear zone where it daylights in the west-facing hillside. Rockfall and debris are accumulating on the slope below the shear, including the irrigation bench and the road side ditch. The amount of talus debris on the west-facing hillside slope on March 22 was estimated between 200 and 400 cubic yards. In addition, scattered rocks had reached the county road, totaling about 5 to 10 cubic yards. The largest rock that fell from the west bulge is a 3-foot boulder that came to rest on the irrigation bench. Other rockfalls are boulders less than 1.5-foot size. Rocks reaching the county road had runout distances of up to 25 feet. At this time, no rockfall has crossed the county road and reached the ROW of the interstate highway (this distance ranges from 50 to 75 feet).

Forecasting Landslide Displacement and Risk Monitoring

This slow translating block slide will likely move for a long period of time until it reaches a state of balance or is mitigated against further movement. Raveling, rockfalls and sloughing will continue as the slide moves. Larger events than the recent occurrences may also develop; however, they are anticipated to be of limited frequency due to the fractured rock conditions in the slide mass and the geometry of the natural and highwall slopes. The risk of catastrophic large-scale slide movement appears very low at this slide due to the rock characteristics of the slide, absence of groundwater pressures, and gentle inclination of the slide movement vector angle.

During December when slide movement velocity was increasing slightly each week, inverse-velocity graphs were plotted in the event trends indicated a rapid failure event. The inverse-velocity prediction method is described in a recent paper by Carla et al *(7)*, citing other related research and publications since 1985. If a large slide event were to occur, it's trend of the inverse-velocity with time would converge on the time axis of the graph, indicating a mass movement event may be imminent. An example of an inverse-velocity plot for the Rattlesnake Hills Landslide is presented in Figure 13A, prepared in Mid-December 2017. However, when the velocity became constant, the inverse-velocity changed from a linear sloped line to a horizontal line, which would not converge on the time axis and therefore would no longer indicate imminent rapid failure, as shown on Figure 13B. The method of Inverse-Velocity to predict failure is intended for brittle behavior, which this slide had experienced in the first few months. By the end of December 2017, the landslide reached a fully residual strength condition, which was indicated by constant movement velocity.

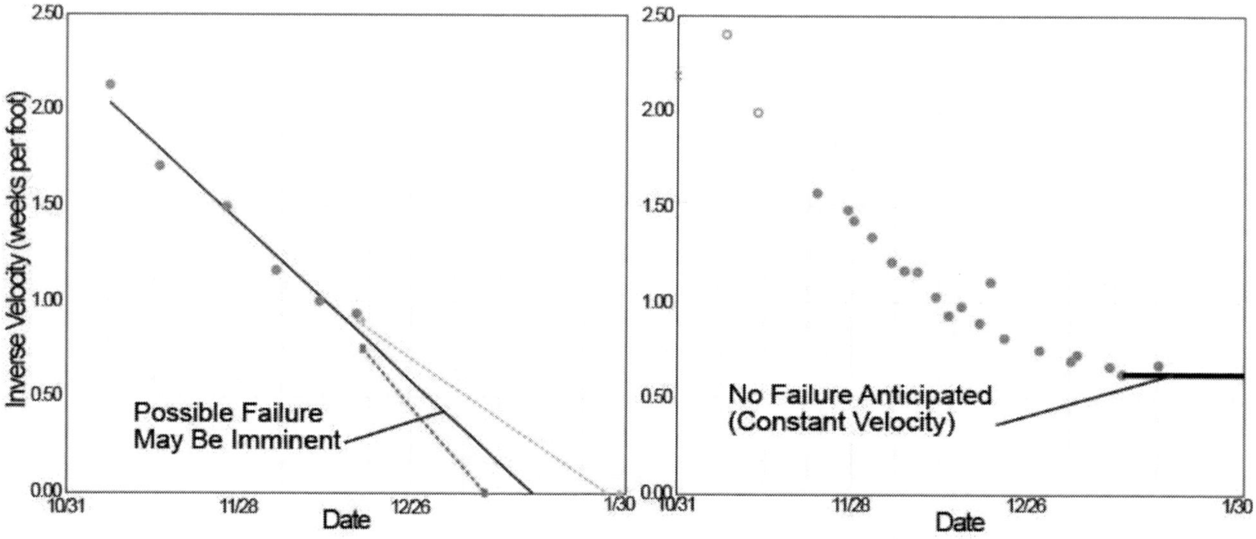

Figures 13A & 13B: Inverse-Velocity Plots

Estimating runout distances for this type of landslide should be based on rockfall modeling and observations of actual rockfall at the site. Runout modeling also exists for rapid brittle slides: however, this is not applicable for the Rattlesnake Hills Landslide. In the author's opinion, runout distances at this landslide are controlled by the characteristics of individual rockfall and shallow debris sloughing. Observations of landslide movements, including rockfall and debris events, indicate that displaced rocks generally stop on the slopes prior to reaching the county road and quarry floor until talus fans are developed. While many of the displaced rocks roll downslope and add to the accumulation of talus, few rocks have bounced and runout onto the quarry floor and county road.

SUMMARY

The Rattlesnake Hills Slide has been sufficiently instrumented and monitored to perform necessary evaluations of landslide movements and potential impacts and risks to nearby roads and facilities. The cooperation and participation by various stakeholders and researchers has provided extensive mapping, imaging and monitoring of landslide features.

Monitoring has quantified the characteristics of the landslide, including the relatively constant rate of slide movement and the gradual raveling of rockfall and development of talus downslope. Rapid slide movement is highly unlikely due to the low angle geometry of the landslide shear zone, the well-drained rock slide mass, and the absence of groundwater. The use of inverse-velocity plots can be helpful for predicting time to failure for slides that are accelerating, and when slides move at constant or reduced velocities, these plots would indicate that a failure event is no longer imminent. Rockfall runout characteristics have been monitored with drone images and video cameras.

Slide movement and rockfall activity have become relatively predictable, allowing reliable management of surrounding facilities and activities. The county road remains closed due to rockfall risks, while the interstate highway is unaffected. Nearby residents have been allowed by the County to remain, considering the risk of rockfall reaching the homes is highly unlikely.

REFERENCES:

1. Washington State Department of Natural Resources, *The Rattlesnake Hills Landslide, Yakima County, WA*, Fact Sheet 1.19.2018, Washington Geological Survey, January 19, 2018.
2. Bentley, R.D., Campbell, N.P., and Powell, J.E., *Geologic maps of part of the Yakima fold belt, northeastern Yakima County, Washington*, Washington Division of Geology and Earth Resources, Open File Report 93-3, scale 1:31,680, 1993.
3. Schuster, J.E., *Geologic map of the east half of the Yakima, 1:100,000 quadrangle, Washington*, Washington Division of Geology and Earth Resources, Open File Report 94-12., 1994.
4. Washington State Department of Natural Resources, *Washington Lidar Portal*, Division of Geology and Earth Resources, http://lidarportal.dnr.wa.gov/. Accessed October 2017.
5. Cornforth, D.H. *Landslides in Practice: Investigation, Analysis, and Remedial/ Preventative Options in Slopes*. John Wiley & Sons, Hoboken NJ. 2005. 596 p.
6. Norrish, N.I. (Wyllie & Norrish Rock Engineers, Inc.), *I-82, MP 36 to 44, Rattlesnake Hills Landslide Evaluation*, Report prepared for Washington Department of Transportation and Department of Natural Resources (Washington Geological Survey), January 30, 2018.
7. Carla, T., Intrieri, E., De Traglia, F., Nolesini, T., Gigli, G., and Casagli, N., *Guidelines on the Use of Inverse Velocity Method as a Tool for Setting Alarm Thresholds and Forecasting Landslides and Structure Collapses*, Landslides (2017) 14:517-534, 2017.

Adding Another Dimension to Rock Cut Slope Evaluations: Looking Out as Well as Up

Chris Ruppen, P.G.
Michael Baker International, Inc.
Moon Township, PA
724-495-4079
E-mail: cruppen@mbakerintl.com

Don Gaffney, P.G.
Michael Baker International, Inc.
Moon Township, PA
412-269-2967
E-mail: dgaffney@mbakerintl.com

Joel Borrelli, P.E.
Michael Baker International, Inc.
Moon Township, PA
412-269-2081
E-mail: jborrelli@mbakerintl.com

Prepared for the 69th Highway Geology Symposium, September 10-13, 2018

Acknowledgements

The authors would like to thank the Pennsylvania Department of Transportation and the Pennsylvania Turnpike Commission for calling our attention to the topic of this paper.

Disclaimer

Statements and views presented in this paper are strictly those of the authors, and do not necessarily reflect positions held by their affiliations, the Highway Geology Symposium (HGS), or others acknowledged above. The mention of trade names for commercial products does not imply the approval or endorsement by HGS.

ABSTRACT

As part of our rock cut slope evaluation work in Pennsylvania, Michael Baker International's Geotechnical Practice recently became aware of a traffic safety issue related to establishment of clear zones to allow vehicles to safely traverse areas outside the paved roadway shoulders.

PennDOT Publication 13M (DM-2), 2015 Edition – Change #2, Chapter 12.1.C. states the following in its discussion of clear zones (underlined text is in the original document):

> When a highway is located in a cut section, the backslope may be traversable depending upon its relative smoothness and the presence of fixed obstacles. If the fore-slope between the roadway and the base of the backslope is traversable (1V:3H or flatter) and the backslope is obstacle-free, it may not be a potential concern, regardless of its distance from the roadway. On the other hand, a steep, rough-sided rock cut should normally begin outside the clear zone or be shielded. A rock cut is normally considered to be rough-sided when the face can cause excessive vehicle snagging rather than provide relatively smooth redirection.

For interstate highways in Pennsylvania, the minimum clear zone is 30 feet. When evaluating cut slope safety hazards within the 30-foot vehicle clear zone, we resolved that we should include loose boulders in swales and irregular slope faces that could be snagged by an errant vehicle in our evaluations. This is a different type of hazard than has been considered during previous ratings, but has become a safety concern.

We consider adding three or four lines to the rockfall hazard rating system (RHRS) form to rate the distance from the edge of travel lane to the toe of cut slope compared to the clear zone and rate the character of the slope (if within the 30-foot clear zone) from the toe of slope to eye-level (nominal 6 feet). These ratings are scaled to be comparable to other rating factors on the RHRS form.

For cut slopes where the ratings were high on this factor, Table 1.1 of the cited edition of DM-2 lists several "Low Cost Safety Improvement Measures" for steep side slopes and roadside obstructions: object markings, slope flattening, ditch rounding, obstruction removal, breakaway safety hardware, and guide rail. Marking the worst objects could be a first step. Making sure maintenance is aware of the need to remove all boulders within the clear zone and not just on the pavement is another easy step. If it won't cause other stability issues, lower rock slopes could be trimmed smooth with hoe-rams on excavators during milling and paving or other roadway contracts. In some areas, guide rail or single-faced barrier would be a simple solution.

However, this is clearly seen as a broader roadway safety issue instead of simply a geotechnical issue. Consequently, the RHRS form has not yet been revised. We have started a multi-disciplined approach to this safety issue, so it can be acknowledged and addressed in a systematic manner as part of overall asset management by transportation agencies. Publicly-available photography shows the safety concern and assists in spreading awareness to appropriate professionals.

INTRODUCTION

When most of us see a rock cut along a highway, our first reaction is to look up. We look to judge its character, its beauty, and its safety. We've also been asked from time to time to design rock cut slopes. When we do, our primary concern is for rockfalls or other stability concerns; we want to make sure no rock ends up in the travel lanes. We have numerous factors and options to consider, including rock type, weathering and discontinuities, slope height and available right-of-way, slope angle, size and configuration of a drop zone at the base of the slope, and the need for other measures to stop the rocks from crossing the white (or yellow) line.

For roadways that are open to the motoring public, we continue to be involved with considering rock cuts through the process of asset management. We evaluate the rock cuts periodically to consider how gracefully they are aging, and the level of risk to the motorists. Again, we typically look over the entire slope area and evaluate whether those rocks will cross the line.

Recently, we've become aware of another risk to motorists. Instead of rocks coming out to meet the vehicles, vehicles sometimes want to get up close and personal with the rock cut. The area between the bottom of the cut proper and that line delineating the edge of the through traffic lane is what this paper is about. We know from our design experience that zone where the rocks come out to meet cars and cars can go meet the rocks can take a variety of different configurations. As we begin, note that we are focusing on interstate-type highways, we are speaking generically about those highways, and we cite examples using only publicly-available images pulled from Google Earth. Slope ratios are given as horizontal to vertical (H:V), except where taken directly from a referenced source.

CLEAR ZONES

Definitions

Most of us are aware of general safety considerations outside the through lanes of traffic. For steep embankment slopes that start within some distance of the through lanes, vehicles typically are protected (shielded) in the form of guiderail or barrier. For some distance from the white line, poles and posts either have a breakaway design or provide vehicle protection by guiderail. Bridge piers and abutments within some distance of the through lanes of traffic also provide vehicle protection by guiderail or barrier. That "some distance" is known as the clear zone. The various protected features are identified as objects, obstacles, or obstructions. The clear zone has a standard definition, and typically features both a foreslope and a backslope:

<u>Clear Zone</u> - The unobstructed, traversable area provided beyond the edge of the through traveled way for the recovery of errant vehicles. The clear zone includes shoulders, bike lanes, and auxiliary lanes, except those auxiliary lanes that function like through lanes.[1]

<u>Foreslope</u> - Area parallel to the flow of traffic that's identified as recoverable, non-recoverable, or critical.[2]

- Recoverable foreslopes are 4:1 (H:V) or flatter. Motorists who encroach on recoverable foreslopes generally can stop their vehicles or slow them enough to return to the roadway safely.

- Non-recoverable foreslopes are defined as traversable but from which most vehicles will not be able to stop or return to the roadway easily. Vehicles on such slopes typically can be expected to reach the bottom. Foreslopes between 3:1 and 4:1 generally fall into this category.
- Critical foreslopes are those which an errant vehicle has a higher propensity to overturn. Foreslopes steeper than 3:1 generally fall into this category. If a foreslope steeper than 3:1 begins closer to the edge of the traveled way than the suggested clear-zone distance for that specific roadway, a barrier might be recommended if the slope cannot readily be flattened.

Backslope - Area parallel to the flow of traffic beyond the foreslope that projects on an upward slope. For this study, the foreslope may be the base of the rock cut slope, a talus slope in front of the rock slope, or a separately designed slope provided to define a drainage channel between the cut and the roadway.

FHWA Guidance

FHWA references the current edition of the AASHTO *Roadside Design Guide* (RDG, 2011) and the AASHTO *A Policy on Geometric Design of Highways and Streets* (Green Book, 2011) for information on the latest practice in roadside safety.[3]

Table 3-1, "Suggested Clear-Zone Distances from Edge of Through Traveled Lane," in the AASHTO *Roadside Design Guide* provides suggested clear-zone distances based on traffic volumes, speeds, and foreslopes. Table 3-1 provides only a general approximation of the needed clear-zone distance based on limited data and is intended as a guide to aid a designer in determining whether an obstruction constitutes an obstacle to an errant motorist that is significant enough to justify action. The distances obtained from Table 3-1 suggest a range to be considered and not a precise distance to be held as absolute. The designer should keep in mind site-specific conditions, design speeds, rural versus urban locations, and practicality. The clear-zone distances in Table 3-1 may be modified with adjustment factors to account for horizontal curvature (sight distance), however these modifications are normally only considered when crash histories indicate a need to do so.

For roadways with interstate type characteristics (high volume and high speed), Table 3-1 suggests the following clear zone ranges:

- 30 to 34 feet for foreslopes 6:1 or flatter
- 38 to 46 feet for foreslopes 4:1 to 5:1
- Fixed obstacles should not be within foreslopes 3:1 and a clear area for vehicle recovery should be provided at the toe of slope
- 28 to 30 feet for backslopes 6:1 or flatter
- 26 to 30 feet for backslopes 4:1 to 5:1
- 22 to 24 feet for backslopes 3:1 to 4:1

For backslopes in a cut section, the AASHTO *Roadside Design Guide* states that the backslope may be traversable depending on its relative smoothness and the presence of a fixed obstacles, and that if the foreslope between the roadway and the base of the backslope is traversable (3:1 or flatter) and the backslope is obstacle-free, it may not be a potential concern, regardless of its distance from the roadway. On the other hand, a steep, rough-sided rock cut normally should begin outside the clear zone or be shielded. A rock cut normally is considered to be rough-sided when the face will cause excessive vehicle snagging rather than provide relatively smooth redirection.[4]

Pennsylvania has called attention to these last two sentences in its Design Manual (DM-2, Publication 13M) by underlining them:

> On the other hand, a steep, rough-sided rock cut normally should begin outside the clear zone or be shielded. A rock cut normally is considered to be rough-sided when the face will cause excessive vehicle snagging rather than provide relatively smooth redirection.

It was this reference that got us re-thinking our approach to cut slope evaluations to include consideration of what can happen at the bottom of rock cuts. Rough-sided rock cuts are considered obstacles from a vehicle safety perspective.

Table 5-2, "Barrier Guidelines for Non-Traversable Terrain and Roadside Obstacles," in the AASHTO *Roadside Design Guide* provides guidance for addressing obstacles within the suggested clear zone. Within Table 5-2, the guidance for smooth foreslopes and backslopes is that shielding is generally not needed. The guidance for rough foreslopes and backslopes is that a judgment decision should be made based on likelihood of impact.

One of the examples[5] in the AASHTO *Roadside Design Guide* provides guidance for a high speed road where a rock cut is within the suggested clear zone:

EXAMPLE 3-I
Design ADT: 3000
Design Speed: 100 km/h [60 mph]
Suggested clear-zone distance for 1V:6H foreslope: 8.0 to 9.0 m [26 to 30 ft] (from Table 3-1)

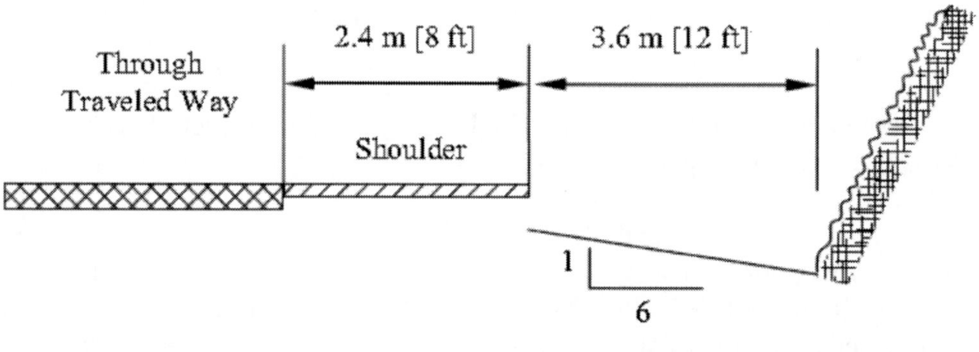

The related discussion for this example states that the rock cut is within the given suggested clear-zone distance but would probably not warrant removal or shielding unless the potential for snagging, pocketing, or overturning a vehicle is high. Steep backslopes are clearly visible to motorists during the day, thus lessening the risk of encroachments and roadside delineation of sharper than average curves through cut sections can be an effective countermeasure at locations having a significant crash history or potential.

A key in this discussion is "unless the potential for snagging, pocketing, or overturning a vehicle is high." This again is subjective. Typically, the bases of rock cut slopes are located within clear zones and are not remediated or shielded on our interstates. The vast majority of them appear to meet the guidance criteria for smooth slopes. However, as part of comprehensive asset management, this other dimension should be considered.

EXISTING ROCK SLOPES

Variations in Design

When we stop looking up, and just look horizontally from the white (or yellow) line and the rock cut – from eye level down – we are reminded quickly of the great variety of geometries we have put at the base of those cuts. With the help of gravity, the lower slope might be covered with rocks, talus, fallen vegetation or other debris. There might be a drainage swale between the edge of pavement and the base of the slope. The cut may have been designed with a catchment area by extending the foreslope and taking advantage of the clear zone. (This reminds us that once rock has fallen into the clear zone, it also becomes an obstacle presenting a safety hazard.) A formal drop zone may have been designed with a guiderail or barrier between the drop zone and through traffic lanes.

We can quickly identify the most common variations of these clear zones. Each of these configurations pose their own individual potential safety concerns within the clear zone.

Cut slope at the edge of pavement

This is the most restrictive condition, but still typically is acceptable for smooth cut slopes. However, there are several concerns. Mass excavation for the cut slope typically does not require the level of accuracy that is found in dimensioning of the shoulder. Therefore, the actual shoulder width typically will vary by a foot or more, and its width should be checked against the minimum requirement. Rock, talus, and other debris typically accumulates at the toe of the slope, and further reduces the effective shoulder width. There is no additional clear zone, so the bottom of the cut slope needs to be maintained in a clean condition.

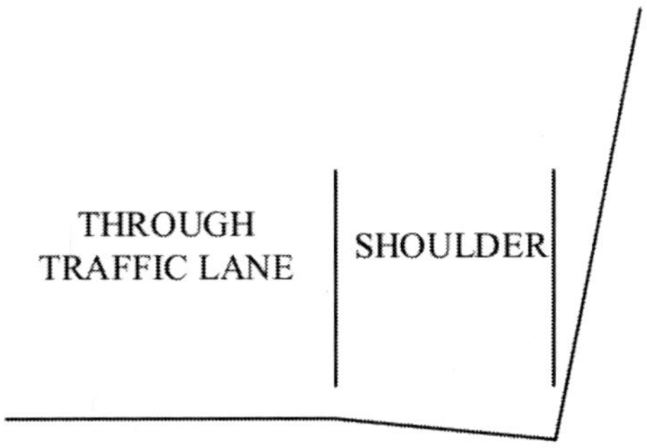

CUT SLOPE AT EDGE OF PAVED SHOULDER

Cut slope behind a barrier at edge of shoulder

The addition of a barrier at the edge of the shoulder can be result of establishing a wider shoulder without slope treatment, extension of a barrier for an adjacent obstruction (such as a bridge abutment), minor realignment of the roadway, or other factors. It may appear to be integral with the toe of the slope, or it may be offset from the toe enough to catch talus and some rock or debris, but not enough to be considered a drop zone designed for rockfall protection. In this section, the barrier practically serves to delay the need for maintenance, while providing a smooth face for errant vehicles.

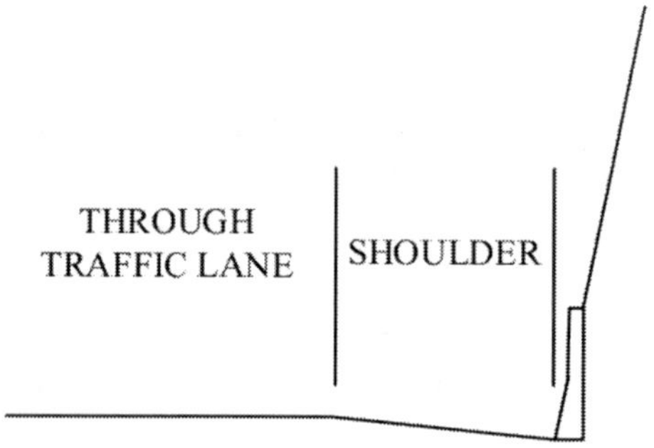

CUT SLOPE AT EDGE OF PAVED SHOULDER WITH BARRIER

Cut slope with a foreslope to the edge of shoulder

With this design, the foreslope has been lengthened, typically to create a drop zone for rock, talus and debris. Alternatively, it may have been lengthened to create a drainage swale at the base of the rock slope. Its width should be compared to the clear zone requirement. If the slope is within the clear zone, it should have a smooth face such that a vehicle will slide along its face and come to rest against it. The accumulation of talus and debris at the toe of the slope, as well as disrupted drainage, should be evaluated for their effects on errant vehicles.

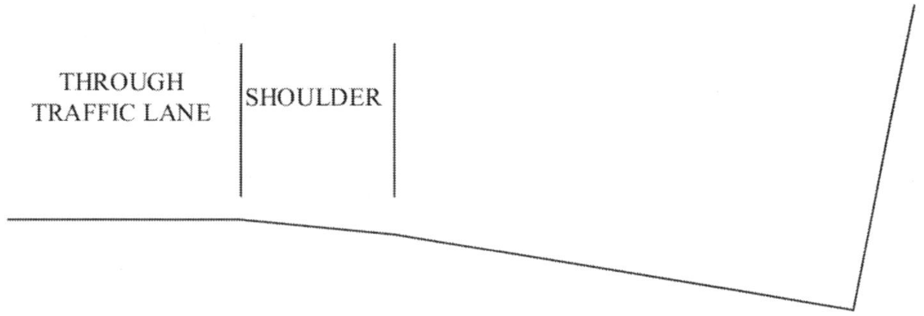

CUT SLOPE WITH FORESLOPE TO SHOULDER

Cut slope with a drainage swale (foreslope and backslope) to the edge of shoulder

This design may reflect only drainage considerations for the widths of the foreslope and backslope, or may include a lengthened foreslope to accommodate rockfall or talus considerations. Its actual total width relative to the clear zone requirement is the primary concern, followed closely by the maintenance of the area to keep it clear of rock, talus, or other debris. Maintenance of positive drainage in the swale is another concern, especially if there are inlets within this area.

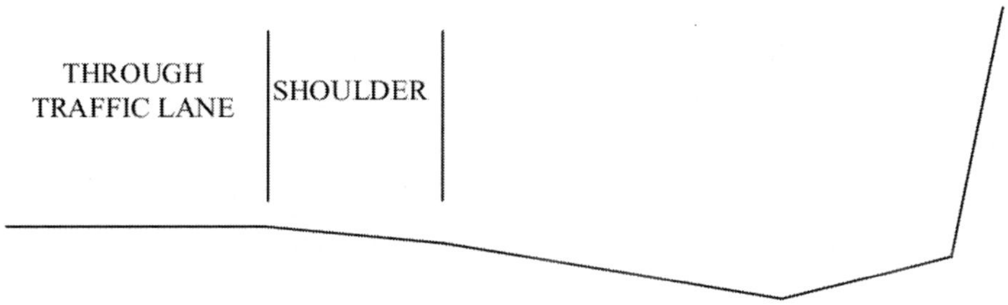

CUT SLOPE WITH DRAINAGE SWALE

Cut slope with a drop zone and barrier

In most respects, this is the cleanest typical section. In the best of circumstances, both the shoulder and drop zone have been designed to minimum requirements. However, there may be other concerns such as right-of-way that override the design minimums for these components. In that case, maintenance may become more critical. Even in the best of cases, the barrier needs to be maintained.

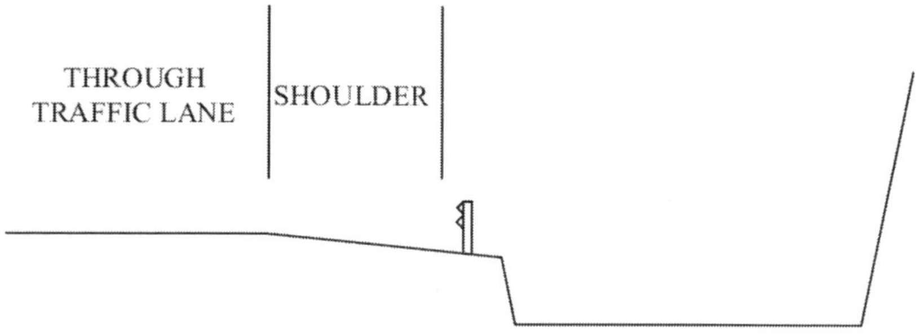

CUT SLOPE WITH DROP ZONE AND BARRIER

Evaluation Considerations

As we start to think about evaluating this aspect of rock cut slopes systematically, we look to the guidance of the FHWA standardized rockfall hazard rating system and its form.[6] That form has four rating levels with a standard spread of point values.

The primary consideration is the location of the base of the rock cut relative to the clear zone for the roadway. Fortunately, the limits of the through traffic lanes are well marked by white lines on the right / shoulder edge and yellow lines on the left / median edge. A relatively simple and straightforward horizontal measurement from the toe of slope to edge of travel line could incorporate this safety hazard into the rating:

- Less than eight feet: 81 points
- Eight feet to 12 feet: 27 points
- 12 feet to 20 feet: 9 points
- 20 feet to 30 feet: 3 points

A second critical consideration is the character of the cut slope face from eye level (nominal six feet) to the toe of slope. Recognizing that slope character is a qualitative assessment, it helps to keep in mind the image of what probably would happen if an errant vehicle were to run into the slope. Are there rough protrusions from the slope that would tend to snag the vehicle? We would suggest a line on the form that considers:

- Numerous rock protrusions extending from slope: 81 points
- Isolated rock protrusions extending from slope: 27 points
- Slope ragged / jagged: 9 points
- Slope relatively smooth / vegetated: 3 points

Another consideration would be to characterize the space between the through traffic lane and the base of the cut slope: the presence and extent of rockfalls, talus, vegetation, or other debris, and the area's general condition. At least one state agency does this now with three lines, identifying the percentages of various-sized fallen blocks, the quantity of fallen material present, and the offset of the rockfall from the slope. Again, these are qualitative assessments.

We are seeing the increased use of concrete barriers placed at the toe of rock cuts, with or without catchment areas behind them. They are being placed for a variety of reasons, from shoulder widening with minimal cut or right-of-way impact to defining the limit of a drop / rollout zone. Sometimes the space between the barrier and the slope is meant to be maintained by periodic removal of debris; sometimes the barrier appears to be integral to the bottom of the cut.

We would propose two additional lines, only one of which would be completed depending on the presence or absence of a barrier between the through traffic lane and the cut slope:

Talus / Rockfall Zone Condition – No Barrier

- Fallen rock on shoulder: 81 points
- Large rock blocks within clear zone: 27 points
- Small rock blocks within clear zone:9 points
- Talus at toe of cut slope: 3 points

Talus / Rockfall Zone Condition – Barrier Present

- Barrier failed to stop rock block(s): 81 points
- Area behind barrier full: 27 points
- Debris immediately behind barrier: 9 points
- Debris in drop zone: 3 points

The movement from casual observation to critical evaluation is a first step in realizing and assessing potential risk. There are other steps that can be taken.

SAFETY OF THE MOTORING PUBLIC

Steps Taken and Options Offered

AASHTO, in its guidance document, notes one simple step: roadside delineation of sharper than average curves in front of rock cut slopes. Other steps have included simply painting or otherwise safety marking protruding rock in the lower cut face and routine, periodic removal of fallen rock and talus between the cut face and the through lanes.

PennDOT lists several "Low Cost Safety Improvement Measures" for steep side slopes and roadside obstructions: object markings, slope flattening, ditch rounding, obstruction removal, breakaway safety hardware, and guide rail. Several of these apply directly to rock cuts. Object marking and obstruction removal are two measures already common and easily handled by maintenance forces. Hoe-ram work to remove protrusions from otherwise stable, smooth rock faces is another obstruction removal technique. This work can be contracted as part of roadway rehabilitation or reconstruction projects. In some areas, guide rail or single-faced concrete barrier would be a simple, cost-effective solution. These barriers are typically applied to the leading edges of bridge piers and abutments within cuts, and their extension to encompass the entire cut slope would help to improve safety.

Going Forward

As we became aware of the potential for this type of safety risk, it didn't take long to realize this risk may be appropriate to consider in many states. Pennsylvania is addressing this issue.

While performing condition assessments as part of asset management, this risk may appear in a gray-zone between disciplines. For example, geotechnical professionals may assess the condition of the cut slopes for rockfalls, and highway or traffic engineers may assess fixed obstacles and roadside barriers / shields. The condition of the lower rock cut slope, including both smoothness and debris, may inadvertently be overlooked by both groups. We have found it beneficial to engage other highway and traffic engineers in both the recognition and improvement of these conditions.

---

[1] AASHTO, Roadside Design Guide, 4th Edition (2011), 3-1
[2] AASHTO, Roadside Design Guide, 4th Edition (2011), 3-4,5
[3] https://www.fhwa.dot.gov/programadmin/clearzone.cfm
[4] AASHTO, Roadside Design Guide, 4th Edition (2011), 3-6
[5] AASHTO, Roadside Design Guide, 4th Edition (2011), 3-24
[6] FHWA, Rockfall Hazard Rating System, NHI Course No. 130220, Participant's Manual, SA-93-057. 1993, 26

Practical Aspects of Using Structure from Motion Photogrammetry Techniques for Characterizing and Monitoring of Rock Slopes

Randy Post, PE
Golder Associates, Inc.
4730 N. Oracle Road, Suite 210
Tucson, AZ 85705
520-888-8818
rpost@golder.com

Roger Pihl, PG
Golder Associates, Inc.
44 Union Boulevard, Suite 300
Lakewood, CO 80228
970-379-5341
Roger_Pihl@Golder.com

Alex Brown, EIT
Golder Associates, Inc.
44 Union Boulevard, Suite 300
Lakewood, CO 80228
303-980-0540
Alex_Brown@Golder.com

Ty Ortiz, PE
Colorado Department of Transportation
4670 Holly Street, Unit A
Denver, CO 80216
303-398-6601
Ty.Ortiz@state.co.us

Prepared for the 69th Highway Geology Symposium, September, 2018

Acknowledgements

The authors would like to thank the following individuals (listed in alphabetical order) for their contributions in the work described:

Kevin Carpenter – Golder Associates Inc.
Bob Group, PE – Colorado Department of Transportation
Nicole Oester, PG – Colorado Department of Transportation
Ryan Preston, P. Eng – Golder Associates Ltd.
Gwynneth Smith, EIT – Golder Associates Inc.
Beau Taylor – Colorado Department of Transportation
Henok Tiruneh, PhD, RG – Golder Associates Inc.
Ty Truelove – Aerial Applications

Disclaimer

Statements and views presented in this paper are strictly those of the author(s), and do not necessarily reflect positions held by their affiliations, the Highway Geology Symposium (HGS), or others acknowledged above. The mention of trade names for commercial products does not imply the approval or endorsement by HGS.

Copyright Notice

ABSTRACT

Historically, DOT's have allocated funding for geohazard response and mitigation (e.g. rockfalls and slope failures) as a reaction to specific events or perceived threats. Asset management utilized at the federal and state level for other highway features provides the framework for new and more proactive approaches to managing geological hazards that negatively impact user safety and mobility as well as maintenance budgets. Remote sensing techniques can be used to supplement and improve likelihood estimations in calculating risk and can facilitate data driven decision making and more efficient funding allocation.

Structure from Motion (SfM) photogrammetry from UAV-collected aerial imagery provides an invaluable tool to characterize geohazard sites in 3D. As equipment costs come down and the technology becomes more accessible, UAV-based lidar will also have a role in data collection at these sites. This paper presents a comparison of these two UAV data collection methods and compares them with more traditional methods such as terrestrial lidar scanning, and collection methods involving full-sized manned aircraft. Through data acquisition and analysis at a combination of semi-controlled test sites, live geohazard sites, and project sites in Colorado, the authors have implemented algorithms and procedures that can ultimately be scaled up to a corridor and even state level. In addition to UAV assisted emergency response to geohazard events, change detection, and remote extraction of geological data, the paper discusses other techniques and tools that provide semi-quantitative, site specific evaluation of risk that is fundamental to meaningful application of asset management principles. The insights gained from these studies have been used to refine the methods of data collection, test new UAV hardware, and to perform comparison with other types of remote sensing data, such as aerial or terrestrial lidar. The SfM tools and techniques are not without their limitations and challenges, including logistical constraints and the management of terabytes of data that must be transferred, processed, and stored.

INTRODUCTION

Historically, DOT's have allocated funding for geohazard response and mitigation (e.g. rockfalls and slope failures) as a reaction to specific events or perceived threats. Asset management utilized at the federal and state level for other highway features provides the framework for new and more proactive approaches to managing geological hazards that negatively impact user safety and mobility as well as maintenance budgets. Remote sensing techniques can be used to supplement and improve likelihood estimations in calculating risk and can facilitate data driven decision making and more efficient funding allocation.

Structure from Motion (SfM) photogrammetry from unmanned aerial vehicle (UAV)-collected imagery provides an invaluable tool to characterize geohazard sites in 3D. Through data acquisition and analysis at a combination of semi-controlled test sites, live geohazard sites, and project sites in Colorado, the authors have implemented algorithms and procedures that can ultimately be scaled up to a corridor and even state level. This paper focuses on important, practical lessons learned during the collection and analysis of UAV data at these sites and provides brief case studies illustrating specific applications of the concepts discussed.

Structure from Motion (SfM) Photogrammetry

SfM relies on the basic photogrammetry principles that 3D positions can be derived from a series of overlapping, offset images. The primary differences between SfM and conventional photogrammetry is that no 3D positions are required a priori, and SfM utilizes more images with a high degree of overlap and from different angles. Camera locations and scene geometry are reconstructed simultaneously and refined iteratively using an optimization algorithm as the software processes additional images (Westoby et al., 2012). SfM can be generated using imagery that is collected from UAVs, from full-size aircraft, from satellites, or even from the ground.

Uses of UAVs and SfM Data

The authors have been able to use UAVs as part of emergency response for multiple geohazard events and project sites in western Colorado. This typically involves having the UAV vendor mobilize to the site (within several hours of being notified in the case of emergency projects) and collect oblique photos of the project area that are sufficient to produce a baseline 3D point cloud using SfM. The photos themselves are also valuable resources for evaluating the hazard, and when the responding engineers and geologists are on-site while the UAV is present, real-time images and video can be viewed from the road level, avoiding the safety issues associated with hiking or using rope-access techniques to physically inspect an area. The 3D point clouds also allow physical comparison with previous models of the site using change detection techniques to evaluate changes and compute volumes in the case of rockfall and landslide events.

Change Detection

Change detection refers to the process of mathematically comparing one 3D data set to another to determine missing and accumulated material or objects. Numerous authors have performed change detection on rockfall slopes, most frequently using terrestrial lidar scanning (TLS) data (Abellán et al., 2009, 2010, 2011; Lato et al., 2009; van Veen, 2016). A complete review of this subject is beyond the scope of this paper, but a general overview is important in the context of the geohazard work described herein. The change detection workflow involves first coarsely aligning the two point clouds, often by manually picking common points in both clouds. Then a fine alignment is performed using the Iterative Closest Point (ICP) algorithm that is essentially an optimization algorithm, iteratively adjusting the orientation (and optionally, scale) of the cloud being aligned to the reference cloud to minimize the difference error between the two point clouds. Ideally this step is performed using only areas of the point clouds that are not changing, or several applications of the algorithm can be applied, screening out high-movement areas in between steps.

Once the point clouds are aligned, the difference between the point clouds is computed using the Multiscale Model to Model Cloud Comparison (M3C2) algorithm. The limits of detection (LOD) are defined as the 95 percent confidence limits assumed to be plus or minus two times the standard deviation of the difference computed for stable areas of the model. Movement within these bounds is within the noise range and should be ignored. The authors implemented these workflows using one of the more common software programs cited in literature for this analysis, CloudCompare (CloudCompare 2018). The final steps for the change detection workflow involve utilizing a noise reduction algorithm and a clustering algorithm (Tohini and Abellan, 2014) to turn the difference cloud data into a set of features (such as rockfall events, debris flow deposition, or landslides) readily stored in a spreadsheet or database. Relevant data includes the centroid coordinates of the feature, dimensions, volume, and other features.

Extraction of Geological Data

Abellan et al. (2014), Lato et al. (2015a) and Sturzenegger et al. (2011) describe various aspects of extracting geomechanical properties from 3D data. The authors have utilized the UAV-derived SfM point clouds and related work products to extract geological/geomechanical data on one roadway re-alignment project and several other sites for general knowledge and understanding. Commercially available software allows a proficient user to map joints on the virtual outcrop and generate stereonet data, as well as collect information on fracture trace length and joint spacing information. Using this technology it is possible to collect data in areas that cannot be accessed safely or easily in the field.

SELECTING THE RIGHT TECHNOLOGY FOR ROCK SLOPE DATA COLLECTION

The primary focus of this paper is the use of UAV technology for characterizing rock slopes. However, a brief overview of related data collection technologies and techniques is warranted in the context of selecting the right method for a given project. Over the past several years, the authors have collected rock slope data using lidar and SfM from a variety of terrestrial, UAS, and full-size manned aircraft platforms. There are advantages, constraints, and safety concerns related to all of these.

The authors have summarized preliminary cost information in Table 1 for the various data collection methods based on the work performed to date and information from literature. This information is presented for preliminary planning purposes only.

Table 1: Comparison of Cost and Data Collection Time			
Method	**Data Collection (hours/km^2)**	**Data Acquisition Cost (per km^2)[a]**	**Additional Processing Cost (per km^2)[b]**
UAV Lidar	4-16[d]	\$5,000[e] - \$10,000[d]	\$0
Helicopter SfM	1[c,d]	\$2,250[c,d]	\$800[d]
UAV SfM	4-8[d]	\$2,700-5,000[d]	\$800[d]

Notes:
a – Includes vendor costs and CDOT or additional field costs (excluding traffic control).
b – Includes additional costs by CDOT, Golder, or other consultant to process the data.
c – Source: Lato, Gauthier and Hutchinson (2015)
d – Source: Golder calculations/estimate.
e – Source: Cozart (2017)

One of the primary deciding factors for which technology to select is the required point density for the anticipated use of the data. Point clouds with a point density of at least 400 ppm^2 are recommended where change detection and geologic/engineering interpretation of the rock slope are desired. Table 2 presents a summary of the range of point densities as well as advantages and disadvantages that can be expected from the various data acquisition technologies.

Table 2: Comparison of Geohazard Remote Sensing Acquisition Methods			
Method	**Typical Point Cloud Density (ppm^2)**	**Advantages**	**Disadvantages**
Aerial Lidar	2-20	• Large coverage area • Bare-earth model	• Low spatial resolution • Conventional not good for steep slopes • Helicopter logistics and safety
Terrestrial Lidar	400 to 10,000	• High accuracy and precision • High point density	• Access not always available for scan locations • Occlusion zones
Mobile Lidar	50 to 500	• Can cover long segments of highway • Moderate point density • Relatively high accuracy and precision	• Specialized vendors • Occlusion zones could be significant for rugged terrain
UAV Lidar	75 to 300	• Occlusion zones minimized • Moderate point density • Relatively high accuracy	• Specialized vendors • No colorized point cloud or inferior colorized point cloud • Can't fly in bad weather

		and precision • Bare-earth model possible	• Relatively high cost of instruments
Helicopter SfM	>200	• High point density possible • True-color point cloud • Can use high-quality lenses and cameras • Can cover entire corridor	• Lower LOD than UAV lidar and TLS • Safety risk in canyons • More occlusion zones than UAV SfM • Removal of vegetation results in interpolation of the surface • Seasonally limited availability
UAV SfM	400 – 4,000	• Very high point density possible • True-color point cloud • Many possible vendors • Occlusion zones minimized	• Lower LOD than UAV lidar and TLS • Can't fly in bad weather • Line-of-sight limitations • Removal of vegetation results in interpolation of the surface

LESSONS LEARNED FROM UAV PROGRAM TO DATE

The authors have been involved in systematically collecting and analyzing UAV data for CDOT geohazard sites since February of 2017. Since then, over 6 TB of imagery, point clouds, and other 3D data files have been collected and generated. This section highlights some of the lessons learned during that time which may be useful in planning similar programs.

Data Collection

Camera Settings and Flight Parameters

- Most rock slopes of interest are relatively steep, so angling the UAV's camera gimbal at approximately 45 degrees and flying multiple overlapping lines parallel to the strike of the slope produced the best results. For mission planning purposes, a given point on the ground should appear in a minimum of 8 images.
- Most UAV cameras can collect images in RAW or JPG format. The authors experience is that the higher-quality RAW image format did not improve the quality of the point cloud models. In theory, RAW images can be post-processed to correct issues such as over- or under-exposure. But practically speaking, manually adjusting hundreds of images for a given data set is not feasible. Additionally, the file sizes are much larger for RAW images.
- UAV cameras can collect images with different aspect ratios. The authors recommend selecting the format that uses all available pixels on the camera's sensor to maximize the potential overlap between adjacent images.

Ground Control Points

Each UAV photo has metadata containing the GPS coordinate of the UAV at the time the photo was taken. Although individual GPS values can be inaccurate, each image dataset contains hundreds of such coordinates. SfM software utilizes these coordinates along with internal algorithms to geo-reference the resulting point cloud. Based on controlled field experiments where point clouds were created both with and without surveyed ground targets, the authors determined that the models without ground control were within approximately 2 meters of the actual location in the horizontal plane. The elevation of the model can be off by as much as 20 to 30 meters. For most applications of interest to the authors, the absolute positional accuracy of the model is not as critical as the relative accuracy and the resolution. There are also significant logistical problems with deploying and maintaining ground control points in steep terrain. So in most cases, UAV surveys without ground control points were acceptable. The exception would be if a single model is desired for a large area or long corridor, or if the data will be used for detailed design in the future, in which case ground control and check points should be used with the help of a licensed surveyor.

Safety

Common safety concerns during collection of data by UAS platforms include working near traffic and hazards to ground personnel and traffic; however the authors' use of UAS platforms in canyons presented additional hazards such as: steep rock slopes, poor stopping site distances due to roadway curves, and narrow shoulder and ditch sections. Finding safe pull offs for staging, operating, and landing presented a common challenge. FAA regulations address many of these hazards, but care should be taken when planning flights. Traffic control, warning signs, operational modification, flight scheduling, flight planning, and/or personal protective equipment may be justified.

Data Processing

The authors utilize the commercial software program Agisoft PhotoScan (Agisoft, 2018) for all SfM processing. Some of the following lessons learned are specific to that program.

Lens Calibration

The spherical nature of a camera lens introduces distortion into images that can affect the accuracy of the point cloud generated by the SfM process if not corrected. PhotoScan uses Brown's distortion model (Agisoft, 2018) to adjust for these errors. The calibration parameters are calculated automatically by the software during the point cloud generation process. Alternatively, the user can utilize Agisoft's built-in lens calibration feature to collect multiple images of a checkerboard pattern and compute the calibration parameters separately. The authors found that calibrations were better when PhotoScan computed the fits from the image data sets compared to using the manual lens calibration procedure.

Processing Time and Resources

There are different quality settings in PhotoScan that greatly impact the output product as well as the processing time. Most point cloud models created by the authors were done at "high" accuracy for the alignment phase, and "high" quality for the dense point cloud phase of the processing. The amount of processing time to generate a dense point cloud with these settings seemed to correlate best with the number of aligned images. The processing time on a computer similar to our benchmark system (Intel® Core™ i7-6820HQ CPU @ 2.70GHz, 6th Generation, 4 cores, 8 threads, 32 GB of RAM, NVIDIA Quadro M5000M Graphics Card, 2 GB GDDR5 Memory, memory bandwidth 76.8 GB/s, 320 CUDA Cores, OpenGL 4.0, Microsoft DirectX 11, 512 GB Solid-State Hard Drive) can be estimated as 3.4 minutes per image to be aligned. Accordingly, large data sets of hundreds of images can take more than a day to process. Breaking up large data sets into smaller "chunks" can be expected to reduce the processing time by approximately 30 percent based on the authors testing but requires more manual input. Some large data sets must be processed at lower resolution to be manageable, but portions of the model where change is noted can be reprocessed at higher resolutions if necessary.

The size of files associated with 3D point cloud data and analysis is enough to tax most current standard file storage devices and computers. The authors recommend assuming file storage needs on the order of 100 GB per mile of rock slope imaged for each temporal data set.

Segmenting Large Sites and Multi-Scale Analysis for Change Detection

Most rockfall site models the authors worked on are in the range of 100 to 1,000 meters long when measured along the roadway. When there is more than one point cloud for a site, the first step of the change detection process is to use manual and automated methods to align the two models. Many point cloud models, particularly those derived from SfM, are observed to have minor distortions that prevent alignment from being perfect between the two scans over the entire scene. But by segmenting the site into maximum lengths of 150 to 300 meters, the effect of the model distortion can be minimized. Performing change detection on this size model will allow an overview of rockfall, erosion, and debris accumulation to be observed for the entire site or a large portion thereof. But frequently it is desirable to further crop the model to focus on the areas of greatest change and re-align the two models to ensure the lowest possible LOD for the change detection. A realistic limit of detection between two high-quality SfM point clouds is on the order of 10 to 30 cm, meaning that positive and negative changes on the slope with magnitudes less than the LOD are in the noise range. The LOD for UAV-based lidar derived data sets is generally lower, on the order of 3 to 8 cm.

Data Management and Logistics

Cloud Storage and Processing

As the authors have noted, simply storing and processing the large data sets for this type of work can be a logistical challenge. Virtual machines located in the "cloud" are being investigated as the primary location of file storage and processing. The advantages of this approach are scalability, virtually unlimited file storage capacity, easier offsite backup solutions, and the potential of configuring faster processing machines (faster processers with more cores, multiple graphics cards, large amounts of RAM) that can greatly reduce the amount of processing time. This approach requires a very fast internet connection to initially upload the raw data as well as to download the finished products. Additionally, much of the analysis work involves manually manipulating 3D models. Doing this over a remote connection can have significant lag if the internet connection is not fast enough.

Tracking Data and Sharing Results

Managing 3D data for a single rock slope is straightforward. However, managing 3D rock slope data and analysis for multiple slopes each with multiple datasets rapidly becomes an issue. Data of interest that must be tracked includes parameters used for data collection of each data set, areal limits, raw point cloud properties, settings used during point cloud processing, settings used during change detection analysis, resulting rockfall or other slope changes, and more. The fact that the data files are so large means that they will be stored in a separate location and not in the tracking database itself. This is truly a "big data" problem.

Another challenge the authors have observed is the logistical challenges of sharing the various raw and processed 3D data products with colleagues. The 3D data can be a tremendous tool to geologists and engineers, but that requires getting it into the hands of the people who need it. It is not as simple as emailing a file attachment, there is the file transfer issue, and the fact that not everyone has access to the often-expensive software needed to fully manipulate 3D data. Adobe 3D PDFs are one potential way to manage this, but these files can only support fairly low-resolution models. The authors have had some success sharing raw point cloud and mesh products using the free Agisoft PhotoScan Viewer software, but that only works if the files are generated in PhotoScan and the ability to annotate, and communicate information on the 3D data is lacking.

ASSET MANAGEMENT

Application to GMP

CDOT's Geohazard Management Plan (GMP, CDOT, 2017) is a work-in-progress that comprises part of the Department's overall Risk Based Asset Management Plan, which resulted from 2013 Federal Legislation that promotes performance and risk-based approaches. Assets within the CDOT GMP include excavated rock and soil slopes, embankments, natural slopes that produce rockfall, debris events, sink holes, and problematic soils. Remote sensing techniques have application to all of the assets within the class to varying degrees. This paper focuses on the application of SfM and lidar methods for excavated rock slopes, natural slopes that produce rockfall and unstable rock features. For example, change detection using point clouds from SfM or lidar could be applied to landslides, sinkholes, pavements, or embankments, where measurements of the location, direction and magnitude of ground movement can result in better Level of Risk (LOR), deterioration rates and estimation of life cycle cost for use in benefit/cost analysis. The ability to assess the condition of a slope more frequently or precisely results in more accurate assessment of likelihood by providing data to partially quantify asset condition and historic event frequency.

Risk in the GMP is the product of likelihood and consequence. Likelihood is represented as an annual probability which is determined by the condition of the asset and the historic number of events over the past 30 years, based on incomplete data. The framework monetizes safety risk, mobility risk and maintenance risk and combines them into a LOR per 0.1 mile segment of highway. Use of remote sensing techniques can improve the accuracy and consistency of the initial calculation of LOR as well as the measurement of performance of the asset. Remote sensing techniques can dramatically improve the precision and efficiency of collecting input data for the likelihood side of the risk equation but have little to do with the consequence side of the equation.

The unequalled visual perspective and detailed images and point clouds available through UAV collected photogrammetry and lidar provides an opportunity to improve precision and consistency of condition assessment. The ability to produce 3D models and extract geologic structure data from dense point clouds makes it possible to analyze stability of excavated rock slopes and unstable features. Use of commercially available software can provide a summary of major joint orientation and spacing, which can be used as input for stability analysis or in rock mass characterization based on existing rating systems.

Measuring and visualizing changes in subsequent point clouds from remotely scanned unstable features, excavated rock slopes and natural slopes with rockfall can be used to improve and quantify the likelihood of future events by increased knowledge of past events. Since the historical number of events for a segment is a factor in calculation of annual probability, more accurate accounting of these events will result in more accurate, consistent and supportable probabilities. Change detection can provide more accurate data by searching for rock accumulations in the ditch, near or below the road, as well as searching for missing rocks on the slope.

The ability to produce terrain models directly with mapping software or from decimated point clouds inherently simplifies construction of profiles for use in rockfall trajectory analysis. The authors used data from SfM to run several two-dimensional rockfall simulation programs during the study. Presumably, an even greater convenience and accuracy could be realized by establishing terrain for a three-dimensional simulation program. Rockfall simulations can be used in life-cycle and benefit/cost analyses.

In Colorado, there are many steep, natural slopes that produce rockfall and are the source of extremely high-energy, damaging events. Many times, slopes that fall into this category are beyond CDOT Right of Way, but if they impact the highway system, CDOT experiences the consequences by reduced system performance, high costs, and poor public perception. Most of the natural slopes that produce rockfall currently do not have an annual probability or LOR grade, but UAV photogrammetry in combination with lidar can be used to accomplish this task and supplement the GMP. There is recognition that naturally occurring rockfall from natural, undisturbed slopes is highly problematic, especially in corridors such as I-70 through Glenwood Canyon, De Beque Canyon, and US 550 at Red Mountain Pass. Four of the largest rockfall events in the Glenwood Canyon corridor have been natural events that cumulatively caused millions of dollars in direct damages with mobility costs estimated in the tens of millions of dollars.

Remote sensing techniques can provide superior visual perspective along with new tools to assess or rate natural slopes that have historically impacted state highways. Visual, subjective assessment of potential source areas, overall surface roughness, rockfall frequency, size, and other factors can be used for preliminary evaluation. A set of likelihood factors, unique to natural slopes, will need to be developed and supported by remote sensing and other data. Unstable rock features on excavated or natural slopes present a special case where the consequence of an event is high. Large features that may release as a rock slide instead of individual pieces are not well handled by existing rating systems developed for rock cuts. Detailed data will be warranted for the most critical features to assess stability or deterioration with the highest degree of precision. Wire frame geometric analysis, extraction of geological data, and change detection can be applied to sufficiently dense point clouds of natural slopes and unstable features so they can be addressed in the GMP.

Life cycle cost, risk management, and asset deterioration are key components of the GMP. The ability to collect and produce detailed terrain and geological data for analysis, preliminary design, and cost estimating of mitigation options will produce more meaningful life cycle costs through a more accurate estimate of capital investment. Risk management approaches that use cash flow diagrams and rates of return on capital also benefit greatly from more accurate initial data because cost inaccuracies are compounded with time. More abstractly, greater efficiency and more accurate initial data facilitates consideration of a wider array of options.

CASE STUDIES

Emergency Response

Over the past three years, the authors have responded to multiple rockfall events that closed the highway or caused property damage and required emergency evaluation of stability. Traditional, long and dangerous hikes up the fall line were avoided at four of the locations by using UAS to view and photograph the area around the source. Proximity of the UAS to the slope, high-quality cameras, and the ability to view and zoom from any perspective make knowledge of the site superior to that gained by traditional methods such as rope and helicopter access, surpassed only by viewing the site from a crane basket.

SH 133, MP 53.5, Redstone Rockfall Site

As part of emergency response to a rockfall at this location in February, 2017, a UAV was mobilized to provide stability assessment of the remaining rock in the source area and the resulting debris field. The site is located in the Maroon Formation on the west side of the Crystal River Valley just north of Redstone, Colorado. The source for the several hundred cubic yard (CY) rockfall event is approximately 1,500 feet above the highway (8,500 feet above mean sea level [amsl]) and the total height of the slope at the rim of the valley is approximately 1,800 feet above the road (8,800 feet amsl). The steep, sandstone cliffs in the upper part have slope angles ranging from 80 degrees to near vertical. The lower half of the slope is characterized by colluvial soils, rockfall debris and alluvial fan materials deposited by frequent debris flows over Maroon Formation bedrock. The average slope gradient for the lower, colluvial slopes is approximately 35 degrees. The AOI is sparsely vegetated with conifers with an estimated ground cover of less than 10%. Figures 3 (a) and (b) show the source location and one of the large boulders displaced from the site at Redstone.

Figure 3: (a) source of MP 53.5 rockfall event (b) large boulder displaced at site. The boulder traveled across both lanes of SH 133 and came to rest above the neighboring creek.

SH 133, MP 55.2, Rockfall Site

Emergency response to this site was initiated in March, 2018 by a road closure when an over-steepened talus slope produced enough material to fill the 25-foot wide ditch, which caused additional cobbles and boulders to enter the road. This traditionally problematic site consists of a talus slope that was excavated without stabilization when the road was constructed many years ago. The unstable talus extends over 100 feet above the road and slope failures commonly result in many cubic yards of material being deposited in the barrier lined ditch. The site was partially mitigated several years ago by improving the volume and configuration of the ditch supplemented by the addition of specially reinforced and modified portable concrete barrier. A UAS was used approximately one hour after the road was closed to determine that unsafe conditions remained and that the road should not be reopened until the ditch was re-established. Figure 4 shows the ditch cleaning efforts following the event.

Figure 4: Ditch cleaning efforts following March 2018 rockfall event at SH 133 MP 55.2

SH 133, MP 30.0, Rockfall Site

While commuting to another study site in April 2018, one of the authors came across a large rockfall that had occurred less than an hour before at MP 30.00. A rock block approximately 6-8 CY in volume fell from the upper slope brow, impacted a ledge 20-30 feet below and bounced into westbound lane of the highway, leaving an impact crater approximately 2-feet deep and 4-feet in diameter (Figure 5). A UAV survey of the area was performed immediately, and it was determined that the road could be safely re-opened but that subsequent scaling was advisable. UAV photography was used to provide information to specialty contractors and CDOT staff. Scaling was successfully completed the following two days.

Figure 5: Largest block from April 24, 2018 rockfall event at SH 133 MP 30.0

I-70, MP 48.8 De Beque Canyon Rockfall Site

During a widening project at this location, a potentially unstable rock was brought to the attention of project staff by the contractor. Several hours later, the site was evaluated using up-close, real time photography collected from a UAV. The evaluation resulted in recommendations for work to continue during analysis of options for future actions. While a six-inch crack was visible when viewed from the road, the unique, multiple perspectives from the UAV allowed observers to determine that there was sufficient intactness along other surfaces so that unnecessary work stoppages or road closures were avoided.

Project Development / Remediation Design

In conjunction with the remote sensing and on-call contract, the authors assisted with several roadway improvement and geohazard remediation designs. These designs included rockfall prevention, roadway re-alignment, and embankment failures. Several of these projects are discussed below.

SH67 MP 90.5

The State Highway 67 projects is a roadway alignment adjustment around a sharp, blind curve 35 miles south of Sedalia on State Highway 67. The curve has been the cause of multiple traffic fatalities over the past years, including three motorcycle deaths within a 1-year span. In an effort to increase safety along this section of roadway, plans designed to increase stopping sight distance and decrease rockfall into the roadway were developed.

A UAV survey was performed by licensed surveyors. In addition to providing topography and base mapping in accordance with CDOT project development requirements, the UAV was also used to collect oblique imagery of the proposed rock cut on the project. The authors used the resulting 3D point cloud to extract geologic properties including fracture spacing and orientation. This data was used to perform preliminary geological engineering analysis and design of the cut slope for the realignment. This type of synergy between project development and geohazard assessment will likely become more common in the future.

I 70 Exit 49

The authors were asked to design a soil nail wall and rock fence to allow widening of a deceleration lane. Photos taken from a helicopter with a DSLR camera and 50 mm lens were used to produce a 3D point cloud to supplement the project topographic data and generate cross sections used for wall layout and for rockfall simulations to determine the necessary height and capacity of the fence.

SH 133 MP 48

The authors coordinated a UAV and GPS survey through an engineering subconsultant. The subconsultant collected photo data using a DJI Phantom 4 Professional quadcopter and provided CDOT with raw (DNG) and JPG imagery, a high-resolution orthophoto, a DWG CAD file containing 1-foot contours, and a point cloud in a proprietary file format. The authors re-processed the aerial imagery to generate a high-quality 3D point cloud using Photoscan and noted that the file size was much larger than the point cloud submitted by the subconsultant. That fact along with anecdotal information from other surveys indicates that the typical point cloud resolution used for topographic mapping purposes is much lower than the ideal resolution for geohazard assessment.

SH 133 MP 21.7

CDOT requested that the authors provide remediation design support for an active embankment failure along State Highway 133 at milepost 21.7. The embankment failure has been an ongoing challenge for the local CDOT maintenance staff over several years. The slope consists of shallow colluvium overlying sandstone and shale bedrock of the upper Mancos Formation and the lower Mesa Verde Formation. The highway platform was constructed with typical cut and fill techniques leaving steep soil and rock slopes on the uphill side and soil embankments on the downhill side. Groundwater seeps are present in the uphill ditch. The slide area is reactivated each spring during the snowmelt, resulting in movement of inches to feet each year. Standard practice by the maintenance crew involves filling the depressed roadway areas with cold patch or milling and repaving the highway between the limits of the failure area (approximately 300 linear feet).

Three UAV surveys were performed: two photogrammetric and one lidar extending from the river at 6,435 feet to more than 200 feet above the road. For all three surveys, heavy vegetation (nearly 100% cover) consisting of Gambel Oak, Pinion, Juniper, native grasses, and typical undergrowth on the downslope portion of the survey created data processing challenges. Software designed for processing photogrammetric surveys was used to develop 3D point clouds and subsequent DEMs of bare earth terrain. One photogrammetric survey was used to develop a cost comparison for various design recommendations and a construction bid package for the site.

Change Detection

As discussed in previous sections, a significant motivation for the authors' use of UAV photogrammetry is to perform change detection on both natural and excavated rock slopes. Sites evaluated to date have yielded encouraging results, showing that a high degree of accuracy is possible. This change detection can be used during emergency response to evaluate rock which may have been impacted by the initial event; to create a record of event frequencies and volumes which may be applied to an asset management plan; or, as limits of detection become smaller with improved hardware and software processing, to measure erosion and deposition of material that may ultimately lead to a future event.

I-70, MP 53 Data Collection and Quality Test Site (Lat. 39.228/Long. -108.260)

The site at MP 53 in De Beque Canyon was chosen as representative of typical locations where data collection will be necessary if a larger scale operation is implemented. It is a steep-walled, Mesa Verde Formation sandstone canyon approximately 800 to 1,000 feet deep. The site has large loose rock blocks partially attached to the multiple sandstone cliffs which are interrupted by steep grass covered slopes with blocks on the surface.

The authors organized initial collection of photogrammetric data with two vendors, but only one was able to collect as a result of camera and mechanical failure. A UAV-lidar vendor scanned the same area. Two data sets were collected from each vendor within the span of 1 to 2 days, so there were presumably no changes to the site between the various data collection events. The data were used to evaluate a variety of variables associated with data processing, such as camera calibration, use of GCPs, and processing parameters. Change detection was performed on the two data sets from each data collection method to determine the realistic limits of detection for the methodology since essentially no change had occurred between the two data sets (see Figure 6). Additional photogrammetry data sets have been collected for this location using the lessons learned and produced some of the highest density point clouds of the study. A second UAV lidar vendor also performed a data collection at the site, flying from a higher altitude with more advanced lidar equipment.

SH 133 MP 24.25

While on-site at MP 21.7 in January 2018, one of the authors was made aware of a new, small rock-fall that occurred three miles east at MP 24.25. The rockfall was quickly removed by traffic control crews, but it allowed for a focused change detection search of the area. A UAV survey had been performed two months prior to the date of the incident, and a subsequent survey was performed the day following the incident. Figures 7 and 8 below show the change detection results performed using CloudCompare (2018) and source images of the area in question.

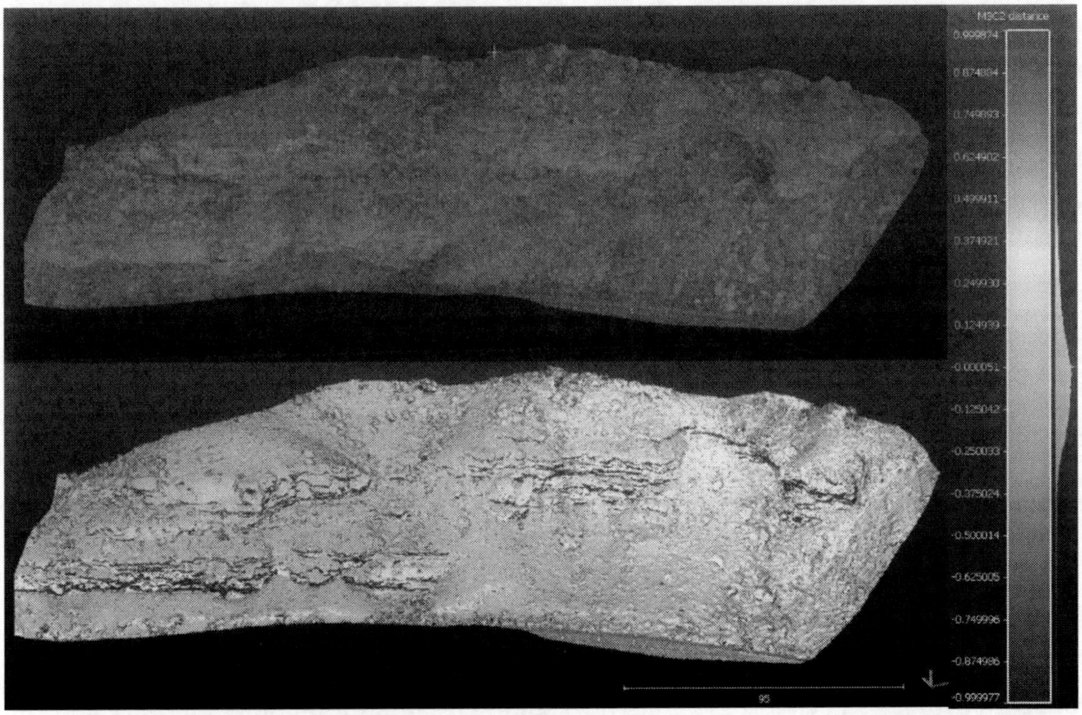

Figure 6: M3C2 histogram plot showing the comparison of the UAV lidar point cloud and the UAV photogrammetry point cloud for the I70 MP 53 test site surveys in April 2017. The mean for this comparison was 0.02 m with a standard deviation of 0.25 m. The variation between the models, flown within 24 hours of each other, is representative of the lower LOD values capable with lidar models.

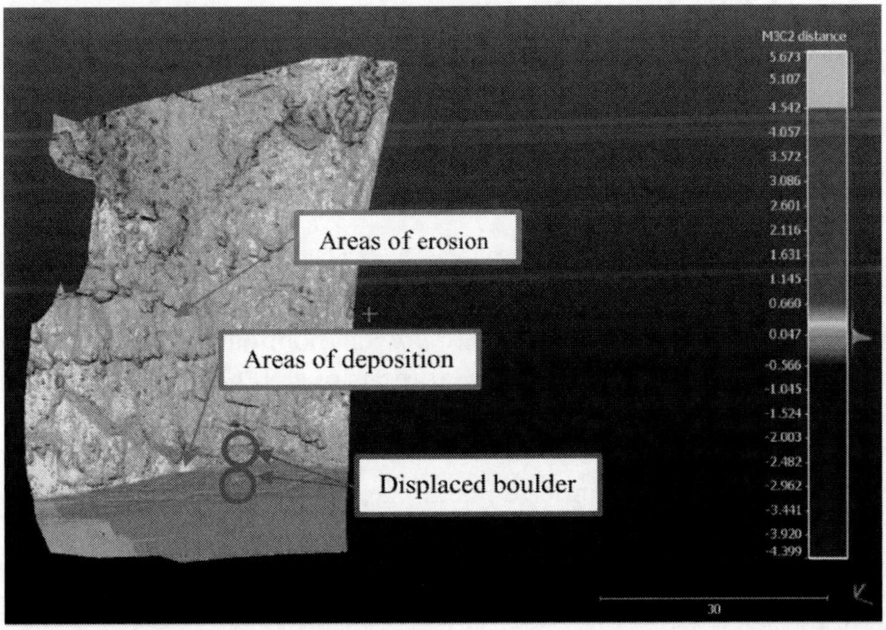

Figure 7: Change detection showing displaced rock from slope and erosion/deposition of soil

Figure 8: (a) November 2017 image showing rock prior to fall (b) January 2018 image showing rock following fall

CONCLUSIONS

The benefits of using UAV-collected lidar and SfM point cloud data are significant for rock slope and geohazard projects. Technological advances, increased equipment availability and rule changes by the FAA will increase the efficacy of UAV usage. The authors have shared a variety of brief case studies where this technology has been applied to emergency geohazard response, project development / remediation, and to perform change detection between two or more 3D data sets. UAV-based SfM has become the preferred data collection method for these projects because of the relatively low technological (and cost) requirements, the ability to collect complete data in areas where TLS would have significant occlusion zones and data gaps, high point density of the models relative to UAV-based lidar, and the ability to use colorized point clouds and the original photos for geological engineering analysis from the office. A collection of lessons learned related to SfM data collection, processing, and change detection analysis have been presented, and are summarized below.

Summary of SfM Conclusions and Recommendations:

- Plan photo collection for a minimum effective overlap of eight, four in direction of flight, at 50% vertical overlap.
- Extend flights several images past extents to prevent poor data at edges of desired AOI.
- No apparent benefit in storing RAW image formats over JPG.
- UAV's gimbal set to approximately 45 degrees below horizontal except when targeting overhangs.
- Set image resolution and aspect ratio to maximize the use of the camera sensor and image overlap.
- Separate camera calibration step not needed with PhotoScan (may not be the case with other software).
- Plan flights for a point cloud density of at least 400 ppm^2.
- Consider TLS for critical locations where possible but use UAV SfM for most locations. Use UAV-based lidar where absolute positional accuracy and vegetation are issues.

- Collect data without ground control points for most locations.
- Flight planning software can present a safety hazard in rugged terrain and should be used with caution or not at all.

Summary of Change Detection Processing Conclusions and Recommendations:

- Alignment portion of change detection is critical and controls limits of detection (LOD).
- Preliminary LOD for UAV-based lidar is approximately 6 cm.
- Preliminary LOD for UAV-based SfM is approximately 30 cm (expected to improve with better processing techniques).
- Evaluate the possibility of dividing areas into smaller segments (0.1 centerline miles) for processing/alignment efficiency and improved LOD.

The use of remote sensing to facilitate management of geological hazards shows great promise but significant additional work will be required to determine what techniques should be applied to specific situations and needs. The additional, better data made available through the use of UAV SfM and lidar can enhance the ability of geohazards professionals to approach the estimation of likelihood with greater confidence, efficiency and consistency between sites and corridors. Using UAVs for emergency response to geological hazards provides unique visual perspectives that reduce risk to the traveling public and to CDOT staff and others involved in the response.

REFERENCES

1. Westoby, M.J., J. Brasington, N.F. Glasser, M.J. Hambrey, and J.M. Reynolds. Structure-from-Motion'photogrammetry: A Low-Cost, Effective Tool for Geoscience Applications. *Geomorphology*, 179, 2012, 300–314.

2. Abellán, A., M. Jaboyedoff, T. Oppikofer, and J. M. Vilaplana. Detection of Millimetric Deformation Using a Terrestrial Laser Scanner: Experiment and Application to a Rockfall Event. *Natural Hazards and Earth System Science*, 9 (2), 2009, 365–72.

3. Abellán, Antonio, Jaume Calvet, Joan Manuel Vilaplana, and Julien Blanchard. Detection and Spatial Prediction of Rockfalls by Means of Terrestrial Laser Scanner Monitoring. *Geomorphology*, 119 (3), 2010, 162–171.

4. Abellán, A., J. M. Vilaplana, J. Calvet, D. García-Sellés, and E. Asensio. Rockfall Monitoring by Terrestrial Laser Scanning – Case Study of the Basaltic Rock Face at Castellfollit de La Roca (Catalonia, Spain). *Natural Hazards and Earth System Science*, 11 (3), 2011, 829–41.

5. Lato, M., J. Hutchinson, M. Diederichs, D. Ball, and R. Harrap. Engineering Monitoring of Rockfall Hazards along Transportation Corridors: Using Mobile Terrestrial LiDAR. Natural *Hazards and Earth System Science*, 9 (3), 2009, 935–46.

6. van Veen, Megan J. *Building a Rockfall Database Using Remote Sensing: Techniques for Hazard Management in Canadian Rail Corridors*. MASc Thesis, Kingston, Ontario, Canada: Queen's University, 2016.

7. *CloudCompare* [Computer Software]. 2018. Version 2.9.1. Retrieved from cloudcompare.org.

8. Tonini, Marj, and Antonio Abellan. Rockfall Detection from Terrestrial LiDAR Point Clouds: A Clustering Approach Using R. *Journal of Spatial Information Science*, 2014 (8), 2014, 95–110.

9. Abellán, Antonio, Thierry Oppikofer, Michel Jaboyedoff, Nicholas J. Rosser, Michael Lim, and Matthew J. Lato. Terrestrial Laser Scanning of Rock Slope Instabilities: STATE-OF-SCIENCE (TERRESTRIAL LiDAR VS. ROCK SLOPE INSTABILITIES). *Earth Surface Processes and Landforms*, 39 (1), 2014, 80–97.

10. Lato, Matthew J., Dave Gauthier, and D. Jean Hutchinson. Rock Slopes Asset Management: Selecting the Optimal Three-Dimensional Remote Sensing Technology. *Transportation Research Record: Journal of the Transportation Research Board*, 2510 (November), 2015, 7–14.

11. Sturzenegger, M, D Stead, and D Elmo. Terrestrial Remote Sensing-Based Estimation of Mean Trace Length, Trace Intensity and Block Size/Shape. *Engineering Geology*, 119 (3), 2011, 96–111.

12. Cozart, Jeff. *The Effectiveness of Drone-Based LiDAR*. LiDAR Magazine, 2017.

13. Agisoft, LLC. *Agisoft Photoscan*. Version 1.4.2 [Computer Software], 2018. Retrieved from www.agisoft.com.

14. Agisoft, LLC. *Agisoft PhotoScan User Manual. Professional Edition*, Version 1.4, 2018. Agisoft LLC.

15. Colorado Department of Transportation (CDOT). FY2018 *Geohazards Management Plan*. Working Draft. 2017

16. Hoek, E. *Practical Rock Engineering*. Course notes, 2007. https://www.rocscience.com/learning/hoek-s-corner/books.

The use of Google Earth/ Google Street View combined with high resolution digital surface modes (DSMs) for rockfall hazard rating

Yonathan Admassu

Department of Geology and Environmental Science
James Madison University, Harrisonburg, VA 22807
540 568 5016
admassyx@jmu.edu

Prepared for the 69th Highway Geology Symposium, September, 2018

Disclaimer

Statements and views presented in this paper are strictly those of the author(s), and do not necessarily reflect positions held by their affiliations, the Highway Geology Symposium (HGS), or others acknowledged above. The mention of trade names for commercial products does not imply the approval or endorsement by HGS.

Copyright Notice

ABSTRACT

One of the common types of slope failures that affect slope cuts along roadways are rockfalls. The main causes of rockfalls are geologic factors including unfavorable orientations of discontinuities, undercutting of weak sedimentary units, and presence of boulders in unconsolidated materials. Various US departments of transportation use proactive rockfall management methodology such as the rockfall hazard rating system (RHRS). RHRS is along other rockfall hazard rating systems rate slope dimension, geologic characteristics, climate, and rockfall history. The most important hazard parameters, geologic characteristics require time consuming data collection in the field. Remote sensing methods such as Google Earth/Google Street View and high resolution digital surface models (DSMs) derived from LiDAR/photogrammetry offer a fast and efficient methodology for rockfall hazard rating. This research proposes a two-staged process where during stage 1) slope cuts are semi - quantitatively rated based on geometric/geologic parameters measureable in Google Earth and visible in Google Street View. During stage 2, slope cuts will further be evaluated using DSMs extracted from photogrammetry or LiDAR. On DSMs, discontinuity surfaces indicative of future likelihoods and past failure will be quantified. The combined result of the two stages will provide a quantitative evaluation of rockfall hazard of slope cuts that may be used as a preliminary rockfall hazard rating.

INTRODUCTION

Rockfalls pose hazards to motorists and cause enormous damage to roadways. The main causes of rockfalls are the presence of unfavorably oriented discontinuities and undercutting of strong rock units by underlying weak rocks. Discontinuities are natural breaks (bedding planes, joints, foliation and fault planes) in rock or their intersections that may daylight on the slope face leading to the generation of rockfalls. Such discontinuity orientation controlled failures are subdivided into plane (Figure 1), wedge and toppling failures. Plane failure occurs when a discontinuity sub parallel to the slope dips gentler than the slope face causes daylighting conditions (Hoek and Bray, 1981). Wedge failure on the other hand results when the line of intersections of discontinuities daylight onto slope surface (Hoek and Bray, 1981). On toppling failures, Goodman (1989) wrote "If layers have an angle of friction Φj, slip will occur only if the direction of the applied compression makes an angle greater than the friction angle with the normal to the layers. Thus, toppling failure with a slope inclined α degrees with the horizontal and discontinuities dipping at σ can occur if $(90 - \sigma) + \Phi j < \alpha$". To evaluate the kinematic possibility of each of these structurally controlled failures, orientations of discontinuities and the slope face along with friction angle circles are plotted on a stereonet. The other common mode of slope failure leading to rockfalls is the result of differential weathering of weak rocks interlayered with strong units leading to undercutting and eventual failure of the latter (Figure 2). These failures are common where the geology is characterized by interlayered strong and weak sedimentary layers.

Figure 1: Structurally controlled failure (plane failure).

Figure 2: Undercutting-induced rockfalls.

Various US state departments of transportation (DOTs) employ proactive assessment of rock slope cuts with respect to rockfall generation. The rockfall hazard rating system (RHRS), first introduced by the Oregon DOT is a system of rating rock slope cuts hazard potentials based on slope height, catchment ditch effectiveness, average vehicle risk, percent of decision sight distance, roadway width, geologic characteristics, rock block size, climate conditions/presence of water, and rockfall history. Each of these parameters are scored based on numerical exponential scoring of 3, 9, 27, and 81. The RHRS is useful for the decision making process for designing rockfall warning systems and identifying slope cuts that have higher potentials for generating rockfalls and are in need of remediation (Pierson and Van Vickle, 1993). Many state DOTs have adopted the RHRS or modified it to fit their respective geologic conditions.

Geologic characteristics are the most important parameters as they are the main causes of rockfall generation. Equal weighing of all parameters is cited as a problem with the RHRS as it underestimates the important role of unfavorable geologic conditions (Russel et al., 2009). Geologic conditions include structural unfavorability, degree of interbedding, degree of undercutting, and average block size. These parameters characterize presence of unfavourably orientated discontinuities and undercutting susceptible interlayered weak/strong rock units. RHRS and other hazard rating systems' evaluation of geologic characteristics require physical investigations through field visits. As an alternative to field data collection, road level remote-sensing methods such as videography, photogrammetry, and terrestrial LiDAR scans (TLS) have become popular. The Missouri DOT uses scaled video captured from a vehicle to measure slope length, slope height, ditch width, ditch depth, rock height, rock length (Maerz and Youssef, 2012). Terrestrial LiDAR acquired from a vehicle can be used to collect discontinuity parameters for kinematic analysis of rock slopes along transport corridors (Lato et al., 2009). This research explores a two stage approach for preliminary rockfall hazard evaluation whereby the first stage involves using the freely available Google Earth/Google Street View to geometrically (slope height, length, aspect) describe slope cuts and identify the mode of failure (either discontinuity orientation controlled or undercutting-induced). Once the mode of failure affecting a slope cut is identified during the first stage, the second stage will quantify rockfall hazard from digital surface models (DSMs) extracted from a 3D point cloud generated from TLS or 3D photogrammetry.

ROCKFALL HAZARD RATING WITH GOOGLE EARTH/GOOGLE STREET VIEW

Swanger and Admassu (In Press) has modified the RHRS chart for parameters that are measurable in Google Earth/Google Street View makes (Table 1). These parameters include semi-quantitatively estimated slope profile (slope height, slope length, slope aspect, slope inclination, and slope roughness), geologic characteristics (structural condition, degree of interbedding, degree of undercutting, and average block size) and impact factors (sight distance and catchment ditch width). Slope height can be measured from elevation differences between slope crest and road level in Google Earth. Linear measurement tools in Google Earth can be used to measure slope length and decision sight distance. The aspect of a slope can be determined with Google Earth's ruler tool which shows azimuth of a line drawn perpendicular to a slope. Google Earth's path tool can be used to construct an elevation profile from which slope height and slope inclination can be measured. On the other hand, Google Street View provides seamless 360^0 street level photographs. These present a close up view of slope cuts adjoining roadways allowing visual inspection of geologic conditions. From Google Street View, one can estimate the size of rock blocks in catchment ditches and rockfall voids by making visual comparisons with objects of known dimensions such as road lane width, height of guard rails or road signs. The presence of daylighting discontinuities and interbeded weak/strong rock units can also be visually inspected.

Table 1: Rockfal hazard parameters measurable in Google Earth/Google Street View.

Rockfall Hazard Rating System					
Factor		**Score**			
		3 Points	**9 Points**	**27 Points**	**81 Points**
Slope Profile	Slope Height	25 to 50 ft	50 to 75 ft	75 to 100 ft	>100 ft
	Slope Length	0 to 250 ft	250 to 500 ft	500 to 750 ft	>750 ft
	Slope Aspect	N	E, W, NE, NW	SE, SW	S
	Slope Inclination	15 to 25 degrees	25 to35 degrees	35 to 50 degrees	>50 degrees
	Slope Roughness	Possible launching features	Some minor launching features	Many launching features	Major rock launching features
Geologic Characteristics	Structural Condition	Discontinuous or Continuous fractures, favorable orientation	Discontinuous or Continuous fractures, random orientation	Discontinuous fractures, adverse orientation	Continuous fractures, adverse orientation
	Degree of Interbedding	1 to 2 weak interbeds, <6 in	1 to 2 weak interbeds, >6 in	>2 weak interbeds, <6 in	>2 weak interbeds, >6 in
	Degree of Undercutting	0 to 1 ft	1 to 2 ft	2 to 4 ft	>4 ft
	Average Block Size	6 to 12 in.	1 to 2 ft	2 to 5 ft	>5 ft
Impact Factors	Sight Distance	>80%	60% to 80%	40% to 60%	<40%
	Catchment Ditch Effectiveness	100%-95%	95%-65%	65%-30%	30%-0

ROCKFALL HAZARD RATING WITH DIGITAL SURFACE MODELS (DSMS)

DSMs from TLS and 3D photogrammetry for discontinuity characterization have been shown to be useful to characterize discontinuity orientation, spacing, roughness (Lato et al., 2009[a]; Lato and Voge, 2012; Nguen et al., 2011, Sturzenegger and Stead 2009) and rockfall hazard evaluation (Lato et al., 2009[b], Nguen et al., 2011; Kemeney and Turner, 2008). Lato et al. (2009[b]). LiDAR scans from different time periods can be used to identify sites of rockfall release based on change detection (Lato et al., 2009[b] and Nguen et al., 2011).

High resolution DSMs are mainly derived from terrestrial LiDAR (TLS) and 3D photogrammetry. Both methods generate a cloud of points (each having x, y, z values as well as RGB values) representing the surface of a scanned object, which in this case is a slope cut (Figure 3). The LiDAR scanner shoots a laser onto a target and records the returned laser signal to calculate x,y,z coordinates of every point from which the laser beam bounces. 3D photogrammetric methods such as structure from motion (sfm), can also generate a point cloud from overlapping photographs. From the 3D point cloud, a digital surface model (DSM) made up of triangulated mesh can be generated (Figure 3). Geologic discontinuities can be identified as flat surfaces on the DSM. Split FX (www.spliteng.com) or Maptek I-Studio (https://www.maptek.com/products/i-site/i-site_studio.html) can be used to identify discontinuities on DSMs and measure their orientations. The Split FX software can also calculate area, elevation, centroid location in space of discontinuities. This data can be exported as .txt file. The ability of TLS and sfm to scan steep inaccessible slope cuts, reduce the risk to personnel, and create a permanent record of slope surface (Higgins and Andrew, 2012).

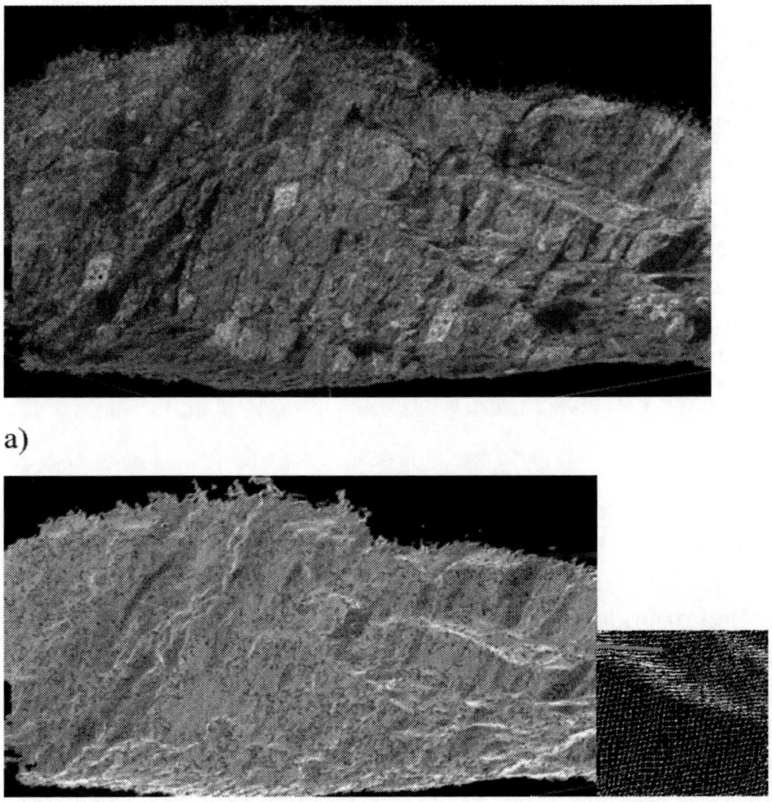

a)

b)

Figure 3: a) point cloud from sfm photogrammetry and b) DSM made from triangulated mesh surface.

INTEGRATING GOOGLE EARTH/GOOGLE STREET VIEW WITH DSM

Various rockfall hazard parameters can semi-quantitatively be determined using the methods described above. This research proposes a streamlined rockfall hazard methodology that integrates evaluation of selected parameters from the Google Earth/Google Street View images and further quantifying geologic parameters using DSMs. In Google Earth and Google Street View, slope cuts can geometrically and geologically be characterized. Once the dominant geologic control is identified, the occurrence of past rockfalls and likelihood of future events can be quantified from DSMs. Therefore, a two staged approach for road level remote sensing based rockfall hazard evaluation is proposed.

Stage 1- Google Earth/Google Street View:

The objective of this initial stage is to characterize cut slopes geometrically (slope height, length, aspect) and identify geologic controls of rockfall generation. Slope profile parameters such as slope height, slope length, slope inclination, and slope aspect can be collected from Google Earth (Table 2). In Google Street View, the main geologic control (rockfalls are resulting from unfavorable orientation of discontinuities or induced by undercutting) potentially causing rockfalls can be visually identified (Table 2). Further, the type of structurally controlled failure, plane, wedge or toppling can be identified (Table 2). From rockfall voids, orientations of discontinuities bounding rockfalls can be estimated using slope aspect (determined in Google Earth) as reference. Rockfall voids are bounded by at least three discontinuity surfaces. The orientations of at least three surfaces (J_1, J_2, J_3) bounding a rockfall void need to be estimated from Google Street View (Figure 4). One of the three bounding surfaces, J_2, is an overhang oriented nearly parallel to the slope face. The presence of J_2 indicates past rockfalls and a possible future release of rockfalls. Similarly for slopes affected by undercutting, the orientation of exposed undercut surfaces, which in most cases are subhorizontal also indicate past rockfalls and imminent rockfall. The estimate of orientation of undercut subhorizontal surfaces and joint spacing controlling depth of undercutting can be estimated. Finally, a geodatabase in Google Earth or Arc GIS can be built to organize the geometric and geologic attributes of each slope cut as well as Google Street View images (Table 2).

Stage 2: Digital Surface Models (DSMs)

Once the preliminary slope profile and geologic data have been collected in Google Earth/ Google Street View and a geodatabase is created, the second step is quantifying geologic parameters that control rockfall release. DSMs can be used to quantify the two main geologic controls, which are structurally-controlled or undercutting-induced failures. Depending on which of the two controls are prevalent, two different approaches may be used.

Structurally Controlled Failures

From DSMs, discontinuities can be identified as exposed flat surfaces and their orientation can be determined using software such as Split FX and Maptek I-Studio. Each discontinuity centroid location, area, and elevation can also be determined in Split FX and exported as a text file. In the case of structurally controlled failures, rockfall source sites are defined as exposed voids that form after the release of rockfalls. As discussed above, such voids are bounded by at least three discontinuity surfaces (J_1, J_2, J_3). More rockfall voids indicate more frequent past rockfall activity and a higher future likelihood of rockfalls. One of the three bounding surfaces, J_2, is an overhang oriented nearly parallel to the slope face serving as a release surface for a previous rockfall (Figure 4). The presence of J_2 indicates the presence of past rockfall events and possibility of future a future release.

Table 2: An example of data table to record stage 1 and 2 data.

Sight information					
Date	6/19/2017				
Location	West Rockingham County-VA/Route 33/Shendoah Mt.				
Site Designation	Rt.33-20				
Lithology/Age	Interbedded Sandstone and Shale/Miss-Devonian(320-410Ma)				
Comments					
Slope Profile	*Values*				
Slope Height	21 ft				
Slope Length	355 ft				
Slope Aspect	75°				
Slope Roughness					
Slope Inclination	>50°				
Geologic Characteristics	*Values*				
Structurally controlled failures (plane, wedge, topple)	None				
Orientation of release surfaces (Dip Dir/Dip)	None				
Undercutting Induced failures	One undercutting unit, undercutting depth 1 – 2 ft				
Orientation of undercut surfaces	Horizontal				
Degree of Undercutting	1-2 ft				
DSM Parameters	*Values*				
Parameter A					
Parameter B					

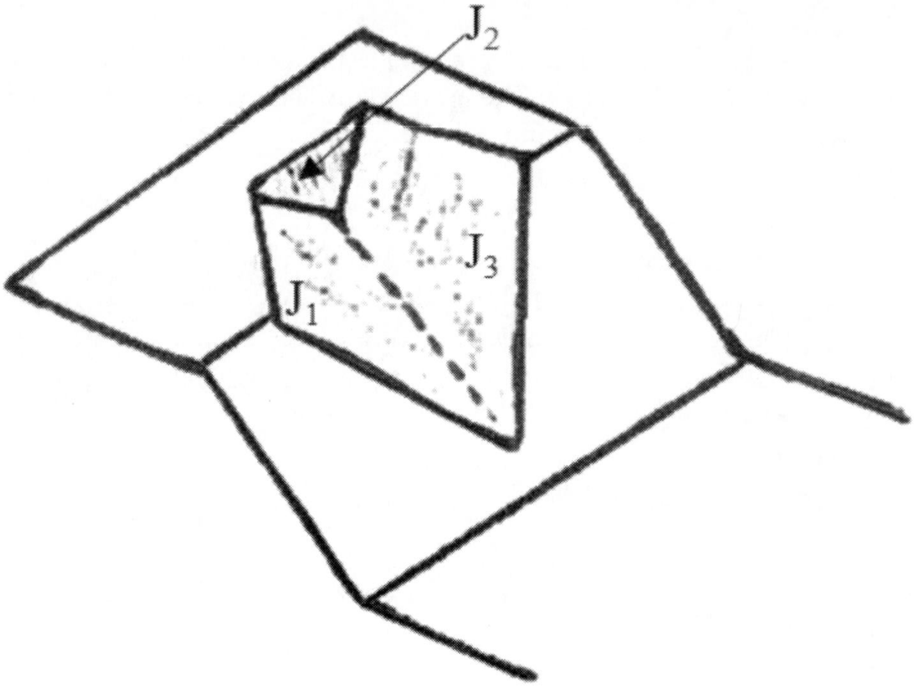

Figure 4: Sketch showing a rockfall void and bounding surfaces (J_1, J_2, and J_3).

In this research, such surfaces are termed as rockfall release surfaces and are quantified as Parameter A. Parameter A is weighted by elevation to give higher weight for high elevation release surfaces.

Parameter A = [Σ (J_{2ai} * Z_i/100) / Slope length*Slope Height] * 100

Where J_{2ai} is the area of the rockfall bounding surface parallel with slope face, and Z_i is the centroid elevation in feet of an individual release surface in feet.

Parameter A score is added to Table 2.

Undercutting Induced Failures

The area of individual subhorizontal discontinuities (undercut overhangs) is proportional to the depth of undercutting. The maximum depth of undercutting is controlled by spacing of discontinuities. Therefore, the area of subhorizontal surfaces should be compared with discontinuity spacing to evaluate how close a rockfall is to being released. An estimate of the orientation of undercut surfaces and joint spacings should come from Google Street View measurements or estimates. Parameter B evaluates how close undercut surfaces are to failing based on maximum depth of possible undercutting before a rock block is released. Parameter B compares the areas of exposed undercut surfaces to the maximum possible area of undercut surfaces before a rockfall is released. As undercutting from higher elevation can generate more energetic rockfalls, Parameter B is weighted based on elevation.

Parameter B = $[\Sigma (A_{hs\,i} * Z_i/100) / \Sigma ([A_{hs\,i} / W_{und}] * W_{und})] * 100$

Where A_{hs} is area of individual horizontal surfaces determined from discontinuity data measured from DSMs

Z_i is centroid the elevation of the flat discontinuity in feet normalized to a 100 ft height,

W_{und} is the average maximum depth of undercutting based on joint spacing estimated in Google Street View,

$A_{hs\,I} / W_{und}$ is an approximate length of individual horizontal surfaces.

Parameter B score is added to Table 2.

CASE STUDIES

Afton Mountain Cut – I-64

The Afton mountain is a slope cut located on the westbound section of Interstate 64 in Virginia (mile marker 101) with a slope aspect of 165^0 (Figure 5). The rock unit is a late Proterozoic meta-basalt with a well-developed highly persistent south dipping foliation and subvertical orthogonal joints. It is evident from Google Street View that rockfalls due to plane failures along the southeast dipping foliation are prevalent.

Figure 5: Google Street View image of Afton Mountain cut on I-64 west bound.

Parameter A

Several areas of past rockfall release sites are observed on the slope face (Figure 5). Release surfaces were visually identified using Google Street View. Judging from the slope aspect determined in Google Earth, rockfall release surfaces (J_2) are near vertical (> 65^0) with dip directions between 135 and 175. Discontinuities with such orientations were selected on the DSM (Figure 6). The Σ (J2ai * $Z_{i\ in\ ft}$ /100) was calculated to be 0.17 m^2 and the Slope length * Slope Height is 311 m^2.

Parameter A = [Σ (J2ai * Z_i /100) / Slope length*Slope Height] * 100

Parameter A = (0.17 m^2/311 m^2) * 100 = 0.05 (the maximum can be 100)

US 33 Cut

A slope cut on US 33 consisting of interlayered sandstone and shale units belonging to the Hampshire formation of Devonian age is chosen as a site representing slope cuts with undercutting-induced rockfall problems (Figure 7). The section of US 33 where it crosses the Allegheny Mountains in western Virginia is prone to releasing undercutting-induced rockfalls due to interlayering of strong layers with soft erodible layers. There is no evidence for discontinuity orientation-controlled slope failures.

Parameter B

At the US 33 cut, 29 subhorizontal surfaces that are sites of undercutting have been identified (Figure 7). Using Google Street View, the average spacing of joints parallel to the slope face (W_{und}) that control the depth of undercutting was estimated to be 0.5 m. The area and elevation of discontinuities was determined using Split FX.

Therefore,

Parameter B = Σ ($A_{hs\ i}$* Z_i/100)/ Σ ([$A_{hs\ i}$ / W_{und}] * W_{und})

Σ ($A_{hs\ i}$* Z_i/100) = 0.19 m^2

Σ ([$A_{hs\ i}$ / W_{und}] * W_{und} = 4.9 m^2

B = 0.04 (the maximum can be 100)

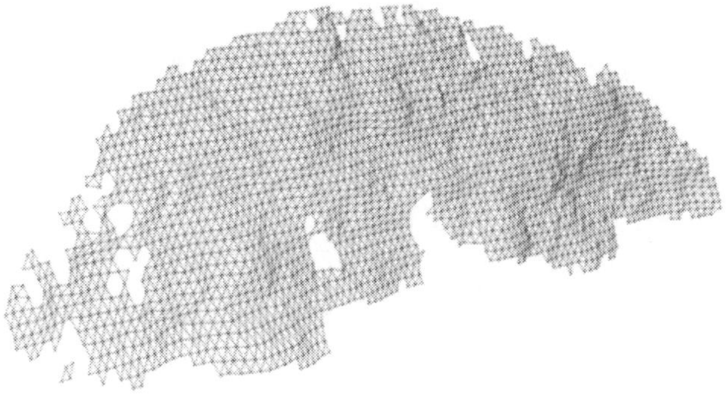

a)

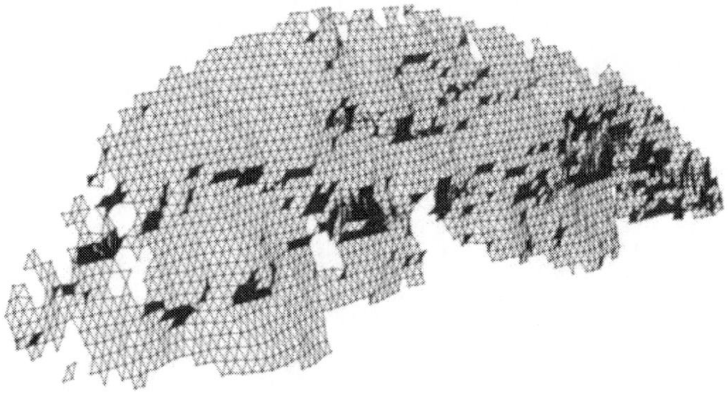

b)

Figure 6: a) Triangulated mesh from cut slope along I-64 highway in Virginia and b) selected release surfaces.

a)

b)

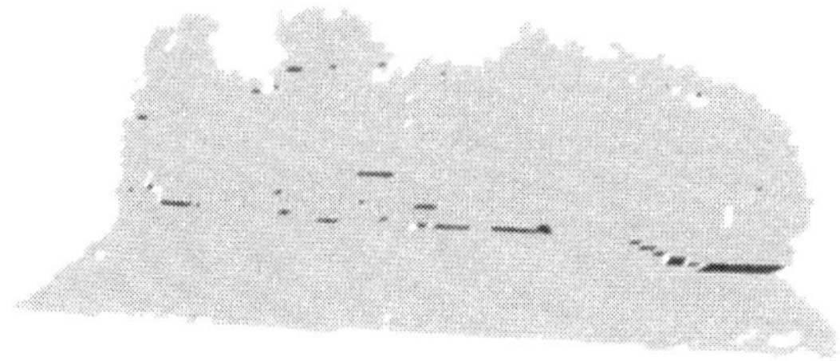

c)

Figure 7: a) Google Earth Street View of a road cut along US 33 highway in Virginia showing undercutting, b) mesh created from a LiDAR derived.

SUMMARY AND CONCLUSIONS

The increasing availability of road level remote sensing methods such as Google Street View and high resolution DSMs offer a great opportunity for support of rockfall hazard rating. Geologic parameters can be semi-quantitatively described by visual inspection. Once the orientations of release surfaces or undercutting surfaces have been estimated in Google Street View, DSMs can easily be used to and repeatedly to quantify rockfall hazard. More sites with geological variation should be tested to refine the method. In conclusion the proposed method:

1) Demonstrates the use of road level remote sensing methods for safe and efficient rockfall hazard rating.
2) Leads to automated rockfall hazard rating from DSMs derived from vehicles.
3) Allows time lapsed monitoring of slope cuts to evaluate long term performance of slope cuts.
4) Leads to digital geospatial management of slope cut information.

In addition to the methods discussed in this research, the use of Google Earth to identify rockfall generation potentials from the back slope (natural slopes above cut slopes) that is not visible from road level should be explored. The back slope is not visible from road level but Google Earth aerial photos can provide preliminary information for further LiDAR or 3D photogrammetric investigation. Finally, it should be noted that the proposed method or other road level remote sensing methods should not replace field-based rockfall hazard analysis but used as a preliminary evaluation technique.

REFERENCES:

Goodman, R. E. 1989. *Introduction to Rock Mechanics.* , New York, USA, 1989, 56 2p.

Hoek, E. and Bray, J. W., 1981. *Rock Slope Engineering*, London, England, 1981, 358 pp.

Kemeny, J., and Turner K. 2008. *Ground-based Lidar: rock slope mapping and assessment. Central Federal Lands Highway Technology Deployment Initiatives and Partnership Program*: Federal Highway Administration, Lakewood, Colorado, 2008.

Lato, M., Diederichs, M. S., Hutchinson, D. J., and Harrap. Optimization of Lidar scanning and processing for automated structural evaluation of discontinuities in rockmasses *International Journal of Rock Mechanics and Mining Sciences*, *46*(1), 2009[a], pp. 194-199.

Lato, M., Hutchinson, J., Diederichs, M., Ball, D., & Harrap. Engineering monitoring of rockfall hazards along transportation corridors: using mobile terrestrial Lidar. *Natural. Hazards Earth Systems Science* 9(3), 2009 [b], pp. 935-946.

Lato, M. J., and Vöge, M. Automated mapping of rock discontinuities in 3d Lidar and photogrammetry models. *International Journal of Rock Mechanics and Mining Sciences* 54, 2012, pp. 150-158.

Maerz, N. H., Youssef, A., Frennessey, T.W. New risk-consequence rockfall hazard rating system for Missouri highways using digital image analysis. *Environmental and Engineering Geoscience* Vol. 11 (3), 2005, pp 229-249.

Nguyen, H. T., Fernandez-Steeger, T. M., Wiatr, T., Rodrigues, D., and Azzam, R. "Use of terrestrial laser scanning for engineering geological applications on volcanic rock slopes–an example from madeira island (Portugal). *Natural Hazards Earth Systems* Science 11(3), 2011, pp. 807-817.

Pierson, L. A. *The Rockfall Hazard Rating System, Oregon Department of Transportation*. Final Report FHWA-OR-GT-92-05, 1991.

Russell, C.P., Santi, P.M., and Higgins, J.D. *Modification and statistical analysis of the Colorado rockfall hazard rating system*. Colorado Department of Transportation – DTD Applied Research and Innovation Branch, No. CDOT-2008-7, 2008.

Sturzenegger, M.,and Stead, D. Close-range terrestrial digital photogrammetry and terrestrial laser scanning for discontinuity characterization on rock cuts. *Engineering Geology, 106*(3), 2009, pp. 163-182.

Swanger, W. and Admassu, Y (in Press). Using google earth and google street view to rate rock slope hazards. *Environmental and Engineering Geoscience*, 2017.

Application of High-Speed Photogrammetry for Rock Cut Assessment

Angus MacPhail
Department of Geological Sciences and Geological Engineering
Queen's University
Kingston, ON, Canada
K7L 3N6
angus.macphail@outlook.com
Phone: 613-985-3656

Dave Gauthier, Ph.D., P.Eng., P.Geo.
BGC Engineering Inc.
Kingston, ON Canada
dgauthier@bgcengineering.ca
613-893-4920

D. Jean Hutchinson, Ph.D., P.Eng., FEIC
Department of Geological Sciences and Geological Engineering
Queen's University
Kingston, ON, Canada
K7L 3N6
hutchinj@queensu.ca
613-533-3388

Prepared for the 69th Highway Geology Symposium, September, 2018

Acknowledgements

The author(s) would like to thank the following individuals for their contributions in developing a new workflow for implementing mobile terrestrial photogrammetry
Mr. David Bonneau – Queen's University

The authors gratefully acknowledge the funding support provided by NSERC and the Ministry of Transportation Ontario.

Disclaimer

Statements and views presented in this paper are strictly those of the author(s), and do not necessarily reflect positions held by their affiliations, the Highway Geology Symposium (HGS), or others acknowledged above. The mention of trade names for commercial products does not imply the approval or endorsement by HGS.

Copyright Notice

ABSTRACT

Ontario has over 16,500 km of provincial highways, many of which are subject to rockfall hazards associated with slopes and rock cuts adjacent to the roadways. The recurrent nature of rockfall hazards necessitates frequent monitoring and a method of prioritization for remedial efforts. To date within Ontario this has been carried out by the application of the Ontario Rockfall Hazard Rating System (RHRON) which acts to quantify relative hazard at a given site, estimate the cost of remediation, and provides a basis for comparison between sites.

The RHRON System relies on on-site manual measurements, exposing employees to some measure of traffic and rockfall hazard. In addition, manual measurements of the rock face itself are limited to portions of the face which can be quickly and safely accessed. In order to minimize the risk to employees, decrease the time required, and increase spatial coverage, mobile terrestrial photogrammetry has been employed to generate 3D models using the Structure from Motion Multi-View Stereo methodology. Using this technique, photographs are gathered at highway speed along a pre-planned survey route and used as inputs to develop three-dimensional photogrammetric models of the slope, from which many of the RHRON parameters can be extracted.

While the use of terrestrial photogrammetry to complement site investigations is well established, performing surveys from a mobile platform allows users to increase spatial coverage per time period while limiting their exposure. Analysis of individual models allows the extraction of many of the RHRON parameters. By collecting multi-temporal data, users are also able to detect changes over time, identify prevalent rock fall source and accumulation zones, and better assess the effects of previously employed remedial actions. A review of the methodology and results for several rock cuts monitored in the South Eastern region of Ontario will be presented.

INTRODUCTION

Rock cuts adjacent to Ministry of Transportation of Ontario (MTO) infrastructure represent a geohazard which must be monitored to manage the ongoing safety of MTO personnel and highway users. Currently, monitoring is accomplished primarily through site inspections with the application of the Ontario Rockfall Hazard Rating System (RHRON) which rates each rock cut to allow prioritization of mitigation efforts (*1, 2*). The inspection and subsequent RHRON rating of each rock cut is either done by MTO Engineers or contracted out to consultants qualified in the use of the system. The RHRON rating level dictates the time between inspections. However, manually rating each site presents a number of hazards and limitations. Even though the surveys are conducted with traffic controls present, performing the ratings manually subjects workers to traffic hazards near high volume highways, and rock fall hazards when working beneath steep slopes. These potential hazards could be reduced by collecting data from a remote platform. In addition, remote sensing facilitates the inspection of areas on a slope which are difficult and potentially hazardous to access in person. The use of remote sensing technologies, which can collect high resolution three-dimensional (3D) data, captures the topography of the slope at a unique point in time. Collection of multi-temporal datasets permits analysis of the changes in geometry, thereby documenting changes occurring on the slope over time (*3, 4*). From this information, the locations of slope activity, the volume and size of material moving or accumulating, and the time period over which the activity is occurring can be assessed. When combined with climate data, this information provides insights into the potential causes of rockfall events.

There are a number of proven remote sensing techniques including photogrammetry, LiDAR, and GB-InSAR which can be used for rock hazard assessment. Photogrammetry has been chosen for further investigation in this study as it is capable of providing high quality results and widespread coverage at a low cost.

Digital photogrammetry involves the reconstruction of 3D models from sets of photographs covering a target feature. Using a 'Structure from Motion Multi-View Stereo' (SfM-MVS) approach, conjugate points between two or more overlapping photographs are identified and 3D coordinates are assigned to each point. While not discussed exhaustively in this paper, readers are referred to James and Robson, 2012, Westoby et al, 2012, and Smith et al., 2016 for a more detailed explanation of SfM-MVS and the generation of 3D models from photographs (*5, 6, 7*). SfM-MVS photogrammetry represents a proven, fast, and low-cost technology which can complement current practices for road cut inspections.

In addition to supporting change detection analysis, photogrammetry models can produce point clouds which can confirm or aid in the extraction of RHRON parameters, as well as providing a medium for information transfer. Once processed, a point cloud is easily transferable, allowing personnel to retroactively analyse sections of the rock face which were difficult to access in the field as well as to manipulate the model in 3D space. Experienced users can quickly and accurately identify hazardous segments, as stipulated in the RHRON workflow, from these models. Within these segments, RHRON parameters such as the crest angle, clear zone width, height, joint orientation, joint persistence, block size, and volumetric rockfall quantities can be extracted, both as single values and as a measure of their variability across the segment.

The RHRON system has been applied to generate ratings for each rock cut in Ontario, which are labelled as Class A, B or C, with Class A being the most hazardous. It is not necessary to collect or process photogrammetric data for every rock cut within the 16,500km of highway monitored by the MTO. Rather, data collection can be focused on Class A and Class B sites which may present recurring problems. In addition, retroactive construction and analysis of models can be completed if sites begin to present problems in the future.

High-speed photogrammetry involves the collection of photographs from a vehicle moving at highway speeds. The photographs are then used as inputs to the construction of 3D models. Many of the benefits described in this paper including change detection, remote slope assessment, and data extraction from point clouds are achievable through other remote sensing technologies or the application of photogrammetry from a different platform. However, mobile terrestrial photogrammetry can allow data collection to cover large areas at low cost without impeding the flow of traffic or acquiring permits for UAV and helicopter flights over transportation corridors.

OVERVIEW

This paper presents some preliminary results of incorporating mobile terrestrial photogrammetry into rock cut inspections. Testing began in the summer of 2017 with the intention of assessing the feasibility of collecting photogrammetric data at highway speeds ranging from 50km/h, on highway off ramps, to 100km/h on 400 Series highways. It is also permissible to collect data at speeds as low as 25km/h where required (4). Eleven rock cuts adjacent to highways were selected as test sites based on the range of different geometries they offered (Figure 1), and their proximity to Kingston, Ontario. Ten of the rock cuts chosen are located on a five kilometer long segment of Highway 15, labelled as Study Area 1 in Figure 1, to the Northeast of Kingston. Models from this segment were used to establish the feasibility of high speed photogrammetry and vary from 2-10m in height and 10-120m long. Only four of the outcrops in this area are rated using the RHRON system as the remainder are too small to pose a hazard to users of the roadway. Study Area 2 was chosen as it represents a more complex site, and is rated as a Class A site. The outcrop reaches 12 meters tall and is 100m meters long. It also displays significant curvature as the site is located on an off ramp connecting Highway 137 to the 1000 Islands Parkway. Surveys were completed on a bi-weekly basis throughout June and July 2017 to assess whether photographs captured at highway speeds were able to produce high quality models capable of being used in a detailed assessment of a slope. After establishing feasibility, subsequent analysis refocused on optimizing camera settings, lens type, and camera orientation for future surveys.

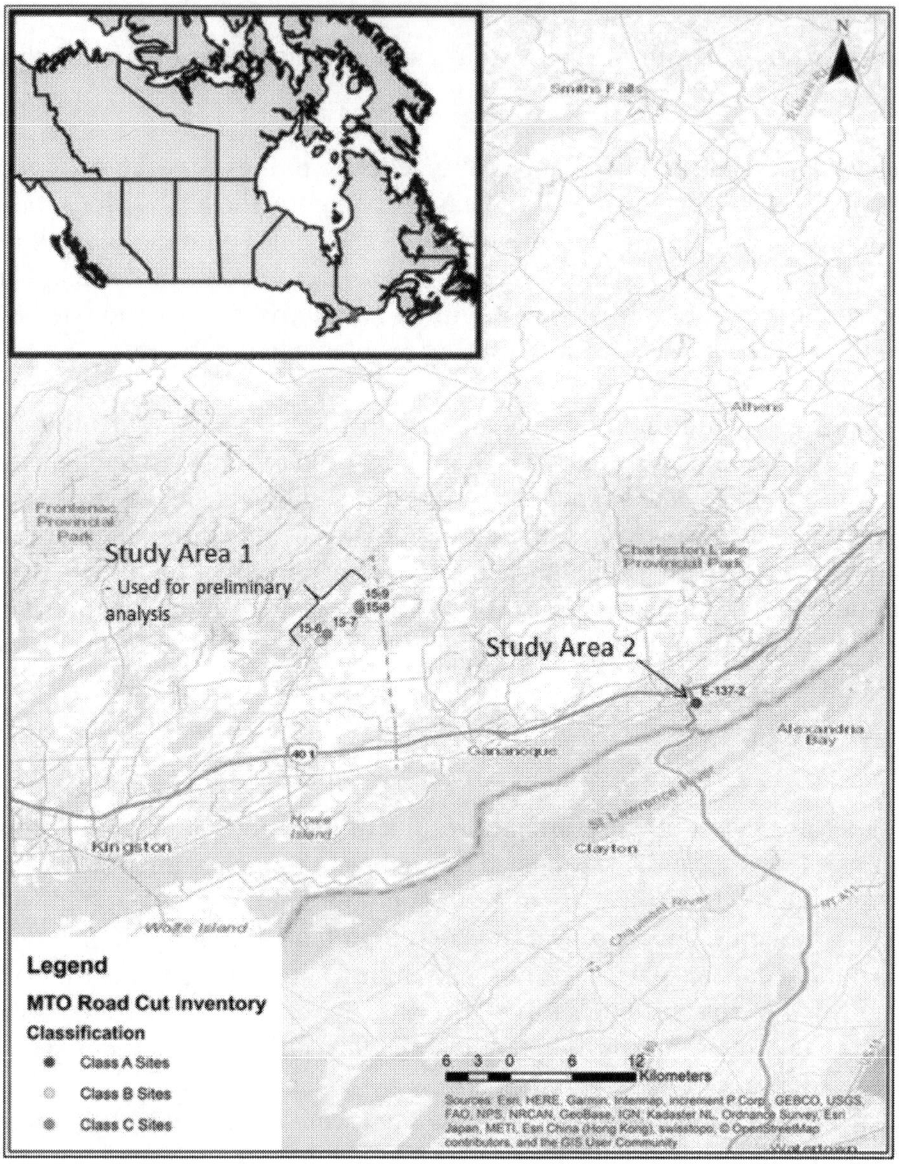

Figure 1: Project location with rock cut hazard ratings based on a 2015 RHRON survey

EQUIPMENT

One of the primary advantages of photogrammetry over other remote sensing technologies is the portability, low cost and low quantity of required equipment. The equipment used in this study consisted of a Nikon D5300 camera (24 MegaPixels), a Nikon AFS 35mm DX rectangular frame lens (f 1.8), a Nikon DX 10.5mm fisheye lens (f 2.8), and a Solmeta N3 global positioning system (GPS) which was mounted on the camera. It is also important to note that the Nikon D5300 has a DX (crop) sensor effectively magnifying the focal length of each lens by a factor of 1.5. Due to constraints discussed later in this paper, photographs were captured in 'fine JPEG' format. For the purposes of change detection, LiDAR scans were captured using a FARO Focus 3D X130 laser scanner (*8*).

METHODOLOGY

Data capture is intended to be performed by two people with one designated driver and one photographer. A survey route is first designed to maximize the number of sites which can be visited over a certain time period. When traveling at speeds of greater than 80km/h it is often necessary for the vehicle to collect data over two passes of the rock face which must be taken into account during survey planning. Alternatively, a single pass may be adequate so long as two cameras are used to acquire data. The exact number of photographs required is dependent on the speed of the vehicle, the distance from the rock cut to the camera, the field of view of the lens and the desired quality of the resultant model. A typical data acquisition campaign for the slopes shown in this paper would attempt to capture a minimum of one photograph for every 2-4m travelled along a rock face, although more photographs will typically result in a more complete model. The weather must also be monitored prior to the specified day of the survey to ensure adequate natural lighting during the data collection work. Ideal weather for photogrammetry is bright but cast-over days, ideally producing uniform shadowing on rock faces. Time of day and orientation of the slope should also be considered where possible, to provide maximum natural lighting on the slope.

During the approach to study sites, the photographer manually optimizes the camera settings based on the geometry and lighting of the upcoming slope. This involves manually altering the three primary camera settings, shutter speed, aperture, and ISO, to ensure the resulting photographs are well exposed, sharp and in focus. This is necessary as the lighting conditions on a rock face can change over short distances depending on the relationship between the aspect of the slope and the position of the sun. Vegetation coverage, especially in the form of trees surrounding the slope, can also have significant impact on the lighting conditions on a rock face. Darker and more uniform colouring in photographs can negatively impact the ability of the software to identify conjugate points between photographs resulting in lower quality models. The most important camera setting to consider with regard to high speed photogrammetry is the shutter speed. When moving at highway speeds it is crucial to ensure the shutter speed is adequate to produce sharp photographs minimally affected by motion blur. Typically a minimum shutter speed of 1/4000 was found to be necessary when travelling at speeds of over 60km/h.

As the vehicle comes adjacent to the beginning of a target feature, data collection is initiated with the camera oriented orthogonal to the overall strike of the slope. The photographer should attempt to maximize the number of photographs gathered as the vehicle drives past a site maintaining a minimum horizontal overlap between photographs of 50% (*9*).

PROCESSING WORKFLOW

The construction of photogrammetric models throughout this study was done using the commercial SfM-MVS software Agisoft Photoscan Pro, Version 1.2.4 (*9*). Figure 2 describes a generalized workflow for photogrammetric model construction (modified from *10*) which mirrors the processing steps in Photoscan.

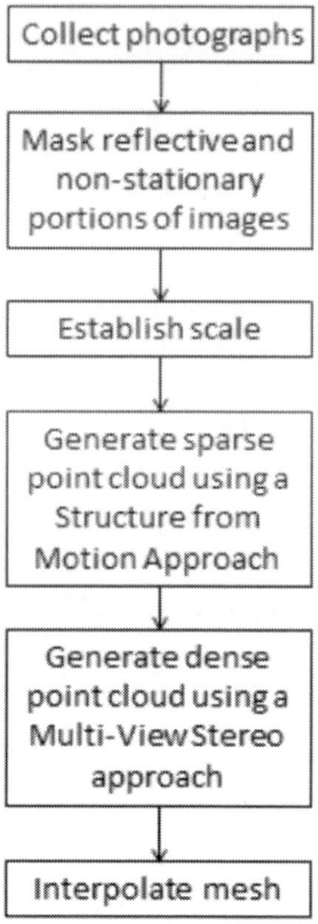

Figure 2: Processing workflow within Agisoft Photoscan Pro, Version 1.2.4

While not necessary for model generation, model scaling was done using a geotagger attached to the camera which adds GPS co-ordinates to the EXIF data for each photograph. This enables the calculation of accurate distance and volume measurements in later processing steps. Other potential scaling options include LiDAR point clouds, georeferenced survey targets fixed to the slope, and locating objects of known dimensions on the slope itself prior to collecting data. Previous work has shown that the addition of geotagger data generally scales models to within 3% of true scale although there are occasionally problems resolving the orientation of the models in 3D space (*11*). These orientation issues can be exacerbated when moving at highway speeds as the precision of the geotagger decreases with the rapid movement of the vehicle.

Models were typically taken to the 'dense cloud' stage of processing where multi-temporal comparisons could be made using a vector based algorithm. In some cases, a mesh was also interpolated through the dense point cloud and used as the basis for model comparison.

Change Detection

Two sites were chosen to test the achievable limit of detection for models produced using mobile terrestrial photogrammetry. As all of the sites investigated in this study are well maintained by the MTO, a number of rocks were actively placed or moved on safely accessible areas of the rock face to determine the limits of detection for change. The overall procedure followed was the same for both study sites and involved collecting baseline LiDAR scans, collecting a 'before change' photogrammetry model, modifying the slope, collecting an 'after change' photogrammetry model, and comparing the results. The LiDAR scans were assumed to be representative of the true shape and orientation of each site and were used to scale and validate the photogrammetry models.

Processing of the photographs was done in Agisoft Photoscan V 1.2.4 (*9*), following the procedure described above, to produce a dense cloud end product. The dense point clouds were then aligned and scaled using the open source software package 'CloudCompare' (*12*). Change detection was primarily performed in CloudCompare using the vector based M3C2 algorithm (*13*). However, in some cases a mesh for the before and after change models was interpolated using the program Polyworks (*14*). In these cases the two resultant meshes were compared using a shortest distance approach within Polyworks.

RESULTS AND DISCUSSION

Models

Figure 3 below a typical series of end products from a high speed photogrammetry survey processed in Agisoft Photoscan. Once produced, dense clouds are typically used during change detection due to their compatibility with the M3C2 algorithm. However, a mesh can also be interpolated based on the dense cloud points which can reveal valuable information relating to the structure of the rock mass. Mesh generation is done using built-in methods in Photoscan which employs a Poisson Surface Reconstruction algorithm to produce a solid polygonal surface (*9*).

In addition to enabling geometric and volumetric calculations of sections of the slope, capturing photographs allows one to reconstruct a high quality scene representative of a site at a given point in time. This can be extremely useful model, allowing personnel who were unable to visit a site in the field to gain a more complete understanding of the geometry, failure mechanism and active processes in the area.

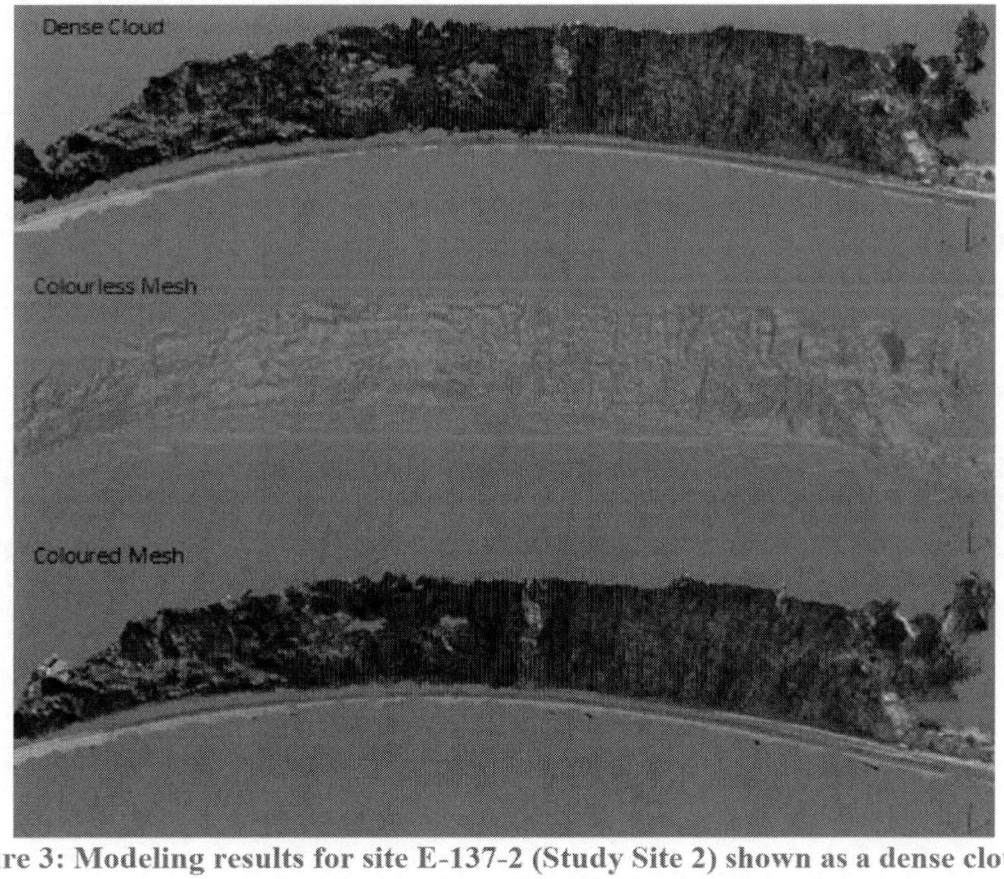

Figure 3: Modeling results for site E-137-2 (Study Site 2) shown as a dense cloud, a colourless mesh, and a coloured mesh.

Lens Type

One of the benefits of using Photoscan Pro is that the software is equipped to support different lens types, including a calibration for both fisheye and rectangular frame lenses. This allowed a comparison of the two lens types used in this study, a 35mm rectangular frame and a 10.5mm fisheye lens, showing that the 10.5mm fisheye was the optimal lens to use in mobile photogrammetry. The primary benefit of using a fisheye lens was to allow a greater field of view to be captured in each photograph. While the focal lengths of the two lenses are different, we found models produced using a fisheye lens tended to be closer to the true shape of the rock cut when compared to LiDAR scans. Figure 5 shows a comparison of models produced using both a fisheye and rectangular frame lens to a LiDAR model of the site. It was found that the shape of models produced using the fisheye lens were closer representations of the true shape of the site as indicated by LiDAR models. The radial field of view using the fisheye lens also allowed photographs to capture debris in the ditch as well as oblique joint faces which would be difficult to capture using a rectangular frame lens. An example of the field of view of both lenses from the nearside highway lane is shown in Figure 4 below.

Figure 4: Field of view for a 10.5mm fisheye lens (Left) and a 35mm rectangular frame lens (right).

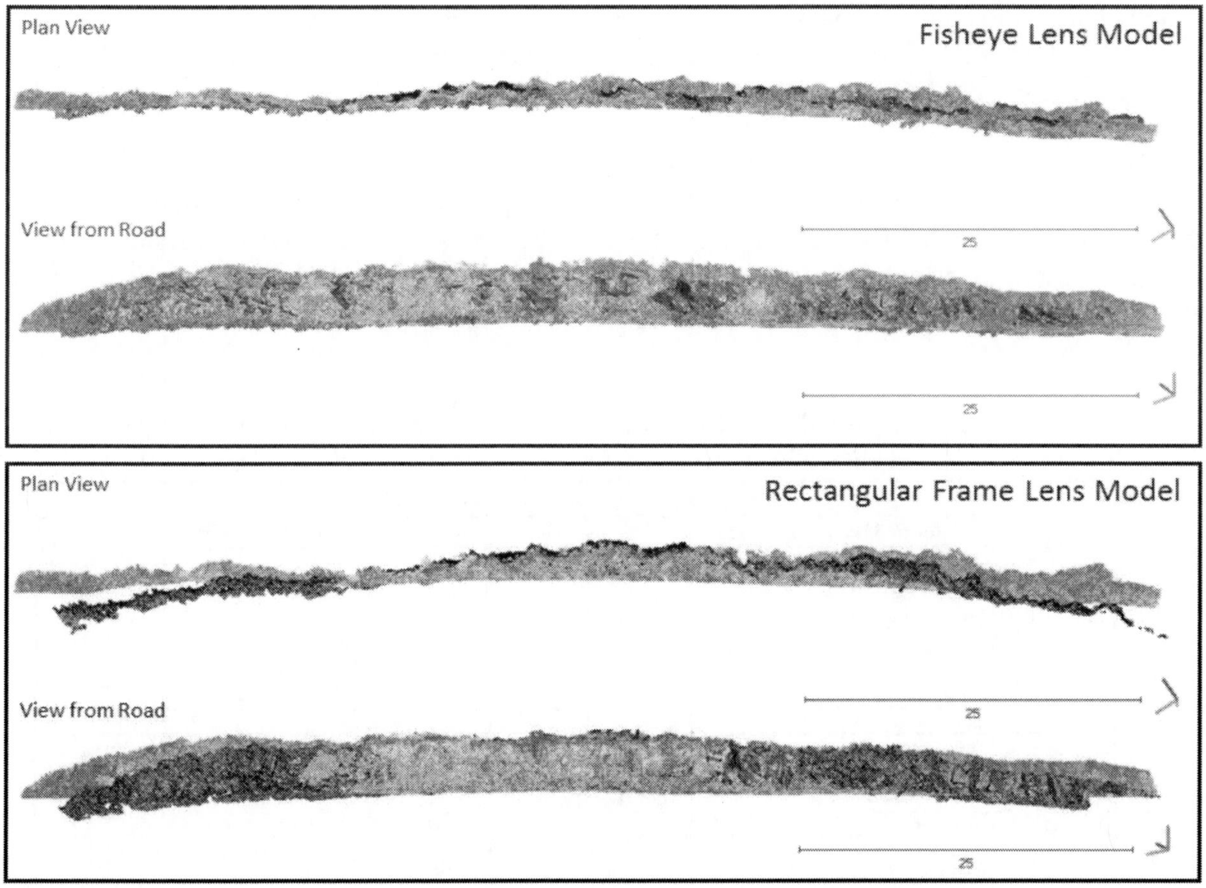

Figure 5: (Top) Photogrammetry model produced using a 10.5mm fisheye lens shown in true colour. (Bottom) Photogrammetry model produced using a 35mm rectangular frame lens shown in true colour. An aligned LiDAR model is shown in both images in orange and yellow.

RHRON Input Extraction

The RHRON system is based on a Rockfall Hazard Rating System (RHRS) developed by the Oregon Department of Transportation which was modified for use in Ontario (*1, 15*). It provides a mechanism for rating and comparing the hazard of rock cuts adjacent to highways throughout Ontario. The system has two separate levels of application depending on the degree of hazard posed by the rock cut. The lowest level is a preliminary assessment which results in a 'Basic RHRON score'. A Basic RHRON assessment consists of evaluating four factors which are rated from 0 (good) to 9 (bad) and averaged to calculate the Basic RHRON score. The four factors include:

F1 – Magnitude – The quantity of potentially unstable rock on a slope
F2 – Instability – An estimate of the temporal probability and frequency of rockfall
F3 – Reach – An estimate of the probability a falling rock will reach the highway
F4 – Consequences – An estimate of the potential consequences of rockfall

The Basic RHRON score is then combined with the crest angle, defined as the upward angle between the edge of pavement and the highest unstable rock, to classify the slope as Class A, B, or C. Reassessment of all sites is conducted every five years. Sites historically labelled as Class C are assessed from a vehicle while Class A and B sites require inspectors to approach the slope on foot (*1*).

Given their higher hazard rating, Class A sites are then evaluated more thoroughly by the application of a Detailed RHRON rating scheme. The first step in a detailed evaluation is to separate the length of the slope into segments which display similar failure mechanisms and degree of hazard. Twenty different parameters are then analysed to determine the detailed RHRON score (*1*).

Table 1: Detailed RHRON sections and parameters including an indication of the possibility of extraction from photogrammetric models

Detailed RHRON Assesment Factors					
Rockfall History and Quantities			Face Geometry		
RHRON Factor	Description	Can Be Extracted from Models?	RHRON Factor	Description	Can Be Extracted from Models?
P1	History of Rock Falls	With multitemporal modeling	P12	Height	yes
P2	Largest Expected Fall or slide volume	Yes	P13	Crest Angle	yes
P3	Total quantity of potential falls or slides	Yes	P14	Clear Zone Width	yes
P4	Face Irregularity	Yes			
P5	Face Looseness	?			
Rock Mass Properties			Traffic Data		
RHRON Factor	Description	Can Be Extracted from Models?	RHRON Factor	Description	Can Be Extracted from Models?
P6	Joint Orientation/Persistance	Yes	P15	Ditch Effectiveness	?
P7	Intact Strength	no	P16	Overspill	?
P8	Shear Strength	no	P17	Average Vehicle Risk	no
P9	Block Size	yes	P18	Decision Sight Distance	no
P10	Slake-Durability Index	no	P19	Available Paved Width	no
P11	Water Table	?	P20	Remediation Cost	no

The input factors for a detailed RHRON analysis are listed in Table 1. In addition to allowing retroactive visual inspection of sites and change detection, model generation can allow users to accurately extract a number of the basic and detailed RHRON parameters (Table 1). Capturing regular, sequential models can increase the number of RHRON factors users can assess to include the rockfall history, ditch effectiveness, overspill, and seasonal water table height. Extracting detailed RHRON parameters from models could decrease the time required for personnel on site and increase the accuracy of measurements in the field.

Due to time constraints, personnel availability, site accessibility, and the length of road which must be monitored, many of the volumetric and geometric parameters are currently visually estimated in the RHRON system. While this has proved effective, the ability to extract these parameters from point clouds when considering remediation options would be useful. The use of photogrammetry to monitor sections of highway could provide a platform to accurately measure many of the parameters which are currently difficult to directly or quickly estimate on site. Specifically, photogrammetry models permit users to extract the parameters contained within the 'Rockfall History and Quantities' and 'Face Geometry' sections of a detailed RHRON assessment.

Figure 6 below shows two examples of point clouds being used to extract data relevant to an engineering evaluation of a rock face. The upper image shows the estimation of the crest angle, defined as the upward angle from the edge of the pavement to the uppermost point of instability. Geometric factors such as the angle of the slope, ditch depth, and clear zone width are important in RHRON ratings and can be swiftly extracted from point cloud data. The lower image in Figure 6 shows an example of joint set extraction and stereonet construction from point cloud data. In some rock cuts there are site accessibility constraints which can add unnecessary hazard to manual measurements being taken on the rock face. Point clouds can provide a mechanism to safely and accurately extract joint set orientation data and analyse potential failure modes.

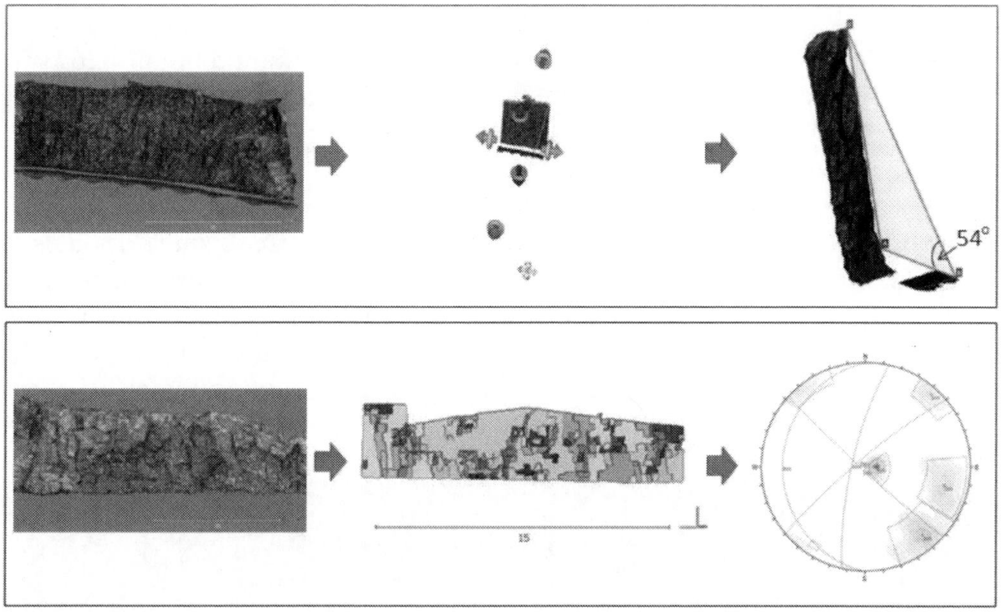

Figure 6: (Top) Cross section of a point cloud used to estimate the Crest Angle. (Bottom) Extraction of joint orientation data from a point cloud to a stereonet.

Change Detection

Analysis of time sequential models permits the detection of change, whether due to loss of material from the slope face or gain of material on the slope face or at the toe and beyond. It is important to detect as small a change as possible, while removing the effects of noise from the analysis. The limit of detection (LOD) represents the threshold separating noise and true change. The limit depends on the quality of the models and the quality of the alignment. When considering photogrammetry models there are a range of variables affecting model quality including the distance from the camera to the target feature, the camera's sensor size and pixel count, the focal length of the lens, and environmental factors such as the lighting on the rock face (*16*).

In order to examine the limit of detection, 'before change' and 'after change' models of two slopes were constructed using photographs taken minutes apart with the same camera and lens, and with minor changes introduced by moving blocks on the slope. The two resulting models should therefore be identical with the exception of the small changes made to the slope between the two surveys. As the two models should be almost identical, any changes preventing a perfect alignment can be interpreted as noise produced during the modeling process or registration errors produced while running the iterative closest point (ICP) algorithm used to align the two point clouds. The practical LOD referenced for Site 1 and Site 2 below is therefore quantified as two times the Root Mean Squared (RMS) of the alignment between the 'before change' cloud and the 'after change' cloud (*11*). One of the limitations of using this method to assess the LOD is that the RMS error represents an average of the registration error across the entire rock face but the quality of the alignment between the two models may be spatially variable.

Site 1

Site 1 is a rock cut approximately 3m tall and 85m in length (Figure 6) adjacent to a highway with a posted 80km/h speed limit (Latitude: 44.448N Longitude: -76.267W). Figure 7 below shows the 'before change' model with the sizes and location of the rock moved between surveys.

For this survey, the rocks were initially part of the slope during the 'before change' models and were removed before capturing the photos for the 'after change' models. As such, the change detection below primarily reveals a loss of material from these locations. Change detection at this site was performed using the program Polyworks which compares two meshed models using a shortest distance approach (*14*). Figure 8 below shows the results of the change detection at Site 1 where the limit of detection was found to be approximately 10cm.

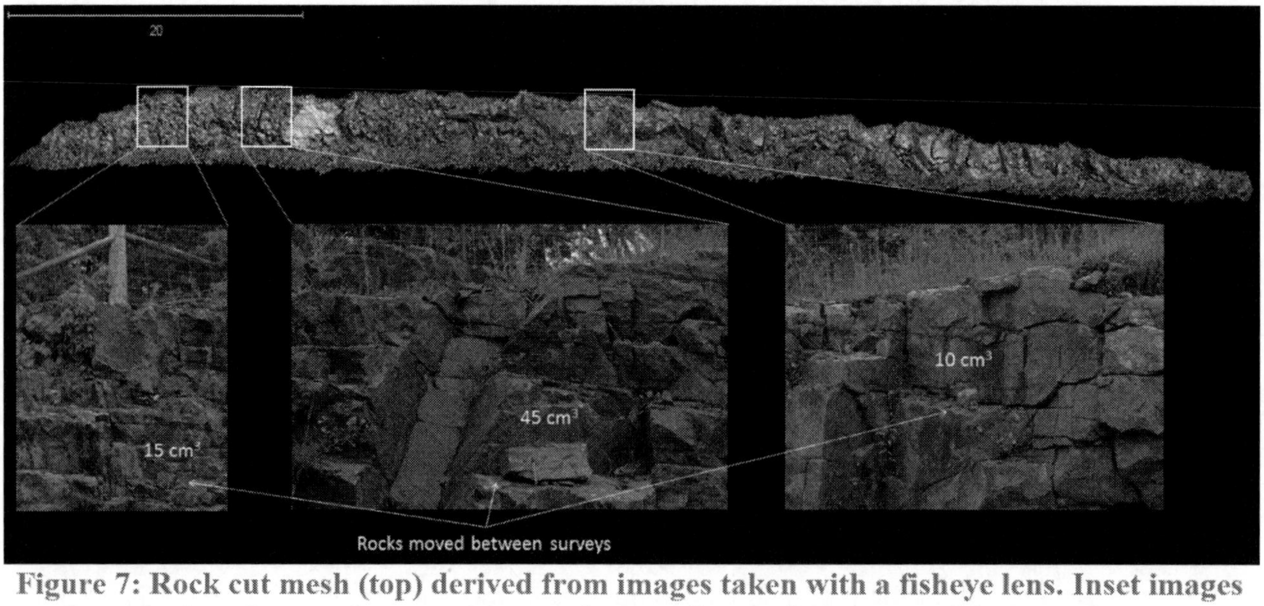

Figure 7: Rock cut mesh (top) derived from images taken with a fisheye lens. Inset images show the locations and approximate volumes of rocks being moved between surveys.

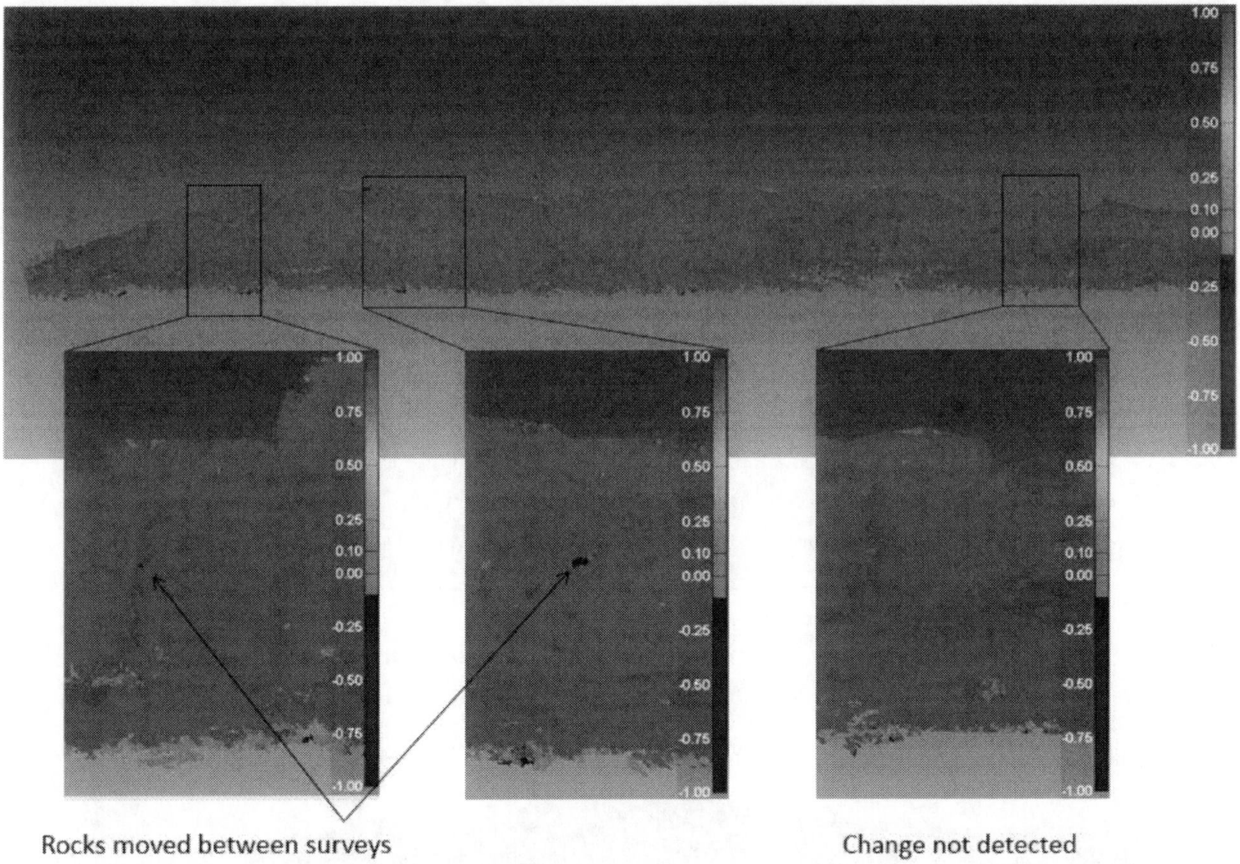

Figure 8: Change detection map highlighting the locations of rocks moved between surveys. Scale bar units are in meters; limit of detection +/- 10cm as shown by grey colour.

Site 2

The outcrop at Site 2 is approximately 15m tall and 110m in length and is located on a highway off-ramp with a speed limit of 50km/h (Latitude: 44.369N Longitude: -75.977W). All of the photos taken during this survey were taken using a Nikon 10.5mm DX fisheye lens which was able to capture the full height of the rock cut in every photo. After the 'before change' survey the rock face was approached and a number of rocks were repositioned. In addition, sections of accumulated talus were removed from areas near the base of the rock slope. All of the induced changes were concentrated on the steeply dipping face outlined in Figure 8 below.

At a speed of 50km/h the limit of detection was found to be 5cm. Differences between models at this site were compared using the program CloudCompare (*12)* and calculated using the vector-based M3C2 algorithm (*13)*. Each of the changes induced on site were identified within the 5cm LOD in the resultant change map (Figure 10). Error caused by model alignment is also visible as loss (blue) in the upper right corner of Figure 10. While alignment errors are not always easy to detect, they are typically located in areas with low point densities or areas where data is not available due to occlusions.

Figure 9: Photogrammetry model (top) and LiDAR model (bottom). The section of the face where modifications were made between surveys is outlined by the black box.

Figure 10: Change detection map of E-137-2 (Top) with the bottom image showing the locations on the slope where modifications were made between surveys. The degree and distribution of change is indicated in the histogram to the right of the colour scale. Units on the scale bar are reported in meters with the 5cm LOD shown in grey. The blocks moved on the slope range in volume from 10-45cm^3.

LIMITATIONS

Despite the success found in being able to generate models at highway speeds, a number of key limitations were also discovered over the course of the study.

Decreasing Image Capture Rate Over Time

Cameras are rated for a certain value of 'frames per second' (fps) which is inherent to a certain camera model. A typical consumer digital single lens reflex (DSLR) camera is capable of capturing approximately 5 fps which is adequate for high speed data collection. However, cameras use a 'buffer space' to handle images. Each time a camera takes a photograph, the data is temporarily stored in the buffer before being stored permanently on the SD card. When continuously capturing images, the buffer can fill, lowering the fps of the camera to below its reported value. For rock cuts greater than 100m in length, the size of the buffer begins to become important as it can prevent users from capturing data with the overlap required by processing software. For the Nikon D5300 used in this study we found that the fps decreased from 5fps to 3.3fps when continuously capturing photographs for over 10 seconds. The buffer is also affected by the file size of the photograph being stored. While it would generally be ideal to capture minimally processed RAW images to prevent data loss, they represent a larger file size. The Nikon D5300 used for this study was only able to capture 1 RAW photograph per second when continuously capturing images for more than 10 seconds. Therefore, JPEG images were collected in order to allow adequate overlap between photographs.

Rock Cut Geometry

Generating models using remote sensing techniques typically requires moving the data acquisition unit to a number of different vantage points in order to limit line-of-sight occlusions in the final integrated model. One of the primary drawbacks of capturing photographs from a vehicle is the limited line of sight available from the highway. As all of the photographs are collected from a uniform height, areas obscured from the perspective of the highway will be occluded in the final model.

Vegetation

Vegetated surfaces present a number of challenges from a modeling perspective. These challenges are not specific to high speed photogrammetry but are important considerations when assessing model quality and limitations. The primary impact of vegetation in mobile surveys relates to the inability of the processing software to identify conjugate points in vegetated areas. As vegetation can move between photographs and is typically of uniform colour, the point density in vegetated sections of rock faces is often orders of magnitude lower than point density on solid rock faces devoid of vegetation. When using data collected at highway speeds, vegetated sections of the slope are often occluded in the final model because the software cannot identify points in these regions. This problem is present when using other photogrammetry platforms (e.g. Unmanned Aerial Vehicle (UAV), helicopter, handheld) but is exacerbated by the speed of the vehicle. The second problem posed by vegetation relates to its impact on change detection. As vegetation grows and diminishes seasonally, the comparison of multi-temporal surveys can show false, spurious change resulting from vegetation growth or loss. This can be resolved through investigation of the two photosets, and removal of data points collected from vegetation, but adds time to the processing workflow.

CONCLUSIONS

This study tested the feasibility and potential applications of gathering and processing photogrammetric data from a mobile terrestrial platform. High speed data collection has proved capable of producing high quality models which can be used as part of an engineering assessment of rock slopes. The platform, methodology, and required training to enable data capture are simplistic when compared to other remote sensing techniques. In addition, high speed photogrammetry can be implemented at low cost, requiring only a consumer grade camera and lens in addition to processing software. A number of the key points discussed in this paper relating to the deployment of high-speed photogrammetry are summarized below:

- Data collection requires a minimum of two people, a camera, and a vehicle.
- User experience with photography is necessary using the current methodology as camera parameters such as the shutter speed, aperture, and ISO must be manually altered to account for variable lighting conditions on rock faces over the course of a survey.
- Fisheye lenses are more typically more effective than rectangular frame lenses due to their larger, and more radial, field of view.
- When travelling over 80km/h, photographs must often be collected over two passes to ensure adequate coverage. A minimum of one photograph every four meters is recommended during data collection.
- High-speed photogrammetry can be effectively employed on sites where hazards are visible from the line of sight available from the passenger window of a car.
- Collecting multi-temporal data can allow the construction of change maps revealing and quantifying changes to the slope over time.
- Models can be used to extract data pertinent to an RHRON hazard assessment without exposing workers to unnecessary hazards.

Many of the benefits of generating 3D models of sites are achievable through the application of other remote sensing technologies or by employing photogrammetry from a different platform. However, high-speed photogrammetry offers an easily deployable option to quickly collect data capable of producing high quality results. While not without its limitations, collecting data at highways speeds has the potential to allow systemic coverage of large areas at a low cost. Even if the data is not processed at the time of collection, photographs can be stored and used as model inputs in the future to provide multi-temporal data describing sites of interest.

Work on this project is currently ongoing at Queen's University, with the support of the MTO, to further define the advantages and limitations of high speed terrestrial photogrammetry.

REFERENCES

1) Franklin, J., Wood D, Senior S, Blair, J. *RHRON: Ontario Rockfall Hazard Rating System - Field Procedures Manual*. Downsview, ON : Ministry of Transportation, Materials Engineering and Research Office Report. MERO-043 , ISBN 978-1-4435-9148-5, 2012

2) Franklin J, Wood D, Senior S, Blair, J. *RHRON: Ontario Rockfall Hazard Rating System: Field Procedures Manual*. Technical Report, 2012

3) Abellán, A., Oppikofer, T., Jaboyedoff, M., Rosser, N.J., Lim, M., & Lato, M.J. Terrestrial laser scanning of rock slope instabilities. *Earth surface processes and landforms*. 39.1, 2014, pp. 80-97.

4) Kromer, R. A., Dutchinson, D. J., Lato, M. J., Gauthier, D., & Edwards, T. Identifying rock slope failure precursors using LiDAR for transportation corridor hazard management. *Engineering Geology* 195, 2015, pp. 93-103.

5) James, M. R., & Robson, S. Straightforward reconstruction of 3D surfaces and topography with a camera: Accuracy and geoscience application. *Journal of Geophysical Research: Earth Surface* 117.F3, 2012.

6) Westoby, M. J., Brasington, J., Glasser, N. F., Hambrey, M. J., & Reynolds, J. M. Structure-from-Motion' photogrammetry: A low-cost, effective tool for geoscience applications. *Geomorphology* 179, 2012, pp. 300-314.

7) Smith, M. W., J. L. Carrivick, and D. J. Quincey. Structure from motion photogrammetry in physical geography. *Progress in Physical Geography*40.2, 2016, pp. 247-275.

8) *FARO Laser Scanner Focus X 130 User Manual*. February, 2015. https://faro.app.box.com/s/r45cyjqengcts8vnh5kawemgsvdfxt8. Accessed June 4, 2018.

9) *Agisoft PhotoScan User Manual, Professional Edition V1.3.0.*, 2017. St. Petersburg, Russia. (http://www.agisoft.com/pdf/photoscan-pro_1_3_en.pdf).

10) Bemis, Sean P., Micklethwaite, S., Turner, D., James, M. R., Akciz, S., Thiele, S. T., & Bangash, H. A. Ground-based and UAV-based photogrammetry: A multi-scale, high-resolution mapping tool for structural geology and paleoseismology. *Journal of Structural Geology* 69, 2014, pp. 163-178.

11) Gauthier, D., Hutchinson, J., Lato, M., Edwards, T., Bunce, C., & Wood, D. On the precision, accuracy, and utility of oblique aerial photogrammetry (OAP) for rock slope monitoring and assessment. *68th Canadian Geotechnical Conference, GeoQuebec*. 2015.

12) *Cloud Compare, V2.9*. http://www.cloudcompare.org/. 2015. Accessed June 2, 2018.

13) Lague, Dimitri, Nicolas Brodu, & Jérôme Leroux. Accurate 3D comparison of complex topography with terrestrial laser scanner: Application to the Rangitikei canyon (NZ). *ISPRS journal of photogrammetry and remote sensing* 82, 2013, pp. 10-26.

14) *InnovMetric*. 2016. PolyWorks V16. Quebec City, Canada.

15) Pierson, L.A., Davis, S.A. & Vickle, R.P.G., 1990. *Rockfall Hazard Rating System Implementation Manual*. USA 1990 average unit bid prices, northwest USA.

16) Tannant, Dwayne D. Review of photogrammetry-based techniques for characterization and hazard assessment of rock faces. *International Journal of Georesources and Environment-IJGE* 1.2, 2015, pp. 76-87.

3-D Geo-View of Subsurface Conditions for Rapid Roadway Stability Assessment

Joel Daniel
Olson Engineering
7529 Standish Place, Suite 102
Rockville, MD 20855
(240) 477-7738
joel.daniel@olsonengineering.com

Phil Sirles
Olson Engineering
12401 W. 49th Ave.
Wheat Ridge, CO 80033
(303) 423-1212
phil.sirles@olsonengineering.com

Dustin Robbins
Federal Highway Administration
Central Federal Lands
12300 West Dakota Avenue
Lakewood, CO
(720) 963-3500
dustin.robbins@dot.gov

Khamis Haramy
Federal Highway Administration
Central Federal Lands
12300 West Dakota Avenue
Lakewood, CO
(720) 963-3500
khamis.haramy@dot.gov

Prepared for the 69th Highway Geology Symposium, September, 2018

Disclaimer

Copyright

ABSTRACT

Following a June, 2017 5.5-magnitude earthquake that shook the Big Island of Hawaii, voids and several cracks were observed at various locations along Chain of Craters Road, Hawaiian Volcano National Park, Big Island, Hawaii. This road is the primary route to the Kilauea volcanic crater rim, which is a major tourist attraction. Accordingly, roadway stability is a key component to assuring public safety within the Park.

Rapid roadway stability assessment within the volcanic geology of the Hawaiian Islands is highly dependent on accurate 3D imaging of subsurface features such as lava tubes and large cracks migrating to the roadway surface. The safety of Park visitors is highly dependent on imaging accuracy since seismic activities or other ground altering events may lead to sudden catastrophic failures. Accurate 3D subsurface views are therefore required to facilitate the best engineering analysis, operations procedures and mitigation methods.

This presentation presents a case study where both 3D ground penetrating radar and 3D seismic tomography were utilized in concurrence to image the subsurface at two impacted sites in the park. The two sites include one where two cracks and a slight depression formed within the pavement, and the second site is where cracks occurred within the pavement and one 4-foot diameter hole had developed at the ground surface along the road shoulder. The 3D ground penetrating radar equipment utilized a series of eight antennas running concurrently to develop a depth profile with high resolution. The 3D seismic tomography data was processed using a newly developed tomography code that provides high resolution images to large depths. Data collected with both methods was used to develop remediation measures to assure Park visitor's safety.

INTRODUCTION

Chain of Craters Road is an approximately 19-mile paved road that passes through the East Rift Zone and coastal area of the Hawaii Volcanoes National Park, on the big island of Hawaii (Figure 1). The original road was built in 1928, and connected Crater Rim Drive around Kilauea crater to Makaophui Crater. During the 1960s, the road was lengthened to reach the small town of Kalapana on the coast. The Kilauea volcano has been erupting continuously since 1983, and since 1986 lava flows have repeatedly covered portions of the road, most recently in 2003 at the coastal end of the road beyond the Hōlei Sea Arch (*1*).

Figure 1 – Map of Chain of Craters Road Area.

Along with having several active volcanoes, the island of Hawaii is no stranger to earthquake activity as magma constantly migrates beneath the land surface. While the magnitudes of such earthquakes typically are not significant, one such earthquake on June 8, 2017 measured 5.5 on the Richter scale. This was the largest earthquake in magnitude since a 1975 quake that measured 7.2 on the Richter scale. One result of the 2017 quake was the development of a hole along the shoulder of the Chain of Craters Road near a natural feature known as Devil's Throat. This area is located within the East Rift Zone (ERZ).

The East Rift Zone (ERZ) is one of two rift zones associated with the Kilauea volcano (Figure 2). The rift zones are locations where the volcano is breaking apart, and thus can be preferential locations for "non-caldera" magma extrusions onto the ground surface.

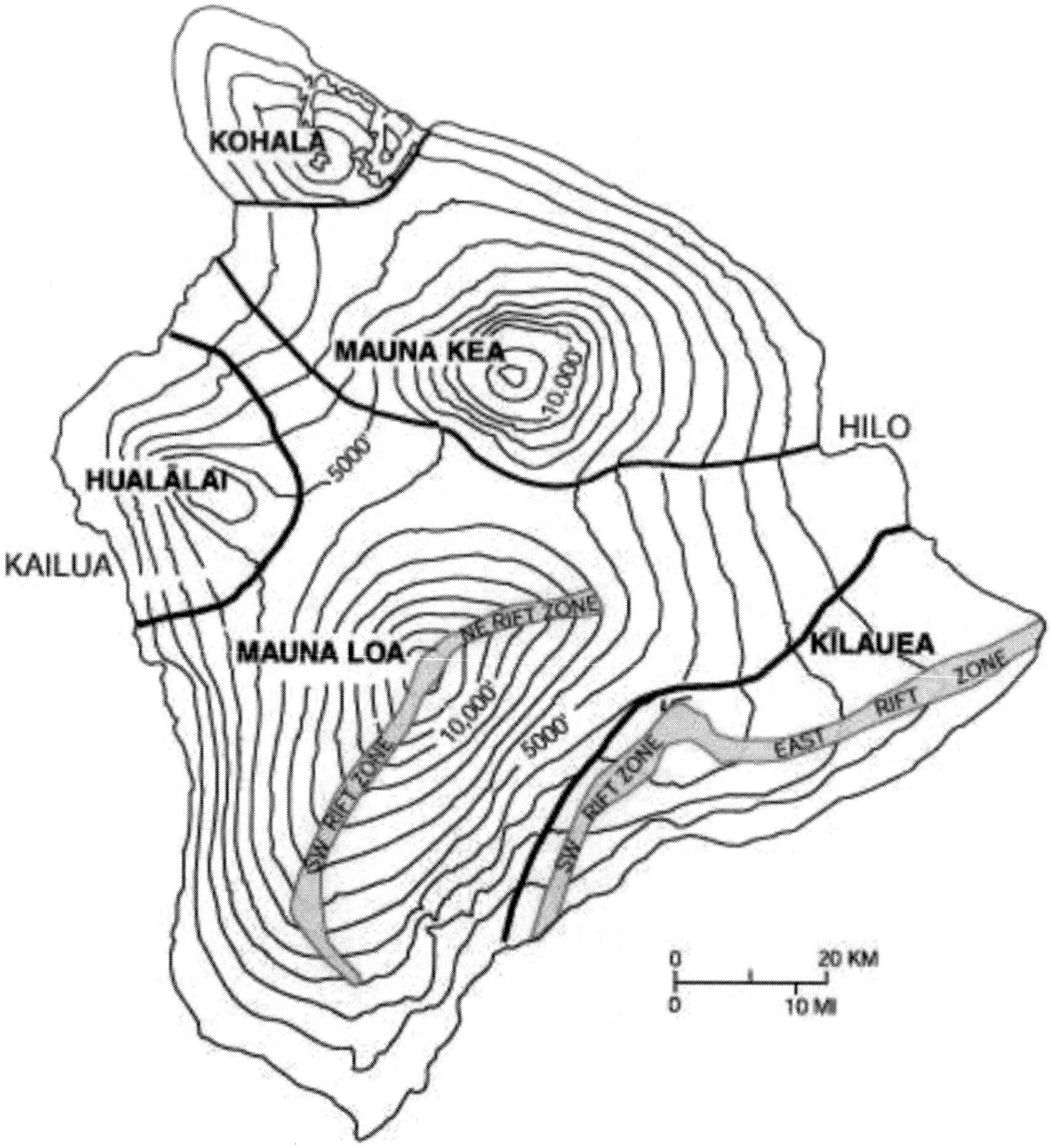

Figure 2 – Map of Rift Zones, Big Island of Hawaii.

A common surficial feature along the rift zones are "pit craters", of which Devil's Throat is one, which are the result of collapses of surficial material into deeper openings that form over time due to the rifting process (*2*). These craters typically range in size from 8 meters to 1,140 meters (26 feet to 3,740 feet), with depths of 6 meters to 186 meters (20 feet to 610 feet). The craters are typically located astride a single rift zone fracture or between a pair of rift fractures, and the fractures may be prominent in the crater walls. The craters are not known to have formed due to eruptive events, although lava has subsequently filled some of the craters. Devil's Throat is the best-exposed pit crater along the East Rift Zone (*2*), and it is situated at the junction of two northeast-southwest oriented fractures (Figure 3). Also nearby are two northeast-southwest trending faults. The geophysical surveys performed at Site 1 and Site 2, shown on Figure 3, are the subject of this paper.

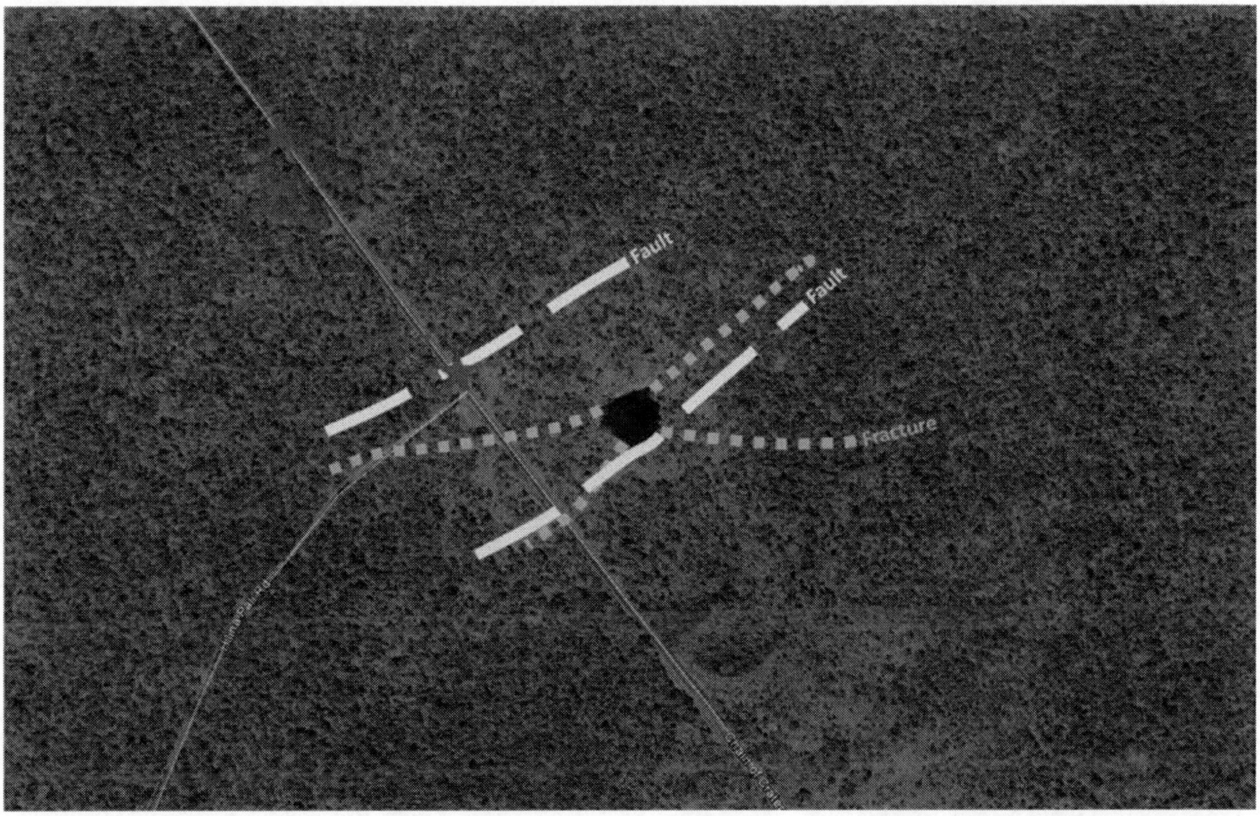

Figure 3 – Devil's Throat with Faults and Fractures.

The conditions at Site 1 included two cracks extending across the asphalt paved roadway, and about ½" of settlement within the middle half of the roadway. Figure 4 depicts the conditions at Site 1. The conditions at Site 2 included one crack extending across the asphalt paved roadway, with a slight depression across the north-bound lane. A 4-foot diameter hole was present within the southbound shoulder of the road, near the western end of the crack. The depth of this hole was greater than 25 feet. In addition to the roadway and shoulder features, a nearby rock face included large fractures of about 8 feet in depth, and ground cracks extending westward away from the roadway measured 4 feet wide and up to 18 feet deep. Figure 5 depicts the conditions at Site 2.

Figure 4 – Site 1 Features.

Figure 5 – Site 2 Features.

The Federal Highway Administration – Central Federal Lands Highway Division (CFL) requested non-invasive geophysical surveys to assess the subsurface conditions at the two subject sites, along with three additional sites further south along the road *(3)*. CFL required that at least two geophysical methods be used for the assessments, and that the methods and equipment must be capable of imaging the subsurface to a minimum depth of about 30 feet. Lastly, the timeframe from receiving notice-to-proceed to providing a draft report was approximately one month. Therefore, efficient methods of obtaining and providing reliable data would be crucial.

GEOPHYSICAL INVESTIGATION

Olson Engineering, Inc. (Olson) selected 3D Seismic Refraction Tomography (SRT) and 3D Ground Penetrating Radar (GPR) to perform the requested assessments. The presence of possible voids and wide fractures is similar to conditions encountered in karst environments, and Olson has performed numerous studies in those environments across the United States using 3D SRT. The benefit of also using 3D GPR is that advancements in technology allow for covering large areas such as roadways more efficient than ever before. The field work was performed between August 29 and August 30, 2017 *(4)*.

3D Seismic Refraction Tomography

In an SRT survey, an impulse (shot) is imparted to the ground (e.g. via a sledge hammer) and the seismic waves generated by the impulse are detected along an array of receivers (geophones). The propagation of seismic waves is governed by the stiffness of overburden soils or the hardness of rock formations. The variability of the soil deposits can be mapped laterally and depth to bedrock can be imaged.

Data were collected using Geometrics Geode 24-channel seismographs with seventy-two (72) 30-Hz gimbaled geophones. Figure 6 shows the seismographs and typical geophones. The geophones were spaced at 1-meter intervals. Data were recorded on a Panasonic Toughbook laptop. The position and orientation of each line was measured using a Trimble GPS unit capable of sub-meter precision. SRT acquisition parameters consisted of 0.5-second records sampled at a 0.03125 millisecond (ms) rate. Shots were collected at every third geophone. This lead to a shot pattern similar to the conceptual layout shown in Figure 7. The seismic energy used for this survey was generated with a sledge hammer impacting a plastic strike plate.

The seismic lines were positioned and oriented in the field based on the recommendations of Federal Highway Administration and National Park Service personnel and safety/accessibility constraints. The location of each geophone was recorded with a Trimble GeoXH series GPS unit capable of sub-meter horizontal precision.

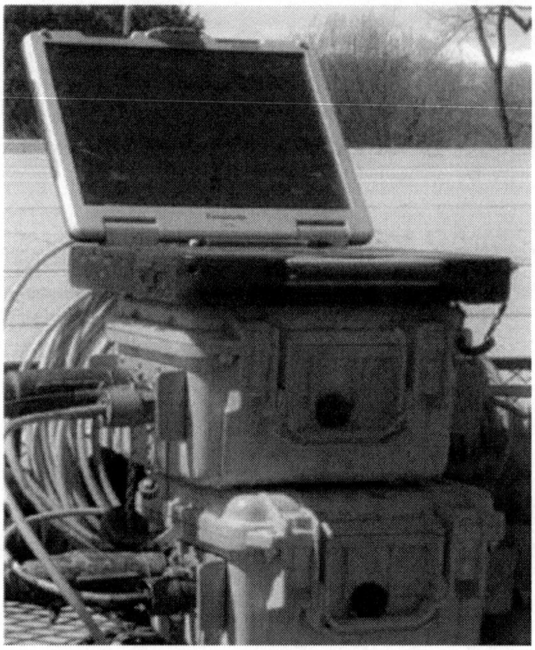

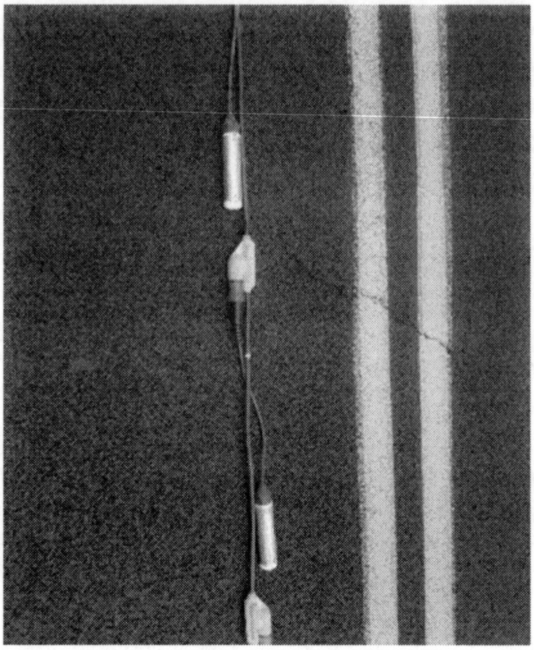

Figure 6: Left – Seismograph and Laptop. Right – Typical Geophone Configuration.

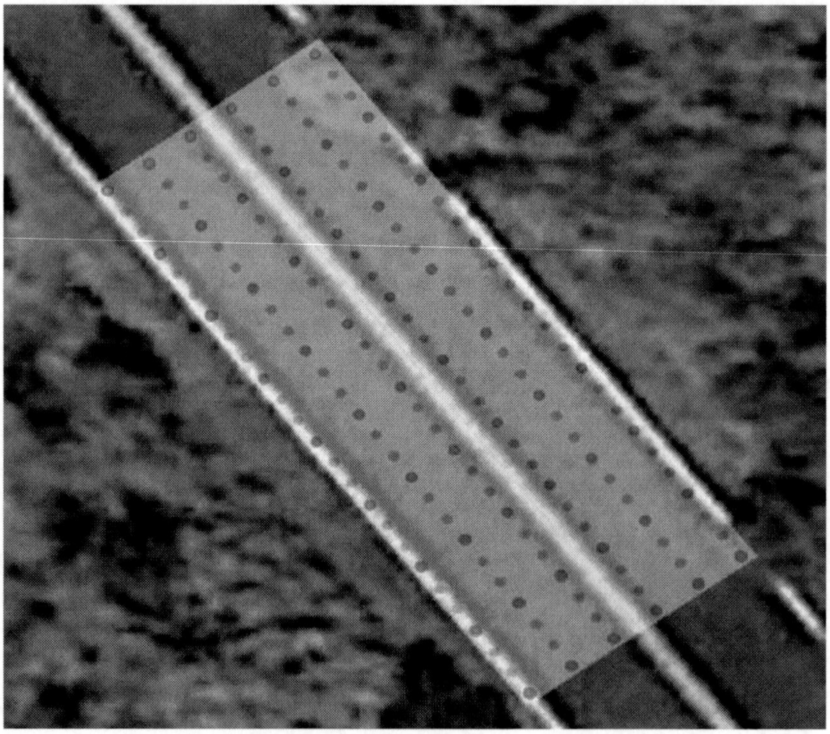

Figure 7: Conceptual 3D seismic layout.
Blue circles represent geophone locations, red circles represent shot locations.

3D Ground Penetrating Radar

For this study, the GPR survey was performed by Diversified Infrastructure Systems (DIS), as a subconsultant to Olson *(5)*. The GPR method is a wave propagation technique that transmits and receives electromagnetic waves (EM or radio waves). When the transmitted energy encounters materials of differing dielectric permittivity, it is reflected back to the surface. Reflection strength depends primarily on the contrast of dielectric and conductivity properties between the materials. GPR for ground imaging is commonly used for mapping underground utilities, karst features such as cavities and voids, soil and/or rock layer thickness, and lava tubes, among other items.

In GPR, the lower frequency range allows for a greater depth-of-investigation but at the cost of target resolution, whereas higher frequencies have higher resolution but are limited in depth-of-investigation. The GPR system used by DIS for this survey was a 3D-Radar DXG 0908 multi-channel system. Unlike conventional GPR systems, this system steps through a wide range of operating frequencies (200 MHz - 3,000 MHz, with 4MHz steps), allowing it to have high resolution in the near surface while still achieving maximum depth of investigation. This system is comprised of 8 antennas spaced 3 inches apart. This gives a very high sample rate, which allows for a 3D volume of data to be collected along each pass of the system. These individual passes are then stitched together to generate larger 3D volumes. For this study, the antennas were configured to collect data every 1.53 inches in the survey direction, with data recorded by a Geoscope MkIV controller. The GPR system was mounted on a cart along with a Distance Measurement Instrument that provided precise distance measurements for collecting data. Differential GPS data were collected along with the GPR data, using a Trimble R8 GNSS/GPS. Figure 8 shows the typical setup for the system.

Figure 8. 3D GPR System.

Data Processing

SRT

The SRT data were processed using Rayfract, version 3.35, by Intelligent Resources Inc., and internal 3D refraction software developed by Olson Engineering (Tomogram). The two major processing steps involved with SRT are first arrival picking and data inversion. The first arrival picking step consists of picking the time for each trace (signal) where the first arrival of wave energy is observed at the geophone position. First arrival picking was conducted in Rayfract. The 3D inversions were then performed using the Tomogram software.

GPR

While onsite, DIS performed preliminary data processing to ensure data integrity and that acceptable coverage was obtained. The full processing generally involved estimating the dielectric constant of the subsurface material by hyperbola matching with data features, and then performing several processing steps including converting from frequency domain to time domain in order to estimate depths to features below the ground surface. Upon completion of the processing for each data set, both vertical and planar depth slices were produced to show features of interest.

GPR Results and Discussion

For the GPR figures, crack features are represented as red lines, whereas the large hole observed at Site 2 is indicated by a red-filled circle. Unknown features as determined from the GPR data are shown as blue lines and polygons. The items outlined in blue do not represent every feature that could be noted in the GPR results. Instead, these are features that Olson considers to be the highest priority. The following paragraphs will summarize the anomalies for each site.

Figure 9 shows an example GPR depth slice of about 2.5 feet for Site 1. The anomaly outlined in blue is likely a critical feature to investigate further for this site. The location is the approximate middle of the roadway. This is not a large anomaly, but is relatively near to the surface and, based on the location, could be related to the southern of the two road cracks that were noted.

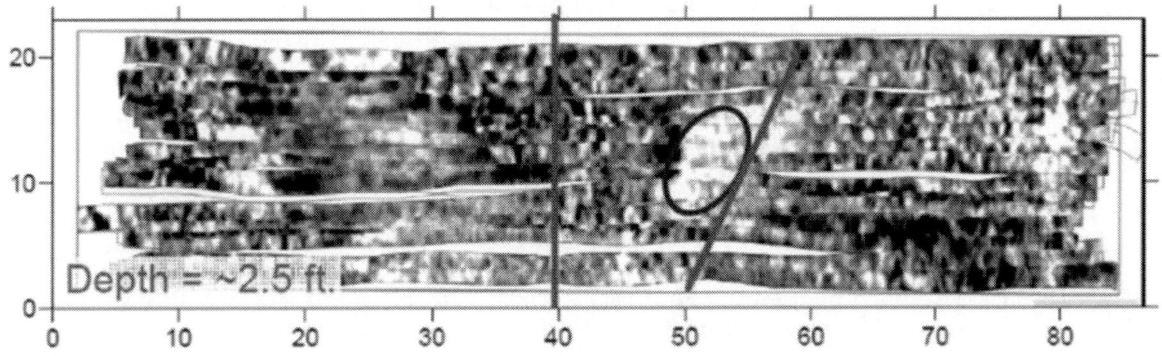

Figure 9. GPR Depth Slice from Site 1.

Figure 10 and Figure 11 show example GPR depth slices of about 3 feet and 5 feet, respectively, for Site 2. The anomaly outlined in blue in both slices is large, and it is likely a high priority anomaly that should be further investigated. The red circle is the 4-foot diameter hole.

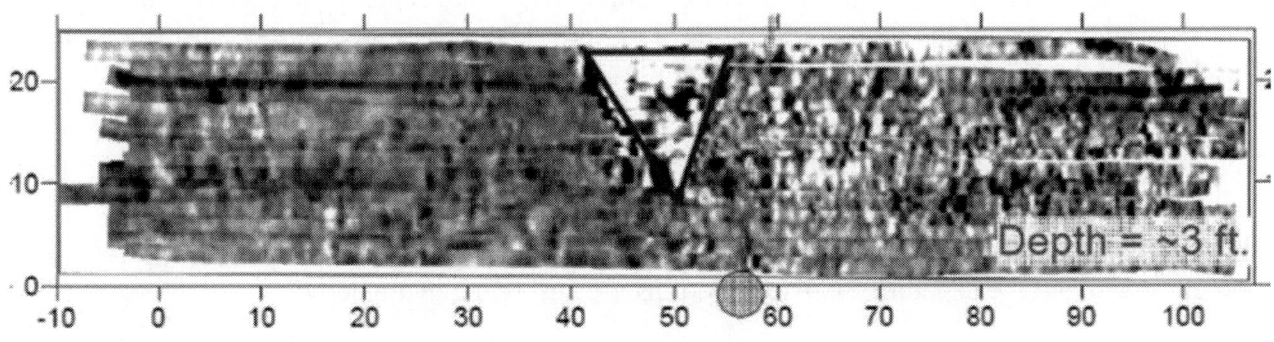

Figure 10. 3-Ft GPR Depth Slice from Site 2.

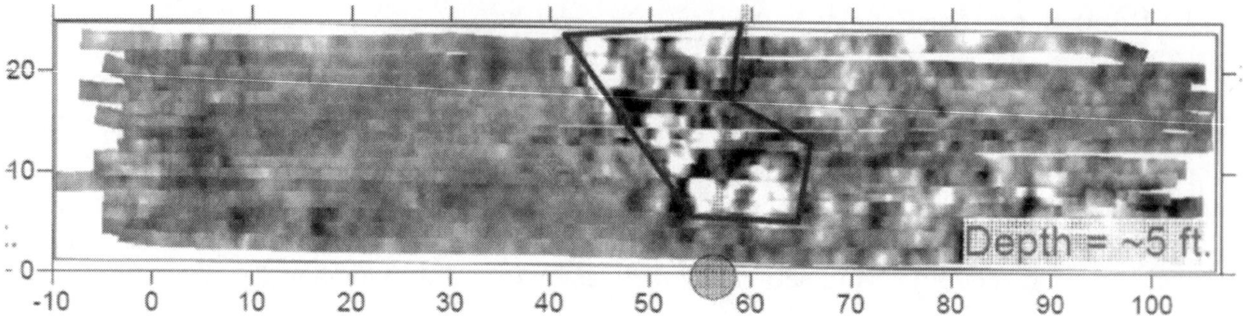

Figure 11. 5-Ft GPR Depth Slice from Site 2.

SRT Results and Discussion

The seismic data processing was complicated by pavement of the road. It is common when collecting on pavement for there to be a high frequency high velocity signal that is caused by the seismic energy that travels along the road pavement instead of the subsurface. If this energy is mistaken for a true first arrival, the seismic model generated from the data will show no depth of penetration and be representative only of the pavement velocity. Therefore, much caution should be used when picking in these situations. The data from this site has an especially strong pavement effect.

Because the first arrival picks nearest the shot location are related to the velocity in the shallow sub-surface, this strong pavement break will greatly limit the accuracy of the seismic results in the top 10-15 feet. Additionally, it is possible that velocity variations within the top 10-15 feet that are not accurately imaged will cause inaccuracies in the results from deeper because their effect is not correctly accounted for in the shallow part of the velocity section.

Three dimensional SRT results for sites 1 and 2 are presented as Figure 12 and 13, respectively. All velocity values are presented using the same color scales, with 'cool' colors (e.g., blue) representing lower velocity values and 'warm' colors (e.g., red) representing higher velocity values. The left side of the figures shows the skewed, 3-D configuration of the subsurface. The right side of the figures shows a 2-D configuration with annotations.

It is very important to understand that these results are generated using Tomogram, a software package developed at Olson Engineering. This package has the possibility of imaging the subsurface in more detail than commercially available software. However, it is still in development and the capabilities and limitations are still being evaluated.

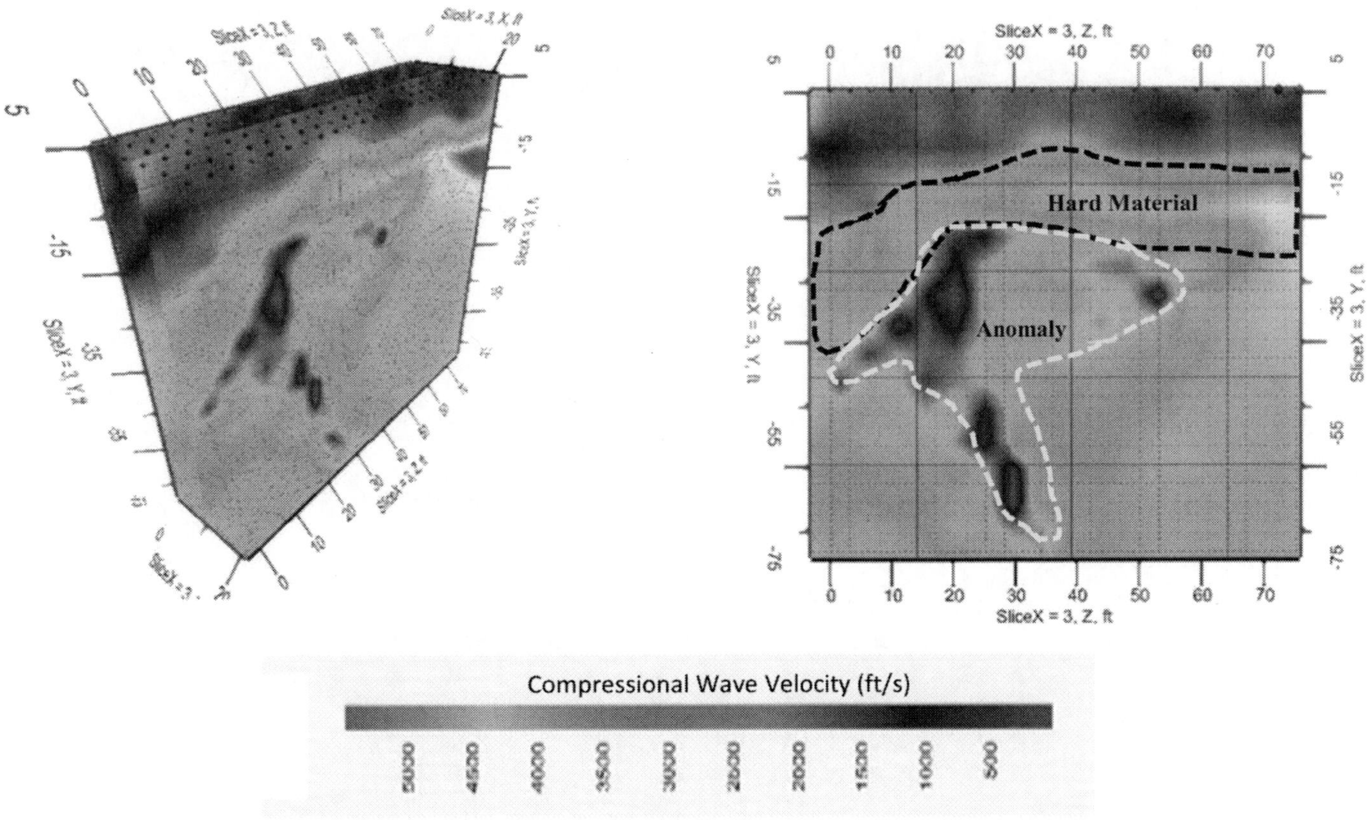

Figure 12. 3D SRT Results for Site 1 (Skewed and Unskewed).

The results for Site 1 (Figure 12) show that the top 10 feet has a low overall velocity value (<2000 ft/s). This could indicate that the top 10 feet is highly fractured basalt. This does not mean that the overall stability or stiffness of this material is low. High levels of fracture can cause low velocity values even when the fractures are small, and the overall properties of the rock demonstrate high stiffness. From 10 feet down to a depth of 75 feet the overall velocity is higher, with values generally greater than 3000 ft/s. However, there are some features within this section where the velocity is low (<1500 ft/s). These low velocities could be related to larger cracks that are attributed to rifting. The geometry does not seem consistent with what would be expected from a lava tube, although this cannot be ruled out based on these results.

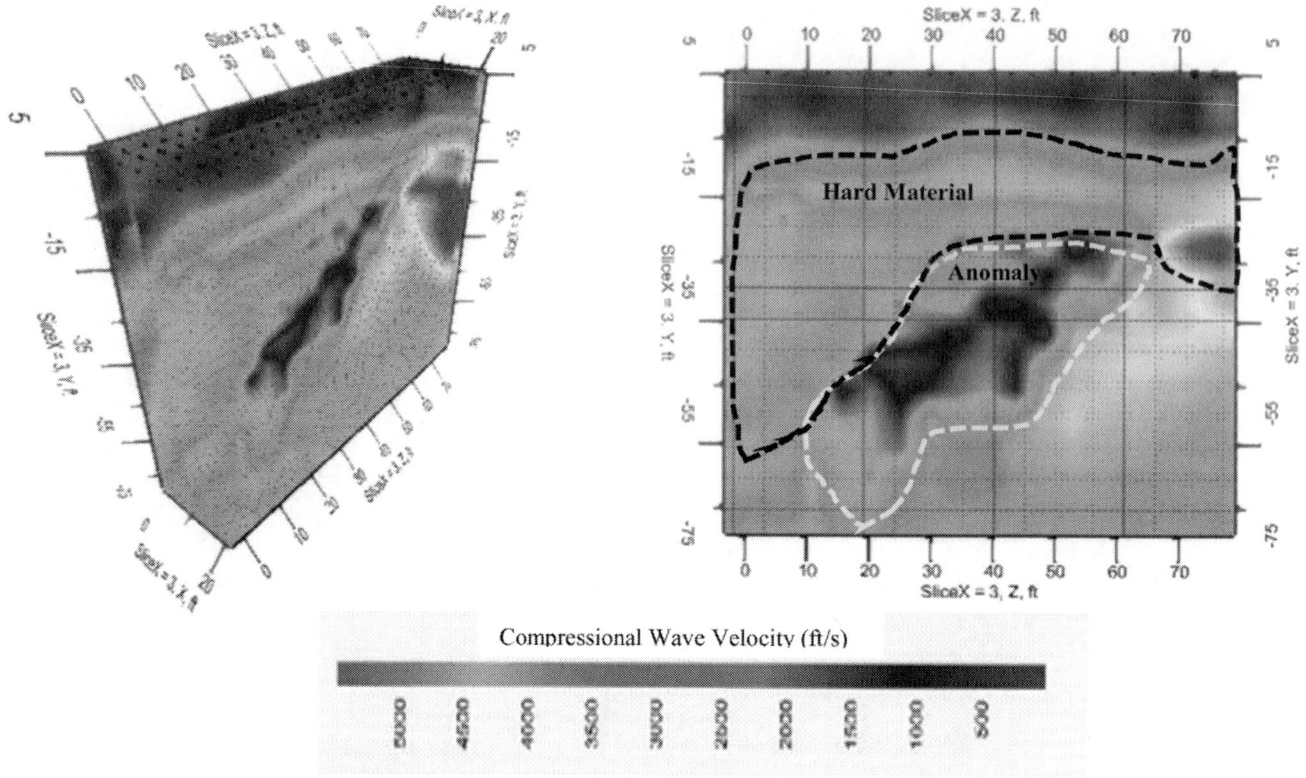

Figure 13. 3D SRT Results for Site 2 (Skewed and Unskewed).

The seismic results for Site 2 (Figure 13) show very similar results to those from Site 1. As with Site 1, there is a 10-foot thick low velocity zone overlaying a generally higher velocity zone from 10 feet down to 75 feet. There are once again low velocity zones within this higher velocity zone. In this case, the feature starts at a depth of about 25 feet and deepens to the southeast. Even more than for site one, this low velocity zone has the appearance of a large crack in the rock.

Qualification of Geophysical Results

For all of the sites assessed for this project, several anomalies were detected. These were generally different in number, size and character for each site. CFL categorized the anomalies into three severity classes based on the quality and reliability of the geophysical data, the risk of near-future distress, and public safety. These classes are listed below.

- Class 1 (High Severity): Immediate remediation is a high priority.

- Class 2 (Medium Severity): Immediate remediation is beneficial. Site should be visually monitored regularly.
- Class 3 (Low Severity): Immediate remediation unnecessary. Occasional visual monitoring should be performed.

The anomaly classifications are then applied to the anomalies previously presented in this paper, as shown in the following figures.

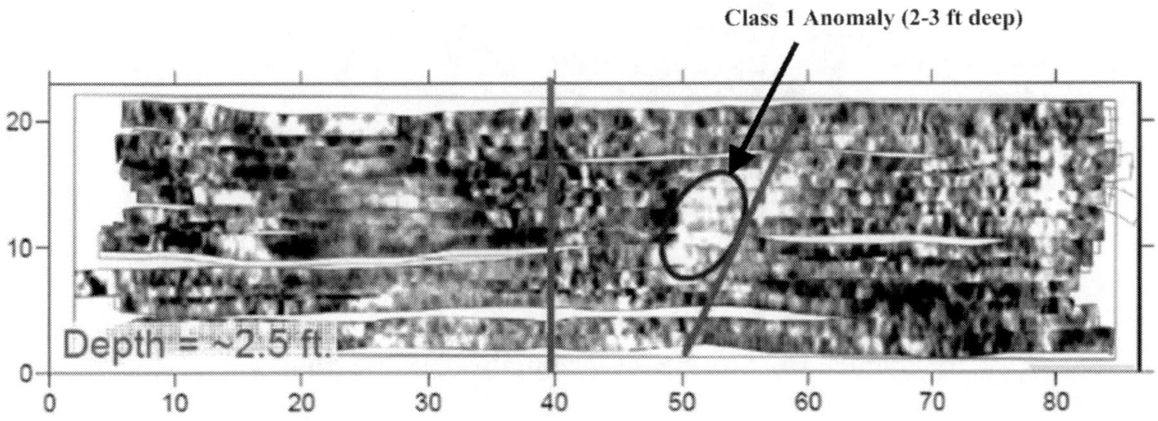

Figure 14. Site 1 GPR Anomaly Classification.

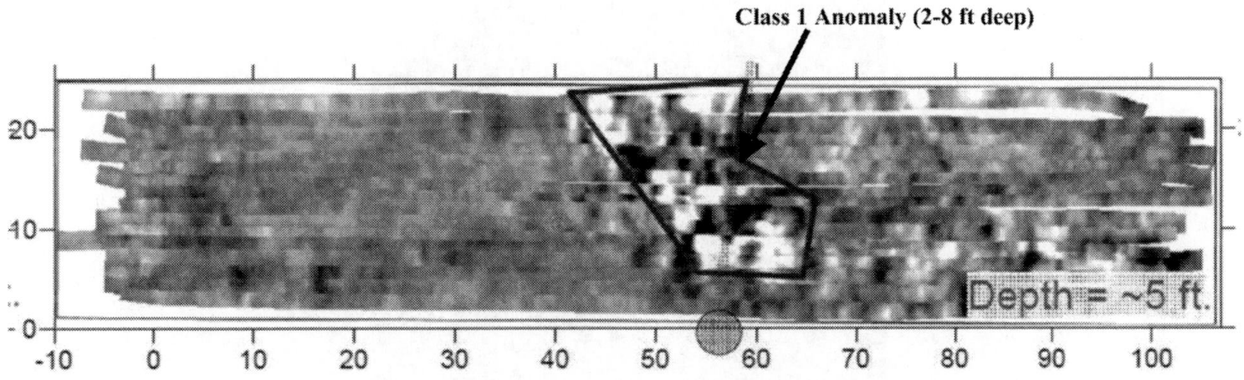

Figure 15. Site 2 GPR Anomaly Classification.

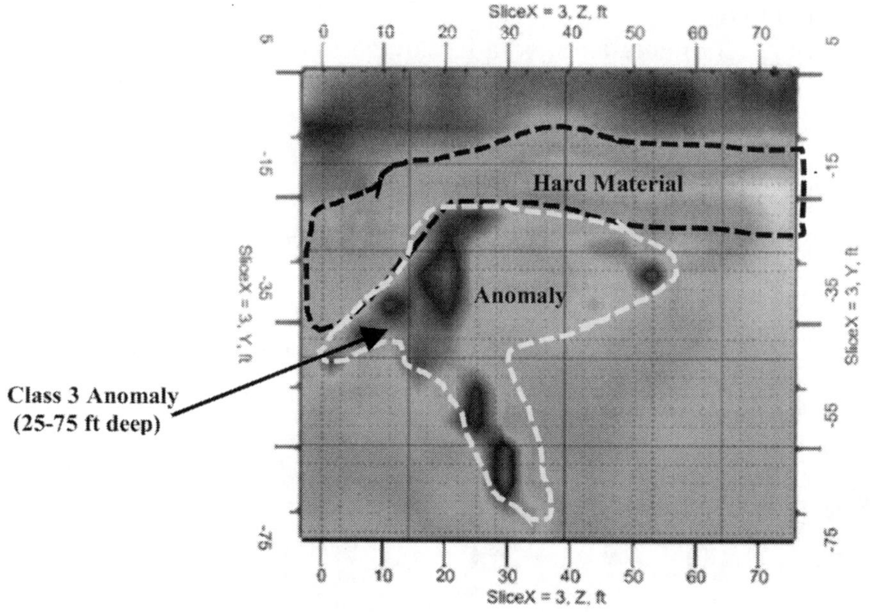

Figure 16. Site 1 SRT Anomaly Classification.

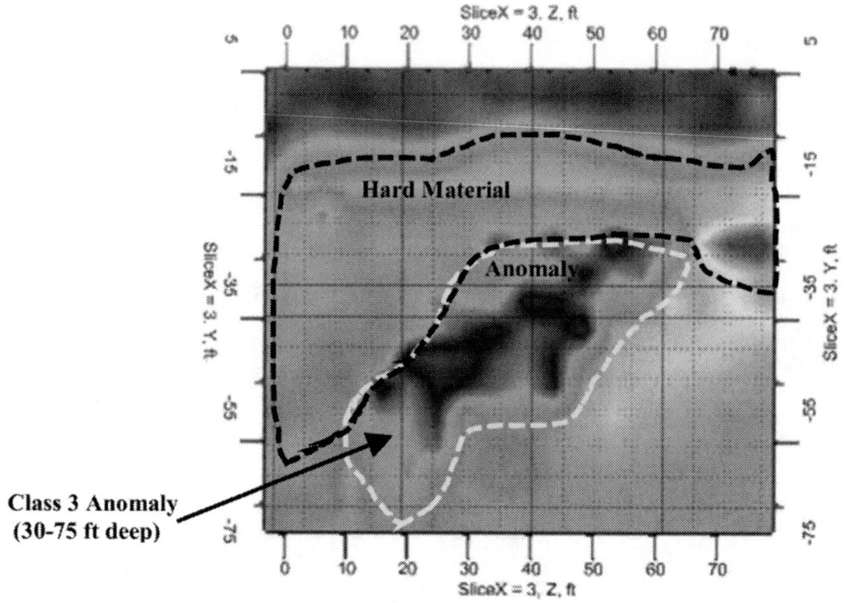

Figure 17. Site 2 SRT Anomaly Classification.

Conclusions

Both of the geophysical methods were successful at detecting anomalies, with each method showing promise for fast data acquisition, data processing, and delivery of useful results within days of data acquisition. The GPR method was generally successful at identifying very shallow features beneath the roadway, to nominal depths of 7-9 feet. The SRT method could not resolve data shallower than about 5 feet below the roadway, based on the pavement issue discussed previously, but was able to identify anomalies to depths of at least 50-75 feet.

CFL was able to make mitigation recommendations without additional geotechnical investigation, with the installation of concrete slabs over the shallow anomalous zones being the preferred solution.

REFERENCES

1. National Park Service. *Hawai'i Volcanoes*. National Park Service, December 3, 2016. Accessed May 28, 2018.

2. Okubo, C. H. and S. J. Martel. Pit Crater Formation on Kilauea Volcano, Hawaii. *Journal of Volcanology and Geothermal Research* vol. 86, 1998, pp. 1-18.

3. Federal Highway Administration – Central Lands Highway Division. *Draft Scope of Work – Geophysical Investigation – Chain of Craters Road – Volcano National Park*. FHWA, August 2, 2017.

4. Olson Engineering, Inc. *Geophysical Investigation Report – Chain of Craters Road*. Olson Engineering, Inc., September 27, 2017.

5. Diversified Infrastructure Services, Inc. *Geophysical Investigation of 'Chain of Craters Road' Using GPR.* Diversified Infrastructure Services, Inc., September 12, 2017.

What are the Benefits of Geotechnical Data Interchange?

Scott L. Deaton, Ph.D.
Dataforensics
2310 Parklake Dr #525.
Atlanta, GA 30345
678-406-0106
sdeaton@dataforensics.net

Prepared for the 69th Highway Geology Symposium, September, 2018

Acknowledgements

The author would like to thank the personnel at Ohio DOT who were instrumental in funding, coordinating the project described herein. In particular:

Chris Merklin, Ohio DOT
Steve Taliaferro, Ohio DOT

Disclaimer

Statements and views presented in this paper are strictly those of the author(s), and do not necessarily reflect positions held by their affiliations, the Highway Geology Symposium (HGS), or others acknowledged above. The mention of trade names for commercial products does not imply the approval or endorsement by HGS.

ABSTRACT

Ohio DOT has been at the forefront of streamlining geotechnical data management for the last decade. Due to serious deficiencies in requiring the exchange of data via gINT project files, in 2017 Ohio DOT engaged Dataforensics to perform an assessment of their in-house processes as well as the processes for several geotechnical consultants that provide data to Ohio DOT. The goals of this assessment were:

- Review how DIGGS can streamline and improve consultant's workflow for managing geotechnical data.
- Identify how Ohio DOT can further improve their internal processes using DIGGS.
- Develop a roadmap for Ohio DOT to realize their ultimate goal of obtaining geotechnical data from their consultants using DIGGS instead of PDF borehole logs containing information.

This paper provides a summary of the findings of this report with significant focus on how DIGGS (data interchange for geotechnical and geo-environmental specialists) benefits both consultants, DOTs and contractors. The typical consultant workflow for subsurface geotechnical data will be discussed in detail identifying the limitations, inefficiencies and opportunities for error that can be eliminated using DIGGS. Additionally, several examples of organizations already using data interchange around the world will be presented to highlight advantages these organizations have because of the widespread usage of geotechnical data interchange in their countries.

INTRODUCTION

Data mining, big data, artificial intelligence are buzzwords describing the rapid evolution of technology related to data and how it is affecting our day to day lives. Whether it is Amazon Alexa, Siri, or Google Assistant, revolutionary changes that are driven by data are occurring in the world around us, yet the standard deliverable for geotechnical and geologic data from site investigations remains a borehole log. The only evolution of this deliverable over the last 50 years is that it has changed from a paper-based deliverable to a PDF (digital version of paper) deliverable. The primary disadvantage of this standard deliverable is that it is ***not*** data and significant value has been removed for the receiver of this deliverable (typically the owner of the project) who is requiring this antiquated communication method.

Revolutionary changes in software, hardware, and cloud-based technology are affecting geotechnical engineers, geologists, hydrogeologists, and engineering geologists as well. These data processing advancements are providing tremendous opportunities for organizations that embrace technology and the ability to properly collect, manage, analyze and visualize data. Subsurface data can be an incredible asset for organizations that are managing this data properly as well as for organizations that are paying for this data to be collected properly such as departments of transportation, the U.S. Army Corps of Engineers, utility companies, and other large owner organizations. To maximize its value, data must be collected, managed, and transmitted as data, not as information.

Data Versus Information

Throughout our industry there is a common misconception about what is geotechnical data. Geotechnical engineers and geologists typically say things like "our geotechnical data archive consists of PDF logs on a server, or PDF logs available on a map or maybe even available in Google Earth." Unfortunately, this is not data. It is valuable and useful information, but it is not data.

The British Standards have codified the definition of geotechnical data in BS 8574:2014 Code of practice for the management of geotechnical data for ground engineering projects [1]. Specifically, they define geotechnical data as: "facts or figures obtained from all phases of a geotechnical project, including derivations from other data. Facts and figures might include text, numbers and formulae." Dataforensics and Keynetix have refined this definition as:

"If you can process it into one or more formats without re-inputting it or using multiple cut and paste operations, you have data; otherwise you have information." Two typical examples of deliverables that are NOT data:1) a paper or PDF borehole log report and 2) a particle size distribution report, shown below in Figure 1.

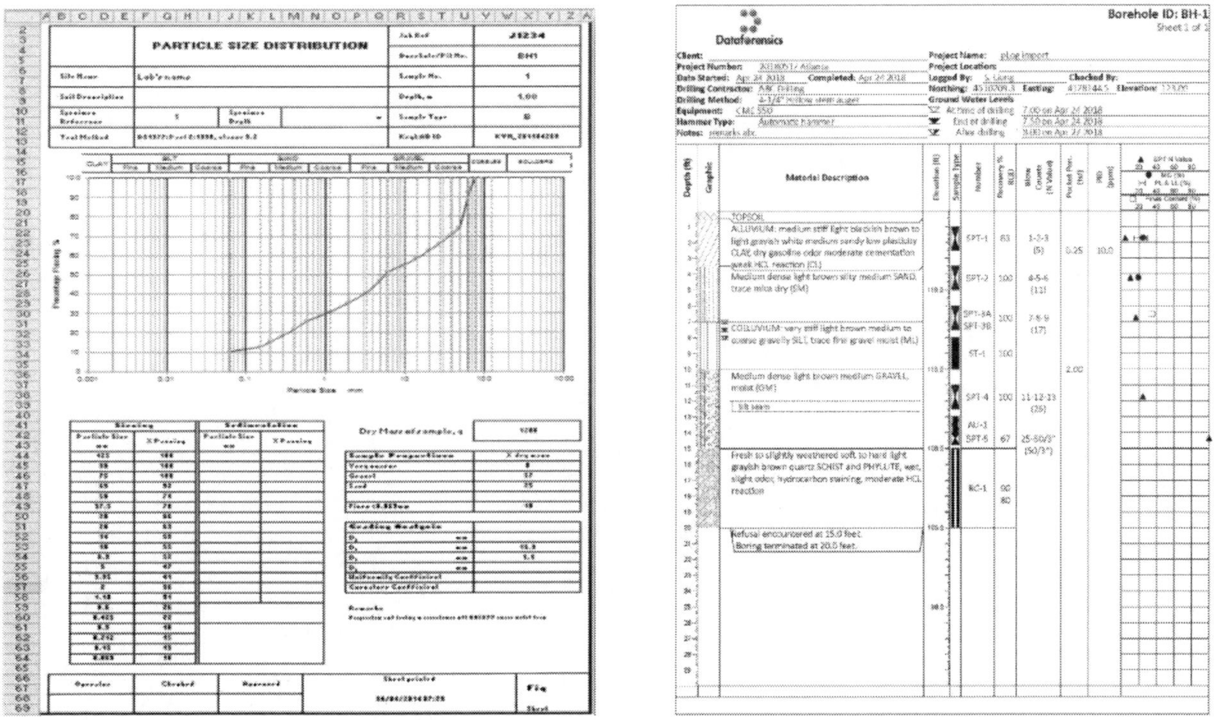

Figure 1 – Standard Information-based Deliverables

There is a tremendous amount of valuable information on these reports; however, it is rendered unusable by this deliverable format without someone reinputting the data again. Having to repeatedly re-input data is a source of tremendous wasted time and resources in our industry and a significant source of errors. A client that routinely works on large highway projects who has been actively promoting proper data management within his organization for several years but is often thwarted by management and others reluctant to "change" their process has repeatedly said to me, *if we transcribe 10,000 laboratory test results from Excel to our laboratory data management software there are a significant number of these results that are incorrect that we may never find.* Researchers at the University of Hawaii have concluded that a typical "mechanical" error rate is 0.5% [2], which would mean that of those 10,000 laboratory test results, 50 would be incorrect.

First Golden Rule of Data Entry

As a result of the inefficiencies, inaccuracies in traditional data management, and inability to use data as identified above, Dataforensics and Keynetix have defined several Golden Rules for Data Entry. The First Golden Rule of Data Entry is "Only Input Data Once". This may seem like an obvious pillar of proper data management, however there are very few organizations that are achieving this goal. In 2017, Dataforensics was engaged by the Ohio Department of Transportation Office of Geotechnical Engineering to assess areas for process improvement within the Ohio DOT as well as within consultants that are performing site investigation projects for the Ohio DOT. Dataforensics found that the typical consultant is re-inputting subsets of the same data between 10 and 15 times per project.

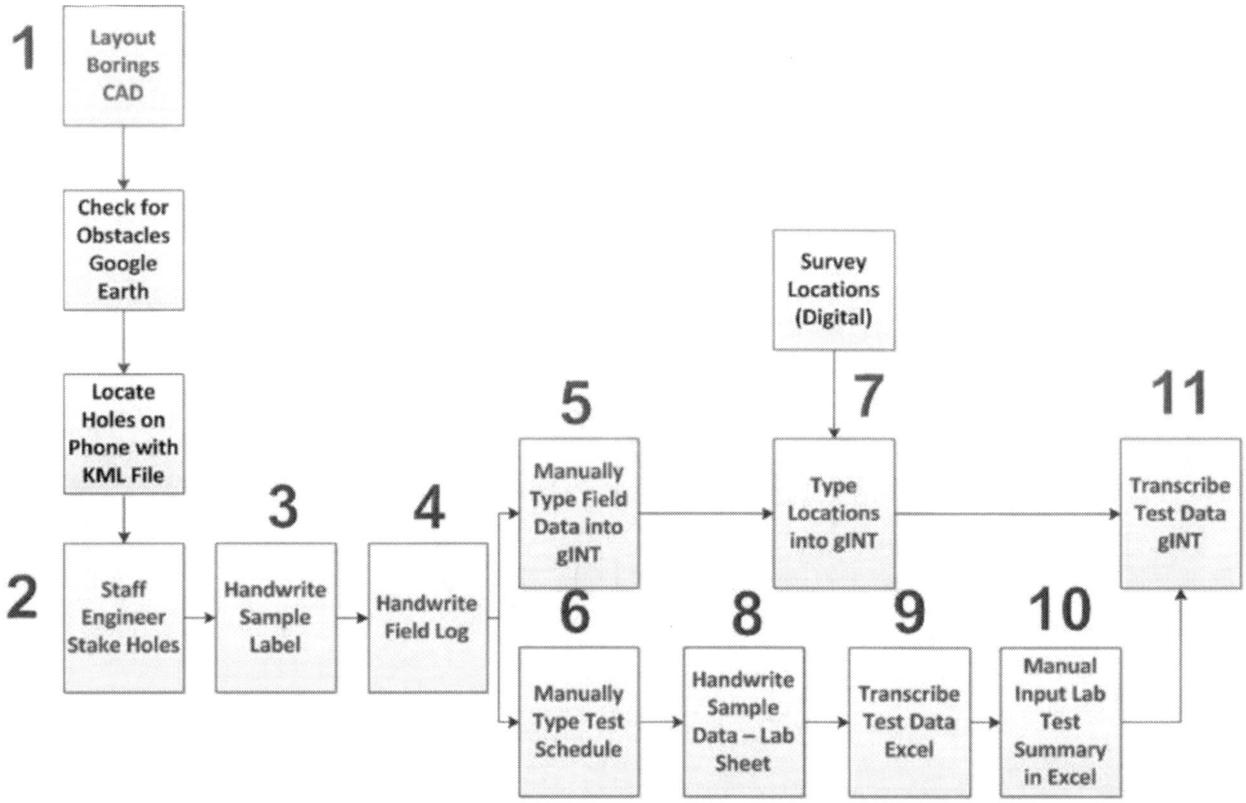

Figure 2 – Typical Workflow for Consultants Performing Subsurface Investigations

The process utilized by consultants managing data related to subsurface investigations is shown in Figure 2, where the boxes in red show re-inputting data that has already been input once before, violating the First Golden Rule of Data Entry.

1. The typical process starts with developing the borehole location plan in CAD. This step may involve two different people, an engineer or a geologist and a CAD professional and in some cases involves transcribing handwritten details from a paper-based site plan into CAD.

   Once in CAD, the locations are exported to a KML file that can be opened in Google Earth to check for obstacles.

   If the person planning the investigation is satisfied with the locations, that same KML file can be transmitted to a field personnel's phone where they can locate the borehole locations. Note that steps 2 and 3 (shown in black) are already properly using a data interchange standard (KML).

2. Once the the Staff Engineer/Geologist is in the field, the locations can be staked and they write on the stake the Borehole ID. This is the second time the Borehole ID has now been recorded.

3. As they are logging the borehole, they handwrite sample labels that include the Project ID, Borehole ID, Sample ID, and Depth.

4. The Staff Engineer/Geologist re-writes this same data (Project ID, Borehole ID, Sample ID, and Depth) on the handwritten field log along with many other pieces of data.

5. Once back in the office, someone manually inputs the field log into software such as gINT, HoleBASE, LogDraft or LogPlot. Essentially all data recorded in step 4 is now being re-input.

6. In order to inform the laboratory about which lab tests should be performed on each sample, a test schedule is developed where the user must once again re-write the Project ID, Borehole IDs, Sample IDs, and Depths for each sample in addition to the necessary lab tests for each sample.

7. Meanwhile, the surveyor using digital surveying technology (Total Station/GPS), performs the survey and transmits these locations to the Staff Engineer/Geologist generating the borehole logs. Note the work is being performed digitally and the data is actually being provided to the end-user digitally, but it is in a format that cannot be imported into gINT automatically. So, this data must be re-typed into gINT.

8. While the Staff Engineer is diligently typing latitude and longitude values that include up to 6 decimal places of precision (a process that is somewhat error prone), lab technicians are handwriting the same details (Project ID, Borehole ID, Sample ID, and Depth) on <u>each</u> laboratory test sheet for <u>each</u> lab test on each sample. So, if an Atterberg Limit, Particle Size Distribution (sieve) and Hydrometer are being performed on a particular sample, you are really inputting this same data three more times.

9. Once the lab test(s) have been completed, the test data is transcribed into Excel and the test results are automatically calculated. In this step all of the data recorded in step 8, along with the test data recorded are re-input again.

10. To summarize the test results, the same data (Project ID, Borehole ID, Sample ID, and Depth) plus the results from each test are tabulated in an Excel spreadsheet.

11. Some of the lab test results (but not all) are then transcribed into gINT so they can be printed on the borehole log.

From this discussion, the inefficiencies in the process are readily identified, and it is easy to imagine all the steps where human errors associated with transcription can be introduced into the traditional process that does not utilize digital data interchange standards.

Second Golden Rule of Data Entry

The Second Golden Rule is "Get Someone Else to Do It". This is really the same as the first rule because if you are only inputting data once then naturally when you need to use a piece of data that has already been created in the process previously, it must have already been input by someone else. A typical example of this is not having to re-input the Project ID, Borehole ID, Sample ID, Sample Depth throughout the project workflow.

Third Golden Rule of Data Entry

The Third Golden Rule is "Store Data in Your Database not Information". A variation of this rule was suggested by a data manager at Golder Associates who was attempting to migrate their data from their antiquated software into a modern enterprise data management system, who suggested that "Data" should never be stored in a comments field. Storing multiple pieces of data in a single field violates the first rule of database design, called First Normal Form, which is each field should store a single atomic (or indivisible) value, essentially a single piece of data [3].

Organizations often store many different types of data in a Comments field that will pri in a column called Notes on their log report. An example of this approach is shown below in tl table where users input the data for Atterberg Limits, Natural Moisture Content and Depth Related Notes. The related column from a log report is shown in Figure 3 as well. This approach for managing information is an example of a "Reportbase" not a database. A "Reportbase" is when you create your database structure based on what your report needs to look like, not based on the physical reality of the data and the relationships between the data. This approach works well for this one scenario (generating the borehole log), however this is not managing data, it is managing information. If the user needs to plot the Atterberg Limit Results on the Casagrande chart, or on a summary table, or plot only the Liquid Limit and Plasticity Index on a cross section the results must be re-input. A good metric to identify this scenario is when you cannot selectively report the data in a different format, in this case, you are not managing data.

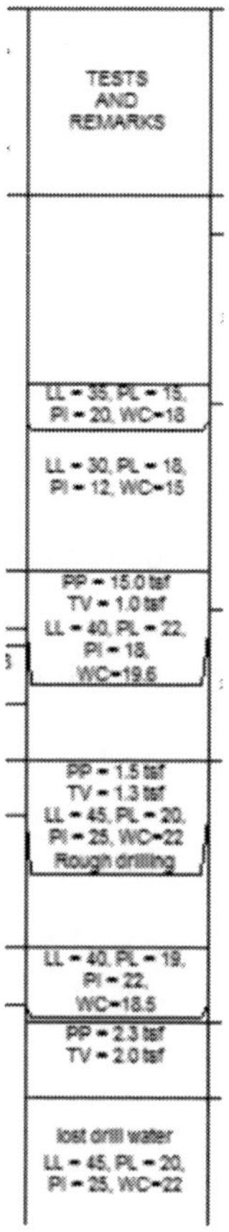

Depth (ft)	Description
5	LL = 35, PL = 15, PI = 20, WC=18
7.5	LL = 30, PL = 18, PI = 12, WC=15
10	LL = 40, PL = 22, PI = 18, WC=19.6
15	LL = 45, PL = 20, PI = 25, WC=22
15.1	Rough drilling
20	LL = 40, PL = 19, PI = 22, WC=18.5
25	lost drill water
26	LL = 45, PL = 20, PI = 25, WC=22
*	

Figure 3 – Example "Reportbase" Information Management

Ultimately all geotechnical data should be stored in accordance with standard database practice following the 1st, 2nd and 3rd Normal Forms for Database Design [3]. When these rules are not followed, inefficiencies in the workflow result, inaccuracies are likely to occur, automation of calculations is impossible, integration with other systems that manage data (e.g. laboratory management or field data collection) is difficult or impossible, data interchange is not possible, and ultimately the ability to use the data in any other system is not practical without re-inputting it. If you have data, it can be reported however you need to report it, transmitted to other systems, analyzed, visualized, summarized and ultimately provided as a deliverable for others to utilize.

In the example shown above in Figure 3, an error was purposely made to highlight how easy it is to have errors because of not storing and managing the data properly.

Data Producers

Data Producers are personnel who are generating the geotechnical data. Typical examples of data producers are field personnel logging boreholes and lab personnel performing testing. Standardized Geotechnical Data Interchange, allows data producers to utilize a system that best fits their needs while providing DATA to users downstream, such as providing data to Data Consumers.

Data Consumers

Data Consumers of geotechnical data can be a variety of different people. In DOT's we often find that various design sections such as Bridge Design, Pavement Design, Culvert Design, Geotechnical Design and Pavement Management all have systems that need to use geotechnical data generated in the subsurface investigation process. Consultants who need to receive historical data from DOT's are also consumers. Consultants doing the site investigation are also consumers of the data produced by the field and lab personnel on new projects. Contractors are data consumers as well, although today they almost never receive any data. No software fits the needs of all producers and consumers. Therefore, software must communicate data to facilitate the work of data producers and data consumers. Data Interchange must be software vendor independent.

Benefits of Data Interchange for Owners (DOTs)

On a typical project, there are often five stages of data transfer between different groups of consumers and/or producers as shown in Figure 4.

1. The Consultant performing the work provides instructions to the drill crew and the drill crew provides field data back to the consultant.
2. The consultant sends lab test schedule data to the lab and the lab sends test results back to the consultant.
3. The consultant sends the deliverables to the owner.
4. The owner (hopefully) loads the data into their regional archive of geotechnical data.
5. This then allows any other consumers of data (current consultant or other consultants and contractors) to utilize this data on projects.

What Dataforensics routinely finds when working with organizations is that each Owner has their own requirements for managing geotechnical data that are imposed on the Consultant. As such, the process shown in Figure 4 is duplicated for each additional Owner the Consultant works for, meaning there are now 15 ways of exchanging data as shown in Figure 5. For each additional Owner that the Consultant does site investigations there are 5 additional data exchange processes. So, ultimately Owners mandating Consultants provide data in their specific gINT format causes many inefficiencies in the process for consultants and prevents Consultants from being able to automate and streamline their processes. If Owners mandate simply that data must be provided in a standardized data interchange format and comply with various standards such as ASTM D2488, ASTM D2487, AASHTO Standards, ASTM Lab Testing Standards etc, the Consultant can then optimize their internal process based on one data management approach and still provide owners the data that the owner requires and needs for their internal processes and managing their geotechnical archive of data.

Additionally, by having the ability to re-use historical data without having to re-input the data, it ultimately reduces the cost of projects for the owners. This same concept can be a benefit for consultants who take advantage of the wealth of data that will be available for them in the regional archives of data for other commercial or industrial projects where historical data will be publicly available.

In our industry today, in many cases there is not a compelling reason for consultants to improve their process and become more efficient, because they are getting paid based on time and materials. This is great for the consultants, but owners are potentially overpaying for services that could be done more cost effectively with improved accuracy. For example, the National Economic Development Office (NEDO) in the UK conducted a review of 5000 industrial buildings and found 50% overran by at least a month [4] of which around 37% of the overruns in the projects were due to ground problems. In another report The National Audit Office [5] cites an Office of Government Commerce study which found that 70% of a range of public projects were delivered late, and 73% were over the tender price. Improvement in the data management process therefore can have very significant ramifications on the cost of projects as well as potential delays.

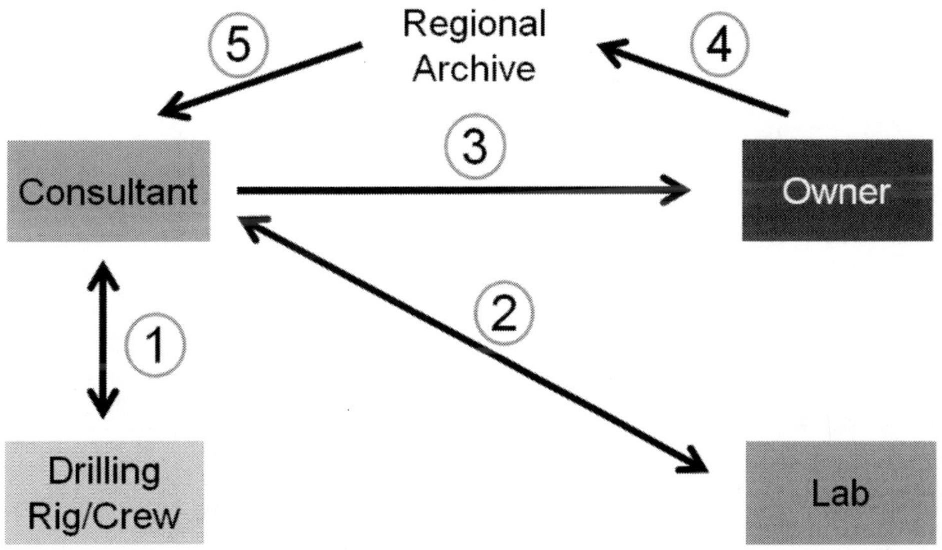

Figure 4 – Five Stages of Data Transfer

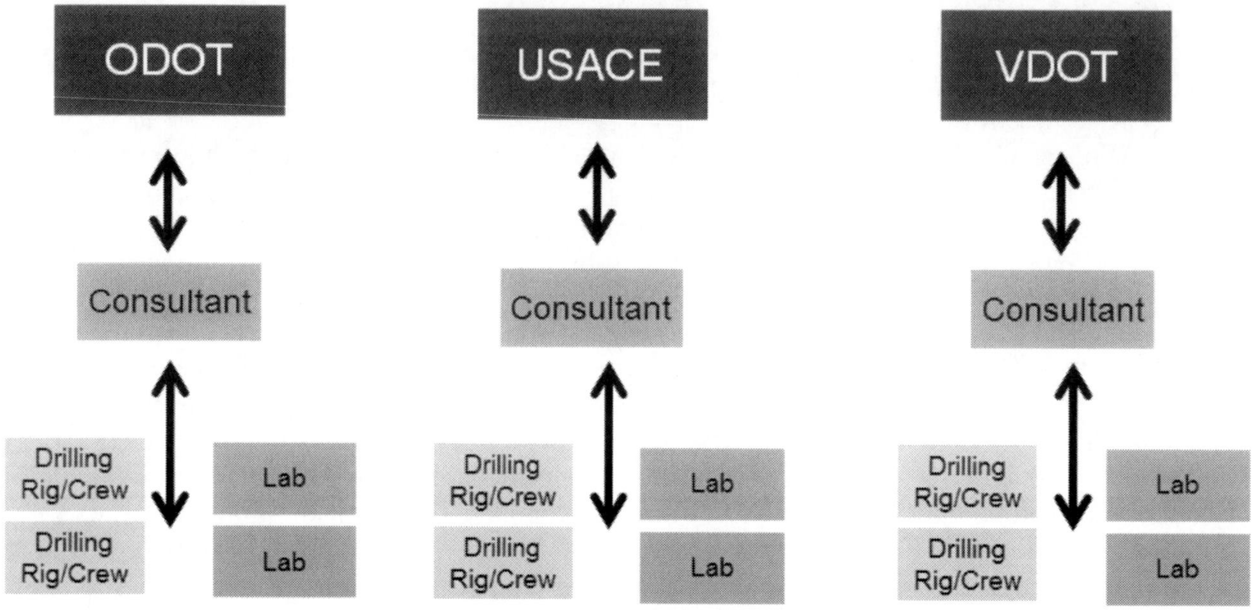

Figure 5 – Fifteen Stages of Data Transfer

Benefits of Data Interchange for Consultants

There are several main benefits for consultants who incorporate digital data interchange based on commonly accepted standards. First, consultants can streamline and improve their internal processes to be more profitable, improve the quality of the data, and reduce their legal exposure. By having a single commonly accepted data interchange standard, all aspects of the workflow for subsurface investigations can be integrated without having to reinvent the wheel for each different owner, as shown in Figure 6.

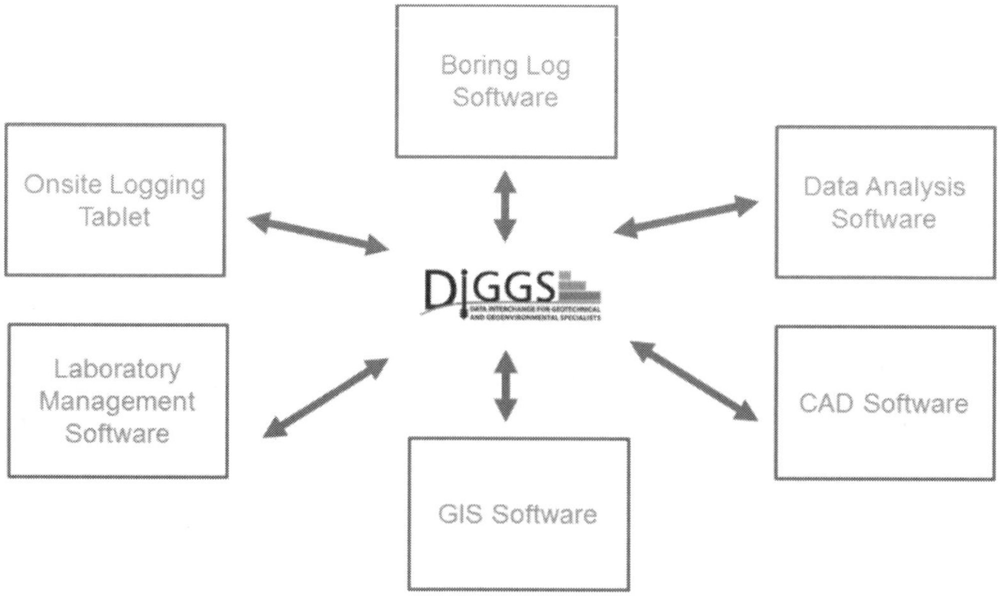

Figure 6 – Workflow with Data Interchange Standard

Furthermore, consultants can better utilize data that today is being managed in separate systems because it is impractical to combine the data in some scenarios. For example, Dataforensics RAPID CPT software for processing CPT data in gINT and HoleBASE SI, allows you to import the CPT data into your data management system from 27 different CPT data file formats. This means Dataforensics has written 27 different importers, because each CPT manufacturer has their own unique data file format. With DIGGS, you would not need Dataforensics software simply to import the CPT data. You might want it for its analysis and visualization capabilities but not everyone needs those. Similarly, for automated data acquisition related to laboratory testing, there are a similar number of equipment manufactures that have different file formats for triaxial, direct shear, consolidation testing, etc. To use this data with your other geotechnical data (borehole, index testing, and in-situ test data), importers for each of these needs to be written in order to have a complete picture of all the test data available on the site.

By simply eliminating the human error associated with transcription of data repetitively decreases the risk and legal exposure for organizations tremendously. Instead of having 15 different ways of managing and communicating data for three different owners, an organization can have a single process that is used for every project as shown in Figure 6. This allows subsurface investigation projects to have a specific and well-defined data management process within an organization. This would be similar to the concept of having an assembly line in manufacturing where a car is built the same way each time, yet different options can be added to the end product. Many of the primary benefits to process standardization whether in manufacturing or in data management are quite similar, improved efficiency, improved consistency, and improved quality [6].

When interviewing each consultant that provides data to ODOT, they indicated that the Owners or clients they work for dictate the format they use for managing their data (e.g. ODOT has their gINT file format, INDOT has their own gINT file format, KYTC has their own gINT file format, USACE has their own gINT file format. Ultimately this leads to the inefficiencies discussed previously, but it also means it makes it difficult for any other organization to use the data in their own systems or with their own internal processes. For example, Dataforensics worked on the Ohio River Bridge project. This project was unique because data was available in the INDOT format, KYTC format, and two different consultants' formats, none of which were compatible without a significant data migration effort.

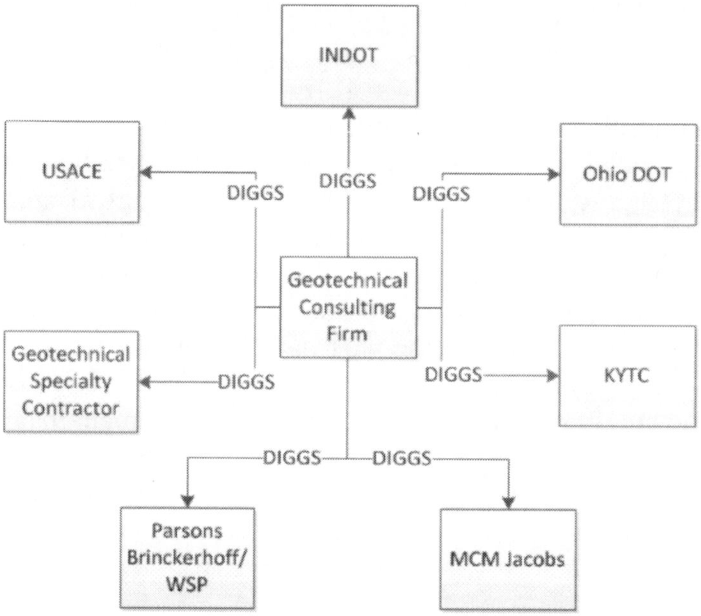

Figure 7 – Data Delivery with Data Interchange Standard

With standardized digital data interchange, organizations simply have to have the ability to export data in this common format and the ability to import data from this common format. So, no matter who the consultant works for, as long as everyone is "speaking the same language" (geotechnical data interchange) they can communicate the data to and from each other easily.

Benefits of Data Interchange for Contractors

Contractors rarely obtain subsurface data that they can use for analyzing, visualizing, and improving the construction process. Instead they are provided geotechnical reports created from significant amounts of useful data, but they must transcribe whichever pieces of data they need to use for their specific design and/or construction process. Whether it is transcribing N-values, moisture contents, depth to rock, shear strength results, water levels, picking key data points off cone penetration test plots, or whatever they need from these reports, a significant amount of effort is expended by contractors to transcribe data. It is potentially error prone, and more importantly incomplete because it is unrealistic to transcribe all the data for each project. Accordingly, they are working with a partial dataset of what is potentially available to them for refining their construction process and design recommendations. This is particularly relevant to specialty contractors who are performing designs or refining designs from consultants. Once contractors have a more complete and accurate picture of the subsurface conditions because they have access to all the relevant geotechnical data, significant efficiencies in the construction process can be realized, ultimately resulting in better construction techniques, and safer infrastructure built at a lower cost.

Does Data Interchange Really Work?

Geotechnical data interchange has been used commonly around the world starting with the AGS Data Interchange standard (Association of Geotechnical and Geo-environmental Specialists). This standard was first developed in the UK in 1992 and has gone through significant enhancements over the last 26 years. Subtle variations to this data interchange standard are used extensively in Australia, New Zealand, and Hong Kong and have been the practice for a couple of decades.

Highways England has a database of hundreds of thousands of boreholes that are available on a map and downloadable as AGS data files and PDF borehole log reports, such that any user can re-use existing historical data without re-inputting it.

Figure 8 – Data Delivery with Data Interchange Standard

Following the Christchurch earthquake in New Zealand, the 'Canterbury Geotechnical Database (CGD) was developed. The CGD enables sharing of geotechnical data collected by various geotechnical firms across the Christchurch area. More than 3928 deep borehole logs and 16407 cone penetrometer tests have been uploaded to the CGD in addition to other test results such shallow Scala penetrometers and test pits.

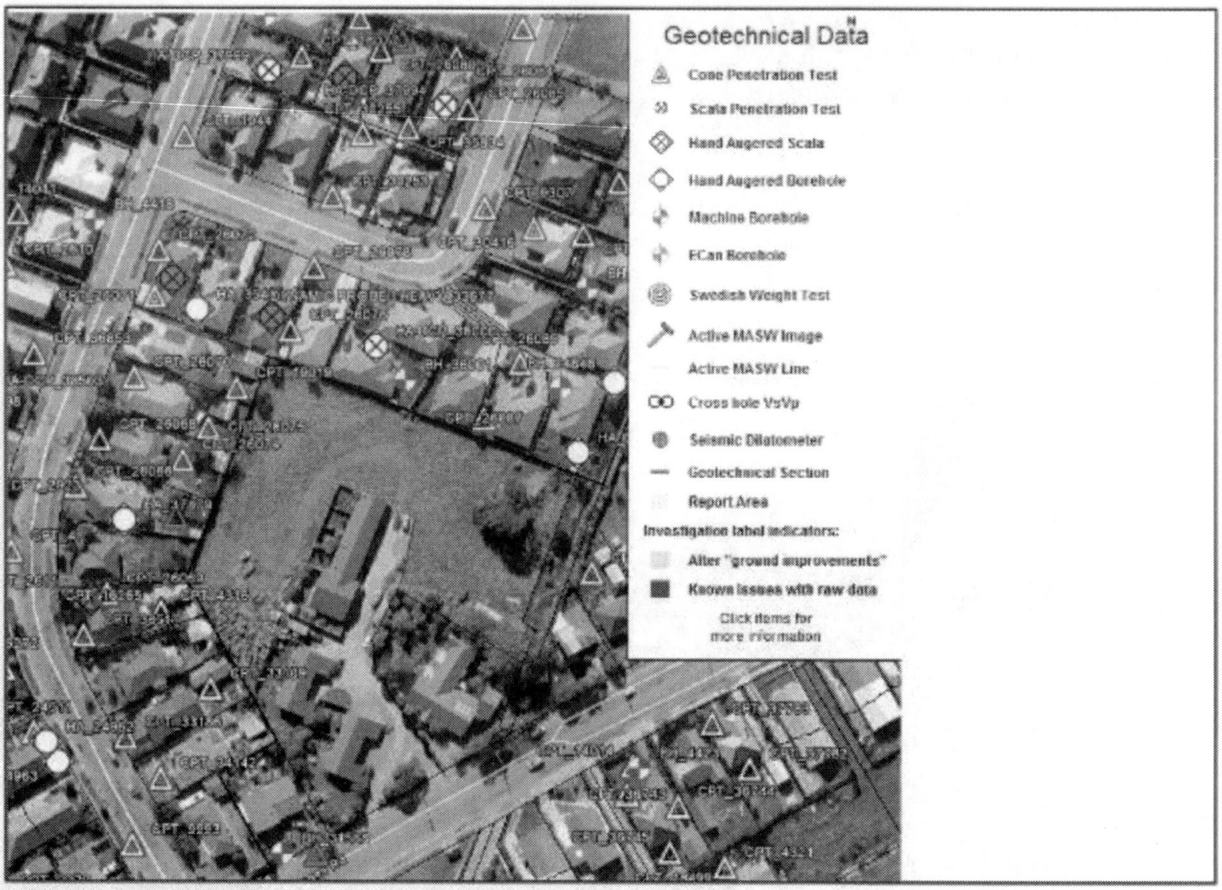

Figure 9 – Data Delivery with Data Interchange Standard

In Hong Kong, Arup utilized thousands of boreholes from AGS data files that had been compiled over the last 20 plus years for designing the new MTR Station as part of the Hong Kong subway system in a 3-D BIM environment [7]. The project involves construction of 1.8km twin railway tunnels, a new underground station with interchange facilities, as well as ventilation buildings and shafts in Wan Chai. Without data interchange standards this level of detail in the design and visualization would not have been possible.

Summary

Standardized digital geotechnical data Interchange using something such as the AGS data interchange standard or the DIGGS data interchange standard, provides tremendous benefits for all aspects of the project lifecycle. Whether the organization using it is an Owner, Consultant, or Contractor, all parties have financial benefits. All parties also have reduced risk as a result of reducing or eliminating vast amounts of data re-entry. Furthermore, more time can be spent performing engineering analysis and design versus simply reinputting the same data. With data interchange, organizations can use available historical data as well as all combine all types of geotechnical data from boreholes, lab testing and in-situ testing together in a single data management system for more advanced analysis, visualization, and data mining.

Digital data interchange is a key part of advancing our industry and keeping it relevant with technological advances. As organizations create vast data repositories of geotechnical data, data mining, artificial intelligence, and other cutting-edge data analytic technologies will allow users of geotechnical data to analyze and visualize data in ways unimaginable today. Organizations such as DOTs will be able to be become proactive regarding potential hazards or potential failures versus being reactive today whether it is predictions related to rockfalls, slope failures, pavement performance, or any other geotechnical or geologic hazard that can impact our transportation infrastructure.

REFERENCES:

1. BS 8574, *Code of Practice for the Management of Geotechnical Data for Ground Engineering Projects*. British Standards, 2014
2. Human Error Website. *Basic Error Rates.* University of Hawaii http://panko.shidler.hawaii.edu/HumanErr/Index.htm. Accessed June 20, 2018
3. Elmasri & Navathe. *Fundamentals of Database Systems.* Addison-Wesley, Reading, Massachusetts, 2000
4. National Economic Development Office, Faster building for commerce, NEDO, London, 1988
5. National Audit Office, Modernising construction, HC87, 11 January, The Stationery Office, 2001
6. Website - What are the Advantages of an Assembly Line? https://www.wisegeek.com/what-are-the-advantages-of-an-assembly-line.htm, Accessed June 20, 2018
7. Arup website. *Arup named top global winner at Autodesk Excellence in Infrastructure Competition.* https://www.arup.com/news-and-events/news/arup-named-top-global-winner-at-autodesk-excellence-in-infrastructure-competition. Accessed June 20, 2018.

Geotechnical Solutions for the I-95 Betsy Ross Bridge Interchange

Structure Alternatives over Soft Soils in Spaghetti Junction

Sarah McInnes, P.E.
Pennsylvania Department of Transportation, Engineering District 6-0
7000 Geerdes Boulevard
King of Prussia, PA 19406
610-205-6544
smcinnes@pa.gov

Geoff Stryker, P.E.
STV Incorporated
1818 Market Street, Suite 1410
Philadelphia, PA 19103-3616
717-471-0600
geoffrey.stryker@stvinc.com

John Pizzi, P.E.
STV Incorporated
1037 Raymond Boulevard, Suite 200
Newark, NJ 07102
973-642-3191
john.pizzi@stvinc.com

Prepared for the 69th Highway Geology Symposium, September 2018

Acknowledgements

The authors would like to thank The U.S. Federal Highway Administration; PennDOT Engineering District 6-0; STV Incorporated; Elastizell Corporation of America; and Menard Group, USA for their work on the various aspects of this project.

Disclaimer

Statements and views presented in this paper are strictly those of the author(s), and do not necessarily reflect positions held by their affiliations, the Highway Geology Symposium (HGS), or others acknowledged above. The mention of trade names for commercial products does not imply the approval or endorsement by HGS.

Geotechnical Solutions for the I-95 Betsy Ross Bridge Interchange

Structure Alternatives over Soft Soils in Spaghetti Junction

Author: Sarah McInnes, P.E., Pennsylvania Department of Transportation, Engineering District 6-0, 7000 Geerdes Boulevard, King of Prussia, PA 19406, phone: 610-205-6544, smcinnes@pa.gov

Coauthor: Geoff Stryker, P.E., STV Incorporated, 1818 Market Street, Suite 1410, Philadelphia, PA 19103-3616, phone: 717-471-0600, geoffrey.stryker@stvinc.com

Coauthor: John Pizzi, P.E., STV Incorporated, 1037 Raymond Boulevard, Suite 200, Newark, NJ 07102, phone: 973-642-3191, john.pizzi@stvinc.com

ABSTRACT

PennDOT's multi-billion-dollar I-95 corridor reconstruction effort in Philadelphia includes replacement of the existing mainline, roadway and ramps in Northeast Philadelphia from Vine Street to the Cottman Avenue Interchange. A major aspect of the corridor reconstruction is the $880 million Betsy Ross Bridge Interchange Reconstruction (referred to as section BRI) in the Bridesburg section of Philadelphia just west of the Betsy Ross Bridge. This complex, three-level interchange, referred to as "Spaghetti Junction," services 160,000 vehicles per day and is the connection point for major Philadelphia arterial routes including Aramingo Avenue and Richmond Street. The first phase of section BRI (referred to as section BR0) connects I-95 more efficiently to the local street network and improves access to the Betsy Ross Bridge into New Jersey.

The present alignment of the mainline and several ramps traverse an area through which the meandering Frankford Creek once passed. During construction in the 1960s the creek was relocated and portions of I-95 were constructed on low-level structure spanning the soft soils comprising the creek deposits beneath. Section BR0 includes Ramp EE-F, which traverses the soft soil. During design, the design team investigated alternatives to bridge structure with the goal of eliminating bridge deck area and reducing construction costs, infrastructure life-cycle costs and future maintenance. The team designed a solution utilizing a combination of compensating fill and column supported embankment for Ramp EE-F, anticipating utilizing one or both applications on future construction of the I-95 mainline and other ramps, resulting in a $160 million savings in construction costs and eliminating 650,000 square feet of bridge deck.

This paper presents the project design, geotechnical challenges, constructability and ultimate successes.

INTRODUCTION & PROJECT DESCRIPTION

PennDOT's multi-billion-dollar I-95 corridor reconstruction effort in PennDOT's Engineering District 6 (comprised of the five-county region in Southeastern Pennsylvania) includes replacement of the existing mainline viaducts, roadway and ramps in Philadelphia from Vine Street to the Cottman Avenue Interchange. One of five sections of the corridor reconstruction is the Betsy Ross Bridge Interchange Reconstruction (referred to as section BRI) northeast of Center City in the Bridesburg and Port Richmond neighborhoods of Philadelphia. This design section, constructed in the 1960s and partially completed in 1973 & 1997, includes the Betsy Ross interchange, which connects I-95 to the Betsy Ross Bridge and New Jersey Route 90, and its associated ramps and connections to surface streets. This geometrically complex, three-tiered "Spaghetti Junction" interchange services 160,000 vehicles per day and is the connection point for the Betsy Ross Bridge and major Philadelphia arterial routes including Aramingo Avenue and Richmond Street. The BRI project section comprises five design sections, BR0 and BR2 through BR5, and the overall goals of the project are to improve traffic movement between I-95 and the Betsy Ross Bridge, complete ramps and connections not previously constructed, increase capacity of the mainline and increase service life of the structures. The first phase of construction, section BR0 and the subject of this paper, connected I-95 to Aramingo Ave. and NJ-90 and focused on interchange ramp construction and improvements to support the connections.

Figure 1: Section BR0 Shown in Green

The present alignment of the mainline and several ramps traverse an area through which a meandering Frankford Creek once passed. During construction in the 1960s the creek was relocated and portions of I-95 were constructed on low-level structure spanning the soft soils comprising the creek deposits beneath. Section BR0 includes Ramp EE-F, which traverses the soft soil. The team designed a solution utilizing a combination of compensating fill and column supported embankment for Ramp EE-F, anticipating utilizing one or both applications on future construction of the I-95 mainline and other ramps, resulting in a $160 million savings in construction costs and eliminating 650,000 square feet of bridge deck in the BRI section.

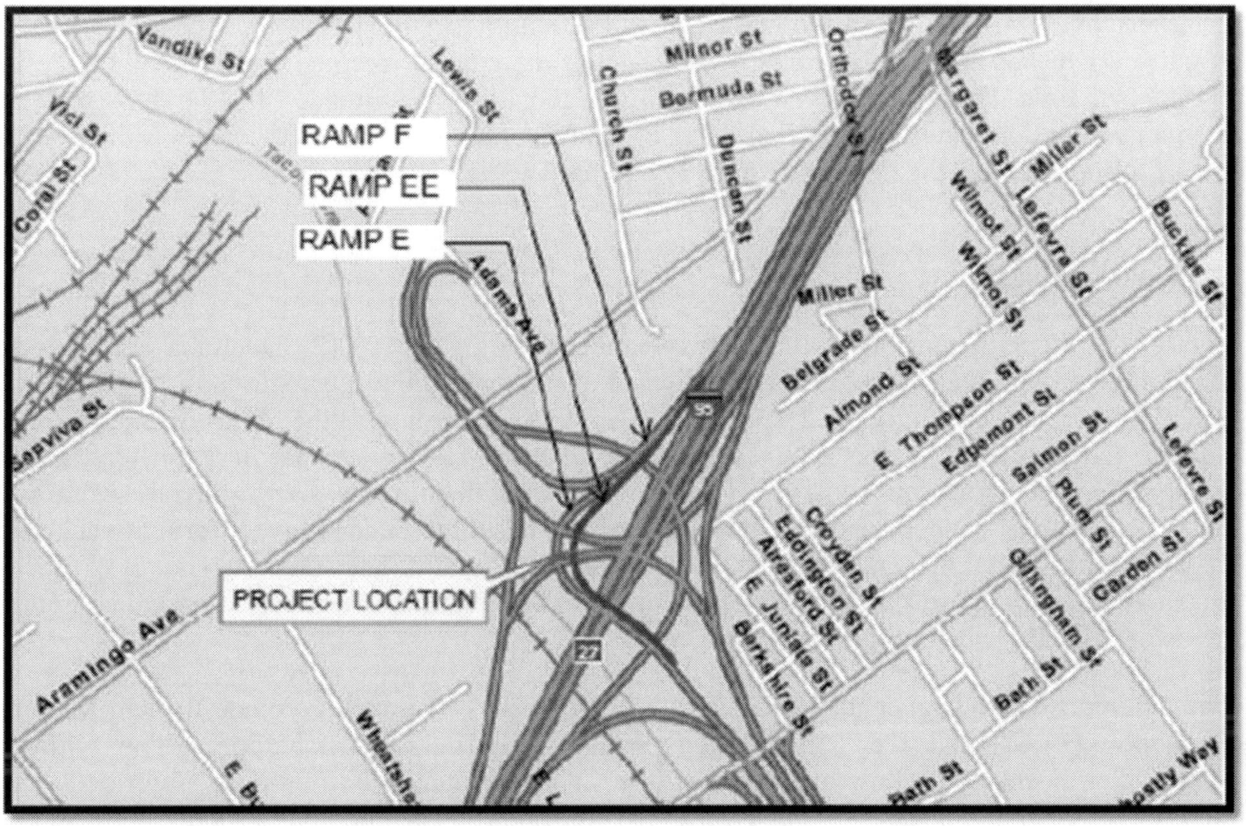

Figure 2: Location of Ramps E, F & EE

SOILS AND GEOLOGY

Physiography & Topography

The project site is located in the Lowland and Intermediate Upland section of the Atlantic Coastal Plain Physiographic Province. It is located near the boundary of the Coastal Plain and the Piedmont Physiographic Province, which is known as the Fall Line. The topography is characterized by flat upper terrace surfaces cut by shallow valleys of very low relief and the Delaware River floodplain. The area is underlain by unconsolidated to poorly consolidated sand and gravel deposits over complexly folded and faulted metamorphosed sedimentary and igneous rocks, primarily schist and gneiss. The drainage patterns are dendritic. Consequently, the project site is underlain by mostly unconsolidated to poorly consolidated sand and gravel.

Soil Survey

The Soil Survey Map of Bucks and Philadelphia Counties indicates the predominant soil within the project area is Urban Land (Ub). This designation represents highly variable and disturbed materials, generally including fill, resulting from previous construction and various land uses over time. Urban structures and works cover so much of this land type that identification of the soils is not practical. Most areas have been smoothed and the original soil material has been disturbed, filled over, or otherwise destroyed over time.

Adjacent to the Delaware River, the project area consists of loose man-made fills of various materials overlying native soils deposited by the action of the Delaware River and locally, Frankford Creek. They consist primarily of granular material intercepted by lenses of clayey and organic silt soils. The uppermost strata are, for the most part, man-made fills. Sand, organic and inorganic silts dominate the stratified deposits, interspersed with lenses of clayey soils. At the site of Ramps EE & F the dominant soils are alluvial and very soft.

Regional Geology

Figure 3 presents a portion of the Pennsylvania Geological Map (Philadelphia and Camden Quadrangles) with the project location indicated. As shown, the project site is mapped as being underlain by the Quaternary-aged Trenton Gravel (Qt). According to the Pennsylvania Geologic Survey and described in the *Engineering Characteristics of the Rocks of Pennsylvania*, the Trenton Gravel formation consists of gray to pale reddish-brown, very gravelly sand with interbedded, cross-bedded sand and clay-silt layers. These interbedded layers form a wedge that begins at the Fall Line and thickens toward the southeast. The Trenton Gravel is deeply weathered and composed of outwash and alluvium that consists of weathered gravel of granite, sandstone, gneiss, siltstone, and quartzite.

The Trenton Gravel generally overlies the Cretaceous System Potomac-Raritan-Magothy Formation (PRM) and the Tertiary-aged Bridgeton Formation (included in Tpb with the Pennsauken Formation) (Low and others, 2002). The Bridgeton Formation is described in the *Geohydrology of Southeastern Pennsylvania* as stratified, feldspathic quartz sand with local beds of fine gravel. Clay and silt beds are rarely present. This formation is described in the *Engineering Characteristics of the Rocks of Pennsylvania* as consisting of extensively crossbedded clayey sand, stained reddish-brown. Underlying this, the sand is either yellow or white and irregularly stained reddish to orange-brown. Beds of gravel are locally present and the gravel is mostly of vein quartz, chert, and quartzite. The maximum thickness of this formation is approximately 30 feet. The Bridgeton and Trenton Gravel Formations have a combined thickness of up to 80 feet (Low and others, 2002).

Oligoclase-mica schist (Xw) of the Wissahickon Formation underlies the unconsolidated formations described above. This metamorphic rock is composed of quartz, feldspar, muscovite, and chlorite mineral constituents. The oligoclase-mica schist variation of the Wissahickon Formation is coarsely crystalline, excessively micaceous, and has abundant feldspar. The estimated thickness of the Wissahickon Formation is 8,000 to 10,000 feet. A sometimes deep saprolitic zone often forms above this rock as the result of weathering.

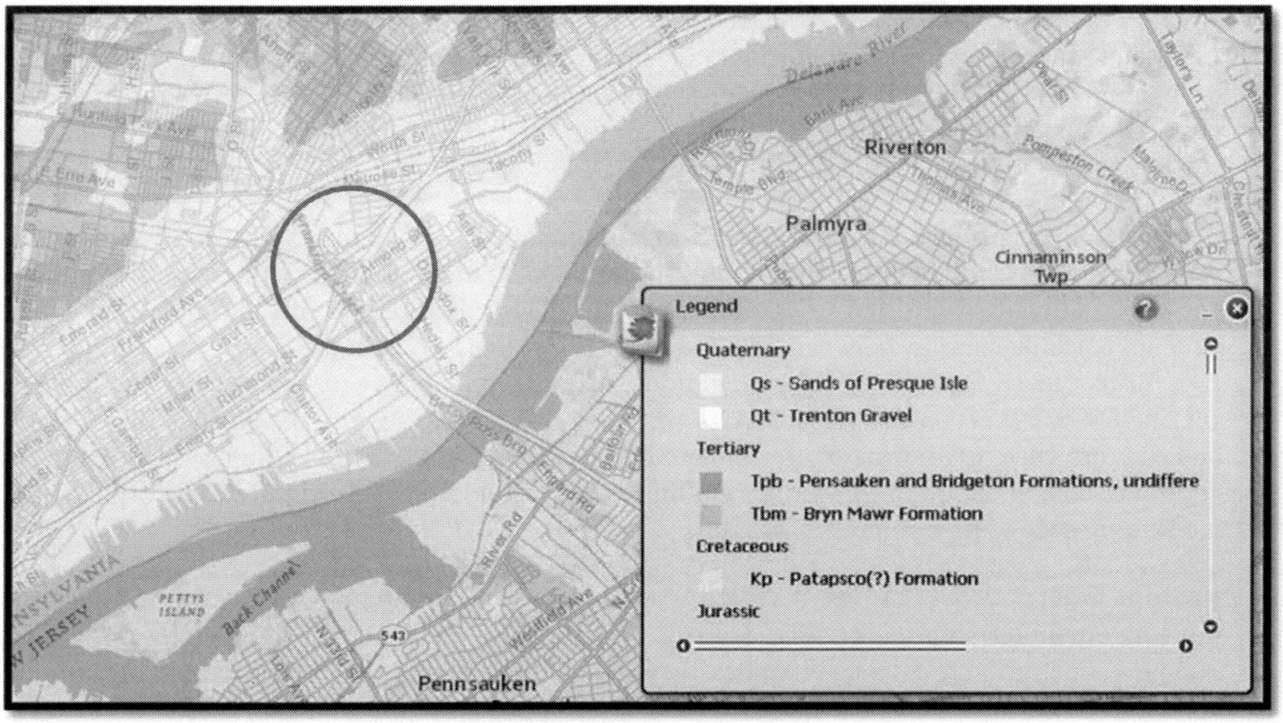

Figure 3: Geology Map

SUBSURFACE INVESTIGATION

The investigation for this project commenced with a review of available data from several historical subsurface investigation programs conducted for this section of I-95 in order to initially characterize the subsurface conditions and guide the design of the supplemental subsurface investigation for the BR0 project. Field observations of the ground below the existing low-level ramp structure indicated that the soils at the site were continuing to settle under their own self-weight, an indication that a careful and thorough evaluation of the strength and settlement properties at the site was required. The final field subsurface investigation program advanced in stages and consisted of 23 SPT soil borings and 12 cone penetrometer soundings. The program also included a series of laboratory testing to confirm the classification of the soils, measure typical index properties, and evaluate compressibility, strength, and corrosion potential. The program also included a series of observation wells installed in selected boreholes and fitted with automated data loggers to record variations in groundwater levels over time. The groundwater data proved invaluable when designing the compensating fill portion of the roadway embankment.

SUBSURFACE CONDITIONS

The following briefly summarizes the major soil strata as represented throughout the project area with increasing depth:

Stratum F (Fill): All the borings encountered a surface stratum of man-made fill material that varies in thickness from 5 feet to 26 feet. The fill material generally consists of medium dense silty sand and gravel with inclusions of brick, glass, plastic, wood, concrete, and refuse throughout. The material typically classifies as A-1-a through A-2-4, with zones of A-4. Standard penetration resistance N-values vary from 1 bpf (very loose) to greater than 50 bpf (very dense).

Stratum O (Organic Silt): All the borings encountered a stratum of medium stiff alluvial brown to gray organic silt underlying the fill material. The material typically classifies as A-7-5 and has a ratio of oven-dried to natural liquid limit less than 0.75, with the measured organic content typically varying between 5% and 10% by weight. The thickness of Stratum O varies from 10 feet to 18 feet. Standard penetration resistance N-values vary from weight of rod (very soft) to 26 bpf (very stiff). The results of the consolidation tests indicate that these alluvial materials are currently normally consolidated under the weight of the existing overburden. Thus any future consolidation of this stratum in the absence of additional loading will be primarily due to on-going secondary compression effects.

Stratum SM1: The borings revealed a relatively thin medium dense silty sand stratum immediately underlying the organic deposits. The maximum thickness of this stratum is 7 feet. Standard penetration resistance N-values vary from 5 bpf (loose) to 25 bpf (medium dense).

Stratum GP: Underlying either Stratum O or Stratum SM1 is a layer of dense sandy gravel across the entire earth fill section footprint. This stratum appears to be of outwash or alluvial origin as evidenced by the rounded gravel particles. The maximum thickness of this stratum is 8 feet. Standard penetration resistance N-values vary from 10 bpf (medium dense) to greater than 50 bpf (very dense).

Stratum SM2: A layer of medium dense silty sand immediately underlies Stratum GP. The thickness varies from 3 feet to 25 feet. Standard penetration resistance N-values vary from 3 bpf (loose) to 43 bpf (dense).

Stratum SM3: A layer of very dense silty sand immediately overlies bedrock throughout this section. The thickness varies from 3 feet to 19 feet. Standard penetration resistance N-values vary from 24 bpf (medium dense) to greater than 50 bpf (very dense), with an average value of greater than 50 bpf (very dense).

Stratum DR: Several borings revealed a zone of highly decomposed bedrock as evidenced by the consistent refusal of the spoon and classifications of samples recovered. The boring log classifies the recovered saprolite samples as micaceous silty sand (residual) followed by very soft biotite gneiss.

Stratum R: The borings reveal that the top of bedrock elevation varies along the alignment from about Elev. –28 feet to Elev. –55 feet. Visual classifications of the bedrock included mica schist, amphibolite, and biotite gneiss. The upper few feet of rock is highly weathered, but generally hard below its weathered crust. Rock core recoveries ranged from 0% to 100%. Measured RQD values ranged from 0% to 100%.

The average groundwater elevations at the site typically range between approximately Elev. +1.6 feet and Elev. +3.6 feet. The observed groundwater elevations in the well closest to Ramp EE-F range from +0.99 to +4.4 feet with an average of +2.2 feet.

Corrosion potential testing indicates that the fill material soil corrosivity rating varies from mildly corrosive to very corrosive. Stratum O generally classifies as very corrosive.

SITE CHARACTERISTICS & CONSTRAINTS

The project site had many challenging site characteristics and constraints. Overall the interchange is located in an urban environment, bounded by AMTRAK's Northeast Corridor to the west, the Delaware River to the East, and surrounding commercial, industrial and residential areas to the north and the south. Site features within the interchange include two Conrail Shared Assets facilities, Frankford Creek, a PennDOT Maintenance Yard, Philadelphia Electric Company and AMTRAK transmission lines, and underground combined sewer outflows.

Historically, Frankford Creek flowed to the Delaware River generally along the alignment of I-95 from south of the Betsy Ross Bridge interchange towards Orthodox Street. In 1955, prior to the construction of I-95, the creek was relocated to its current location parallel to the Conrail Del Air Branch, which can be seen in Figure 4. As part of this relocation, according to the *Report on Flood Control Frankford Creek* (Knappen Engineering Company, 1946), the adopted alternative directed filling the former bed of Frankford Creek in conjunction with the construction of two underground combination sewer outfalls. The fill material consisted primarily of coal waste, cinder and ash, and other various urban fill.

The interchange was constructed in multiple phases after the original interstate construction was completed in the 1960s. Beginning in the early 1970s with the construction of the Betsy Ross Bridge as part of the Pulaski Highway, new connection ramps were built to and from I-95 along with collector/distributor roads along the interstate alignment. In the 1980s, after portions of the original low-level structures began to settle, portions of the mainline were replaced. In the 1990s, interchange connections were completed to Aramingo Ave west of the interchange, which allowed better access to the industrial and commercial areas to the west and south. These ramps relieved truck traffic within the residential areas of Bridesburg, north and east of the interchange between I-95 and the Delaware River.

As a result of the nature of the urban fill that was used during the flood control project and the mixed industrial/commercial land use over the years, soil investigations and testing during design revealed that the soils within the interchange contained various contaminants exceeding regulatory levels that classified them as residual and/or hazardous waste. This significantly impacted the application of embankment alternatives, as the quantity of material generated for disposal would be a significant cost driver. Subsequent groundwater testing during construction revealed contamination, resulting in additional treatment during dewatering operations.

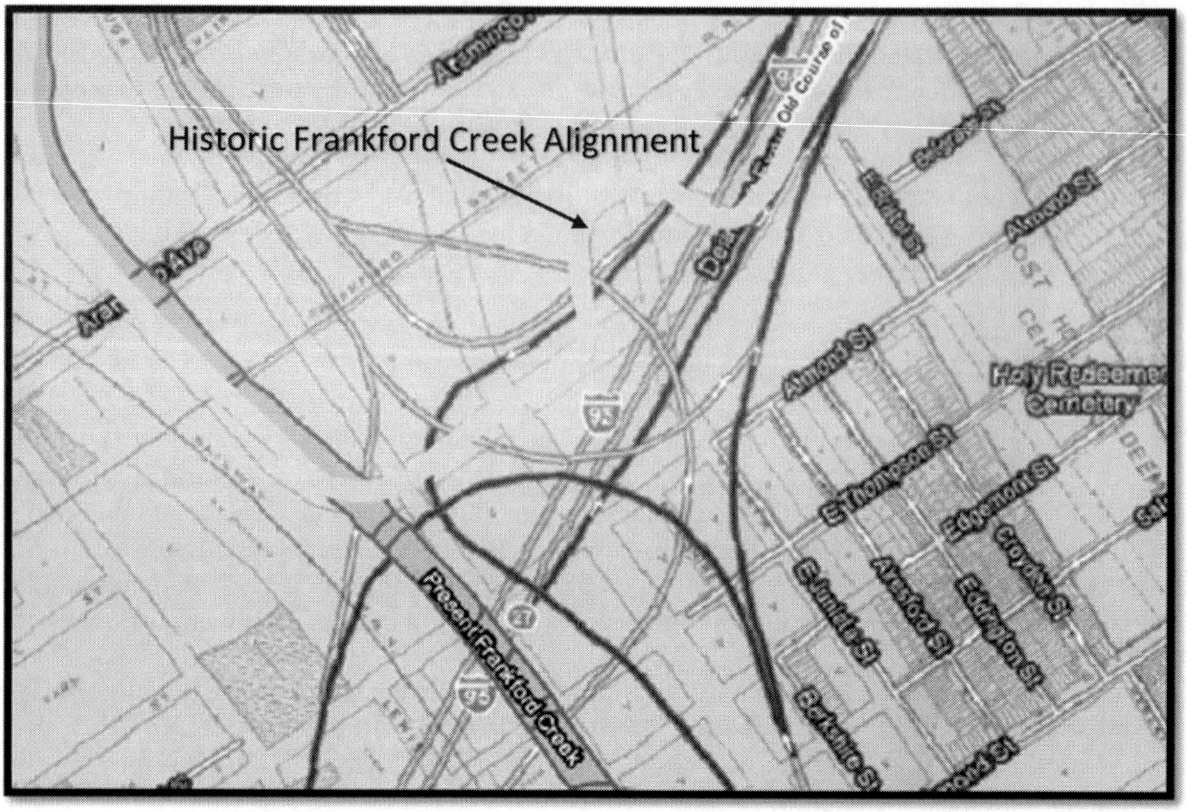

Figure 4: Historic vs. Present Frankford Creek Alignment

This portion of Frankford Creek is part of a detailed FEMA study area, and a large portion of the interchange west of I-95 is within the 100-year flood plain. This means that the increase to the post construction 100-year water surface required additional permitting considerations and a flood map revision. The flood plain limits include the area of application of the embankments for BR0. As a result, any embankment construction within the flood plain needed to not impact the 100-year flood elevations, and buoyancy was considered for light-weight fill alternatives. Both of these factors drove the limits of application of each geotechnical alternative. The flood plain and 100-year water surface elevation also constrained the roadway profiles, so that positive drainage and roadway freeboard during high storm events could be provided. These factors impacted required design heights for the embankments.

As mentioned previously, the flood relief project constructed underground combination storm sewers, on top of which the existing low-lying I-95 structures and the interchange ramps were constructed. As part of the design development of the new embankments future access and maintenance of these sewers by the City of Philadelphia Water Department (PWD) was critical. A reconfiguration of the sewer network and relocation of these facilities outside the footprint of I-95 was developed. While every effort was made to reduce these conflicts, there were several major crossings required which had to be accounted for in the design of the embankments.

As discussed, replacing the low-lying structures within this network of ramps presented a number of constraints during the project. During construction, PennDOT managed to maintain traffic, as well as access to both the maintenance yard and railyard in addition to relocating the underground combination sewers while maintaining service. There were numerous constructability issues during construction. The variability of the urban fill encountered on the project site and the existing structure foundations slowed excavation operations to extract existing piles and remove non-native debris. Pockets of contaminated soils, when discovered, required in-situ soil testing, segregation, and re-handling for disposal. While obstructions were anticipated in the design, drilling operations were slowed when they were encountered requiring rig reset, predrilling or obstruction removal. The variation of soil strata on the project site also presented challenges to ensuring column depths met design requirements.

DESIGN ALTERNATIVES & REQUIREMENTS

Alternative A - Compensating Fill Embankment

In general, compensating fill embankment design and construction utilizes lightweight fill materials in combination with limited excavation of existing subgrade materials to raise the roadway grade without increasing the effective overburden pressures on the existing underlying soil strata. The weight of the new embankment fill is compensated for by replacing some of the existing subgrade material with lightweight engineered fill to produce no net additional, or even a slight reduction, in load.

The compressible soils at this site were determined to be normally consolidated under the weight of the existing fill material placed during the original roadway construction and creek relocation. Because the compensating fill design does not impose additional loading upon the existing subgrade soils, the new embankment construction does not result in any immediate settlement or long term primary consolidation settlement. The new embankment will experience very minor long term settlement due to ongoing secondary compression of the underlying compressible soils which is a residual effect of the historical placement of the existing fill materials. The prediction for this secondary consolidation settlement was less than one inch over a 40- to 50-year roadway life cycle.

During the preliminary design phase, several alternative materials were evaluated for use as a lightweight engineered fill. The materials included fly ash and air-cooled slag (70 to 95 pcf), expanded shale (40 to 65 pcf), lightweight foamed concrete (20 to 50 pcf) and expanded polystyrene (1 to 2 pcf). Given environmental considerations, durability and range of unit weights available, the design team selected lightweight foamed concrete as the best option.

For the purposes of determining excavation depths and thicknesses of lightweight foamed concrete, two classes of lightweight foamed concrete material were considered:

- Class IV is a higher density and higher strength material to better help distribute traffic loads immediately beneath the roadway pavement structure, with a density of 42 pcf, a minimum compressive strength of 120 psi and a fixed thickness of 2 feet.

- Class II is a lower density and lower strength material of variable thickness as fill between the bottom of Class IV and the bottom of excavation with a density of 30 pcf and a minimum compressive strength of 40 psi.

The design for the compensating fill section required an assessment of likely unit weights of the existing subgrade material in order to determine the depth to be excavated to balance the new embankment load. Based on the data available from Standard Penetration Tests, Cone Penetrometer Tests, and laboratory tests, design analyses were performed for unit weights of fill ranging from 90 pcf to 105 pcf in 5 pcf increments. Using the test data, engineering judgment, and an interest in balancing economy with safety, the design team selected a unit weight of existing subgrade material equal to 95 pcf for of the development of the final embankment design sections.

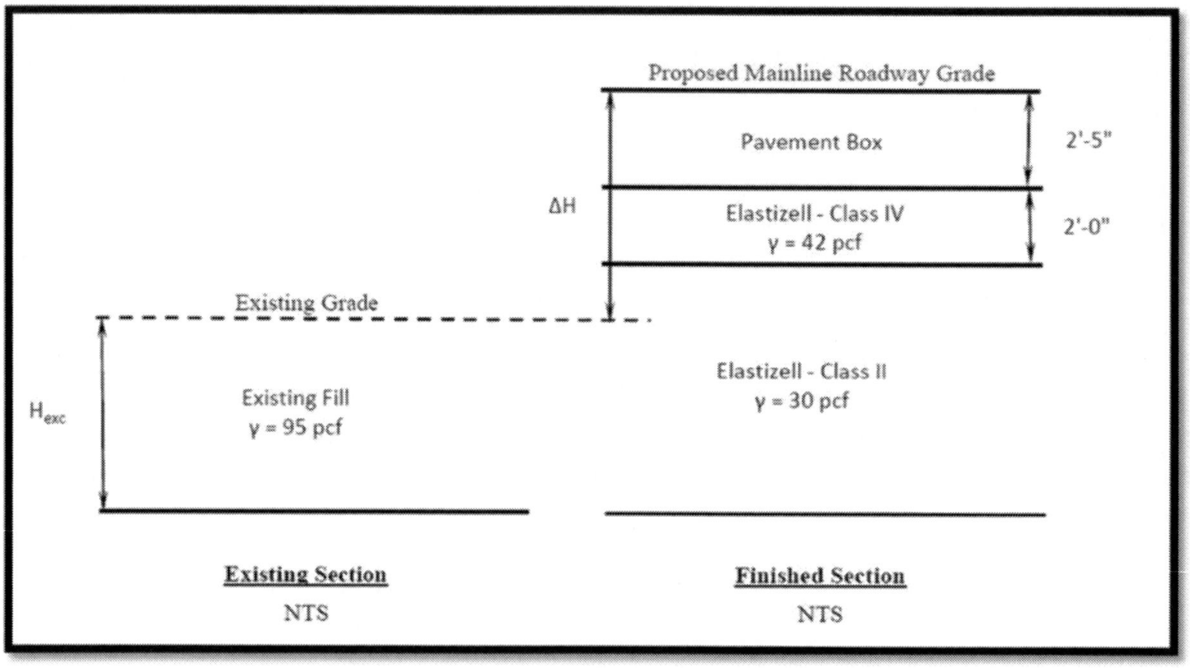

Figure 5: Compensating Fill Embankment Concept

The required excavation elevations ranged from +3.0 feet to +7.5 feet, which maintained the bottom of excavation above the calculated average groundwater elevation of +2.2 feet. However, the 100-year flood water surface elevation for the project site is Elevation +15.2 feet. Because the lightweight foamed concrete engineered fill material is lighter than water, an evaluation of the potential for developing net uplift due to buoyancy during such a flood was necessary. This evaluation consisted of estimating the factor of safety against uplift, defined as the ratio of the total weight of the pavement and engineered fill column to the weight of the column of water between the bottom of excavation and the 100-year flood level. The results of this evaluation yielded an acceptable factor of safety equal to or greater than 1.1 for all roadway segments designated for compensating fill embankment.

Alternative B - Column Supported Embankment

This geotechnical engineered system involves the construction of a roadway embankment supported on a geosynthetic-reinforced granular soil load transfer platform (LTP) that is, in turn, supported on a pattern of vertical columns or inclusions extending from existing subgrade to a suitable bearing stratum at depth. The inclusions may consist of driven or drilled piles, vibro-concrete columns, controlled modulus columns, etc. The approximate total height of embankment supported by the vertical elements on this project, including the load transfer platform and pavement, is approximately 13.5 feet.

A geotechnical design analysis based on the Collin Beam method was performed to design the following elements for the column supported embankment:

- Column spacing
- Column load
- Load Transfer Platform (LTP) thickness
- LTP base reinforcement requirements (lateral spread resistance and catenary support)
- Tensile load requirements for LTP reinforcement

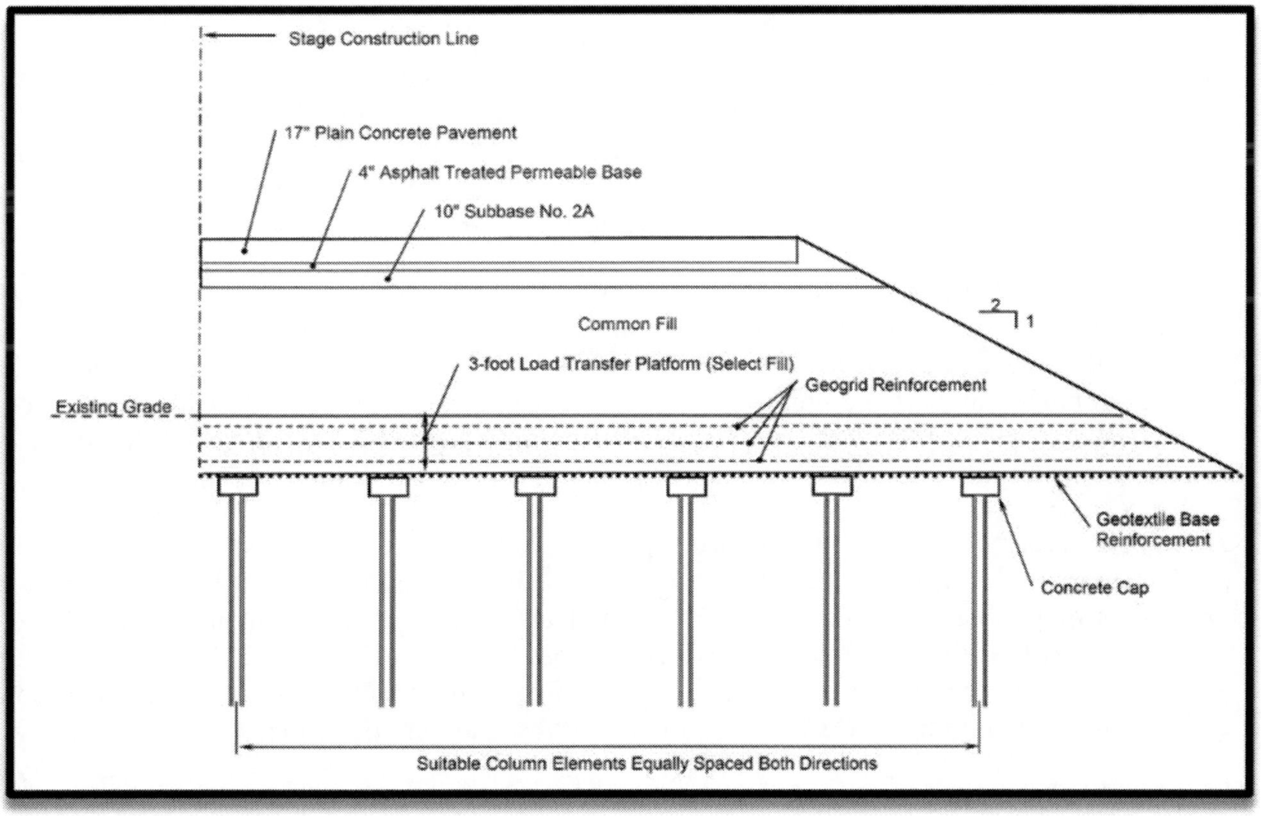

Figure 6: Typical Column Supported Roadway Section

Using this methodology, the minimum load transfer platform thickness is a function of the column spacing and is equal to or greater than one-half the clear span between columns. For the BR0 project, the design team adopted a column spacing equal to 8-foot center-to-center in a square pattern to facilitate layout and construction. Specification of a 2-foot width/diameter cap at the top of the vertical column elements results in a clear span between columns equal to 6 feet. This column spacing and pattern is consistent with the use of a 3-foot thick load transfer platform.

Based on the weight of the pavement section, the embankment material, the weight of the LTP, and an assumed live load equivalent uniform surcharge representing the traffic load equal to 360 psf, the maximum unfactored column load corresponding to this configuration equals 65 tons.

The Contract Drawings presented a design utilizing driven steel H-piles as the vertical supporting elements. Static analysis demonstrated that steel HP12x53 pile sections driven to end-bearing on bedrock would be adequate to support the required column loads. These piles required a 2-foot wide/diameter precast cap installed at the top for proper load distribution and steel tip reinforcement at the toe for seating the pile into bedrock. One factor leading to the selection of end bearing driven piles was that unlike the compensating fill alternate, this column supported embankment alternative would not experience long term settlement due to ongoing secondary compression of the underlying compressible soils. Nevertheless, the specifications permitted contractor-selected alternate designs using prestressed precast concrete piles, vibro concrete columns, and controlled modulus columns for the vertical column elements provided the alternate designs could satisfy a post-construction settlement criteria on the order of one to two inches.

With the Collin Beam method, the load transfer platform included a minimum of three layers of internal geosynthetic reinforcement to stiffen the platform and develop beam-type action in transferring the embankment load to the columns. The LTP fill material consists of select granular material. Analysis demonstrated that three layers of geogrid or geocell reinforcement were adequate to reinforce the load transfer platform for this project.

The 3-foot recommended depth of excavation accounted for removal of existing surface obstructions as well as grading and proof rolling of the subgrade while providing a suitable working platform and adequate embankment height above the load transfer platform to permit full development of soil arching. Bottom of excavation levels remained above the local groundwater table level. In addition, because the column supported embankment utilized select and common soil fill material, buoyancy under 100-year flood conditions was not an issue.

A base reinforcement geotextile provided separation between the subgrade and the select fill to resist lateral spreading of the new embankment fill and to support the soil below the zone of arching through catenary action. Based on analysis, base reinforcement design specified a woven geotextile with a minimum Long Term Design Strength (LTDS) at 3% strain of 4.4 kips per foot width. Based on similar analysis, the specified intermediate reinforcements consisted of geogrids with a minimum LTDS at 3% strain of 360 lbs/lf in both directions.

CONSTRUCTION & SETTLEMENT MONITORING

The Contractor bid the reconstruction of Ramp EE-F on the basis that the compensating fill segment would be constructed in accordance with the design presented in the contract documents with no significant modification.

For the column supported embankment segment, the contractor proposed and bid an alternative design using controlled modulus columns (CMCs) and an alternative LTP designed by Menard USA as the vertical elements. This CMC alternative was approved by PennDOT and successfully implemented.

The CMCs were designed to be installed from a pre-excavated working pad elevation through the Stratum F and Stratum O to achieve an adequate bearing capacity within Stratum SM1 and Stratum GP. The final drilling depth of the production CMCs was based on automated field observations of multiple drilling parameters (torque, rotation speed, vertical speed of penetration, downward thrust, etc.) with control values established after review of the results of successful CMC load tests conducted to confirm vertical bearing capacity of the inclusions. The CMCs were cut-off approximately 12 inches into the working pad. Production rates averaged 40 to 50 CMCs per day.

Following CMC installation, a 2-foot-thick LTP was installed on top of the working pad. Two layers of geogrid were installed – one layer between the top of working pad/bottom of LTP, and another layer 1 foot into the LTP. The Contractor constructed the remainder of the embankment and pavement using standard means and methods.

Because the CMC design did not bear on the underlying bedrock, settlement analyses were necessary to predict immediate and long term settlements. The immediate settlement would occur during construction as load was applied (such as placement of the LTP and construction of the embankment). The long-term settlement would occur over time as the compressible soils which received some load from the vertical inclusions further consolidated. The majority of the estimated settlement was expected to be long-term. Menard used finite element modeling to predict total settlements at the top of the LTP ranging from 1.2 inches to 2.4 inches, with post-construction settlement ranging from 0.8 inches to 1.7 inches.

Construction monitoring included settlement platforms set at maximum 100-foot intervals along each ramp structure segment and survey points on adjacent structures. Final settlement at the completion of embankment construction for the Compensating Fill Embankment varied from 0.00 inches to 0.08 inches and for the Column Supported Embankment settlement was recorded as 0.00 at all settlement plates.

Figure 7: Compensating Fill Construction with Precast Facing Panels

Figure 8: Column Supported Embankment Drilling Operation

Figure 9: Aerial View of Completed Section BR0

CONCLUSIONS

The implementation of both geotechnical embankment alternatives to structure construction contributed to the success of the BR0 project. Section BR0's construction was substantially completed within the required schedule and opened to vehicular traffic in August 2017. There are several recognized benefits of these embankment alternatives to structures along with lessons learned.

- The application of these geotechnical alternatives eliminated more than two percent of District 6-0's approximately 25 million square feet of bridge deck while reducing future maintenance and asset management costs.
- Both Compensating Fill and Column Supported Embankment alternatives are viable applications for the I-95 corridor.
- Application of an embankment alternative to future sections of the BRI project and other design sections on the I-95 corridor will contribute to further short- and long-term cost savings and reduction of bridge deck area and structure maintenance.
- A possible future alternative could include a combination of both applications to achieve higher embankment heights.

- The estimated construction cost of the H-Pile column supported embankment was $3.3M the Alternative CMC Column Supported Embankment item bid price was $1.8M resulting in a $1.5M savings.

- A constructability advantage of CMCs compared with H-Piles is production and construction durations. Production rates for the CMCs were 40 to 50 per day compared with average production rate of five to eight H-Piles per rig per day. The achieved CMC production rate equates to approximately 22,000 square yards of area in a few weeks.

- The Column Supported Embankment alternative is preferred over Compensating Fill for the following reasons:

    - Lack of waste generation – CSE requires much less excavation of contaminated soils

    - Ability to use contaminated waste soils from adjacent design sections

    - Faster construction production rates

    - Little to no support of excavation

    - Construction is not weather sensitive. The compensating fill operation was shut down quite frequently due to rain and unsuitable weather conditions.

With the success of these alternatives on section BR0 there are already plans to use this on two other I-95 project sections and another project in District 6.

REFERENCES

1. Berg and Dodge, "Atlas of Preliminary Geologic Quadrangle Maps of Pennsylvania", Pennsylvania Bureau of Topographic and Geologic Survey, 1981.

2. Geyer, McGlade and Wilshusen: "Engineering Characteristics of the Rocks of Pennsylvania", Pennsylvania Bureau of Topographic and Geologic Survey, 1982.

3. Low, Dennis J., Daniel J. Hippe, and Dawna Yannacci. 2002. Geohydrology of Southeastern Pennsylvania, Water-Resources Investigations Report 00-4166. United States Department of the Interior, U.S. Geological Survey, New Cumberland, PA.

4. Pennsylvania DCNR, "The Geology of Pennsylvania", 1999.

5. STV Incorporated, Preliminary Geotechnical Engineering Report prepared for 0095 Section BRI, February 2011.

6. STV Incorporated, Final Geotechnical Engineering Report prepared for 0095 Section BRI, April 2014.

7. URS Corporation, Foundation Recommendation Reports prepared for 0095 Section GR2, July 2011.

Risk Assessment for Long-Term Performance of US 101 and Alternative Alignments, Del Norte County, California

Scott A. Anderson
(Corresponding Author)
BGC Engineering Inc.
701 12th Street
Golden, CO 80401
720-289-9430
scanderson@bgcengineering.com

Cole Christiansen
(Contact as above)
cchristiansen@bgcengineering.com

Dave Gauthier
(Contact as above)
dgauthier@bgcengineering.com

Sebastian Cohen
Caltrans North Region Construction
1656 Union St.
Eureka, CA 95521
707-496-4096
sebastian.cohen@dot.ca.gov

Prepared for the 69th Highway Geology Symposium, September, 2018

Acknowledgments

The authors would like to thank the following individuals for their contributions to the work described:

Tom Badger – Badger Geotechnics
Scott Burns – Portland State University
George Machan – Landslide Technology
Kenneth Johnson – WSP USA

Jaime Matteoli, Charlie Narwold, and Marietta J. James - Caltrans

Disclaimer

Statements and views presented in this paper are strictly those of the author(s), and do not necessarily reflect positions held by their affiliations, the Highway Geology Symposium (HGS), or others acknowledged above. The mention of trade names for commercial products does not imply the approval or endorsement by HGS.

Copyright Notice

ABSTRACT

US Highway 101 in Del Norte County, California, climbs a grade between Wilson Creek and Crescent City, cutting through redwood forest and, at its height, traversing a steep slope known as the "Last Chance Grade," several hundred feet above the coastline below. For decades, the slope has been the site of landsliding, large and small, and is a perpetual safety concern (1 fatal accident), mobility concern (there are no practical detours, and one-lane traffic control has been in place for 5 years now), and financial concern. Millions of dollars are now spent each year, on average, to maintain the roadway, even in its poor condition. Because of this, Caltrans is interested in an alternative route, possibly involving tunneling, to bypass most of the known sliding areas. Unfortunately, there are no easy solutions, so hard alternatives need to be compared. To do this, Caltrans conducted and expert-based risk assessment to compare the risks of high lifecycle cost, impacts to mobility, and safety, that was based on the judgment of a panel of experts. To get the most from a judgment-based approach like this, especially when alternative designs are no more than conceptual, requires the best communication of existing site conditions possible and the decomposition of the complex problem into smaller, more interpretable segments. This paper will present multiple ways this was done, including tapping the knowledge of Caltrans and other experts, use of oblique aerial photography, and 3-D visualization using augmented reality.

BACKGROUND

It is unclear who gave the name Last Chance Grade to a part of US Highway 101 (US 101) in Del Norte County, and near Crescent City, California, or when they did so, but it is a very appropriate name now. US 101 crosses landslides along Last Chance Grade that have been actively moving and impacting the highway for decades. More recently, the highway has generally been a site of one-way controlled traffic and ongoing structural repairs. The annual maintenance and preservation cost of $2 to $5 million is increasing. Nearly continuous repair efforts have kept some access through the site, but they are not sustainable as a long-term approach per findings of Caltrans and Federal Highway Administration (FHWA) reviews in 2016 and 2017.

A number of years ago, Caltrans suspected they might need an alternate alignment and began exploring options as a 'last chance'. The highway is of critical importance, not just because of its part on the iconic coastal highway and the tourism that brings, but because it's the only connection between Crescent City and the rest of the state. Without it, there would be a 320-mile detour suitable for only some vehicles, to approach Crescent City from the south. The alternative alignments are shown on Figure 1 and have been compared based on considerations that are not geotechnical. Criteria such as length, environmental impact, cultural considerations and construction cost were used to describe the alternatives and are reported in the Project Study Report *(1)*.

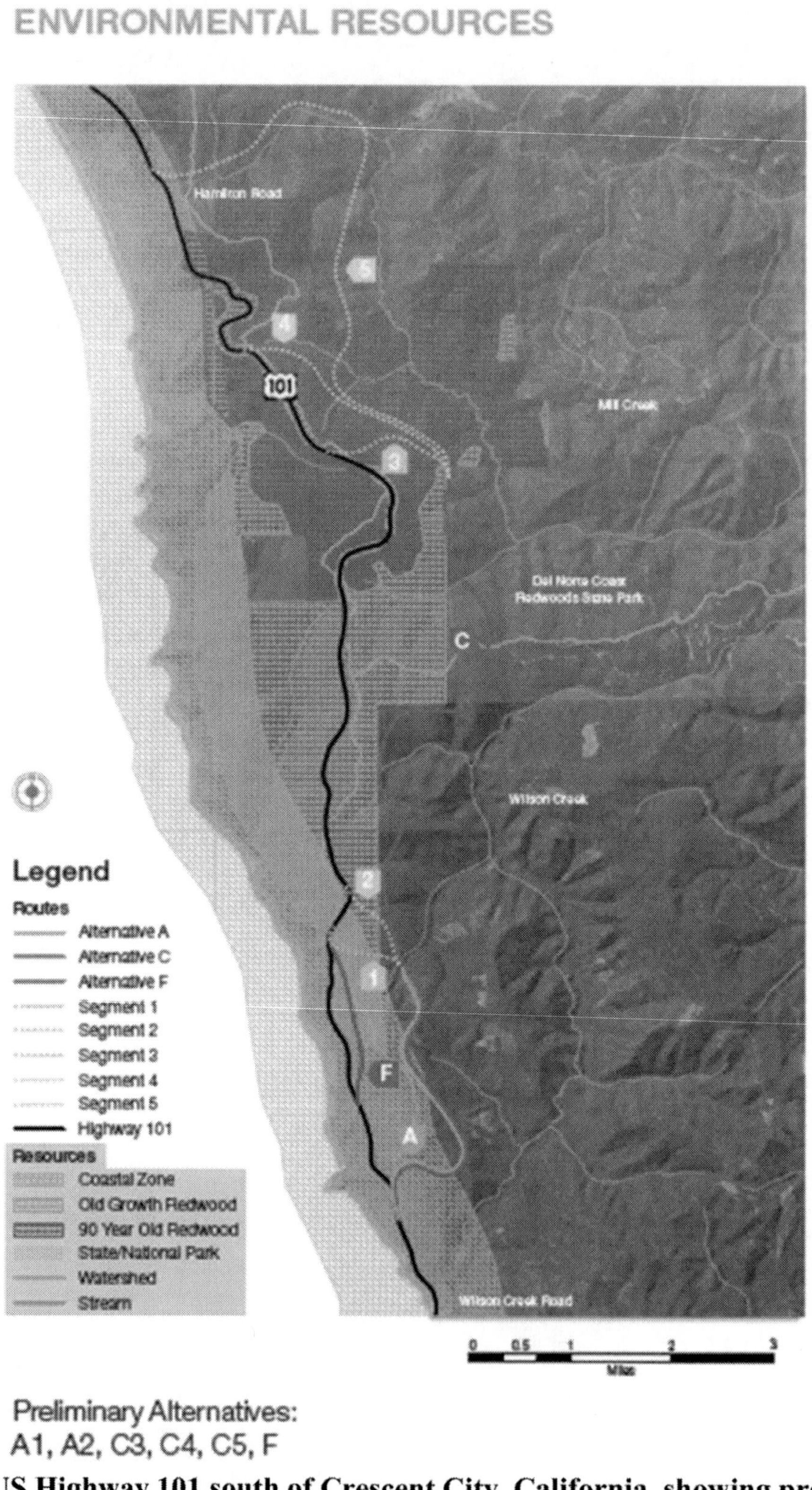

Figure 1 – US Highway 101 south of Crescent City, California, showing proposed alternative alignments and environmental resources *(1)*. Note that this risk assessment also considered two new regrading alternatives that lie near Alternative F, shown here in plan view.

The alternative alignments involve lengths of new highway between 1.3 and 11.7 miles, tunnels on all but one alternative, multi-span bridges on all but one alternative, and cuts and embankment fills exceeding 100 feet in height. At the concept level, the alternatives vary in construction cost from $275 million to 1.25 billion *(1)*. The scale of the new construction is largely due to the fact that where the slopes are not actively sliding, they are part of other important domains, such as land important to tribes of Elk Valley Rancheria, Tolowa Dee-ni' Nation, and the Yurok Tribe; state and national parks; and a UNESCO World Heritage site. Old growth redwoods in this area are so majestic that they have been given numbers and names. No preferred alternative has been identified but there is plenty of information from within the previously recorded criteria to see differences.

Unfortunately, however, the factors that have led to the problems on the highway are not unique to one location or one type of phenomenon. The geology is a problematic one. There is a large earthflow within the shallower slope formed by the Franciscan Mélange (see Figure 2), and there are steep erosional gullies below the highway to the coastline (see Figure 3), and deep-seated landsliding in the generally stronger Franciscan Broken Formation with a headscarp near the ridge top (see Figure 4).

The geology has been mapped by the CGS at a reconnaissance level *(2)* and shows multiple "active" to "dormant-old" landslides throughout the area where alignments are proposed. The following description of geologic units is from the CGS report *(2)*. The units of the Franciscan Complex in the study area are referred to as the "Broken Formation" and "Mélange." Both are composed of intensely sheared and fractured sandstone, siltstone and shale. The Broken Formation is composed mainly of gray, thickly bedded sandstone with siltstone and shale interbeds. The outcrops commonly represent relatively intact blocks of rock bounded by shear zones. The massive, hard sandstone blocks, bounded by weak, sheared zones leads to steep slopes and slides of large intact blocks of rock. The Mélange is composed of dark gray, highly sheared siltstone and shale. Outcrops commonly show highly contorted bedding or rock so sheared that bedding cannot be traced across the outcrop. The Broken Formation can be considered as a mass of hard sandstone blocks separated by shear zones, and the Mélange can be considered essentially a large shear zone containing relatively few intact blocks.

Figure 2 – The Franciscan Mélange and Broken Formation at the Last Chance Grade *(1).*

Figure 3 – Steep erosional gulley below US Highway 101.

Figure 4 – Deep-seated landsliding (Northern and Southern Last Chance Grade Landslides) in the Broken Formation with headscarps near the ridge top.

The reality of this type of geologic setting is that the investigation, design, and construction standards that Caltrans uses, and the expected standard of care practiced by the professionals delivering work to Caltrans, cannot assure the same level of long-term performance for each alternative alignment. There will remain uncertainty with respect to how this geologic terrain will respond to the proposed construction. Some of these alignments may look good on paper, and based on the criteria previously considered they may look preferable to others; however, when it comes to how they perform during a lifetime of operation, they may be quite a disappointment. They may require frequent and costly maintenance and road closures as are being experienced today on the Last Chance Grade, or possibly even worse.

Because of the challenging geologic setting and alignments selected based primarily on other criteria, Caltrans was concerned with making such a large investment and possibly finding themselves with a different set of geotechnical problems that have similar long-term impacts in terms of maintenance cost, impacts to mobility, and road serviceability. Caltrans addressed this concern through a structured risk assessment that used available information to quantify an estimate of geotechnical risks for alternative alignments. The estimated risks involve the cost of ownership and maintenance, the possibility of future short-term closures and impacts to mobility when repairs are being made, and the possibility of a long-term or permanent closure.

ASSESSING THE RISK

Basis of the Risk Assessment

The risk assessment is based on expert opinion and the recognition that expert opinion can be quantified. Similar to probability estimates based on statistics or other logic, subjective probability estimates can be used to estimate risks for complex events. Background for this approach is nicely summarized in the following references, which span 50 years: Role of "calculated risk" in earthwork and foundation engineering – The Terzaghi Lecture, Arthur Casagrande, 1965, ASCE Journal of the Soil Mechanics and Foundation Division; Degrees of Belief – Subjective Probability and Engineering Judgment, Steven G. Vick, 2002, ASCE Press; Risk-Informed Decision Making (RIDM) – Risk Guidelines for Dam Safety, Federal Energy Regulatory Commission, Version 4.1, March 2016.

For complex problems or paths to failure, it is important to be able to "decompose" the problem into smaller steps because this allows a better assessment of probability for each step. The project can then be "recomposed," and the probabilities combined in appropriate ways. Usually, this is done by considering conditional probabilities of failure, but other ways of decomposition for probability estimation are also acceptable.

Risk is the product of a probability and a consequence, and consequence can be defined in different ways. Caltrans' interest in the cost of maintenance, the possibility of having road and lane closures similar to what has been occurring recently on the highway, and the possibility of long-term closure represents three different consequences. If each of these is defined by way of a threshold event, the consequence becomes simply that a threshold is crossed, and exactly what that means in terms of dollars, time, or other measures is tied to the definition of the threshold.

The estimated probability of the event of crossing a threshold is therefore equal to the risk of it occurring (consequence has the value of unity (1.0)).

Therefore, the basis of the risk assessment is expert opinion of the probabilities of crossing well-defined thresholds. The opinion is compiled from a panel of experts, each of whom brings unique and complementary experience and it is elicited in a carefully designed and structured way.

Risk Assessment Design

The risk assessment is designed to get the best possible estimates of the risk to achieving Caltrans' objective to build a low maintenance-cost and reliable highway to replace the existing US 101. This assessment is done for several different alignment alternatives so the risks can be considered with other objectives and used to help inform the selection of a preferred alternative. The quality of the estimate is limited at this stage by the information that is available on the geologic setting, the mechanisms and activity level of known landslides, and the limited, conceptual nature of the alternative alignment designs. Nevertheless, careful design of the process has led to estimates of risk that are meaningful and objective, and helpful for the project.

The opinion of experts is formed by their past experiences as well as their interpretation of the current problem, so they can differ somewhat. A panel of experts with complementary experience is convened to capture this range of opinion and to encourage debate of contributing factors and risk estimates. Informing the panel on the current problem is done through providing access to published studies and information, and having the panel, who are experts in a 'global' sense meet with people who are experts on this project – those that have been working on it extensively. It is important that all panel members have the same understanding of what they are estimating, so clear understandings of objectives and precise definitions are required.

Regarding Caltrans' performance objectives, it was possible to consider cost, mobility, and closure separately and to give them precise definitions by defining four condition states (A through D) and identifying three thresholds that represent the change from one condition state to another. The condition states are described in Table 1. If a condition state changes, then a threshold has been crossed and a risk realized. The change from Condition State A to B means the cost threshold has been crossed and this risk realized; B to C means the mobility threshold has been crossed and this risk realized; and, C to D means the closure threshold has been crossed and this risk realized.

Table 1 – Description of condition states with representative maintenance and preservation actions.

Condition State	Description	Actions
A Routine Maintenance Work/ Average Maintenance Efforts for Type and Location of Highway	Highway segments that require no more than average maintenance for that type of highway lane mile. Average refers to the type, quantity, and frequency of application. Temporary lane and shoulder closures have frequency and duration consistent with other California highways requiring average maintenance, and work is scheduled to be minimally invasive.	Field Maintenance efforts include planned recurring work, such as vegetation, rock, and debris removal; minor ditch excavation; repair and resetting of guardrail; cleaning culverts; minor patching of potholes; repair of pavement sags and small embankment slumps; and other minor or routine work that's expected on California highways, including regularly programmed bridge and tunnel inspection and associated maintenance and traffic interruption.
B Above Routine Maintenance Work/Above Average Maintenance Efforts for Type and Location of Highway	Highway segments that require more than an average amount of resources to keep the highway safe and open. Maintenance and repairs require traffic control and short-duration lane closures and cause interruptions to mobility that are above average for similar California highways.	Requires above average Field Maintenance efforts, which approach or exceed the annual budget allocated or expected. Projects funded from various sources (Programs) often used to repair or construct improvements on a higher than average or expected frequency. Includes minor bridge repairs from ground movement or environmental factors, and portal and tunnel repair from drainage and minor ground movement.
C Significant Damage Repair Work/ Emergency Projects Required	Highway segments that require significant emergency response actions and funding to keep the highway safe and open. Projects are large and substantial. (retaining walls, structures, minor realignments/retreats, bridge and tunnel structure mitigations, etc.). Mobility is impacted by restricted speeds and frequent lane closures, but a minimum of one lane is maintained open a majority of the time. Bridges and tunnels are distressed but still safe to allow traffic (with possibly some restrictions).	Programs of Emergency Relief, Safety and Pavement are accessed. Activities involve building structures, changing drainage, and construction activities that significantly interrupt traffic. Includes structural mitigation of bridges and tunnels/portals due to ground movement. Full, temporary closures from ground movement are rarely experienced. One-way Traffic Control measures, with delays of one to two hours, are sometimes required for damage repair activities. Weight restrictions might be imposed on distressed bridges.
D Long-term Full Closures/ Abandonment	Impractical to keep the highway open via emergency and other programs (Safety, Pavement, etc.), because the costs are too high. Closures that last more than a few weeks and may be permanent. Bridges, walls, and tunnels are significantly distressed and not safe to allow traffic.	Repair or stabilization of road, bridges, walls and/or tunnels require at least extended temporary road closure (traffic safety concerns, and not feasible to mitigate/repair facility/structure under traffic).

Some assumptions are made to facilitate the risk assessment as follows. Caltrans will first invest additional money, above an expected maintenance budget, to maintain their objective of keeping the road open and unrestricted. If necessary, Caltrans' interventions will escalate, and they will take steps that do compromise the mobility objective next, in order to preserve the road and keep it open. It follows from this logic is that there is essentially no mobility risk until money has been spent on unusually heavy maintenance, and there is no closure risk until interventions that have impacted mobility have been exhausted. The probability of a closure risk is therefore assumed conditional on the probability of mobility and cost risks (thresholds) having been realized first, and the progression of changing condition and crossing thresholds could happen quickly or slowly.

This simplification of conditional relationship is valuable and reasonable here and it is a useful way to decompose the problem. It means that progression through the condition states in Table 1 happen in sequence, and without skipping or reversing.

Another way this problem is decomposed for risk estimation is through breaking alternative alignments into construction segments. Construction segments have been selected based on the primary construction type in that part of an alignment – earthwork, bridge or tunnel, and the geologic and topographic setting for a segment of alignment. With these considerations, eleven construction segments have been identified and are the building blocks for the assessment. The panel considers one of these eleven construction segments at a time and thinks only about the performance over 10 to 50 years of that type of construction in that environment and the risk of the construction segment advancing across three thresholds and four condition states.

The formulation of an event tree is used to track the estimates and calculate conditional probabilities, as shown in Figure 5. The segment risk assessments are then combined to build the alignments by treating each alignment as a system of segments.

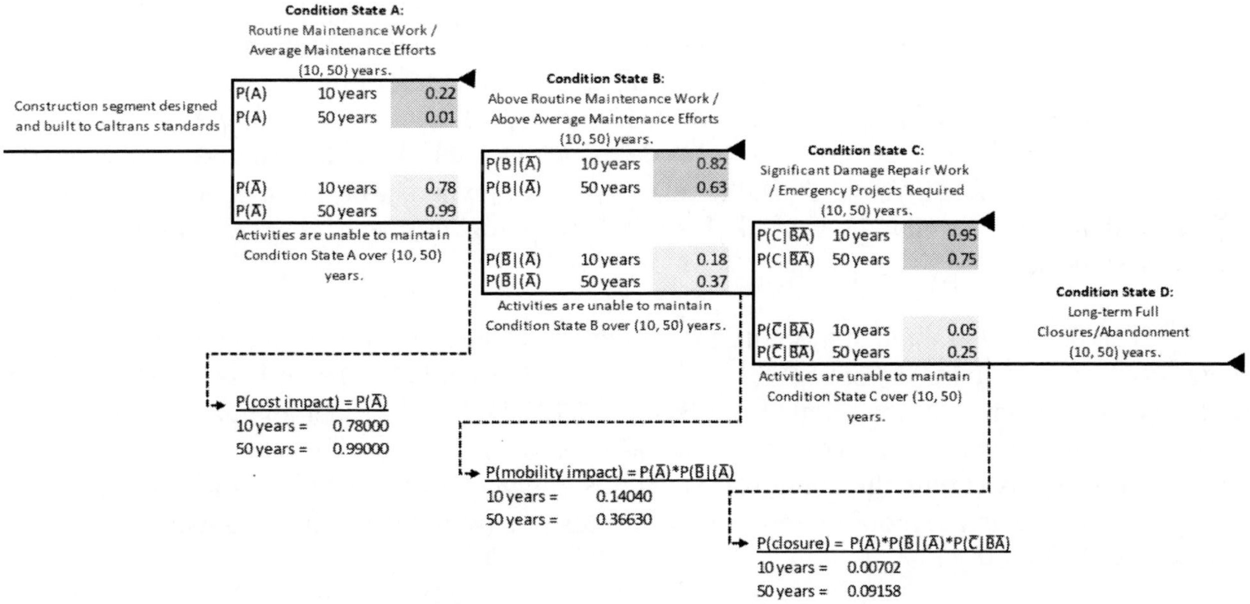

Figure 5 - Representative event tree.

Conduct of the Risk Assessment

Extensive writing on the Last Chance Grade project was edited for what is most important from a geotechnical standpoint and what could be reviewed in an appropriate amount of time and was part of a workbook presented to the panel. In addition, LiDAR data collected in 2011 and 2016 was processed to create a bare-earth terrain model of the site. Creation of this model provided new insight into the geomorphology by allowing the panel to view the underlying landforms. To provide additional understanding, the alternative alignments including the proposed earthwork were merged with the topography to show the footprint of the road prisms. Geology and landslide maps *(2)* were added to the base maps to show where the proposed alignments intersect mapped landslides and where they cross geologic contacts. These data were presented in a series of maps and as 3D visualizations using Microsoft's HoloLens mixed reality headsets and BGC's Ada software. HoloLens headsets enabled the panel to simultaneously view the features from any perspective as they walked around a 100-square-foot virtual model in a meeting room. The panel also viewed a 3-D model of the slope that afforded them a perspective from off the coast. This model was created using oblique aerial photographs of the slope collected from a moving helicopter. Together these powerful tools provided insight and improved the expert's understanding and interpretation of the geologic hazards and their impact on the proposed alternative alignments.

OBSERVATIONS AND CONCLUSIONS

The expert panel was informed by a summary of published materials and project work. This included new conceptual design drawings, oblique aerial photogrammetry models, mixed reality images viewed through the HoloLens, and presentations by Caltrans staff in a panel meeting and in the field. With this understanding, the panel was able to reach a consensus opinion on all estimates of risk in the assessment, and that is the first observation. Initial opinions were sometimes different, but with discussion and reviewing of material, and sometimes another view through the HoloLens, consensus was reached.

The assessment was useful in contrasting six different alignment alternatives and the different performance objectives, especially with respect to the mobility and closure risks. The risk of crossing the cost threshold is very high: it is expected that each alternative alignment would move from Condition State A to B (as described in Table 1) within 10 years and is nearly certain to do so within 50 years. Thus, the relative cost of ownership is not a good differentiator: all alternatives are viewed to be very likely to have costs higher than average for similar construction elsewhere.

The risk of crossing the mobility threshold is also high but shows greater differentiation between the alternatives. Within 50 years' time, only Alternative F (Figure 1) is "less likely than not" to change from Condition State from B to C (Table 1). For the other alternatives, the progression is nearly certain. In other words, based on the available information now, the panel believes it is nearly certain that within 50 years this highway will be in a condition similar to today with respect to the mobility it provides - unless it is routed through the tunnel of Alternative F (Figure 1).

Another important observation is that three of the alternatives (F and A1, as shown in Figure 1, and one of the new regrading alternatives) are very likely to avoid closure (Condition State D) within 10 years, and more likely than not to avoid closure within 50 years. The other alternatives are very likely to result in closure within this time. These results mean that the risks to the performance objectives of low cost, relatively unimpeded mobility, and avoiding closure are high. Indeed, they are higher than one would expect for any new construction. One reason for this is the uncertainty that exists now. As exploration is conducted, the site understanding improved, and concepts developed in recognition of the geotechnical challenges, it is expected the estimated risks will come down.

The results show that alternatives are not equivalent with respect to risks of ownership. In fact, the estimated risks vary by approximately two orders of magnitude between the alternatives. With respect to the risks estimated through this process, Alternative F has the least risk and Alternative C3 has the highest risk. These alternatives are shown in Figure 1. Alternative F mostly uses the existing alignment and includes a high construction cost solution – a large tunnel. Alternative C3 uses the greatest amount of new alignment, including cuts, embankments and bridges, and a tunnel. It also has a high construction cost, but the vulnerability of having so much new work in such a difficult terrain is revealed by the structured process of this risk assessment.

Longer variations of the C Alternative shown in Figure 1 were not part of the risk assessment, but they can be judged to have equal or higher risk by way of their common elements and greater length. The other alternatives considered here have risks that lie between these two extremes and are also expected to have lower construction cost. Caltrans can consider the estimated risks presented here for ownership cost, mobility impacts, and closure, along with estimated construction costs, and other important selection criteria when choosing the best alternative to meet their overall objectives. The findings will also help Caltrans plan site investigations and prepare for ownership of this part of US 101 for many years in the future.

In conclusion, the risk assessment proved to be very valuable for Caltrans. It provided a transparent and objective look at factors not previously considered: the long-term ownership risks and expectations. Caltrans and stakeholders for this project can use these findings to better evaluate a preferred solution to a challenging part of US Highway 101. Additionally, the work has revealed construction segments and features that carry most of the risk. As project development continues on one or more of the alternatives, these observations will be valuable for planning geotechnical explorations and for refining highway designs.

REFERENCES:

1. California Department of Transportation. June 2016. Project Study Report, Permanent Restoration, Last Chance Grade. File: 01-DN-101 PM 12.0/15.5. EA: 01-0F280K/EFIS. EFIS ID: 0115000099. Program Code: 20.XX.201.131. URL: http://www.lastchancegrade.com/files/managed/Document/208/lcg_psr_final_s.pdf
2. Wills, C.J. 2000. Landslides in the Highway 101 Corridor Between Wilson Creek and Crescent City, Del Norte County, California. Department of Conservation, California Geological Survey, Special Report 184.

LANDSLIDE APPLICATION OF THE GEOTECHNICAL OBSERVATIONAL APPROACH

George Machan, PE
Wade Osborne, PE
Chris Carpenter, PE
Charlie Hammond, CEG
Landslide Technology (A Division of Cornforth Consultants, Inc.)
10250 SW Greenburg Road
Portland, OR 97223
(503)-452-1200
georgem@landslidetechnology.com

Philip Wurst, PE
Oregon Department of Transportation
455 Airport Road, SE,
Salem OR 97301
(503) 986-2818
. Philip.Wurst@odot.state.or.us

Prepared for the 69th Highway Geology Symposium, September, 2018

Acknowledgements

The author(s) thank the individuals/entities for their contributions in the work described:

Michael Long – Oregon Department of Transportation
Joe Squire – Oregon Department of Transportation
Derryl James – Oregon Department of Transportation
Gene Wilborn – Oregon Department of Transportation
Steven Schultz – Oregon Department of Transportation
Markus Schaaf – Oregon Department of Transportation
Michael Tardif – Oregon Department of Transportation
Jerry Wolcott – Oregon Department of Transportation
Darren Beckstrand – Landslide Technology/Cornforth Consultants
Brent Black – Landslide Technology/Cornforth Consultants
Dr. Derek Cornforth – Landslide Technology/Cornforth Consultants

Disclaimer

Statements and views presented in this paper are strictly those of the author(s), and do not necessarily reflect positions held by their affiliations, the Highway Geology Symposium (HGS), or others acknowledged above. The mention of trade names for commercial products does not imply the approval or endorsement by HGS.

ABSTRACT

The geotechnical uncertainties and challenges faced by the US Route 20 realignment project that crossed landslide terrain warranted the use of the observational approach to achieve acceptable performance at least cost. Over 12 landslide areas have been instrumented and monitored, primarily using inclinometers, piezometers and surveys, along with visual reconnaissance.

During earlier phases of the project, instrumentation successfully measured groundwater responses due to horizontal drains, removal of toe support during excavations, and added loads due to fill placement. Embankments and cut slopes were designed for the project with estimated Factors of Safety that generally ranged from 1.0 to 1.2.

Preparations for construction included the development of a process to systematically apply the observational approach. Based on prior international experiences, the process utilized three management categories to reflect 1) "GREEN" = anticipated satisfactory conditions, 2) "AMBER" = increased movement possibly causing damage, and 3) "RED" = significant movement that could cause damage. Supplemental mitigation strategies would be considered when "AMBER" conditions occurred, and implementation of mitigations would be recommended when conditions entered the "RED" category.

During construction, one area detected accelerated slide movement in an inclinometer that prompted application of the "AMBER" and "RED" categories. The new process was followed, with additional instruments and ground surveys as conditions went into the "AMBER" category, followed by development of mitigation strategies as conditions appeared to move from "AMBER" to "RED" that were implemented with modifications to design and construction.

INTRODUCTION

The US20 highway realignment project is located between Pioneer Mountain and Eddyville in the Oregon Coast Range, between approximate milepoints 16 and 24. A vicinity map is shown in Figure 1. In western Oregon, US20 links the coastal communities with the Willamette Valley by crossing the Coast Range mountains, and US20 extends further east across the Cascade Range.

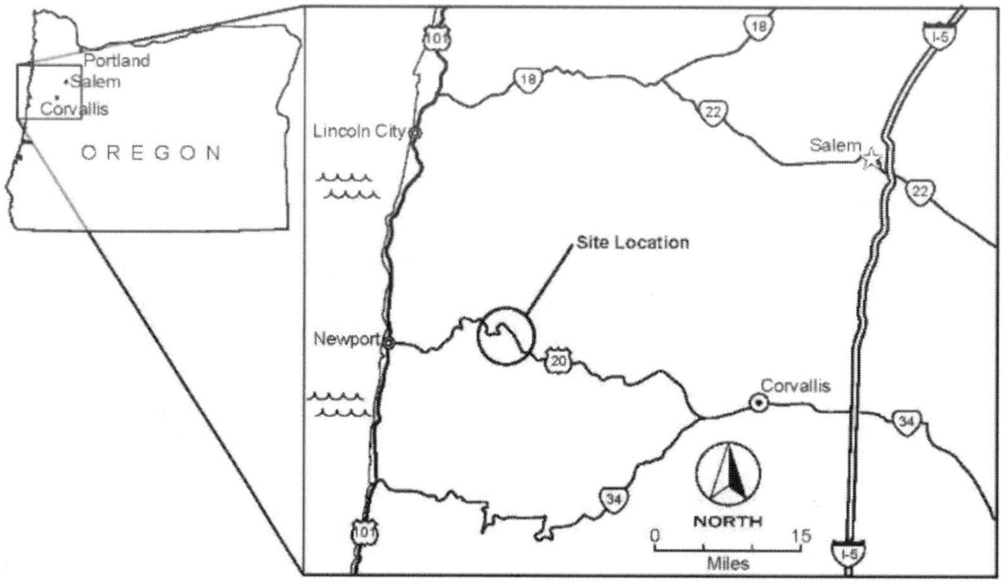

Figure 1 – Vicinity Map

The need for highway realignment was to replace a dangerous section that had a very high accident/injury/fatality rate. The project included substantial cuts and fills up to 200 feet tall to cross hillsides and incised drainages. The project delivery method was initially Design-Build; however, reactivated landslides during construction caused damage to initial earthwork and structures. Subsequently, the project was redesigned by the Oregon Department of Transportation (DOT) to minimize the harmful effects of the landslides. The redesign consisted of four fast-tracked phases of Design-Bid-Build contracts within a five-year program, which is the focus of this paper.

Geotechnical investigations and testing were fast-tracked in parallel with geotechnical analyses and design engineering. Stability analyses and designs were performed understanding there were geologic uncertainties and gaps in geotechnical information and parameters. While analyses and designs were progressing, new data was being obtained, requiring updating of geologic models and redesigns in some cases. Significant interpretation and assumption were necessary in the development of landslide models, including the depths and extent of potential slide shear zones.

The risks associated with fast-tracked design on a complex landslide project is the reliance on a greater number of assumptions and interpretations, and all potential consequences may not be readily understood because the accelerated process bypasses the traditional sequence of investigation, monitoring and design. The geotechnical Observational Method was used to augment design and construction decisions. Consequently, a higher level of geotechnical involvement was necessary during construction to address conditions not fully understood at the time of design.

LANDSLIDE CONDITIONS

The most problematic part of the realignment project was a mountainous 3-mile stretch of the highway across four drainage basins, each with different landslide conditions on both sides of the drainages, as shown on the LiDAR shaded relief site map in Figure 2.

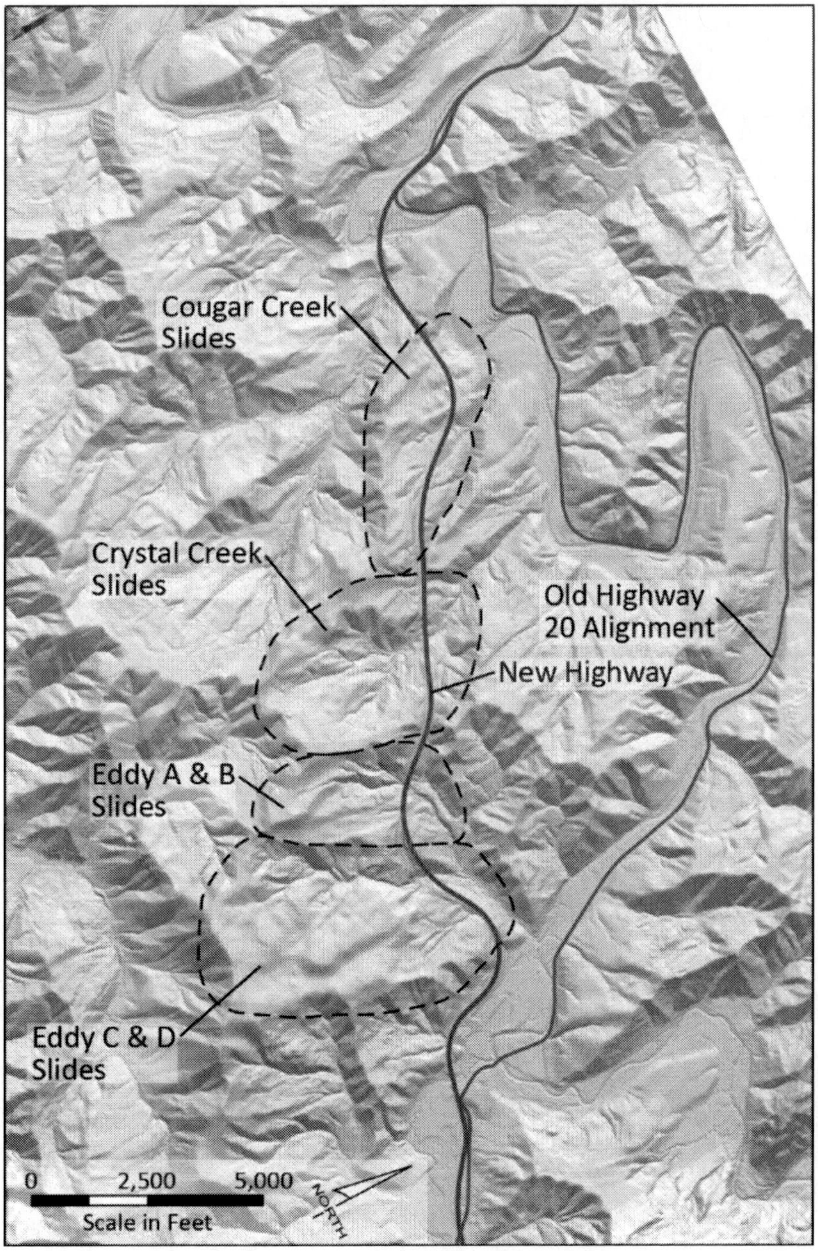

Figure 2 – Site Plan – New highway crossing landslide terrain

The bedrock geology in the project area is primarily rhythmically bedded turbidite siltstone and sandstone of the Tyee Formation *(1)*. Weak rock layers in the Tyee Formation are uplifted, twisted and faulted into blocks and slabs that can separate like a deck of cards. The upper slide materials consisted of colluvium slide debris (USD), underlain by layers of weathered rock slide debris (LSD), as depicted in the photograph and diagram in Figure 3.

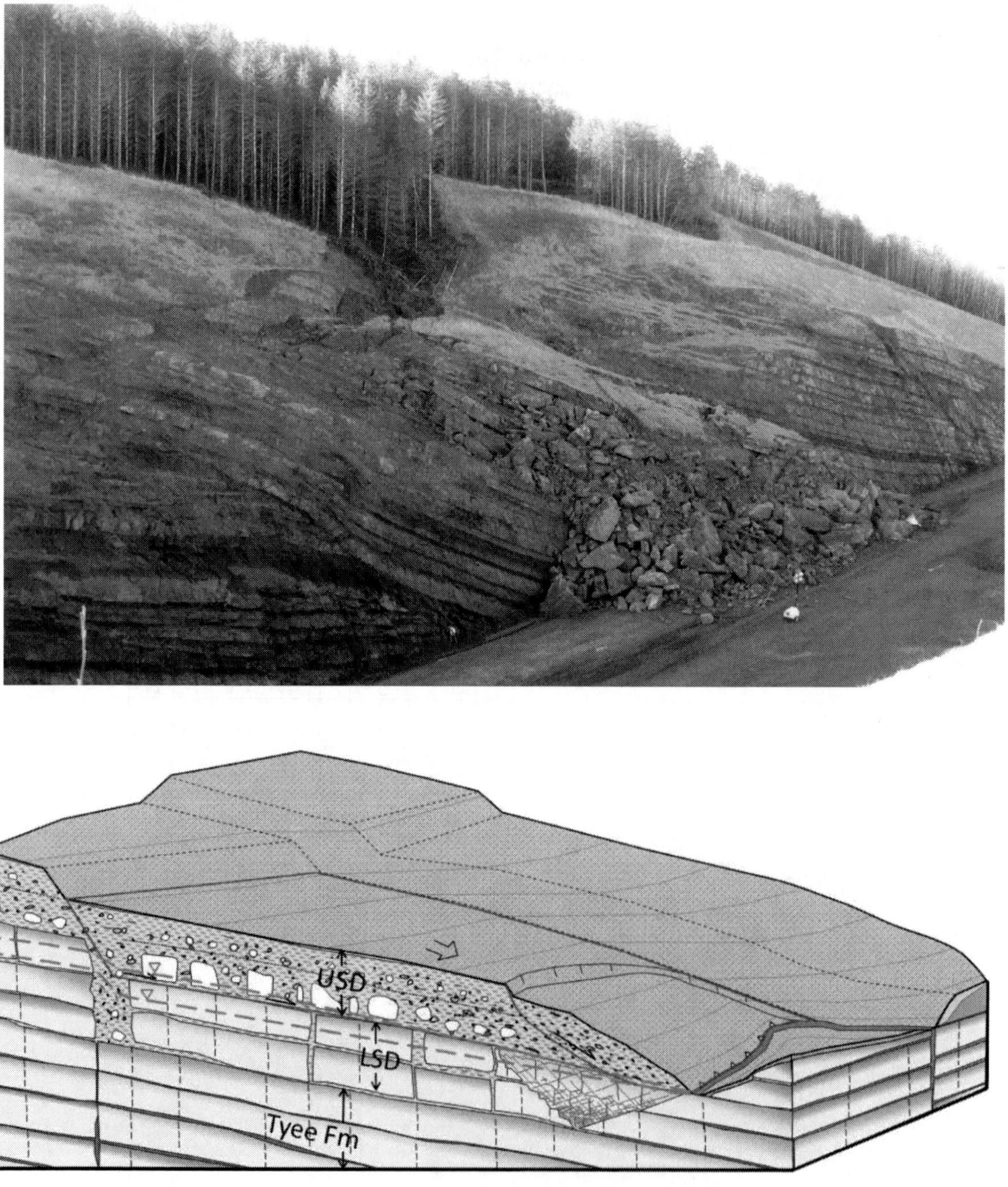

Figure 3 – Landslide Stratigraphy

Geologic materials include low strength layers, some which have reduced to residual shear strength due to past displacement. The project area is subjected to Oregon coastal weather patterns which include moderate summer temperatures, prolonged wet season, intense coastal fall and winter storms, and occasional rain-on-snow events. Seasonal precipitation ranges up to 90 inches per year. Groundwater occurs as flow along bedding and through fractures, and water pressures rise appreciably as a result of individual storms.

Twelve different large landslides in this area exhibit translational geometry and vary in depth from 20 to over 100 feet. Different subsurface conditions existed in each drainage, with variable groundwater levels and overburden soil that varied from clay/silt to boulders to slide blocks as large as buildings. The landslides are often vertically-nested due to the parallel layers of low-strength rock in stratified geology. Landslide geometry and groundwater levels were determined by extensive instrumentation, where approximately 150 inclinometers and 250 piezometers were installed. Most of the piezometers were vibrating wire. Of the inclinometers, 29 were outfitted with in-place sensors. On average, each slide had about 12 inclinometers and 20 piezometers.

ANALYSES AND DESIGN CRITERIA

Observed active ground deformations at several locations along the new highway alignment indicated that landslide mitigation measures were necessary to improve stability. Geotechnical investigations and analyses were performed to evaluate causative factors for the landslides. The landslides occur in tilted beds of the marine sedimentary Tyee Formation, which are generally inclined 15° to the NNW. Primary factors that contributed to instability of the terrain include: inclined stratigraphy, low-strength residual shear zones, groundwater pressures, and erosion. In addition, alterations to the terrain, such as cuts and fills and changes to drainage channels had reduced stability and induced ground deformations.

Slope stability analyses were performed using 2D limit equilibrium computer modeling software. Interpreted geologic cross sections at each landslide (based on subsurface conditions, monitored groundwater levels and slide shear zones) were used to develop models for analysis. Analyses were performed for each of the interpreted shear zones. Interpretations were typically necessary for the locations and orientations of the active and passive wedges since these features were not visible on the ground surface. The back-analysis method was used to determine reasonable values of the average residual shear strength (ϕ'_r) of the shear zone materials, assuming a Factor of Safety (FS) of about 1.0. In general, the residual shear strength (ϕ'_r) ranged 10 to 15°.

The groundwater profile assumed for back-analyses was selected based on measured levels at the time when the landslide stopped/started moving as evidenced by instrumentation monitoring. Mitigation options were analyzed using estimated elevated groundwater levels for 100-year return period storm events. Parametric stability analyses of groundwater levels indicated a FS increase of 3 to 6% for each 10 feet of groundwater lowering. A representative analysis cross section is shown in Figure 4.

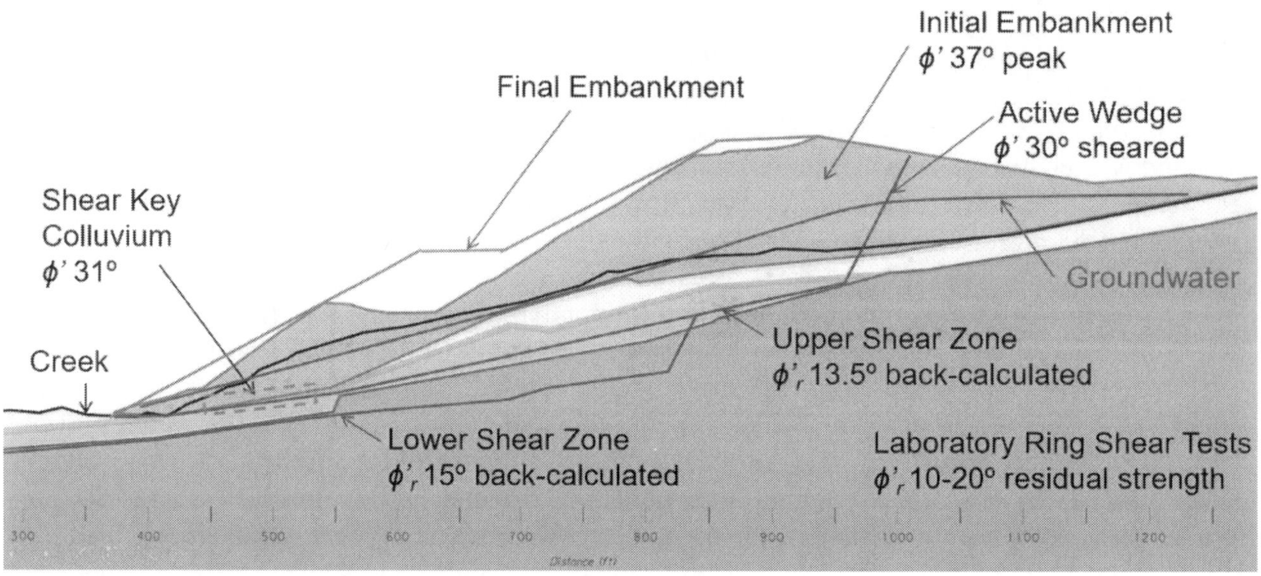

Figure 4 – Parametric Stability Analyses (Cougar Creek Slide)

Initial geotechnical mitigation designs utilized standard DOT criteria for stability; however, it was concluded that extraordinarily high costs would be incurred to achieve such criteria. Influencing the designs was conservativeness associated with geologic and geotechnical uncertainties. The DOT considered relative risks and consequences to determine cost-effective approaches for this project.

Embankment stability FS criteria were modified from DOT standard, and subsequently discussed with Federal Highway Administration (FHWA). The DOT and FHWA concluded that marginal mitigation and risk-acceptance was acceptable when mitigating the large landslides on this project, provided that the Observational Method be employed to verify performance expectations and that unsatisfactory performance would be rapidly detected and supplemental mitigations developed and implemented. The selected criteria for embankment stability employing earthwork mitigation measures for this project were as follows:

- Avoid catastrophic embankment or cut slope failures;
- Attempt to stop slide movement; however, movements up to 3 inches/year was acceptable;
- Design stability FS = 1.2 if it can be attained at reasonable cost;
- Accept marginal stability (FS = 1.1 +/-) where the foregoing FS criteria cannot be achieved within reasonable cost, providing supplemental mitigations could be implemented if necessary to achieve satisfactory performance;
- Use the Observational Method to monitor stability as the embankments and cut slopes are being constructed, and implement supplemental mitigations where performance needs to be improved.

HIGHWAY AND LANDSLIDE MITIGATION DESIGN APPROACH

It is rare that enormous excavations and embankments are constructed in landslide terrain due to the significant risk of triggering movement. However, this major earthwork undertaking was necessary in the selected highway alignment due to the mountainous conditions bisected by numerous drainages and the incised meander of the Yaquina River.

The highway redesign included new embankments at four large drainage crossing areas where bridges (in various stages of completion) were abandoned due to slide movements. The bridges were replaced with embankments and flexible large diameter culverts. The embankment fills are more flexible, can tolerate deformation and can be repaired in the event of landslide movement. Where embankments decreased landslide stability, methods to reduce the impact of embankments were considered, including reducing fill height/weight (by shifting and/or lowering the highway alignment), reducing groundwater pressures, and by adding measures to increase downslope resistance.

Earthwork and drainage mitigation methods were the preferred approach to achieve the stability FS criteria due to their simplicity and flexibility. Embankments could be stabilized by constructing large buttress fills that gain resistance by widening the embankment footprint and gaining support from opposing hillsides across drainages and streams. Environmental constraints restricted the extent of buttressing in some streams and riparian areas, resulting in smaller than desired improvements in the FS.

Slope stability is significantly affected by groundwater and surface water impacts. This was exhibited by seasonal variations in landslide movement and catastrophic failures in cut slopes during intense rainfall events. Therefore, a significant part of the landslide mitigation plan included control/diversion of surface water and subsurface drainage systems to reduce groundwater pressures within the landslides. Subsurface drainage systems included deep horizontal drains and shallow intercepting trench and blanket drains. The horizontal drains were designed following the procedures recommended by Machan and Black *(2)* and Cornforth *(3)*. Over 1,200 horizontal drains were installed, up to 900 feet long each. The horizontal drains reduced the groundwater pressures that were causing the landslides to move and also reduced excess pore water pressures during embankment placement.

Cut slopes were affected by adverse geologic conditions and weak shear zones. Some slides had occurred when excavations were made in north-facing cuts. Slide movement was generally slow and 'ductile' in most cut slope areas. However, localized rapid/catastrophic ('brittle') events had occurred in two locations where the bedding was locally steeper. The stability and relative risk of each cut section was analyzed, and mitigation measures were evaluated, including buttresses, deep horizontal drains and ground anchors. Due to the large size of the slide blocks, the increase in stability was relatively small, typically within marginal limits (FS = 1.0 to 1.1). If the risk of catastrophic failure was low for a particular cut section (gently-inclined shear zones), the risk of constructing a marginal 'ductile' mitigation was considered acceptable. Conversely, cut sections with a greater risk of 'brittle' failure were mitigated to more stringent design criteria (FS=1.2), which necessitated large buttresses where space allowed or structural reinforcement using high-capacity ground anchors.

RISK-INFORMED DECISION MATRIX

The reasonableness of selected landslide mitigations for each landslide area was evaluated with a "Risk Informed Decision Matrix." Scott *(4)* presents methodology for risk-informed decisions to prioritize limited resources. Silva et al. *(5)* describe quantification of expert judgment and semiempirical relationships for conducting probabilistic assessments.

A methodology and chart were developed for this project that compares "Likelihood of Failure Occurrence" with "Consequence or Impact" for selected mitigation options, as shown on Figure 5. The likelihood of failure occurrence is described as either: extremely low, very low, low, moderate, or high. The consequence or impact is a measure of the risk of injury and/or cost and impact to the highway user, and is described as either: very low, low, moderate, high, or very high.

One goal for selecting landslide mitigations is to avoid high-risk conditions that are represented by the red area in Figure 5. Conversely, costly mitigation options that address extremely low risk events (lower left corner) might be considered excessive.

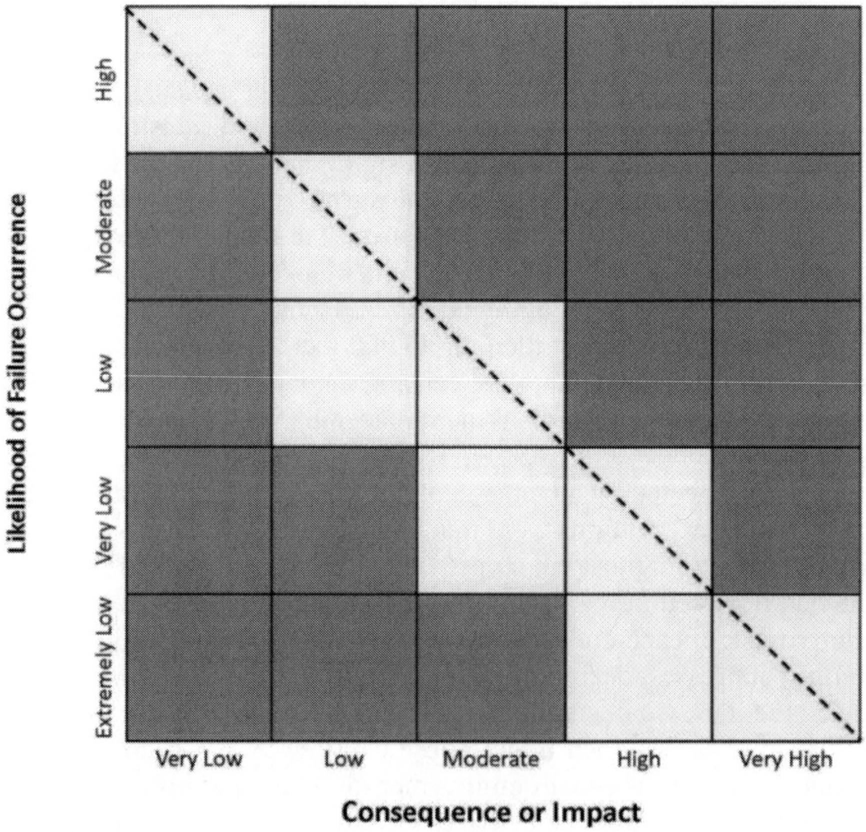

Figure 5 – Risk-informed decision matrix (based on Scott 2012)

The determination of the "Likelihood of Failure Occurrence" and "Consequence or Impact" for each landslide area was based on evaluation of many aspects of the landslide and attributes of highway construction and mitigation options, including the following:

- Landslide movement amount and rate (velocity)
- Seasonal movement trends
- Potential for rapid movement impacting the highway
- Hazard type and consequences
- Bedding and shear zone geometry
- Groundwater influence
- Loss of toe support
- Gaps in engineering and understanding
- Initial cost and maintenance and long-term costs

The sum of the various aspects was qualitatively evaluated to interpret relative levels of "Likelihood of Failure Occurrence" and "Consequence or Impact" for each landslide area. The results were plotted onto risk informed decision matrix charts. The evaluations shown on the risk-informed decision matrices were used to check the suitability and reasonableness of alternative mitigation options. The following risk-informed decision matrix for cut slope slides (Fig. 5) shows how alternative mitigation measures can be evaluated by how much they lower the risk level. For example, the slide area labeled "Cut 5" was rated a moderate likelihood of occurrence with a moderate consequence/impact, which places it in the red higher risk zone. By constructing an infill rockfill buttress against the cut slope, the relative risk would be reduced to an acceptable level as shown in Figure 6.

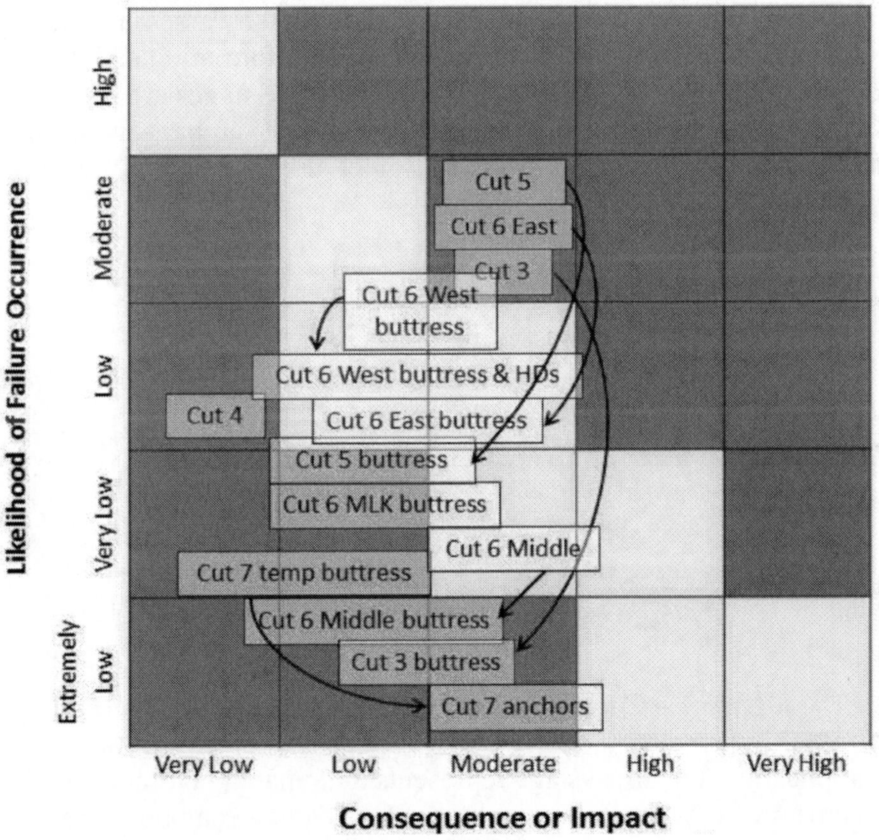

Figure 6 – Risk-informed decision matrix example (cuts)

OBSERVATIONAL METHOD

The "Observational Method" is sometimes used during construction of complex geotechnical projects to deal with uncertainty and risk. Nicholson et al. *(6)* define the Observational Method as "a continuous, managed and integrated process of design, construction control, monitoring and review enabling appropriate, previously-defined modifications to be incorporated during (or after) construction. All these aspects must be demonstrably robust. The objective is to achieve greater overall economy, without compromising safety". The geotechnical use of this method was initially proposed by Karl Terzaghi and later described in the Rankine Lecture paper by his associate Ralph Peck *(7)*. Cornforth *(3)* describes the application of the Observational Method to landslide mitigation.

Use of the geotechnical Observational Method allows consideration of designs at lower calculated FS to reduce construction costs. DOT standards generally require a FS for landslide mitigation of at least 1.25. Factors of Safety between 1.0 and 1.25 are considered "marginal", where there could be greater risk and uncertainty, but it is also recognized that there could be conservativeness in the analyses and that satisfactorily stable slopes could be designed with lower FS. The Observational Method provides a risk management framework which allows observed performance of the mitigation elements to verify suitability of designs and development of supplemental mitigation measures, where needed, to achieve acceptable performance.

A unique process was developed for this project based on principles of the Observational Method. The process includes three stages that progress from acceptable stable conditions to unacceptable unstable conditions. The stages were quantified by assessing the degrees of landslide movement that could cause increasing levels of deformation and damage and actions that should be taken to prevent significant damage. The three stages are represented by green, amber and red coding, resembling a traffic signal (green = go; amber = proceed cautiously; red = stop), based on the approach described by Nicholson et al. *(6)*. The three stages are illustrated in the following diagram, Figure 7.

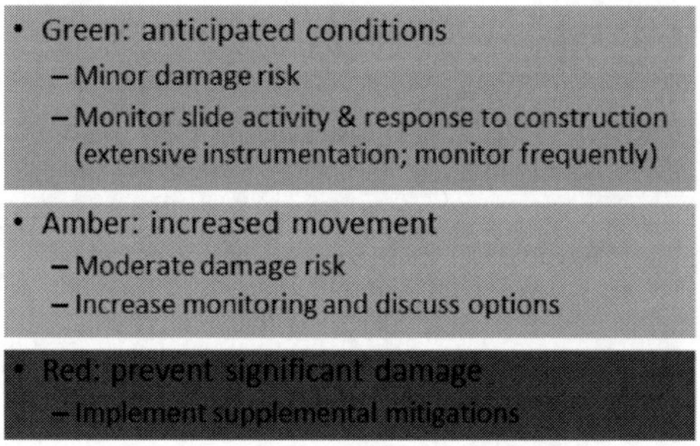

Figure 7 – Condition stages for Observational Method

In this application, the green stage represents anticipated favorable conditions resulting in only small movement where there is only minor risk of damage and normal site monitoring would suffice. The amber stage represents increased movement where moderate damage might occur, which triggers the need for additional geotechnical monitoring and site observations to assess the severity of the condition. The amber stage would include planning potential mitigations and discussions with design and construction representatives to assess possible next steps and options. The red stage represents high risk of significant damage and should be avoided if possible. Supplemental mitigations should be implemented during the latter portion of the amber stage or early in the red stage to prevent significant damage from occurring.

Planning was necessary to apply the Observational Method to this project. Peck *(7)* states "The most serious blunder in applying the observational method is failing to select (in advance) an appropriate course of action for all foreseeable deviations (disclosed by observation) from those assumed in the design. The engineer must devise solutions to all problems which could arise under the least-favorable conditions." For each slide area, the embankment and cut slope designs were evaluated for geotechnical uncertainties and potential consequences. The ranges of potential failure mechanisms were identified and strategies were developed to mitigate these potential failures if they were to occur.

Engineering judgment was used to develop trigger criteria for each stage. Tolerable movements were estimated for the green and amber stages, based on constructed elements in embankments, cuts and ground anchor applications. For example, constructed elements in embankments include culverts, horizontal drains, and trench drains, as illustrated in the cross section in Figure 8. Figures 9 and 10 provide illustrations for cut and ground anchor applications.

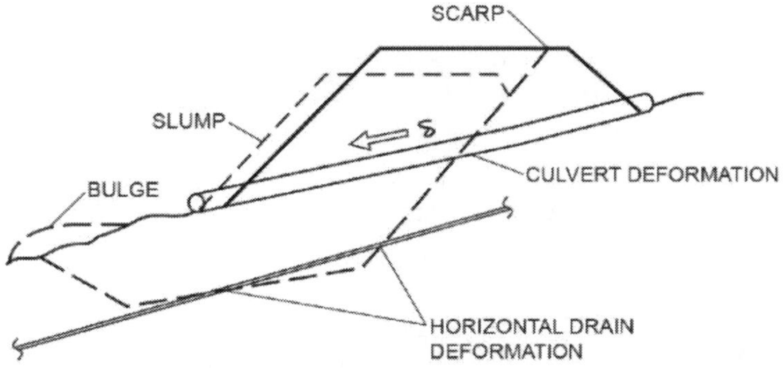

Figure 8 – Illustration of potential damage in embankments

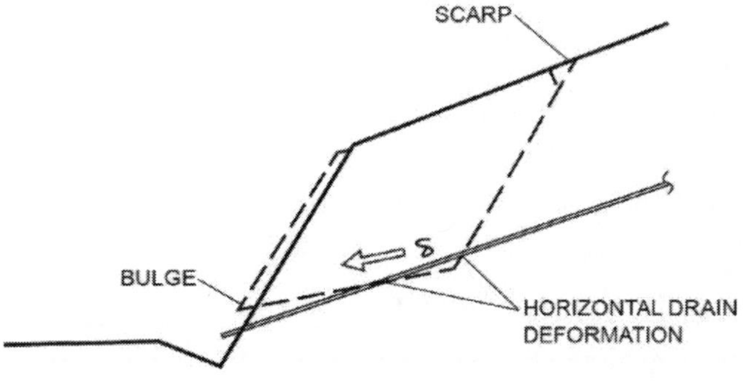

Figure 9 – Illustration of potential damage in cut sections

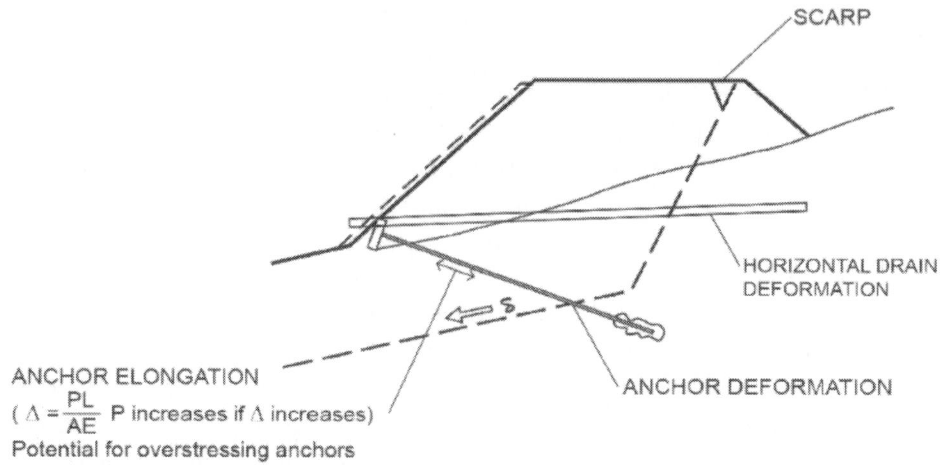

Figure 10 – Illustration of potential damage in ground anchors

Placing fill material on landslides and excavating into slopes could cause slumps and landslides, with movement along shear zones and scarps. Stability could also be impacted by blasting and uncontrolled water. The following lists possible types of damage.

- Deformation or shearing of horizontal drains
- Deformation or shearing of culverts
- Deformation or shearing of trench drains and underdrains
- Elongation or overstressing of ground anchors

The green stage was defined as the range of tolerable movement a constructed element could experience and still retain functionality and effectiveness. The amber stage was defined as the range of increased movement which could potentially result in moderate damage approaching but not likely to cause critical impairment of the feature and/or facility. The estimates of tolerable deformation for embankments, cut slopes and anchored slides are shown in the following three exhibits, Figures 11 through 13, respectively.

EMBANKMENT SECTIONS	Tolerable Movement (acceptable damage)	Increased Movement (moderate damage)	Critical Damage
Embankment	< 5 feet?	5-10 feet?	> 10 feet?
78" Culvert	< 12"?	12-18"?	> 18"?
36" Culvert	< 6"?	6-12"?	> 12"?
1.5" Horizontal Drain pipe	< 1"	1-4"?	> 4"?
Trench drain 6" pipe, 3-ft wide trench	< 6"	6-12"?	> 12"?
Slide scarp or toe bulge forms	No scarps	No crack, to crack beginning to form	Significant Crack

Figure 11 – Tolerable movement criteria for embankments

CUT SLIDE SECTIONS	Tolerable Movement (acceptable damage)	Increased Movement (moderate damage)	Critical Damage Potential
Cut Slide	< 0.5"/yr?	0.5-1.5"/yr?	> 1.5"/yr?
1.5" HD pipe	< 0.75"?	0.75-3"?	> 3"?
Slide scarp or toe bulge forms	No scarps	No crack, to crack beginning to form	Significant Crack

Figure 12 – Tolerable movement criteria for cuts

GROUND ANCHOR SECTIONS	Tolerable Movement (acceptable damage)	Increased Movement (moderate damage)	Critical Damage Potential
Ground Anchor elongation	< 0.7"?	0.7-1.5"?	> 1.5"?
1.5" HD pipe	< 1"?	1-3"?	> 3"?
36" Culvert	< 6"?	6-12"?	> 12"?
Slide scarp or toe bulge forms	No scarps	Impending failure?	Critical, action required

Figure 13 – Tolerable movement criteria for ground anchors

Measuring the movement of landslide shear zones during construction was done to determine the level of concern and the appropriate action to be taken. Instrumentation of the many landslide-prone areas is an integral component of this project through design and construction. Project instrumentation, including inclinometers and piezometers, utilized the methods and principles applicable to landslides described by Machan and Beckstrand *(8)*. Many instruments existed that could be continually monitored during construction. Additional instruments were installed to allow for an adequate distribution in critical construction areas. Monitoring of the performance of installed drainage systems was accomplished by using piezometers to measure groundwater level responses, which allowed for focused design and mitigation efforts at key locations to remove excessive groundwater where it caused destabilization. Groundwater level monitoring was augmented with measurement of horizontal drain discharge flow rates. In each landslide area, critical instruments were identified that would provide early warning, and these instruments were automated with data transmitted by telemetry for rapid viewing and assessment.

One example of a trigger criteria for embankment sections is the evaluation of the tolerance of horizontal drains to landslide movement. A horizontal drain is a 1.5-inch diameter PVC pipe installed in a 4-inch diameter hole with an open annular space. It was reasoned that if a landslide shear zone were to displace across the diameter of the drain, the pipe could move a distance of up to 1.5 inches into the annular space without damage (green stage). The drain might experience slight damage where displacement of up to 4 inches occurs but remain functional to allow collected groundwater to flow through the drain pipe across the damaged zone (amber stage). Once the amber stage is reached, additional monitoring and evaluations would need to be performed to check the severity of the interpreted hazard and to analyze mitigation options. In addition, the construction team would be made aware of the interpreted risks and potential mitigations being considered. If the displacement along the landslide shear zone approaches 4 inches (the drilled diameter of the horizontal drain hole), it could become sheared or obstructed, resulting in a loss of functionality and effectiveness (red stage).

The intent of adequately applying the Observational Method was to identify potential concerns early and to timely implement mitigation strategies to ensure normal serviceability limit states are not exceeded and to avoid catastrophic and expensive consequences. The following diagram (Figure 14) illustrates the progression of this process and risk condition states for an embankment example that eventually requires red stage attention. The vertical scale is the landslide shear zone displacement as measured in nearby inclinometers. This example assumes approximately half the embankment is constructed the first summer season and completed the following summer. The diagram assumes landslide movement decreases in the fall/winter months as pore pressures dissipate from embankment loading. Movement increases again as fill loading resumes during the second construction season.

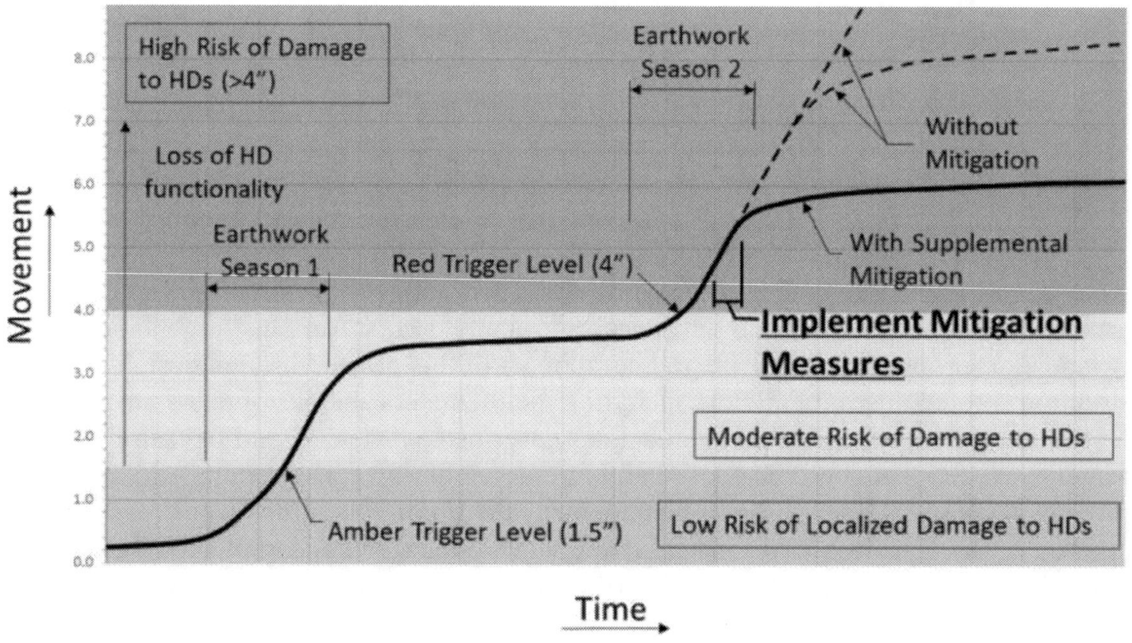

Figure 14 – Illustration of potential movement trends and stages of Observational Method for embankments, including trigger criteria and increased level of monitoring, possibly leading to decision to implement a supplemental mitigation

The planning of the Observational Method for monitoring the integrity of ground anchors installed through an embankment and slide debris and into the underlying Tyee Formation rock was complex because of settlement-induced stress changes (relaxation) in the ground anchors as well as landslide-caused increases in anchor stress. In addition to preventing damage to culverts and horizontal drains, it was also critical that the landslides do not cause excessive straining of the ground anchors. It was reasoned that a ground anchor with a 100-foot free length could be stretched (elongated) by the landslide as much as 0.7 inches and still retain proper functionality well within the elastic range of the steel tendon according to industry-accepted allowable stresses (65% of guaranteed ultimate tensile strength), which would be represented by the green condition stage.

Industry practice is to not allow tensile stresses to exceed 80% of ultimate strength to prevent the steel tendon from experiencing plastic deformation, which could lead to failure of the tendon. Hence, the amber stage was defined where movements could approach 1.5 inches and cause stresses in the steel tendons to approach 75% of ultimate strength. If this level of movement and elongation of the steel tendon was exceeded (red stage), then a supplemental mitigation measure would need to be implemented rapidly to prevent plastic deformation and failure of the ground anchors.

The following diagram (Figure 15) illustrates this process and risk condition states progressing into the critical red stage for an example where ground anchors are used to increase embankment stability. The vertical axis measures anchor load (measured by strain gages on the tendon and load cells at the anchor head) which are related to the identified triggers for the Observational Method. Stress levels in the steel tendons need to be controlled to prevent ground anchors from becoming overstressed. Anchor restressing was performed 14 days after initial stressing to regain load after settlement of the bearing pad occurred, causing relaxation of the anchors. In the event landslide movement occur, anchor loads could increase due to stretching and increased tension in the anchors.

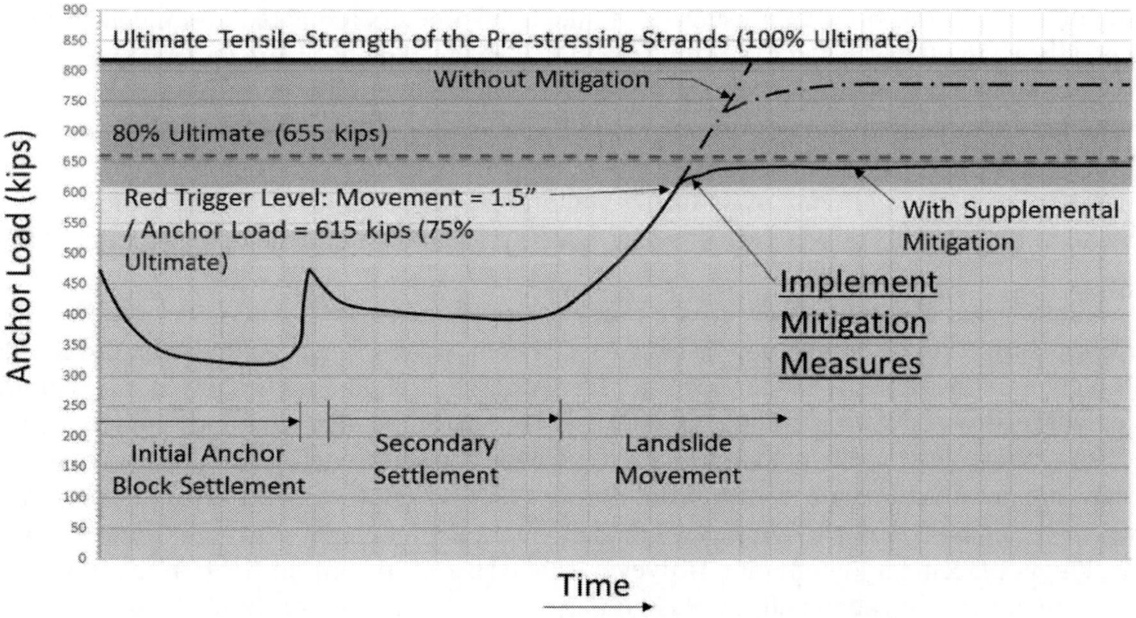

Figure 15 – Illustration of potential movement trends and stages of Observational Method for ground anchors to stabilize embankments, including trigger criteria and increased level of monitoring, possibly leading to decision to implement a supplemental mitigation

CONSTRUCTION CONSIDERATIONS

Constructability reviews were conducted to check the reasonableness and effectiveness of project details and specifications to manage and minimize landslide risks, and to prevent construction activities that might exacerbate landslide conditions. Prospective construction contractors were informed in prebid meetings of potential landslide risks and to anticipate and accommodate design changes if measured performance indicated a need for supplemental mitigation outside those provided in the bid documents. Flexibility in the construction scope and schedule were necessary to modify designs using the geotechnical observational approach.

The contractors were also informed of the extensive instrumentation that they would need to protect and prevent their equipment from inadvertently damaging them. Many of the instruments were converted to digital sensors where cables were extended in conduits to the perimeter of the construction corridor to lessen the risk of accidental damage. Contract requirements included provisions for the contractor to repair or install replacement instruments in the event of damage.

CONSTRUCTION PERFORMANCE

Most landslides performed satisfactorily during construction of the large embankment fills and hillside cuts, as verified by the extensive network of geotechnical instruments. Landslide movement was typically less than ¼ inch. This demonstrated the successful application of designing to marginal Factors of Safety and verifying performance using the Observational Method.

During installation of horizontal drains, groundwater response was measured with piezometers in various borings. Responses in most piezometers indicated groundwater lowering of typically 10 to 20 feet when installing horizontal drains. However, piezometers in several areas, including a critical area of the Crystal Slide beneath the highway embankment, did not respond to the horizontal drains. At the Crystal Slide location, the designed installation consisted of three arrays of horizontal drains (90 drains). To achieve the desired reduction in groundwater pressures, a supplemental array of 45 horizontal drains was added to target the critical area. These supplemental drains were effective in reducing groundwater head approximately 17 feet.

One unstable condition occurred at the west portion of the Crystal Creek embankment toe where accelerated movement along a shear zone was measured in an inclinometer during the 2014 construction season. The toe movement rate approached 0.25 inches/day during active fill placement in July and August. The total shear movement at the toe of the local slide lobe reached 5 inches within a few weeks, which triggered the need for additional mitigation to reduce impacts and possible damage to horizontal drains. Geogrid reinforcement layers were added to reinforce the embankments as a quick method of obtaining an increase in stability to allow embankment placement to continue. However, this was a short-term mitigation to allow uninterrupted fill placement, and further mitigation was necessary to achieve long-term stability.

Stability analyses were fast-tracked to evaluate permanent mitigation options. Two options were determined to be effective and feasible. The embankment slope geometry was modified to reduce the embankment loading in the vicinity of the locally active slide area by creating a 60-foot wide bench mid-height of the embankment slope and steepening the upper slope to 1V:1.5H. While this slope geometry modification improved local stability of the middle portion of the embankment height, the global stability of the final design embankment height was still considered marginal (FS of about 1.0). A supplemental toe buttress was designed to achieve acceptable levels of stability, which required a modification of the environmental permit due to impacts to the stream as well as acquisition of additional right-of-way.

The revised embankment slope design and supplemental buttress have resulted in improved stability FS (approximately 1.15 based on stability analyses). Horizontal drains were subsequently inspected by insertion of a jetting hose and found to be unobstructed and still functional. The 78-inch diameter culvert was inspected by walking inside and confirmed that no elements or bolts were damaged and the cross-sectional area had not visibly deformed. Slide movements have significantly slowed, verifying satisfactory performance.

CONCLUSIONS

Geotechnical solutions moved mountains and drained hillsides. Highway construction required excavation, transport, and placement of 7.3 million cubic yards of earth, including the construction of shear keys and buttresses. Over 1,200 horizontal drains were installed over a 3 mile stretch of new highway, pulling an estimated 5 million gallons of water per day from the hillsides during large storms. Horizontal drains were drilled up to 900 feet long each, totaling in excess of 100 miles. The horizontal drains reduced the groundwater pressures that were causing the landslides to move and also reduced the impact time of storms by rapidly removing groundwater. Stability of the treacherous mountain slopes was improved, enabling the construction of large unprecedented highway cuts and embankments necessary for the selected alignment.

Risk-based evaluations using engineering judgment were essential for weighing landslide risks, anticipated performance of constructed features, and construction cost. Some low risks could be tolerated which achieved a corresponding savings in construction costs without comprising traffic safety. Overly-conservative designs were avoided. This resulted in cost-effective mitigation decisions.

Overall, using the Observational Method to evaluate the project from a performance-based perspective resulted in a net savings compared to a conventional design approach. This project utilized an adaptive design approach confirmed with verifiable performance criteria. Using this approach resulted in a net savings to the DOT, where the geotechnical elements of the construction were completed 20% under budget.

REFERENCES:

1. Hammond, C.M., Meier, D., and Beckstrand, D.L. Paleo-landslides in the Tyee Formation and Highway Construction, Central Oregon Coast Range. In *Volcanoes to Vineyards: Geologic Field Trips through the Dynamic Landscape of the Pacific Northwest*, O'Connor, J.E., Dorsey, R.J., and Madin, I.P. (Eds) Field Guide 15, Geological Society of America, Portland OR. 2009. 14 p.

2. Machan, G and Black, B.A. Horizontal Drains in Landslides: Recent Advances and Experiences. In *Landslides and Engineered Slopes: Protecting Society through Improved Understanding, Proceedings of the 11th International and 2nd North American Symposium on Landslides and Engineered Slopes.* Eberhardt, E., Froese, C., Turner, A.K., and Leroueil, S. (Eds). Banff, Canada. June 2012. pp. 1967–1972.

3. Cornforth, D.H. *Landslides in Practice: Investigation, Analysis, and Remedial/ Preventative Options in Slopes*. John Wiley & Sons, Hoboken NJ. 2005. 596 p.

4. Scott, G.A. Using Risk-Informed Decisions to Prioritize Limited Resources. *Geo-Strata, GeoInstitute.* 2012. pp. 16-22.

5. Silva, F., Lambe, T.W., and Marr, W.A. Probability and Risk of Slope Failure. *ASCE Journal of Geotechnical and Geoenvironmental Engineering.* vol. 134 (12), 2008. pp. 1691-1699.

6. Nicholson, D., Tse, C. and Penny, C. *The Observational Method in Ground Engineering – Principles and Applications, Report 185*. Construction Industry Research and Information Association (CIRIA), London UK. 214, 1999. 214 p.

7. Peck, R.B. Advantages and Limitations of the Observational Method in Applied Soil Mechanics. *Ninth Rankine Lecture, Geotechnique.* vol. 19 (2), 1969. pp. 171-187.

8. Machan, G and Beckstrand, D.L. Practical Considerations for Landslide Instrumentation. In *Landslides and Engineered Slopes: Protecting Society through Improved Understanding, Proceedings of the 11th International and 2nd North American Symposium on Landslides and Engineered Slopes.* Eberhardt, E., Froese, C., Turner, A.K., and Leroueil, S. (Eds). Banff, Canada. June 2012. pp. 1229–1234.

Soil Mixing: An Innovative Solution for Resiliency in a Flood-Prone Canyon

Todd Schlittenhart, P.E.
Yeh and Associates, Inc.
2000 Clay Street, Suite 200
Denver, CO 80211
(303) 781-9590
tschlittenhart@yeh-eng.com

Prepared for the 69th Highway Geology Symposium, September, 2018

Acknowledgements

The author would like to thank the individuals/entities for their contributions in the work described:

James Usher, P.E. – Colorado Department of Transportation
Doug Stremel – Jacobs Engineering
Derek Heflinger – Jacobs Engineering
Eric King – Jacobs Engineering
William DeRosset – Ayres Associates
Harry Koenigs – Kiewit Infrastructure
Khamis Haramy – FHWA/CFL
Marilyn Dodson – FHWA/CFL

Disclaimer

Statements and views presented in this paper are strictly those of the author(s), and do not necessarily reflect positions held by their affiliations, the Highway Geology Symposium (HGS), or others acknowledged above. The mention of trade names for commercial products does not imply the approval or endorsement by HGS.

Copyright

ABSTRACT

US 34 Big Thompson Canyon has experienced two significant flood events in 1976 and 2013, destroying many sections of the 23-mile highway segment between Loveland and Estes Park, in Colorado. Following the 2013 flood event, the Colorado Department of Transportation completed emergency repairs to rebuild the highway infrastructure and restore public and private access. The next phase involved planning and designing permanent repairs to reconstruct US 34 to be more resilient to future flood events and minimize future flood damage so that the highway remains accessible.

Post-flood observations of damaged and undamaged portions of the highway emphasized the natural resiliency and scour resistance of the crystalline bedrock formations found within the canyon. Permanent repair designs included shifting the highway onto a bedrock surface through rock cuts and bridges. However, these designs proved to be too costly to implement throughout the entire canyon and, in many locations, were not feasible to construct due to the steep canyon walls rising over 300 feet above the highway. Innovative solutions were needed to connect the highway to the resilient bedrock found below the pavement surface, which in some locations was within 20 feet. Through the CM/GC (Construction Manager/General Contractor) process, soil mixing was identified as a design method to create a resilient bedrock-like surface between the highway and the bedrock. Soil mixing involves combining a cement slurry with the native materials. This paper presents the subsurface investigation, initial test section construction and investigation, design procedures, variable soil mixing construction techniques, and contracting methods used in over 5000 linear feet of permanent highway repairs.

INTRODUCTION

When it officially opened on May 28, 1938, US Highway 34 through Big Thompson Canyon provided a paved two-lane road connecting Loveland, CO to Estes Park, CO, the gateway to Rocky Mountain National Park. The paved roadway replaced a primitive, single-lane dirt road dating back to 1903. The alignment of US 34 followed that of the historic single-lane road, maintaining the breath-taking views and grandeur of a deep canyon carved out by the Big Thompson River. Completion of US 34 marked the beginning of population and tourist growth in Big Thompson Canyon. With a paved highway open year-round, it was possible to enjoy the tranquility of the river and canyon while still being a short drive from the necessary services of Loveland and Estes Park. Multiple communities and a few small towns sprung up and are now established within the canyon, relying on US 34 as their sole access.

Along the 23-mile portion between Loveland and Estes Park, US 34 winds its way through numerous geologic formations and deposits. As the highway heads west out of Loveland and follows the Big Thompson River, steeply-dipping, north-south trending shale and sandstone formations intersect the highway. The encountered formations within the foothills range in age from Cretaceous at the east end to Pennsylvanian at the mouth of the canyon. The Big Thompson Canyon portion of US 34 is comprised of crystalline rock formations of Pre-Cambrian age, overlain by alluvial and colluvial deposits. The first deep, narrow stretch of highway west of the foothills is known as The Narrows. This 2-mile stretch of canyon formed by the downcutting of the river through hard metamorphosed sedimentary rock formations. In cutting its passage, the river naturally sought the line of least resistance, turning abruptly to pass around resistant rock, and creating an intricate winding course. The canyon walls through The Narrows are characterized by steep, near vertical rock slopes of varying form and height, sometimes rising hundreds of feet above the highway. The base of the canyon is only a little wider than the river, and in many places the roadway was constructed by blasting into the rock slopes or supported on the river side by a series of soldier pile and lagging retaining walls. Farther upstream the granite and pegmatite rocks which intruded into the sedimentary rock are more abundant. Due to varying resistances to erosion within the granites and pegmatites, the canyon walls west of The Narrows vary from near vertical to rocky, open mountain valley slopes. Numerous synclines, anticlines, and faults are part of the forming geomorphology of the canyon. Most notable is the Big Thompson Canyon fault, an east-west trending fault, which the river follows for extended portions of the canyon.

The Big Thompson Canyon has an infamous history of disastrous floods. The July 1976 flood was a devastating flash flood that swept down the steep and narrow canyon, claiming the lives of over 140 people; the deadliest flash flood of Colorado's history. This flood was triggered by a nearly stationary thunderstorm near the upper section of the canyon that dumped over 12 inches of rain in less than 4 hours. A wall of water more than 20 feet high raced down the canyon, destroying cars, homes, and businesses while washing out most of US 34 (see Figure 1). Recordings taken at the mouth of Big Thompson Canyon showed the flow of water at 31,200 cubic feet per second (cfs) at its peak.

Figure 1. Photograph of damage to US 34 after the 1976 flood.

After the 1976 flood, US 34 was repaired and rebuilt in the same position it was, with better drainage control and retaining walls to mitigate erosion. One major improvement occurred in The Narrows. Through this section, the roadway was elevated above the 100-year flood elevation through a series of four soldier pile and lagging retaining walls and a bridge. Construction of the retaining walls and bridge through The Narrows was completed in 1978. US 34 Big Thompson Canyon was again heavily damaged during the September 2013 floods, with many homes and businesses damaged and over 100 air-lifted evacuations. Sustained rainfall over multiple days created watershed runoff combining with flows released from Lake Estes Dam. Surges down the canyon exceeded the 500-year flood event. While not as intense as the 1976 flood, the sustained rainfall of 2013 sent enough water down the canyon to wash out major sections of the highway (see Figure 2). In The Narrows, the retaining wall foundations were undermined in many areas and sections of backfill and pavement were washed out, leaving large voids behind the wall facing (see Figure 3).

Figure 2. Photograph of September 2013 damage.

Figure 3. Photograph of "The Narrows" after the September 2013 flood.

Following the 2013 flood event, emergency repairs restored public access along the highway, reopening the highway to traffic in both directions by November 11, 2013. The next phase involved designing permanent repairs to provide resiliency to future flood events and minimize flood damage so that the highway remains accessible. Governor John Hickenlooper gave the directive to "build back better than before." Following that directive, the Colorado Department of Transportation (CDOT) engaged engineering design professionals to study the hydraulic flow of the river in the canyon and its impact on the road and bridges in its channel.

Post-flood observations of damaged and undamaged portions of the highway emphasized the natural resiliency and scour resistance of the crystalline bedrock formations found within the Big Thompson Canyon. Design concepts sought to maintain at least one travel lane of the highway from being compromised during a flood event. Two of the major design concepts to improve highway resiliency were to relocate sections of the roadway onto competent rock (Figure 4) or add scour resistant elements adjacent or under the roadway along sections particularly vulnerable to high-energy erosion. Relocating the roadway ideally prevents any major damage in future floods, as the roadway moves above 100-year flood elevation and onto a scour resistant bench of metamorphic rock. In locations where it is not feasible to shift the roadway onto bedrock, scour resistant elements help limit damage and destruction in future floods. In the 2013 flood, there were areas where both lanes of the roadway were compromised, limiting emergency access and trapping people within the canyon until helicopter rescues could be coordinated.

Figure 4. Schematic of roadway relocation design.

SCOUR RESISTANT DESIGN CONCEPTS

In June of 2015, Kiewit Infrastructure was awarded a CM/GC (Construction Manager/General Contractor) contract with CDOT. CDOT brought together a design team of engineers in the fields of roadway, hydraulics, structures, and geotechnics led by Jacobs Engineering. Kiewit engineers and construction managers worked alongside the design team to develop scour resistant elements for scour-susceptible areas where shifting the highway onto bedrock under the roadway was not feasible due to cost and/or constructability. Resilient conceptual designs included soil nail walls, tangent pile walls, tangent micropile walls, gravity walls, grouted boulder systems, roller compacted concrete, and many others. The design concepts were evaluated for cost, resiliency, and constructability. Of note, with regards to the soil mixing concept, is the grouted boulder system. This concept built on the design and construction methods used on County Road 43 (CR 43) and the North Fork of the Big Thompson River. There, native overburden materials were excavated to the bedrock surface or below the anticipated scour elevation. Large (greater than 8-foot diameter) boulders were moved into the excavation and cemented together to form a bedrock-like system below the roadway. This concept affectionately became known as "Build Your Own Bedrock" for the US 34 project.

Out of the evaluation of all the conceptual designs, the following elements were selected for implementation as scour resistant designs:

- Bridge structures above the 100-year flood elevation connecting to a rock cut "through-cut" in an area known as The Horseshoe. These bridges bypassed a high energy oxbow bend in the river that scoured out completely in both the 1976 flood and the 2013 flood. Figure 5 shows a computer generated model of the Horseshoe Bridge. The portion of US 34 around the oxbow (left side of the photograph) was abandoned.

Figure 5. Rendering of Horseshoe Bridge.

- Matrix grouted riprap and large clast, mechanically placed riprap buried below the roadway template with an additional layer of riprap ("nuisance riprap") exposed adjacent to the river. The cross-section shown in Figure 6 shows the stout matrix riprap section buried below the roadway and the "nuisance riprap" adjacent to the river. This design concept was implemented in multiple areas due to the relative low cost compared to the other design concepts. During a flood event, the "nuisance riprap" will scour away, allowing the river to widen out and lose energy. The matrix riprap and large riprap protect a single lane of the highway and maintain access.

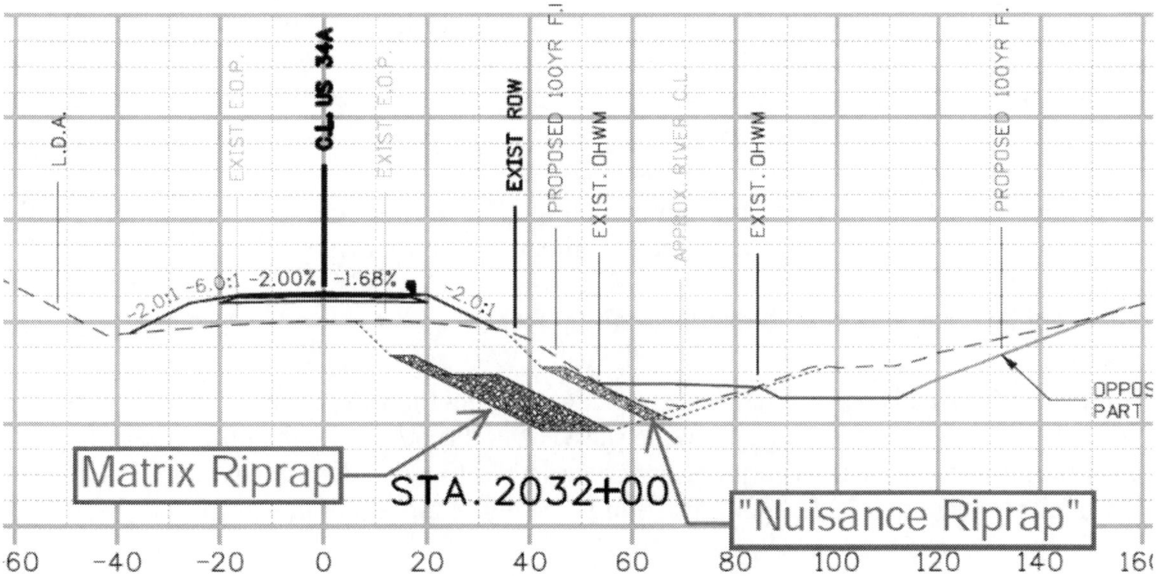

Figure 6. Schematic of Matrix Riprap with Nuisance Riprap.

- Soil mixing below the roadway within the traffic lane furthest away from the river. Soil mixing was implemented in narrow reaches of the river where riprap construction was impractical. The soil cement connects the highway to the crystalline bedrock found below the pavement surface. Figure 7 shows a typical section with the approximate bedrock surface shown as a dashed line and the area of soil mixing shaded in brown.

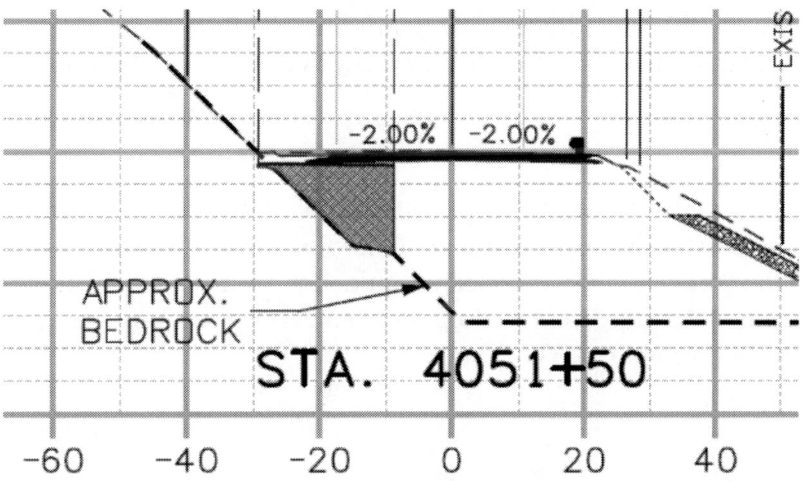

Figure 7. Schematic of Soil Mixing Below Roadway.

SOIL MIXING CONCEPT

Within the CM/GC process, CDOT solicited an independent team of engineers and construction managers to review the US 34 Big Thompson Canyon project for innovative design concepts. Building off the CR 43 and design team's concept of "Build Your Own Bedrock," the independent team identified soil mixing as a more constructible and more affordable option.

Soil mixing is an in-situ ground treatment in which soil is blended with cementitious and/or other binder materials to improve strength, permeability, and/or compressibility characteristics of the native soils. The strength of the soil cement correlates to its durability and scour resistance. Various methods of soil mixing exist to create uniform, dense masses of soil cement.

- The auger method is the most common technique for deep soil mixing. Generally, drill rigs equipped with specialty augers are used to inject and mix a cement slurry with the surrounding soils. The auger method is effective in sandy, silty, and clayey materials that allow for auger penetration to depths up to 60 feet.

- Injection tillers and rotary drum mixers are another method for deep soil mixing and are suitable for the in-situ mixing of a wide range of soil types to depths up to about 15 feet. Injection tillers and rotary drum mixers are attached to a standard excavator through which the cement slurry is injected and mixed in place at the target depth.

- In-situ bucket mixing refers to using an excavator and a specialty excavator bucket to mix the cement slurry with the soil in place. A variety of soils can be mixed through bucket mixing; however, this method typically does not provide a uniform mix in clayey soils. Bucket mixing depths are limited to the reach of the excavator. The cement slurry is added to the soil to obtain a high slump soil cement. The excavator removes several buckets of soil from below the previously mixed soil, deposits the unmixed soil above that previously mixed, and proceeds to uniformly mix the soils.

Bucket mixing was the preferred method for US 34 due to the on-site sands, gravels, and boulders encountered in the subsurface. The depth limitations of bucket mixing required the design team to locate soil mixing areas where the depth to bedrock did not exceed 25 feet.

SOIL MIXING DESIGN

Once the soil mixing design concept was approved, the design team was charged with identifying locations and designing the soil cement. For the US 34 Big Thompson Canyon project, design of the soil cement was based on guidance from Design Guideline 7 (DG-7) of FHWA Hydraulic Engineering Circular No. 23 (*1*) and Chapter 17 of Design Standard No. 13 (U.S. Bureau of Reclamation DS-13) (*2*).

Design Guideline 7

DG-7 reflects bank stabilization guidance provided by the Pima County Department of Transportation (Pima County) in Tucson, Arizona and the Portland Cement Association. When placing soil cement as a countermeasure for scour protection, DG-7 provided the following applicable recommendations to the Big Thompson Canyon project.

- Tie the soil cement into nonerodable surfaces to prevent undermining.

- Design the soil cement with free drainage to reduce hydrostatic pressure behind the soil cement section.

- Construction specifications should include types of materials and equipment, mix design and methods, handling, placing, and curing techniques.

- A 7-day compressive strength of 750 pounds per square inch (psi) should be used for the soil cement mix design.

The performance and resiliency of soil cement bank protection constructed under the guidance of DG-7 has been documented (*3*) following five major floods between 1983 and 2006. One soil cement bank stabilization study, the Salt River Flood of January 1993, provides the closest documented condition to the Big Thompson Canyon. The cement treated alluvium contained clasts greater than 3 inches in diameter and an estimated peak flow of 124,000 cubic feet per second (cfs) containing boulders impacted the soil cement banks. It was noted that the banks performed well when subjected to the high velocity, turbulent flow containing up to boulder-sized clasts.

Design Standard No. 13

Chapter 17 of the U.S. Bureau of Reclamation (USBR) DS-13 provides guidance for soil-cement slope protection measures for the upstream face of embankment dams and dikes. The recommendations applicable to the Big Thompson Canyon project, based on implementation at 22 USBR sites are:

- Avoid the use of plastic soils for making soil cement. A silty sand soil is considered the best suited for making soil cement., however, several USBR projects used coarser materials successfully.

- Design the soil cement with a cement content of approximately 12 percent by dry weight of soil.

- Use minimum compressive strengths of 600 psi at 7 days and 875 psi at 28 days for the soil cement mix design.

SUBSURFACE INVESTIGATION

At the direction of CDOT, a geotechnical subsurface investigation was not performed at each proposed soil mixing location. This decision was based on the cost of such an investigation and the time required to complete the investigation. In place of a geotechnical subsurface investigation, a bedrock surface survey was selected by CDOT and the CM/GC team, which provided data at lower relative costs and within a shorter time period. The bedrock surface survey utilized hard rock hammer drill rigs to advance a 3.5-inch diameter drill bit to the desired depth. M.C. Donegan operated the drill rigs, with Kiewit directing and observing the drilling operation. The planned drilling generally consisted of a grid pattern of drill holes equally spaced within each proposed soil mixing area. Figure 8 shows a plan view of the typical grid pattern drilling.

Figure 8. Plan View of Typical Grid Pattern Drilling.

The desired depth of penetration was a minimum of 5 feet into bedrock in order to distinguish the bedrock surface from boulders within the overburden materials. Subsurface samples were not obtained as part of the bedrock surface survey and the quality of the encountered bedrock was not quantified. Drill hole locations and elevations were surveyed by Kiewit, combined with bedrock depths, and provided to Jacobs for bedrock surface modeling.

STABILITY DESIGN

Based on the bedrock surface model, Jacobs generated cross-sections showing the bedrock surface and areas of soil mixing. Refer to Figure 7 for a typical section of the soil cement overlying dipping bedrock.

Stability of the soil mixing area is critical during and following the extreme flood event, in which the overburden materials are removed due to scour and the soil cement remains on the bedrock surface. The primary failure mechanism for the soil mixing area in the extreme flood event is failure by sliding. Resistance to a sliding failure was checked utilizing three methods.

Plane Failure of a Rock Slope

Utilizing limit equilibrium analysis for a plane failure of a rock slope (*4*), driving forces and resisting forces were used to calculate a Factor of Safety (FS) for critical sections. The driving forces included the weight of the soil cement, traffic surcharge, and pore water pressure. Pore water pressures were neglected since the plans and specifications require installing a drainage system. The resisting force for the plane failure is the shear strength of the sliding surface between the soil cement and the bedrock. A cohesion (*c*) of 200 pounds per square foot (psf) and friction angle (ϕ) of 40° were assumed as input parameters for the shear strength of the sliding surface. A FS greater than 1.2 was considered stable in the extreme event.

Failure by Sliding of a Spread Footing

AASHTO LRFD design specifications (*5*) provide calculations for analyzing failure by sliding of a horizontal spread footing. This design method requires the factored resistance to be greater than the factored loads. A load factor of 1.0 was applied to the dead load weight of the soil cement and a 0.50 load factor was applied to the live load traffic surcharge in accordance with the Extreme Event II limit state and Table 3.4.1-1. Pore water pressures were neglected since the plans and specifications require installing a drainage system. An internal friction angle (ϕ_f) of 45° was assumed for the sliding surface when calculating the nominal sliding resistance. A resistance factor of 0.90 was applied to the sliding resistance based on Table 10.5.5.2.2-1.

Failure of Cement-Bedrock Adhesion

Using the equations for a plane failure of a rock slope as a basis, adhesion between the soil cement and the bedrock is substituted for the shear strength of the sliding surface in the resisting forces. Driving forces are compared to the resisting forces to determine the FS for critical sections. The driving forces include the weight of the soil cement and traffic surcharge, with pore water pressures neglected since the plans and specifications require installing a drainage system. The resisting force of adhesion is the bond strength of the soil cement-bedrock interface. An ultimate adhesion of 2 kips per square foot (ksf) was assumed for this bond strength and a FS greater than 1.2 was considered stable in the extreme event.

TEST SECTIONS

Prior to preparing project plans and specifications, two soil mixing test sections were constructed by Kiewit under the design supervision of Advanced Geosolutions, Inc. (AGI) to verify design assumptions and evaluate constructability.

The first test section was in the shoulder of the eastbound lane of US 34 within The Narrows. This test section was constructed by excavating an area approximately 100 feet long by 15 feet wide. Overburden soils were removed down to the bedrock surface and stockpiled adjacent to the test section. A bentonite-cement grout was delivered by mixer trucks to the site and poured into the excavation. A Kiewit excavator mixed the cement slurry with soil to achieve an approximate 1:1 ratio of soil to cement slurry. Samples were taken for unconfined compressive strength testing.

The second test section was in the westbound lane of US 34 near Idlewild Lane. An area approximately 30 feet long by 35 feet wide was excavated, but soil was left in the bottom half of the excavation to test in-situ mixing methods. A bentonite-cement grout was delivered by mixer trucks to the site and poured into the excavation. A Kiewit excavator mixed the cement slurry with the soil left in the excavation until an approximate 1:1 ratio of soil to cement slurry was attained. The excavator continued mixing until encountering bedrock at depths that matched the bedrock surface model in the area. Samples were taken for unconfined compressive strength testing.

Upon completion of the test sections, trenches were cut through the soil cement to observe the adhesion between the soil cement and the bedrock mass and to verify the constructability of drainage trenches in the soil cement. Drilled core samples (Figure 10) were also taken from each test section to observe the uniformity of soil mixing as well as bedrock adhesion.

Figure 10. Recovered Core Samples of Soil Cement.

Observations and conclusions (*6*) of the test sections included:

- Both test sections achieved 7-day unconfined compressive strengths exceeding 1000 psi when using an approximate 1:1 mix ratio.

- Boulder-sized clasts can be incorporated and mixed into the soil cement. With sufficient in-situ mixing, the boulders did not segregate to the base of the excavation prior to the cement curing.

- Cement slurry was observed to be in direct contact with the bedrock surface, indicating adhesion between the soil cement and the bedrock.

- Appropriately sizing the excavator and excavator bucket is critical to ensuring proper soil mixing along the irregular bedrock surface. Larger boulders could not be mixed with the cement slurry when using undersized excavators.

- The depth of soil mixing is limited by the reach of the excavator. Areas where the depth to bedrock is less than 25 feet are constructible.

PLANS AND SPECIFICATIONS

Soil mixing locations were identified based on limitations to other resilient design concepts and where both lanes of US 34 were destroyed in the flood events.

- Rock cuts to shift traffic onto a bedrock surface were not cost effective in reaches where steep canyon walls (higher than 300 feet in some places) rose above the highway. Additionally, CDOT was advised to avoid rock cuts in areas of known rock slides and adverse geologic features.

- Bridge structures elevated above the anticipated flood event proved costly to construct and required a complete shutdown in narrow sections of the canyon to remove and replace the existing highway with an elevated bridge. Although full highway closures were scheduled, limited access was required to remain for the residents of the canyon and emergency vehicles, thus necessitating at least one lane of the highway remain open.

- Matrix and large clast riprap sections need to toe below the anticipated scour elevation or into the non-erodible bedrock surface and require slopes flatter than 1.5 Horizontal to 1 Vertical on which to be placed. In narrow sections of the canyon, steep rock slopes below the highway template did not allow the riprap to tie into the bedrock surface or below the anticipated scour elevation.

The majority of planned soil mixing was in The Narrows in between retaining wall sections that proved to be vulnerable during the 2013 flood (Figure 11).

Figure 11. Destroyed Section of Highway Between Walls at Dille Dam.

A performance-based specification was created for soil mixing. A specialty contractor was required to construct the soil mixing and to submit a soil cement mix design based on the subsurface materials at each site. During construction, soil cement samples were taken and cast into cylinder molds for compressive strength testing. The soil cement was required to meet a 28-day strength of 750 psi. Full depth core samples were also obtained following completion of each section. The cores were inspected for uniformity of the soil cement as well as adhesion to the bedrock surface.

CONCLUSION

In a steep, narrow mountain canyon affected by floods, soil mixing proved to be a cost-effective, constructible design innovation when compared to structural concepts. Soil mixing was able to meet the goals of the project by providing a resilient roadway design in harmony with the river and ecological systems, maximizing the allotted budget to obtain the greatest benefit, minimizing the inconvenience to the public and residents along the corridor, and following the Governor's directive to "build back better than before."

Much like the four soldier pile and lagging walls built in 1978 through The Narrows, full scale resiliency testing may not occur until the next flood event in Big Thompson Canyon. When that does occur, the resiliency of soil mixing in a flood prone canyon will be tested.

REFERENCES

1) Lagasse, P.F., Clopper, P.E, Pagán-Ortiz, J.E., Zevenbergen, L.W., Arneson, L.A., Schall, J.D., Girard, L.G. *Bridge Scour and Stream Instability Countermeasures: Experience, Selection, and Design Guidance-Third Edition*. Federal Highway Administration Publication No. FHWA-NHI-09-111, Hydraulic Engineering Circular No. 23, 2009.

2) Kiene, A.H. *Reclamation: Managing Water in the West, Design Standards No. 13: Embankment Dams, Chapter 17: Soil-Cement Slope Protection*. U.S. Department of the Interior Bureau of Reclamation, 2013.

3) Hansen, K.D., Richards, D.L., Krebs, M.E. *Performance of Flood-Tested Soil-Cement Protected Levees*. PACE. http://pacewater.com/news-resources/publications-presentations/ Accessed June 22, 2018.

4) Wyllie, D.C.; Mah, C.W. *Rock Slope Engineering*. 4th Edition. Spon Press, London and New York, 2004.

5) American Association of State Highway and Transportation Officials (AASHTO). *LRFD Bridge Design Specifications*. 8th Edition, 2017.

6) Kiewit Memoranda. *US 34 Permanent Flood Repairs, Soil Mixing Test Section Report*. dated April 7, 2017 and June 9, 2017.

INNOVATIVE USE OF HORIZONTAL DIRECTIONAL CONTINUOUS ROCK CORING FOR THE DESIGN OF THE EAST END TUNNELS For THE LOUISVILLE –SOUTHERN INDIANA OHIO RIVER BRIDGES

Craig S. Lee, P.E.
S&ME Inc.
2020 Liberty Road, Suite 105
Lexington KY 40505
859-293-5518
clee@smeinc.com

Prepared for the 69th Highway Geology Symposium, September, 2018

Disclaimer

Statements and views presented in this paper are strictly those of the author(s), and do not necessarily reflect positions held by their affiliations, the Highway Geology Symposium (HGS), or others acknowledged above. The mention of trade names for commercial products does not imply the approval or endorsement by HGS.

Copyright Notice

ABSTRACT

This paper describes the innovative and successful use of horizontal directional continuous rock coring to develop a model of the subsurface conditions along the Pillar and both the North and South bound tunnel alignments. This paper introduces the concept of horizontal directional continuous rock coring to the transportation tunnel design process and presents the unique continuous baseline data generated by horizontal exploration versus traditional vertical borings.

S&ME advanced horizontal directional core borings through the crown of each tunnel and in the pillar section between the tunnel openings and performed laboratory testing on the recovered rock core to assess the engineering properties of the rock and published a Geotechnical Data Report for the tunnel. This was the first use of horizontal directional continuous rock coring as the primary technique for a major transportation tunnel and was the longest horizontal directional rock coring project in the United States.

The tunnel project is part of the Kentucky East End Approach segment. The approximate 2,000 foot long twin tunnels begin about 1,000 feet east of the intersection of Highway 841 North and Route 42. The tunnels have an inside finished width of approximately 60 feet with an inside finished height of approximately 41 feet. The original design of the I-265 extension proposed a conventional open cut roadway through the hillside that includes the Drumanard Estate. The Drumanard Estate was placed in the National Registry of Historic Places and must be preserved. This forced the alignment underground into twin tunnels, a northbound and a southbound tunnel. The design team proposed a 12-foot diameter exploratory (pilot bore) tunnel to evaluate the support conditions along the tunnel alignment but the cost was prohibitive. S&ME proposed horizontal directional core borings as an alternative to the exploratory tunnel.

The horizontal directional coring was a successful alternative approach to exploring the subsurface conditions along the tunnels. It provided a comprehensive model of the subsurface conditions along the tunnel alignments that was used for the final design of the tunnels. This approach saved the Kentucky Transportation Cabinet both money and time compared to the proposed pilot bore tunnel.

INTRODUCTION

The Louisville-Southern Indiana Ohio River Bridges Project is a "priority" national transportation project which addresses long-term, cross-river transportation needs in Louisville, Kentucky and Southern Indiana. It is one of the largest transportation projects in the country and results in safer travel, less congestion and improved access to destinations in the region. The overall project consists of six segments:

1. Kennedy Interchange
2. New Downtown Bridge
3. Downtown Indiana Approach
4. East End River Bridge
5. Kentucky East End Approach
6. Indiana East End Approach

The tunnel project is part of the Kentucky East End Approach segment. The approximate 2,000 foot twin tunnels begins about 1,000 feet east of the intersection of Highway 841 North and Route 42. Each tunnel has an inside finished width of approximately 60 feet with an inside finished height of approximately 41 feet.

The original design of the I-265 extension proposed a conventional open cut roadway through the hillside that includes the Drumanard Estate. During the project design phase, opponents to the project successfully placed the Drumanard Estate on the National Registry of Historic Places and must be preserved.

This forced the alignment underground into twin tunnels, a northbound and a southbound tunnel. The opponents hoped the expense of a tunnel would kill the project. As part of the design phase, the initial tunnel designer proposed to excavate a 12-foot diameter pilot bore through the crown of one of the tunnels to explore the subsurface conditions along the alignment. The pilot bore had three drawbacks. First, the cost. Second was the disruption to the nearby neighborhoods to construct a pilot bore. And lastly, the liability of an open 12-foot diameter tunnel if the project was cancelled.

S&ME proposed horizontal directional continuous rock coring as an alternative to the pilot bore. After the pilot bore bids came in significantly over the Engineer's Estimate, the Kentucky Transportation Cabinet (KYTC) turned to the S&ME alternative approach.

The objectives of our subsurface exploration were to advance horizontal directional core borings through the crown of each tunnel and in the pillar section between the tunnel openings and perform laboratory testing on the recovered rock core to assess the engineering properties of the rock. An assessment of site environmental conditions for the presence or absence of pollutants in the soil, bedrock, surface water, or groundwater along the alignment or on adjacent properties was beyond the scope of this exploration.

Our scope of work included the following:

- Drilling a total of 1,900 feet along the North Bound Tunnel section of the alignment.
- Drilling a total of 2,287 feet along the South Bound Tunnel section of the alignment
- Drilling a total of 2,337 feet along the Pillar section of the tunnel alignment.
- Providing a brief review of our field exploration
- Provide the results of the laboratory testing conducted.
- Review of subsurface rock stratigraphy with pertinent available physical properties along the alignments.
- Hydraulic Conductivity (Packer) of the borehole
- Providing boring logs

PROJECT LOCATION

The project site lies on the east side of the Greater Louisville area near the town of Prospect.

GEOLOGY

The Louisville Bridges Twin Tunnels encountered three rock formations along the alignment. The Silurian aged Louisville Limestone is the uppermost formation at the project site and is comprised of soluble limestone. The Louisville Limestone is mostly thin-bedded gray dolomitic limestone and gray calcitic dolomite, commonly in lumpy or irregular beds. Shale, in partings and very thin beds, constitutes a few percent, and very sparse chert is present in nodules and thin layers. In the project site, the Louisville Limestone is finely crystalline calcitic dolomite; the sparse fossils are dolomitized and include crinoid columnals, brachiopods, horn corals, and colonial corals.

From an engineering perspective, the Louisville Limestone is characterized by solution enlarged joints and bedding planes. Deep weathering and sinkhole formation are common. The primary impact for conventional building and roadway construction is the presence of latent drop-outs and a highly variable top of rock profile. The residuum derived from the Louisville Limestone is predominantly fat clay with limestone slabs and can exhibit problematic shrink and swell characteristics. For the tunnel, the Louisville Limestone presents several potential problems, most associated with the discontinuities such as solution enlarged joints (both horizontal and vertical), solutioning along bedding planes, voids, and sinkholes. The Louisville Limestone can also produce significant groundwater flows after rain events. Water flow is largely along open joints, fractures and bedding planes.

The Waldron Shale is immediately below the Louisville Limestone. The Waldron Shale is composed of greenish-gray shale and minor gray dolomite; typically at least 95 percent is shale. The shale is dolomitic and weathers with angular fracture or crude fissility, eventually producing a plastic clay. The dolomite is clayey and occurs in irregular masses, lumps, and thin discontinuous beds. Fossils, which are sparse in both the shale and the dolomite, include brachiopods, crinoid columnals, gastropods, and bryozoans. At the tunnel site, the Waldron Shale ranges in thickness from 9 to 15 feet. The basal contact with the underlying Laurel Dolomite is conformable and sharp.

The Waldron Shale breaks down when exposed to water and air. This formation is problematic in conventional earthwork construction as those unfamiliar with its properties, mistakenly place the shale as a durable shot rock fill. Over time the shale will degrade causing structurally significant settlement of buildings and roadways. The Waldron Shale presents a challenge to the construction of the tunnel as the shale is prone to delaminating and degrading during construction of the tunnel. In addition, the Shale will undergo a change in its physical properties over time after exposure to the elements.

The Laurel Dolomite underlies the Waldron Shale. The Laurel Dolomite is composed 95 percent or more of gray dolomite with minor greenish-gray shale and sparse gray limestone.

LABORATORY GEOTECHNICAL TESTING PROGRAM

The following strength and index tests were performed on selected rock core specimens in general conformance with ASTM International Standards, Kentucky Methods Manual, or other standards where applicable. The laboratory tests were conducted in the S&ME Knoxville, Tennessee Rock Mechanics laboratory and at the Geotechnical Engineering Center at the University of Texas at Austin.

- Axial and Diametrial Point Load Test (D5731)
- Unconfined compressive strength (D7012)
- Direct Shear (D5607)
- Brazilian Stress/Splitting Tensile Strength (D3967)
- Slake Durability (D4644)
- Cerchar Abrasivity (D7625)
- Huder-Amberg (Axial Swelling)
- Thin Section Petrographic Analysis
- pH
- Saturation and void ratio

The samples collected for testing were selected from the proposed alignment starting at the tunnel face to the termination of each of the three borings. The point load, unconfined compressive strength, and Brazilian Split Tensile tests were selected approximately every 60 feet along the boring starting at a distance of approximately 330 feet, which is the distance to the tunnel face, to the termination of the each boring at 1,900 feet, 2,287 feet and 2,337 feet respectively.

The slake durability samples were selected from the Waldron Shale. These samples were collected approximately every 30 to 35 feet along the borings starting at the contact between the Louisville Limestone and the Waldron Shale extending to the contact with the underlying Laurel Dolomite. Samples collected for the Cerchar Abrasivity, Huder-Amberg, and petrographic analysis were also selected from the Waldron Shale and then sent to the University of Texas at Austin. The pH, saturation and void ratio, and sulfur testing were also selected from the Waldron Shale. The locations of these samples along the alignment were selected by the S&ME geologist in the field based on visual observations and characteristics of the shale.

SUBSURFACE CONDITIONS

Subsurface conditions along the proposed tunnel alignments were explored with horizontal directional continuous core borings. The coordinates for the three boring locations within the tunnel template (at the tunnel face) were provided by Parsons and used by S&ME in planning our boring path profile. We advanced the borings using HQ size core tools to a distance of 122 feet to 163 feet – the section with the tightest curvature of the boring path. The HQ tools cut a

2.5 inch diameter core and a 3.7 inch diameter boring. From a distance of 122 feet to 163 feet NQ size and AQ size core equipment was advanced the remainder of the boring. The NQ tools cut a 1.8 inch diameter core and a 3 inch diameter boring. The AQ core was cut during directional drilling. The directional equipment cut a 3 inch diameter boring and a 1 inch diameter core.

Continuous core samples were collected along each boring during both the conventional (tangent section) and directional phases. Our boring logs, laboratory test sheets, and core boxes referenced the location of the core with respect to "distance" from the boring collar instead of depth. Our Geotechnical Data reports included a table of distance from the collar of each boring as well as project datum coordinates for each rock core sample interval. The field logging was performed by an S&ME geologist and consisted of:

- Measuring and logging the core and describing the physical appearance and lithology of the rock.
- Identifying and documenting the discontinuities, and bedding planes within the formations.
- Measuring the core recovery and Rock Quality Designation (RQD)
- Selecting specimens for laboratory testing
- Photographing the core after placing the recovered core in the labeled core boxes. The rock core photographs are included in Appendix A of this report.
- Assigning project coordinates of the selected rock core specimens

The Devico System used at the Louisville Bridges tunnel job consists of the DeviDrill, the PeeWee tool, and the DeviFlex. The DeviDrill is the steerable core barrel while both the PeeWee and DeviFlex are used to measure the physical parameters of the borehole. The principle behind the DeviDrill core barrel is a drive shaft running through a bushing, offset from the center line of the tool. Expanding pads operated by a differential pressure is keeping the DeviDrill in a fixed tool face while drilling in a curve. The inner assembly carries an inner tube collecting the core, a mule shoe system, and an instrument barrel with the survey tool recording inclination and tool orientation. Data is stored inside the tool and downloaded wirelessly to a PDA after each run.

The PeeWee is a miniature electronic multishot based on the same technology as the DeviTool Standard. The PeeWee uses three high-accuracy magnetometers and accelerometers. It records inclination, azimuth, tool face, temperature, gravity vector, magnetic field vector, magnetic dip angle, and battery status.

DeviFlex is a non- magnetic electronic multishot for surveying inside casings and drill strings by simply using the wireline system. The DeviFlex is less prone to magnetic disturbances. The DeviFlex tool consists of two independent measuring systems. Three accelerometers and four strain gauges are used to calculate inclination and change in azimuth. In addition, the DeviFlex records and stores gravity vector, temperature, and battery capacity.

S&ME partnered with Tech Directional a licenses franchisee of Devico to assist in the directional phases of the project. S&ME maintains an exclusive agreement with Tech Directional to provide directional continuous rock coring on projects in the United States.

PROJECT PLANNING AND EXECUTION

The exploration operations were located within an abandoned exit ramp of Highway 841. The photograph below shows the project staging plan relative to the proposed tunnels and adjacent neighborhood to the north of the staging area.

The North Bound boring was the first boring drilled. We used an LM-75 skid mounted underground drill rig to complete the North Bound boring. The South Bound and Pillar borings were drilled using an LM-90 skid mounted underground drill rig. The drill rigs were powered by Caterpillar generators which were quiet enough you could easily talk with a normal volume standing immediately next to one. This was important as the affluent neighborhood is on the other side of the trees adjacent to the project site as shown in the adjacent photograph.

In order to steer the hole, the boring must be started into a vertical surface. For the North Bound boring we excavated a pit into bedrock as shown in the photograph. We applied shotcrete to the face to control spalling over the project duration. We used a short rock cut face for both the South Bound and Pillar borings, so no pit was needed. Prior to starting each boring, our surveyor placed a PK nail in the shotcrete face at the starting coordinate from our borehole planning. The inclination of the drill steel string was measured by our surveyor to make sure our inclination was correct.

From the boring collars to the face of the tunnel is a distance of approximately 330 feet. In that 330 feet the borings had to descend from elevation 595 feet at the surface to elevation 535 feet – the elevation of the crown at the tunnel face. The Devico Drill was used to steer the boring from the drill setup location down to the crown of the tunnel. A portion of the boring profile plan is reproduced below to illustrate the boring path. Pillar 2 was an alternate second pillar boring that was not drilled.

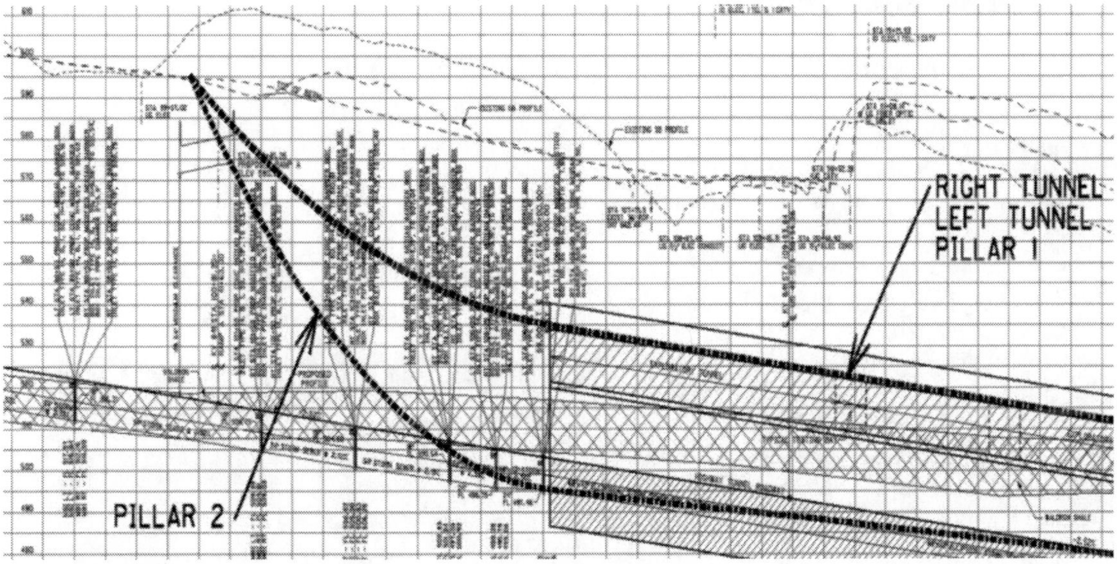

At predetermined intervals the boring is surveyed using the multishot survey tool. The photograph at left shows our engineer downloading the survey data from the multishot tool. He then plots the "as drilled" coordinates against the borehole plan coordinates to assess what steering maneuvers are needed. Surveying the borehole as it advances is important as the drill string tends to wander downward and to the left during the drilling process. The Devico Drill allows us to correct the boring path to stay as near as practical to the planned path.

As each coring run is completed our geologist boxes and logs the core and generates gINT logs. An example of the recovered core Louisville Limestone is shown in the photographs to the right.

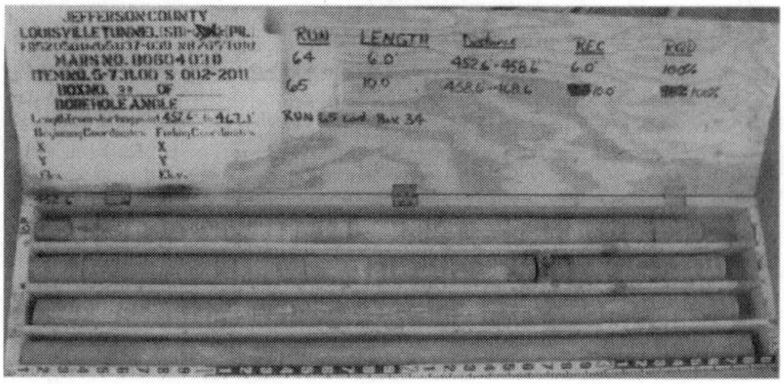

Location / Orientation North Bound, Box 33, 452.6 ft to 467.1 ft

Remarks Louisville Limestone

Our engineer (below) is holding a ten foot long core recovered from the Laurel Dolomite.

Horizontal Boring Relative to the Geologic Units

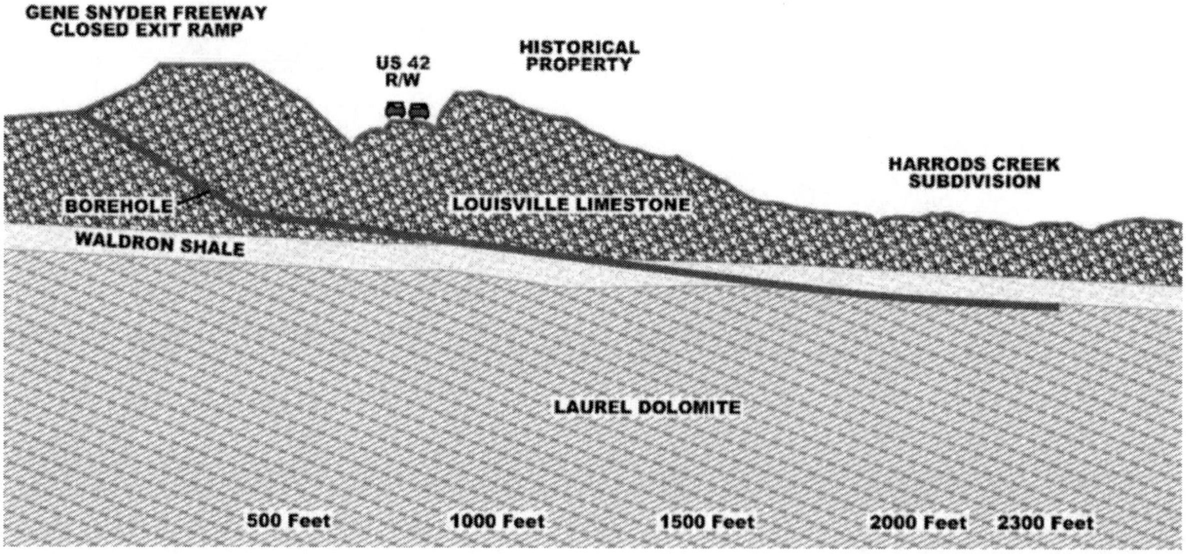

LABORATORY

A horizontal corehole enables the Engineer to generate comprehensive strength and material properties from the laboratory testing program along the entire alignment. Instead of a series of discrete profiles along a traditional vertical corehole spaced hundreds of feet apart, the horizontal corehole allows the Engineer to "see" conditions continuously along the tunnel length. For example, in the plot of Slake Durability (SDI) versus Distance from the Hole Collar you can readily see where the boring encountered the Waldron Shale by noting the drastic drop in the SDI value beginning at about 1650 feet into the alignment.

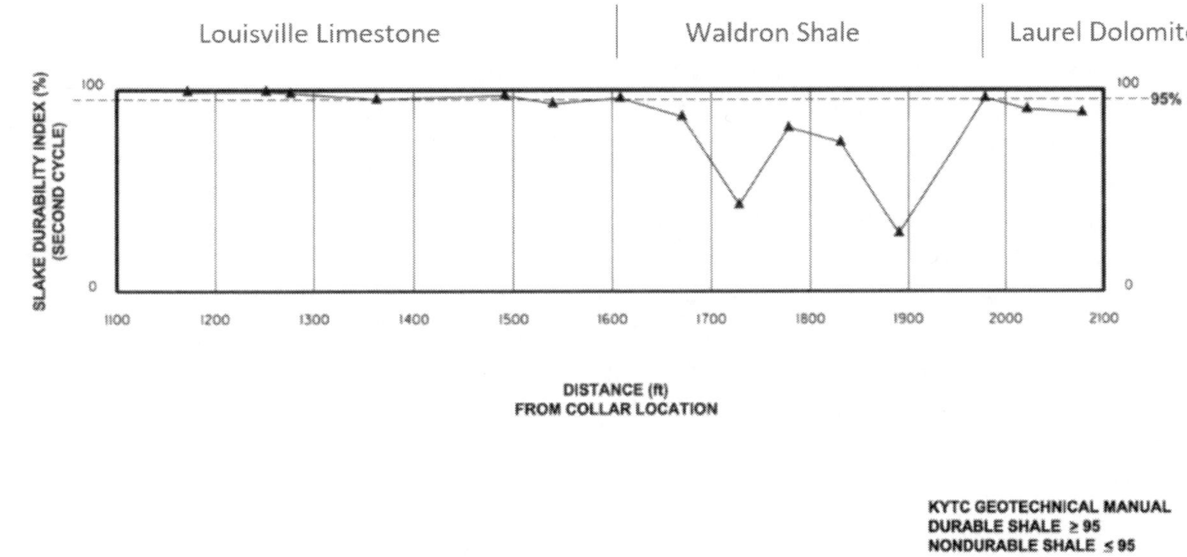

1

designer with the rock unconfined compressive strength along the alignment and not just at discrete vertical locations as in traditional vertical borings. The plot below shows the unconfined compressive strength along the tunnel alignment measured from the collar.

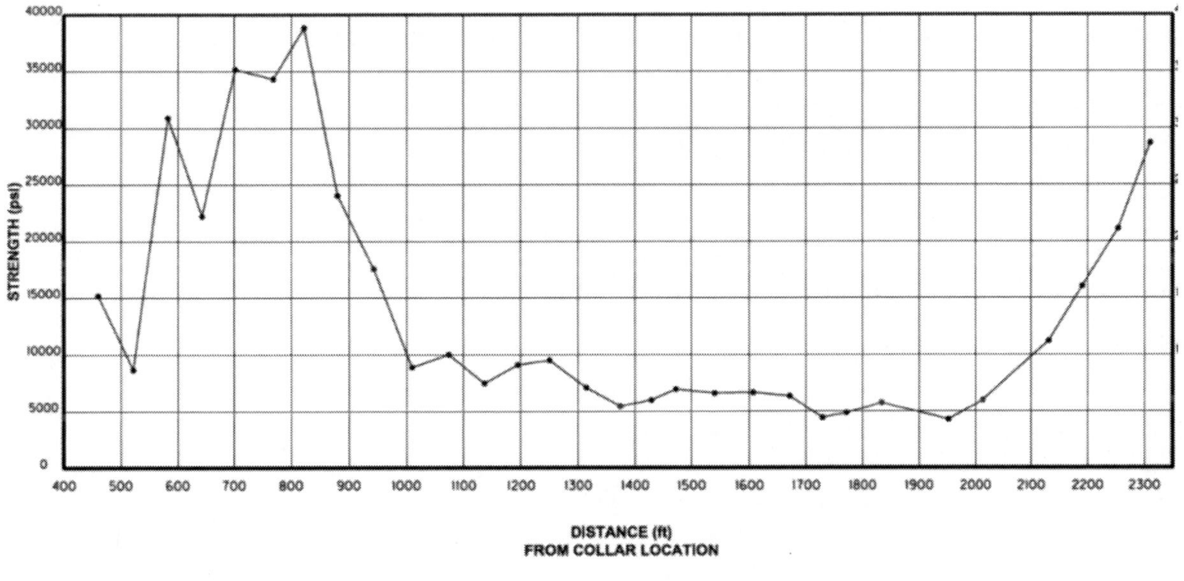

DISCONTINUITIES

The Louisville Limestone can be observed at road cuts along Highway 181 and Highway 42. The Limestone is weathered to light gray with substantial fossils throughout. Solutional weathering can be observed in the exposed rock extending ten to twenty feet deep into the rock. Reddish brown clay exists within the solution channels.

The Louisville Limestone along the alignment was observed to be sound to slightly fractured, hard, and crystalline. No solutional weathering or features were observed within the limestone portion of the alignment.

Water circulation was lost during the drilling process through small fractures and bedding features within the Louisville Limestone and Waldron Shale. The fractures and bedding features encountered in these formations were transporting water from the nearby Pillar Boring, which was being drilled simultaneously, to the North Bound Boring.

The Pillar Boring was located to the west and at a higher elevation to the North Bound Boring. Water gain of approximately 5 to 10 percent was observed during drilling. At the termination of the Pillar Boring the water gain in the North Bound Boring was lost. Complete water return within the North Bound Boring was lost at approximately 1,220 feet at the contact between the Louisville Limestone and the Waldron Shale.

HYDRAULIC CONDUCTIVITY (PACKER) TESTING

Hydraulic conductivity testing (also known as permeability or "packer" testing) was conducted in the North Bound Boring upon completion of coring activities. The test intervals were selected by KYTC and S&ME based on the results of the coring activities and subsurface conditions encountered in the bedrock.

The permeability test results were reported as Lugeon values. The Lugeon unit is commonly used in grouting practice for measuring the permeability and the grout take potential of bedrock. Reporting the permeability test results using this method allows for the evaluation of the permeability characteristics for each stage tested. The equation to calculate permeability in Lugeon units is:

L_u = ((Water take, in gallons ÷ 7.48 gal/ft^3) x (142 ÷ gauge pressure in psi))
divided by (Stage length in feet x test time in minutes x 0.0107620)

The packer system used in the borings was custom built for S&ME by Tam International. The packer system utilized two inflatable packers 2 foot in length and 2 inches in diameter were set 22 feet apart. Solid steel centralizers were placed above each packer to protect them during the placement and retrieval from the boring. Between the packers one inch diameter steel pipes were connected. The central section of the steel pipe contained off set holes to allow the water to fill up the test section between the packers. Above the packer at the top of the boring an In-Situ Incorporated transducer was attached. The transducer provided the pressure level within the boring as the water filled the test section between the packers. The transducer allowed the selected pressure levels to be set and held at each test interval.

The tests were conducted at three pressure intervals with a low pressure of 60 psi and a high pressure of 120 psi. The recorded Lugeon values and the hydraulic conductivity summary sheets were included in the reports. An example is provided to the right.

One Lugeon unit is the type of permeability consistent with sound bedrock. 10 Lugeon units typically indicates a permeable formation in which seepage occurs. 100 Lugeon units is the type of permeability typically observed in heavily jointed bedrock with relatively open joints, or in slightly to moderately jointed bedrock where joints are wide to very widely open (i.e., severe solution zones).

LOUISVILLE TUNNEL PROJECT
GEOTECHNICAL INVESTIGATION
WATER PRESSURE TESTING

S&ME

Field Test Data

Boring : ***South Bound*** Test by: ***N. Peterson***
Elevation: **592.3** Date: ***9/15/2011***

Formula for Lugeon (Lu) calculation:
(water take in gallons ÷ 7.48 gal/ft³) x (142 psi ÷ gauge pressure in psi)
divided by (stage length in feet x test time in minutes x 0.0107620)

Data Entry - Enter Borehole Stage (from & to); Test Time; Gauge Pressure; and Water Take.
Spreadsheet calculates Stage Length and Lugeon Units.

Borehole Stage Interval (ft)	Vertical Stage Interval (ft)	Increment	Stage Length (ft)	Test Time (min.)	Gauge Pressure (psi)	Water Take (gallons)	Lu (incr.)	Lu (stage)
2257.0	2257.0	1	30.0	10	60.0	44.4	4	
2287.0	2287.0	2	30.0	10	90.0	42.3	3	
		3	30.0	10	120.0	29.1	1	1
		4	30.0	10	90.0	18.7	0	
		5	30.0	10	60.0	14.4	0	
2157.0	2157.0	1	30.0	10	60.0	0.1	0	
2187.0	2187.0	2	30.0	10	90.0	0.7	0	
		3	30.0	10	120.0	0.0	0	2
		4	30.0	10	90.0	0.0	0	
		5	30.0	10	60.0	0.0	0	
2057.0	2057.0	1	30.0	10	60.0	28.6	3	
2087.0	2087.0	2	30.0	10	90.0	36.0	2	
		3	30.0	10	120.0	53.5	3	3
		4	30.0	10	90.0	41.3	3	
		5	30.0	10	60.0	28.7	3	
1957.0	1957.0	1	30.0	10	60.0	19.5	2	
1987.0	1987.0	2	30.0	10	90.0	30.2	2	
		3	30.0	10	120.0	41.5	2	4
		4	30.0	10	90.0	30.3	2	
		5	30.0	10	60.0	21.7	2	
1857.0	1857.0	1	30.0	10	60.0	23.5	2	
1887.0	1887.0	2	30.0	10	90.0	30.9	2	
		3	30.0	10	120.0	38.8	2	5
		4	30.0	10	90.0	31.8	2	
		5	30.0	10	60.0	21.8	2	

CONCLUSION

For the East End Tunnels project, the horizontal directional continuous coring was a successful alternate to the proposed pilot bore. Some noted advantages are:

- Provided detailed subsurface and laboratory data continuously along all three alignments instead of the single alignment proposed by the pilot bore.
- The continuous subsurface and laboratory data allows the designer to more confidently assign roof support types along the alignment.
- Generates rock core continuously along the alignment for observation and testing by the designer and as reference to the tunnel contractors.
- Packer testing along the entire alignment to identify opportunities for groundwater inflow into the tunnel construction.
- The continuous core together with the surveyed borehole data correctly predicted conditions that would be problematic for roof support where the Waldron Shale transitioned to roof support.
- A higher probability of identifying vertical Karst features compared to vertical borings. Karst was a critical concern in designing the tunnel support.
- Resulted in three 4-inch diameter boreholes that were grouted closed upon completing the holes. No safety or liability issues for KYTC.
- Was less costly than the proposed pilot bore and much less disruptive to the adjoining neighborhoods.

Geotechnical Seismic Design in New England

Craig W. Coolidge, P.E.
Summit Geoengineering Services, Inc.
173 Pleasant Street
Rockland, ME 04841
(207)-318-7761
ccoolidge@summitgeoeng.com

Prepared for the 69th Highway Geology Symposium, September, 2018

Acknowledgements

For example: The author(s) would like to thank the individuals/entities for their contributions in the work described:

Stephen Marcotte, C.G. P.G., L.S.E. – Summit Geoengineering Services, Inc.
Erika Stewart, P.E. – Summit Geoengineering Services, Inc.
Jenna Gilbert – Summit Geoengineering Services, Inc.

Disclaimer

Statements and views presented in this paper are strictly those of the author(s), and do not necessarily reflect positions held by their affiliations, the Highway Geology Symposium (HGS), or others acknowledged above. The mention of trade names for commercial products does not imply the approval or endorsement by HGS.

Copyright Notice

ABSTRACT

Earthquakes are not commonly associated to New England when compared to more seismically active regions of the United States. However, recent updates to seismic design codes and earthquake hazards mapping have significantly impacted geotechnical design and corresponding site development for projects including highway structures. A point of interest for this paper is the comparison of the current seismic design standards by the American Society of Civil Engineers (ASCE) to that of the American Association of State Highway and Transportation Officials (AASHTO).

Geology within New England is quite diverse including alluvium sand, sensitive marine clay, heterogenous glacial till, and a wide range of sedimentary, metamorphic, and igneous bedrock. Overburden thickness, or depth to bedrock, can also fluctuate greatly within a short distance resulting in a range of subsurface variability and uncertainty.

To investigate subsurface conditions, the current state of practice for geotechnical investigations within New England consist predominately of test borings. To improve geotechnical investigations within New England for seismic design, changes to explorations are necessary. These changes may include the use of geophysical testing such as multichannel analysis of surface waves (MASW) and seismic cone penetration testing ($SCPT_u$). Improved exploration methods will result in less conservatism for evaluating site classification and liquefaction potential and facilitate proper seismic code application to geology within New England. Discussion includes a case history showing comparison of the changes in seismic hazards mapping along with a comparison of exploration methodologies.

INTRODUCTION

Earthquakes are not commonly associated to New England when compared to more seismically active regions of the United States such as the west coast, predominately California. Additional areas of elevated seismic activity include the middle of the country along the Mississippi River valley, coastal areas of South Carolina, and Alaska. Still, seismic activity has and will continue to occur within New England. The challenge for local design code officials, geologists, and engineers is the consideration of to what effect and risk do earthquakes have on existing and future infrastructure within New England?

That answer may still be up to debate. However, continual updates in engineering standards with provisions to seismic design are suddenly impacting local projects. While the provisions to seismic code and design are likely derived from more seismically active areas, the impacts are being felt local to New England. This is particularly important to geotechnical investigations where seismic loads and liquefaction potential were previously considered an afterthought or low risk. Evidence of this includes the widespread presence of large masonry brick buildings present within the heart of many older cities and towns. Most of these are constructed or retrofitted without geotechnical or structural considerations of any seismic loading given their age. For new projects built to updated standards, the seismic design criteria may actually govern the foundation and soil related buildability and construction method considerations previously reserved for bearing capacity and settlement limitations.

So why the issue? One reason is that many original communities in New England are located near the coastline or along river valleys because the early modes of transportation and source of energy were by water. Unfortunately, many of these areas are where the local geology consists of marine sediment such as soft clay or river valley alluvium such as loose sand and silt. The increase in mapped peak ground accelerations, recognized in newer standards, and unfavorable geology has created a challenge for geotechnical consultants within New England.

GEOTECHNICAL SEISMIC DESIGN CODES

The determination of appropriate seismic site classification and associated hazards are often necessary as part of reporting requirements for geotechnical investigations. The results are used for structural design of foundations for bridges, buildings, towers, and other similar structures. Reporting requirements generally include the following:

- Slope Instability
- Liquefaction Potential
- Total and Differential Settlement
- Surface Displacement Due to Faulting, Lateral Spreading, or Lateral Flow

The methodology for determining geotechnical seismic design parameters varies depending on the code applied. For most highway related projects, the AASHTO LRFD Bridge Design Specifications are adopted by individual state departments of transportation. State and city building codes adopt the International Building Code (IBC) which uses the recommendations provided in ASCE 7.

Is there much difference between the standards? Presently, this is one of the bigger challenges for local geotechnical engineers to sort out because seismic design standards vary between editions. Until recently, most states within New England utilized 2009 IBC, which reference the seismic design maps and procedures of ASCE 7-05. The maps and procedures were significantly modified for the recently adopted IBC 2015 which references ASCE 7-10. To date, new changes are established under ASCE 7-16 that will apply to future editions of IBC.

For highway projects the AASHTO Bridge Design Specification 8th Edition was released in 2017 replacing the 2007 AASHTO 4th Edition that was to be adopted by individual state departments of transportation. Additional publications by the Federal Highway Administration include, the LRFD Seismic Analysis and Design of Transportation Geotechnical Features and Structural Foundations released in 2011, and the LRFD Seismic Analysis and Design of Bridges released in 2014. The current design manual editions for the New England States include:

- Maine DOT Bridge Design Guide (August 2003)
- New Hampshire DOT Bridge Design Guide (January 2015)
- Vermont AOT VTrans Structures Design Manual 5th Edition (2010)
- Massachusetts DOT LRFD Bridge Manual (2013)
- Connecticut DOT Bridge Design Manual (2003)
- Rhode Island DOT LRFD Bridge Design Manual (2007)

The use and details of each of these documents are beyond the scope of this paper. In summary, it appears most of the local DOT bridge design manuals still reference older methodology and mapping for determining seismic parameters as compared to the newer standards and mapping provided by ASCE 7 referenced in IBC.

SESIMIC DESIGN MAPPING

The National Earthquake Hazards Reduction Program (NEHRP) is a multi-agency program with focus on reducing losses due to earthquakes in the United States. The seismic design codes used by AASHTO, ASCE, and IBC generally adopt standards established from the NEHRP provisions. Interactive mapping programs for code related seismic design parameters are provided through the United States Geological Survey (USGS) website.

Deterministic Peak Ground Acceleration

The peak ground acceleration mapping by NEHRP includes provisions in 2003, 2009, and 2015 calculated as the largest 84th percentile geometric mean peak ground acceleration. From 2003 to 2015, the peak ground acceleration has increased significantly for sites located within central to southern Maine and New Hampshire and eastern Massachusetts. Below are graphic results of the corresponding changes for a select list of 15 locations within New England:

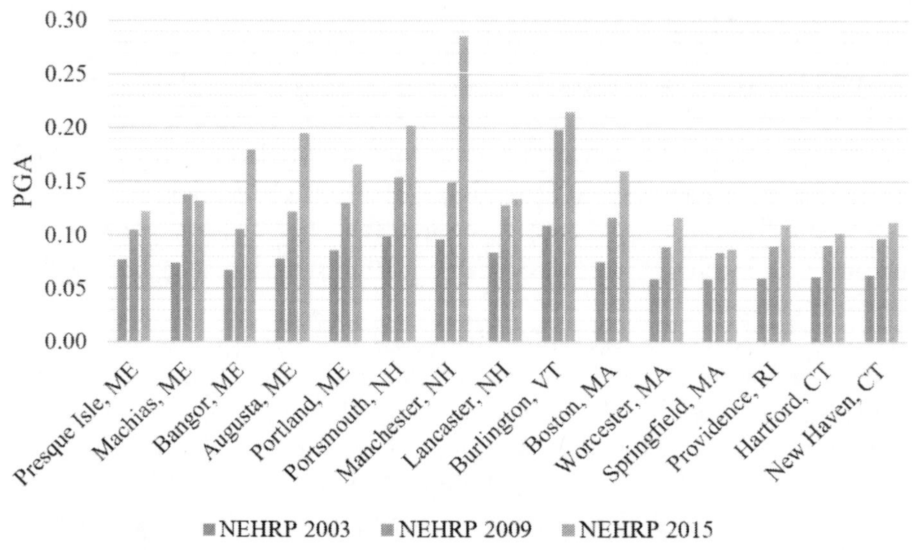

Figure 1a – Peak Ground Acceleration (Site Class B)

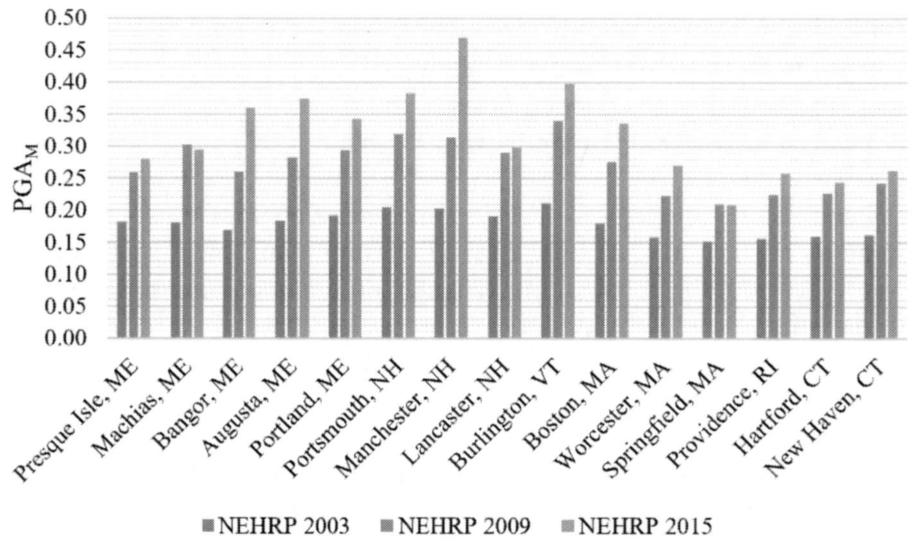

Figure 1b – Peak Ground Acceleration (Site Class E)

When comparing results of Figures 1a and 1b, it is important to note the increase in deterministic peak ground acceleration PGA_M from 2003 to 2015 ranging from 150% to 300%.

Probabilistic Method – Peak Ground Acceleration

The probabilistic geometric mean peak ground acceleration is commonly determined from hazard mapping provided by the United States Geological Survey (USGS). It is common for the designer to select a site specific peak ground acceleration determined as having a 2% probability of exceedance in 50 years.

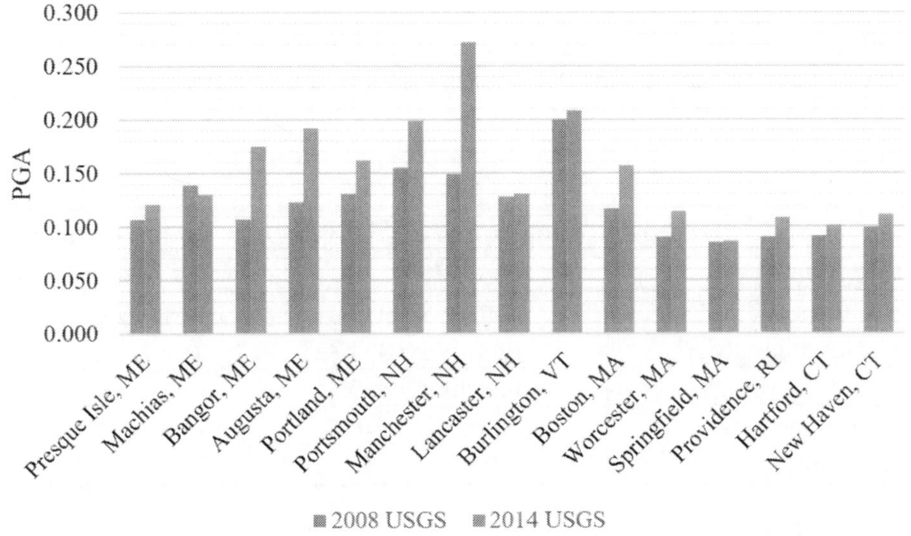

Figure 2 – USGS Hazards Mapping PGA (2% in 50 Years)

The probabilistic peak ground acceleration PGA does not account for soil strata such as shallow bedrock, stiff soils, or soft soils in comparison to the deterministic peak ground acceleration PGA_M which does account for the subsurface soil profile. Comparison of the mapping from 2008 to 2014 shows a 25% to 85% increase in southwestern Maine, southeastern New Hampshire, and northeastern Massachusetts.

Earthquake Magnitude

The earthquake magnitude a quantitative measurement of earthquake size derived from maximum ground shaking measured by a seismograph. The magnitude for a site can be estimated using the USGS Hazards Mapping for a specific location.

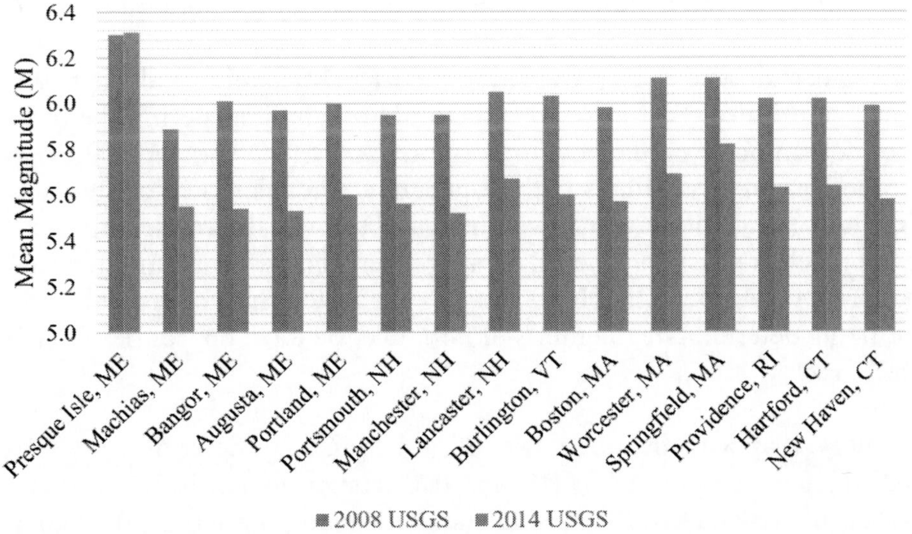

Figure 3 – USGS Hazards Mapping Magnitude (2% in 50 Years)

From observation of Figures 2 and 3, we can see a general increase in peak ground acceleration (PGA) from mapping in 2008 to 2014, however a decrease in earthquake magnitude. One can conclude the updated mapping has recognized in an increase in the ground acceleration intensity but a decrease in the earthquake energy. In general, the mean magnitude has decreased from 6.0 to 5.5 for locations within New England. Historic earthquakes within New England include magnitudes of 5.5 or greater are summarized on Table 1.

Table 1 - Historical Earthquakes (5.5 Magnitude or Greater)

Location	Date	Magnitude
Central, NH	June 11, 1638	6.5
Newbury, MA	November 10, 1727	5.6
Cape Ann, MA	November 18, 1755	6.2
Eastport, ME	March 21, 1904	5.9
Whittier, NH	December 20 & 24, 1940	5.5

It might be reasonable to consider the topography and geology within New England having variability within a short distance. The Appalachian Mountain range extends through the central portion of New England. Glacial till and bedrock reside within higher elevations. River valley and flood plain alluvium reside at lower elevations along with widespread marine sediments towards the coastline extending to the Atlantic Ocean. Thus, the effects of variable topography might shorten duration of earthquake shaking. However, the presence of localized deeper alluvial and marine sediments would increase ground acceleration during earthquake.

Site Specific – Peak Ground Acceleration

In retrofitting existing structures and design of new structures, engineering judgement will be necessary to evaluate risk and application of peak ground accelerations. Presently IBC utilizes the procedure specified under ASCE 7. In determining the site-specific PGA for design, ASCE 7-10 states the designer may use the lesser of the probabilistic PGA (2% in 50 years) or deterministic PGA_M (84th percentile) but not less than 80% of the deterministic PGA_M.

AASHTO design guides, as provided in the LRFD Seismic Analysis and Design of Transportation Geotechnical Features and Structural Foundations, permit the use of either the probabilistic or deterministic methods. Furthermore, AASHTO concludes the probabilistic approach incorporates all possibilities with respect to earthquake location, magnitude, and ground motion attenuation, producing a weighted average to estimate seismic activity. Thus, the probabilistic approach is often considered an appropriate basis for making rational design decisions about risk versus benefit. However, it is suggested that for critical structures of high importance utilizing deterministic methods or both may be most appropriate to evaluate site specific seismic risk.

In summary, it appears the utilization of probabilistic or deterministic PGA is still a fundamental difference between AASTHO and IBC design standards in current practice. To further assess seismic risk, AASTHO design standards incorporate hazard levels for bridge structures ranging from significant to minimal damage with consideration of bridge importance. Consideration needs to be given for an acceptable return period for seismic risk from earthquake.

The return period commonly ranges from 1,000 to 2,500-years with an expected life span ranging from 50 to 100 years for a bridge structure. The selection of a site-specific return period and hazard level for the structure will influence the peak ground acceleration applied for design.

SEISMIC SITE CLASSIFACTION

Determination of the seismic site classification is based on the results of a subsurface investigation using test borings or piezocone penetration testing. Alternatively, site classification is based on geophysical testing such as spectral analysis of surface waves (SASW), multichannel analysis of surface waves (MASW), downhole and/or crosshole shear wave velocity testing. `The procedure for determining site classification adopted by ASCE Standard 7 and AASHTO 2009.

SITE CLASS DEFINITIONS

Site Class	$\bar{v}_s$	$\bar{N}$ or $\bar{N}_{ch}$	$\bar{s}_u$
A. Hard Rock	>5,000 ft/s	N/A	N/A
B. Rock	2,500 to 5,000 ft/s	N/A	N/A
C. Very dense soil and soft rock	1,200 to 2,500 ft/s	>50	>2,000 psf
D. Stiff Soil	600 to 1,200 ft/s	15 to 50	1,000 to 2,000 psf
E. Soft clay soil	<600 ft/s	<15	<1,000 psf
	Any profile with more than 10 ft of soil having the characteristics: • Plasticity index $PI > 20$, • Moisture content $w \geq 40\%$, and • Undrained shear strength $\bar{s}_u < 500$ psf		
F. Soils requiring site response analysis in accordance with Section 21.1	See Section 20.3.1		

For SI: 1ft/s = 0.3048 m/s 1lb/ft² = 0.0479 kN/m²

Figure 4 – Standard for Site Classification

The procedure defines a site as having a profile of 100 feet with classification ranging from A (hard rock) to E (soft soils) utilizing shear wave velocity v_S. Alternative methods for classifications C through E include the use of the standard penetration resistance (N) from test borings and/or undrained shear strength (s_u) from laboratory testing of cohesive soils.

The classification of A and B are assigned where the thickness of soil between bottom of foundation and surface of competent bedrock is 10 feet or less. Classification A is permitted when verified by onsite shear wave velocity testing or with testing and knowledge of similar geology along with confidence in competent bedrock. Classification B is applied where determined by shear wave velocity testing or where bedrock is estimated as competent with moderate weathering and fracturing. Classification of C is applied where bedrock is considered soft or more highly fracture and weathered unless verified by shear wave velocity testing.

The classification of C, D, or E are determined for the site-specific subgrade profile to 100 feet by using one of the three methods; average shear wave velocity, average standard penetration test (granular, cohesive, and rock), or individual use of standard penetration test for cohesionless layers with undrained shear strength for cohesive layers to 100 feet.

Most commonly the subgrade is evaluated using the N method for combined cohesionless, cohesive, and rock conditions for the profile of 100 feet. Value for N is determined for cohesionless and cohesive layers using the standard penetration test. Where rock is encountered, N is applied for the rock layer as being equal to 100.

Alternatively, the subgrade soil can be evaluated using N_{ch} for granular layers and undrained shear strength s_u for cohesive layers. The limit for cohesionless layers is an N value of 100 and for cohesive layers is 5,000 psf. Determination between cohesionless and cohesive soils is a plastic limit value of 20. Classification is determined base on the lesser value of the two individual methods.

Where soft clay soils are present a classification of E is assigned where soil profile 10 feet or greater in thickness includes; a plasticity index greater than 10, moisture content equal to or greater than 40, and an undrained shear strength of less than 500 psf.

The classification of F is applied where any of the following are present:

- Liquefiable soils, quick/highly sensitive clays, and collapsible weakly cemented soils.
- Peats and/or highly organic clays with thickness greater than 10 feet.
- High plastic clays greater than 25 feet with plasticity index greater than 75.
- Thick soft clays greater than 120 feet with undrained shear strength below 1,000 psf.

GEOTECHNCIAL EXPLORATIONS

Test Borings w/SPT Sampling

It is common practice for geotechnical investigations in New England to be conducted using conventional test borings. Test borings are performed by hollow stem auger or by rotary wash with driven casing. Sampling is conducted using the standard penetration test SPT to collect split spoon samples and to estimate the in-situ density of soils. In-situ field vane shear tests can be performed to obtain estimates of undrained shear strength. Thin wall tube samples can be collected for soft cohesive samples to perform laboratory testing in determining undrained shear strength. Rock core samples can be collected to determine rock type and quality.

Geology within New England is quite diverse and includes; loose alluvial sand, sensitive soft marine clay, heterogenous glacial tills, and a wide range of sedimentary, metamorphic, and igneous bedrock. Overburden thickness also fluctuates from loose soil to hard till over bedrock commonly within the upper 100-foot profile. The use of test borings provide versatility by an ability to collect data for a range of subsurface conditions common to New England.

Despite the versatility, the application of data collected from conventional test borings is generally a poor application for site classifications of D, E, and F. When below groundwater, sand can undergo upheave and disturbance during performance of standard penetration testing unless borehole hydrostatic pressure is maintained. This is commonly done with rotary wash and occasionally the use of drilling muds. Additionally, the standard penetration test is a poor measurement of cohesive strength, especially for soft clays common to marine deposits.

Seismic Piezocone Penetration Testing ($SCPT_u$)

An alternative to test borings for investigating the subsurface conditions is the performance of seismic piezocone penetration testing ($SCPT_u$). $SCPT_u$ is performed by a cone on the end of a series of rods pushed into the ground at a constant rate (2 cm/s) to obtain near continuous measurements of the resistance to penetration of the cone. Parameters obtained include cone resistance (q_c), sleeve friction (f_s), and piezocone pore pressure (u_2). The results are interpreted to obtain soil type and soil parameters for engineering design. Shear wave velocity tests are performed at select intervals, typically 1 meter (3-feet).

The in-situ shear wave velocity profile (V_s) can be obtained from shear wave testing performed during $SCPT_u$. Correlation for standard penetration resistance N and undrained shear strength S_u can be obtained independently from the same $SCPT_u$ test from cone penetration resistance for further evaluation of appropriate seismic site classification. The ability to match 3 independent methods of analysis utilizing one exploration provides the engineer with greater accuracy and less conservatism. The near continuous data acquisition by $SCPT_u$ provides enhanced profiling of the soil stratum and better identification of sub-layering.

Multichannel Analysis of Surface Waves (MASW)

Gaining in popularity are multichannel analysis of surface waves (MASW) surveys conducted to measure shear wave velocity profile (V_s) and to map stratum layering. MASW surveys are non-invasive and performed using a series of 24 or more geophones along a straight alignment at the ground surface. An energy source such as a steel plate and sledge hammer with a trigger switch are used to develop surface waves recorded at low frequency (1 to 30 H_z). The results are collected through a data acquisition system and then processed by dispersion properties to determine V_s profiles in 1D for depth and 2D for depth and location.

The advantage of conducting MASW surveys is the direct site measurement of shear wave velocity for determining appropriate seismic site classification. Additionally, approximate stratum layering between soils types and bedrock are possible from MASW data or additional high-resolution reflection and/or refraction surveys. Still, test borings and/or cone penetration testing should be performed to verify soil strata, depth to groundwater and/or bedrock, and for consideration of liquefaction potential.

Laboratory Testing

For most geotechnical investigations, laboratory testing is conducted on samples obtained from test borings. Laboratory tests determine both index and strength properties and can be helpful for estimating seismic site classification. Performance of index testing is required in determining the special requirement of site class E for soft clays where the undrained shear strength for cohesive soils is below 500 psf and thickness is greater than 10 feet.

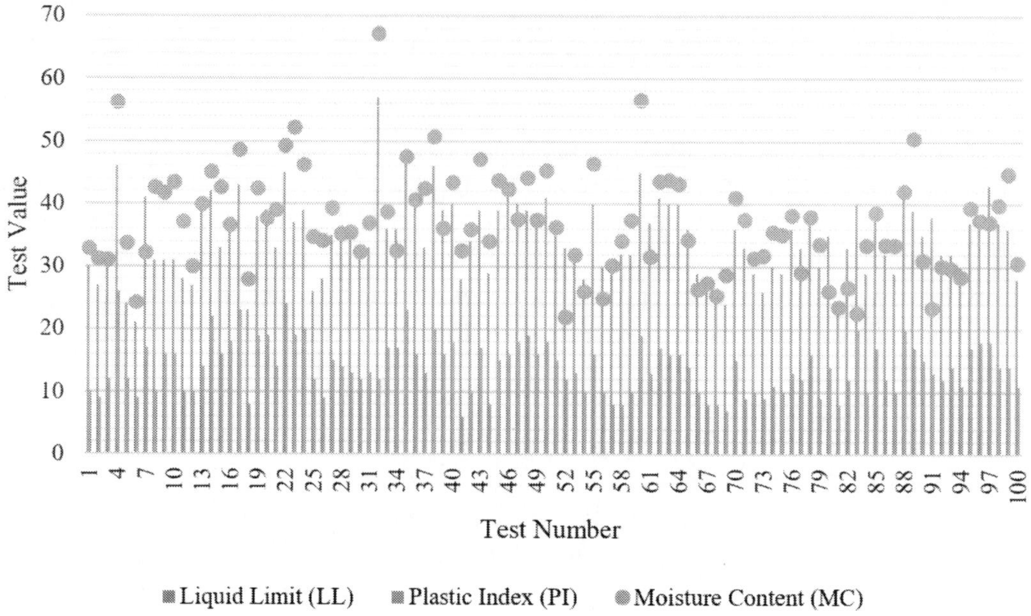

Figure 5 – Atterberg Limit Tests for Presumpscot Formation

Figure 5 shows summary of 100 Atterberg limit tests conducted for marine deposit (soft clay) further described as the Presumspcot Formation for various sites in Maine. The average liquid limit (LL) is 35 with a range of 57 to 21. The average plastic index (PI) is 14 with a range of 26 to 6. The average moisture content is 37 with a range of 67 to 22. In summary, the Presumpscot Formation comprises of lean clay with variable silt classifies as CL in accordance with the Unified Soil Classification System. Further, the lean clay is generally considered to have moderate to high sensitivity.

In comparison to the index requirements for site class E, the range of values are both below and above the criteria for essentially the same geologic formation. Thus, detail profiling of the undrained shear strength and index values is recommended for soft clay deposits.

LIQUEFACTION POTENTIAL

The LRFD Seismic Analysis and Design of Transportation Geotechnical Features and Structural Foundations publication defines sites as having a Seismic Hazard Level as follows:

Table 2 – Seismic Hazard Level

Hazard Level	Using $S_{D1} = F_v S_1$	Using $S_{DS} = F_a S_s$
I	$S_{D1} \leq 0.15$	$S_{DS} \leq 0.15$
II	$0.15 < S_{D1} \leq 0.25$	$0.15 < S_{DS} \leq 0.35$
III	$0.25 < S_{D1} \leq 0.40$	$0.35 < S_{DS} \leq 0.60$
IV	$0.40 < S_{D1}$	$0.60 < S_{DS}$

Table 3 below presents data for 15 select sites within New England mapped with site class E soils using AASHTO 2009 seismic parameters.

Table 3 - AASHTO 2009 (Class E)			
Location	**PGA**	**S_{D1}**	**S_{DS}**
Presque Isle, ME	0.077	0.184	0.439
Machias, ME	0.074	0.138	0.368
Bangor, ME	0.067	0.153	0.362
Augusta, ME	0.078	0.159	0.404
Portland, ME	0.086	0.156	0.429
Portsmouth, NH	0.099	0.155	0.473
Manchester, NH	0.096	0.157	0.466
Lancaster, NH	0.084	0.178	0.447
Burlington, VT	0.109	0.198	0.552
Boston, MA	0.075	0.135	0.377
Worcester, MA	0.059	0.133	0.326
Springfield, MA	0.059	0.132	0.326
Providence, RI	0.060	0.122	0.318
Hartford, CT	0.061	0.128	0.330
New Haven, CT	0.063	0.122	0.331

The seismic accelerations in Table 3 are mapped as hazard level I using the criteria for spectral acceleration of 1 second (S_{D1}) and hazard levels of II and III using criteria for spectral acceleration of 0.2 second (S_{DS}) based on AASHTO mapping of 2009 for site class E. The discrepancy and relatively low peak ground accelerations suggests earthquakes within New England are considered to be short in duration and magnitude but higher in initial intensity as expected for soil class E. The discrepancy also suggests the criteria of Table 2 do not apply well to the mapped accelerations for New England currently published by AASHTO 2009.

For hazard levels I and II, the peak ground acceleration and earthquake magnitudes are 0.14g and 6.0 or less, respectively. Liquefaction potential for hazard levels I and II are considered low thus liquefaction analysis is not required. Further criteria are provided for determining the need for liquefaction analysis of hazard level III. As part of the AASHTO criteria, liquefaction analysis is not required for hazard level III where mean magnitude is less than 6.0. Hazard level IV sites are strictly required to have liquefaction analysis performed but are not mapped within New England. The mean magnitude in New England is generally mapped near 6.0 based on 2008 data and 5.5 using 2014 data by the USGS, as shown on Figure 3. Thus, evaluation of liquefaction potential is not generally considered necessary based on the screening criteria used by the hazard levels of I, II, and III. In comparison, ASCE 7-10 requires all sites having a soil class of D, E, or F be evaluated for liquefaction potential regardless of magnitude.

Screening Criteria of Granular Soils

Sandy soils are defined as being susceptible to liquefaction for sites of level III based on having corrected standard penetration tests (N_{60}) below 30 or normalized cone penetration resistance (q_c) below 160 ksf. Additionally, liquefaction analysis is not needed where groundwater is at a depth of 50 feet below grade or deeper. An exception exists where the mean magnitude is between 6.0 and 6.4 with an SPT N_{60} value below 20.

Screening Criteria of Cohesive Soils

Clayey soils are defined as being highly sensitive and susceptible to liquefaction for sites of level III and IV based on having all of the following:

- Liquid Limit below 40
- Moisture Content/Liquid Limit Ratio > 0.9
- Liquidity Index > 0.6
- SPT N_{60} below 5 or CPT resistance q_c below 50 ksf

The criteria are applicable for seismic hazard levels III and IV which are not common to New England. Still, soils satisfying the screening requirements as having liquefaction potential exist within geology local to New England. As an example, Table 4 shows a summary of 100 Atterberg limit tests conducted for Presumpscot Formation (soft clay) as provided in Figure 5. Results are summarized on Table 4 for comparison to the criteria outlined above.

Table 4 – Summary of Results for Atterberg Limits (Presumpscot Formation)

Index Value	Average	Maximum	Minimum
Moisture Content (MC)	37	67	22
Liquid Limit (LL)	35	57	21
Plastic Index (PI)	14	26	6
Liquid Index (LI)	1.2	2.2	-0.1
Ratio MC/LL	1.1	1.4	0.6

Comparison of the test results suggest possible conformance to the index criteria for clayey soils as being highly sensitive. Based on local experience, SPT N_{60} values for soft clay of the Presumpscot Formation are commonly below 2, CPT resistance q_c below 15 ksf, with an undrained shear strength of 1,000 psf or less. Additionally, the general properties of Presumpscot Formation are similar to Boston Blue Clay and higher elevations of Pleistocene Lake deposits in Vermont and New Hampshire.

In summary, because of the hazard levels determined through mapped accelerations of AASHTO 2009 and the mean magnitudes provided by USGS, liquefaction analysis is not generally triggered through the screening criteria. Still, sandy soil having low fines content and located below groundwater yielding SPT N_{60} values below 30 are commonly present in alluvial deposits prevalent in New England. Additionally, marine clays may also meet the requirements for being highly sensitive yielding liquefaction or potential for shear strength reduction during earthquakes. Thus, despite not satisfying the screening requirements, engineering analysis for liquefaction potential should still perhaps be checked using other published methodologies to better determine risk.

CASE EXAMPLE – RAILWAY FACILITY

The project consisted of a new site development for a 50,000 ft^2 steel frame building within a former railroad yard used for storage and minor maintenance for passenger rail service in Brunswick, Maine. Additional development included approach and descent railroad lines and pavement access drives for the facility. Preliminary geotechnical investigation included the performance of 4 test borings utilizing rotary wash drilling advanced to depths of 30 to 40 feet below ground surface (bgs). Further geotechnical investigation included 24 shallow test borings to depths of 10 to 20 feet bgs along with 4 seismic cone penetration tests ($SCPT_u$) to depths of 45 to 80 feet bgs. The subsurface conditions consisted of the following:

- (0 to 4 ft) *Existing Fill* - Sand with Variable Gravel, Silt, and Coal Ash
- (4 to 30/50 ft) *Marine Regressive Delta Deposit* – Sand with Variable Silt
- (30 to 60 ft CPT-1) (50 to 80 ft CPT-2) *Presumpscot Formation* – Silty Clay
- (4 to 8 ft) *Groundwater Depth* – Sand & Gravel Aquifer at 10 to 50 gpm Yield

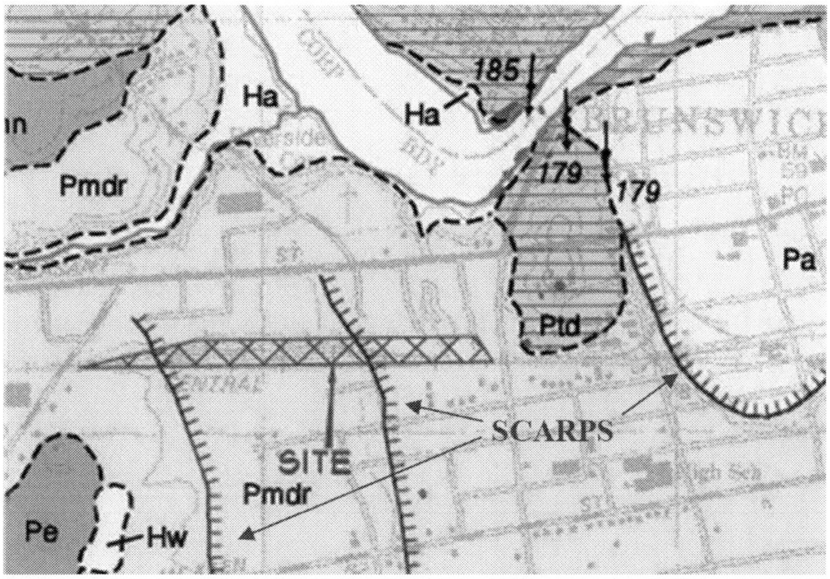

Figure 6 – Geological Mapping by Maine Geological Survey

Geologic scarps are mapped within the marine regressive delta deposit by the Maine Geological Survey as potentially located between the cone penetrometer CPT-1 and CPT-2 tests performed at a horizontal spacing of 200 feet. The geologic scarps represent a shift or division between historic stream channels. The findings suggest a scarp of 20 feet in elevation where the transition of sand to clay shifts from 30 to 60 feet bgs at CPT-1 to 50 to 80 feet bgs at CPT-2.

A challenge for geotechnical design of the facility was the presence of variable sand and silt located below a relatively shallow groundwater table overlying undulating layers of silty clay. Topography for the site was relatively flat requiring minimal grading for cuts or fills. Bearing capacity and estimated settlements of the upper soils were determined sufficient for the building foundations, railway tracks, and pavement areas based on the results of the shallow test borings and soils laboratory testing. Limitation for conventional site development was the determination of seismic site classification along with potential for liquefaction by earthquake.

Results of preliminary test borings conducted using rotary wash with split spoon sampling are compared to the correlated SPT-N_{60} values determined by seismic cone penetration tests ($SCPT_u$) shown below on Figure 7.

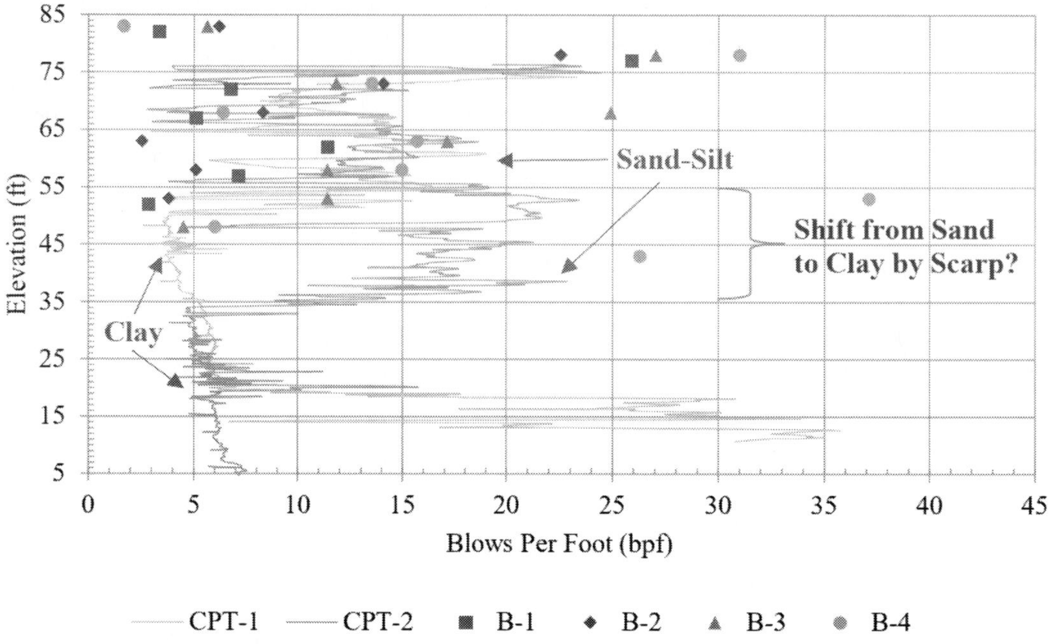

Figure 7 – Standard Penetration Test (SPT N_{60})

The average SPT-N_{60} value obtained from test borings performed from elevations 85 to 45 feet for the upper sand-silt layer is 12. The average correlated N_{60} determined from the cone penetration tests (CPT-1 and CPT-2) is also 12 showing agreement between methodologies. The average SPT-N_{60} for the complete profile from elevation 75 to 5 feet from the cone penetration tests is 12 with a range from 4 to 35.

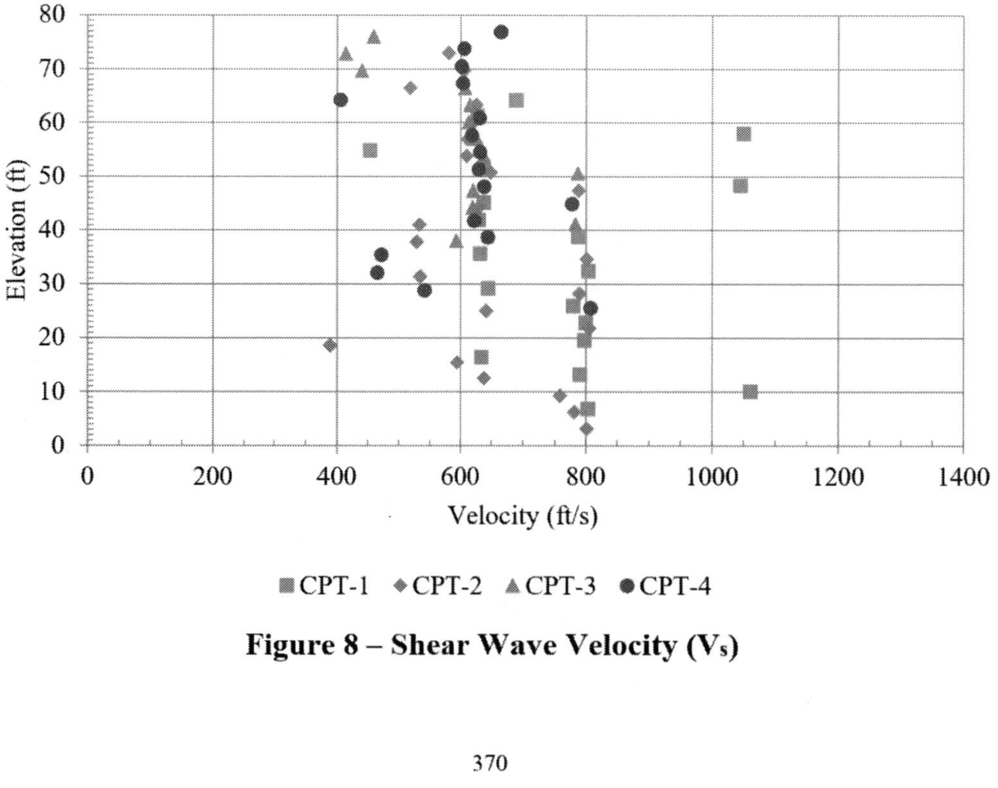

Figure 8 – Shear Wave Velocity (V_s)

The shear wave velocity obtained during performance of seismic cone penetration tests ($SCPT_u$) resulted in a range of 390 to 1,060 ft/s with an average of 665 ft/s. In determining seismic site class per Figure 4, a classification of E is determined by use of the N method (N < 15) where the average N_{60} value for the profile is 12. However, by use of shear wave velocity (V_s) where the average for the profile is 665 ft/s (V_s > 600 ft/s) the site is classified as D. The undrained shear strength (S_u) estimated from cone penetration tests for the underlying Silty clay range from 1,000 to 2,000 psf with an over-consolidation ratio (OCR) of 1.5 to 3.0 precluding the special requirements for soft clay soils (S_u < 500 psf).

Determination between site class D and E can significantly impact the applied peak ground acceleration used for liquefaction analysis. Below is a list of variable mapped peak ground acceleration for the site using deterministic and probabilistic methods.

Table 4 – Deterministic Method for PGA			
Reference	**PGA**	**PGA_M (Class D)**	**PGA_M (Class E)**
2009 AASHTO	0.079	0.104	0.164
2003 NEHRP	0.118	0.123	0.186
2009 NEHRP	0.122	0.189	0.283
2015 NEHRP	0.170	0.248	0.348

Table 5 – Probabilistic Method for PGA		
USGS	**PGA**	**Mean Magnitude**
2008	0.123	6.0
2014	0.168	5.6

At the time of design for the project, local and current codes included ASCE 7-05 along with AASHTO 2009. The peak ground acceleration for site class D determined by 2009 AASHTO was 0.104g and for ASCE 7-05 utilizing NEHRP 2003 was 0.123g. Additionally, the probabilistic peak ground acceleration obtained from the 2008 USGS hazards mapping was 0.123g at a mean earthquake magnitude of 6.0. For design of the project, a peak ground acceleration of 0.123g at a magnitude of 6.0 was used.

Liquefaction potential was evaluated using the results from the seismic cone penetration tests ($SCPT_u$) and methodology provided by Robertson, et al, in Guide to Cone Penetration Testing for Geotechnical Engineering 5th Edition. The methodology utilizes the fundamental equation for liquefaction analysis of balancing the soil strength cyclic resistant ratio (CRR) and the earthquake forces as cyclic stress ratio (CSR). The factor of safety against liquefaction under design earthquake loading is fundamentally determined by the ratio of CRR/CSR. The soil strength is estimated by cone penetration resistance (q_c) along with adjustment based on soil type such as variability of fines and effective stress parameters. The earthquake force is estimated by the design peak ground acceleration along with adjustments of magnitude, deposit thickness, and effective stress parameters. Use of CPT data provides a more rigorous evaluation for liquefaction potential due to the near continuous profiling data along with in-situ measurements for cone resistance, friction ratio, and drainage properties from piezometer pore pressure for estimating soil behavior type.

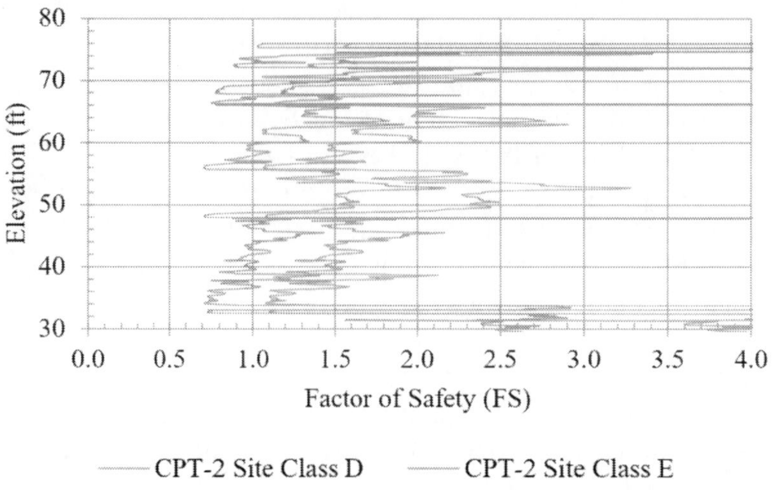

Figure 9a – Shear Wave Velocity (V_s) by NEHRP 2003

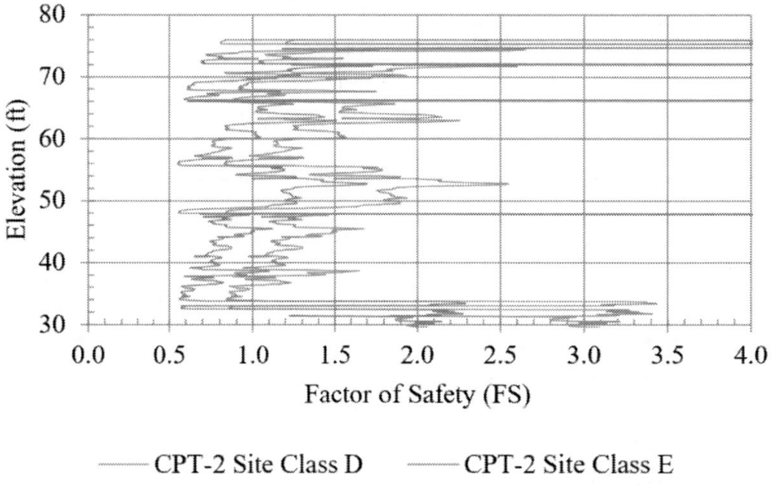

Figure 9b – Shear Wave Velocity (V_s) by NEHRP 2009

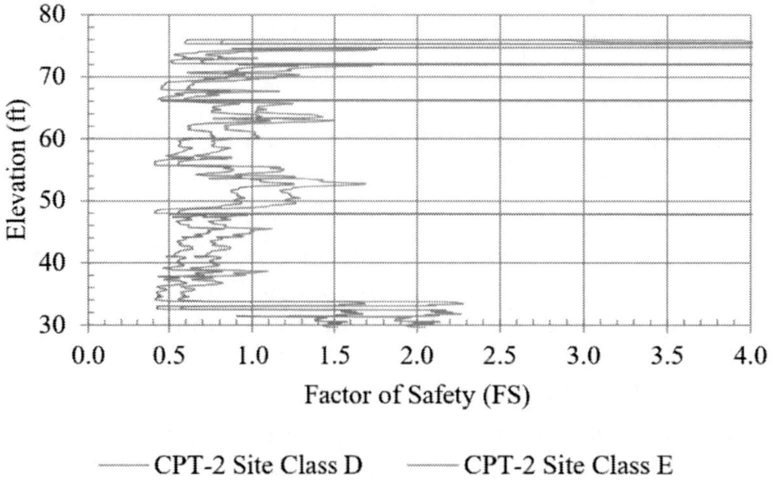

Figure 9c – Shear Wave Velocity (V_s) by NEHRP 2015

Reviewing the results for liquefaction potential, the profile is fully resistant from liquefaction using NEHRP 2003 site class D, and slightly susceptible using site class E. The potential for liquefaction increases using NEHRP 2009 for both site class D and E where the peak ground acceleration is increased but magnitude is decreased from 6.0 to 5.6. The potential for liquefaction becomes widespread using NEHRP 2015 for both sites classed D and E due to the increase peak ground acceleration.

The results compare the same profile of CPT-2 simply adjusting the earthquake parameters for peak ground acceleration and magnitude. Based on this comparison, it appears likely that many sites located within New England that are marginal or slightly above susceptible to liquefaction may eventually, under future seismic mapping and associated codes, be deemed risk to widespread liquefaction which can significantly affect foundation design, construction methods employed, and overall site development feasibility.

CLOSURE

Let's recap to the question of, to what effect and risk do earthquakes have on existing and future infrastructure within New England? The findings of this paper show a comparison of the past, present, and future mapping recommendations by NEHRP that recognizes an increase in peak ground accelerations suggesting an increase in risk. Conversely, the mapping also recognizes a decrease in earthquake magnitude suggesting the general risk of earthquake may actually remain the same and simply the application of seismic design parameters is evolving.

For engineering design, the determination of seismic site classification, hazard risk, and liquefaction potential is dependent on the code applied such as AASHTO, ASCE, and IBC. In current practice, it appears better awareness for evaluating liquefaction potential is provided by ASCE as compared to AASHTO but at perhaps a more conservative approach when using traditional exploration methods such as test borings. Determining a site specific seismic site classification by method of shear wave velocity may reduce conservatism when a result of higher site classification is determined as compared to the traditional method of N from test borings.

The case example further shows the sensitivity of increase peak ground acceleration with an increase liquefaction potential. Because of this, the updated mapping recommendations by NEHRP may result in an increase use of ground improvements such as stone columns, deep dynamic compaction, or even alternative pile support foundations. Improving geotechnical investigations to better predict liquefaction potential and determine site seismic classification will be necessary. In closing, the points of emphases for this paper include the following:

- Updates in seismic mapping by NEHRP will increase design peak ground acceleration.
- Variations in codes will bring discrepancy for geotechnical seismic design risk.
- Techniques such as $SCPT_u$ and MASW should be considered for site class determination.
- Sand deposits for both seismic class D and E may still present liquefaction potential.
- Clay deposits should be evaluated for special case conditions and level of sensitivity.

REFERENCES:

- Text Books
    1. Kramer, S.L. *Geotechnical Earthquake Engineering*, Prentice-Hall Inc., Upper Saddle River, N.J., 1996.
    2. Day, R.W. *Proc., Geotechnical Engineer's Portable Handbook 2nd Edition,* McGraw-Hill Companies, Inc., ASCE New York, 2012, pp. 14.1-14.95
    3. Robertson, P.K. and K.L Cabal *Guide to Cone Penetration Testing for Geotechnical Engineering 5th Edition,* Gregg Drilling & Testing, Inc., Signal Hill, CA, 2012, pp. 95-119
- Publications
    1. ASCE Standard 7-10. *Minimum Design Loads for Buildings and Other Structures*, American Society of Civil Engineers (ASCE), Reston, VA, 2010. pp. 57-68, 203-209.
    2. 2015 International Codes. *International Building Code*, International Code Council, Inc., Country Club Hills, IL, 2014. pp. 417-445.
    3. Bonneville, D.R. and Shuck, A.M. *An Introduction to the 2015 NEHRP Recommended Seismic Provisions for New Buildings and Other Structures*, NEHRP Provisions, San Francisco, CA, 2015.
    4. Berry IV, H.N. *Earthquakes in Maine*, Department of Conservation, Maine Geological Survey, Augusta, ME, 2003.
    5. Park, C.B., *Multichannel Analysis of Surface Waves (MASW) – Active and Passive Methods*, The Leading Edge, Kansas Geological Survey, Lawrence, KS, 2007.
    6. Weddle, T.K., *Surficial Geology for Brunswick Quadrangle, Maine*, Maine Geological Survey, Augusta, ME, 2001.
- Federal Government Reports
    1. Mendez, V.M. *AASHTO LRFD Bridge Design Specifications 4th Edition*, American Association of State Highway and Transportation Officials, Washington, DC, 2007.
    2. Kavazanjian, Jr., E. *LRFD Seismic Analysis and Design of Transportation Geotechnical Features and Structural Foundations Course No. 130094* Publication No. FHWA-NHI-11-032, U.S. Department of Transportation, 2011.
    3. Marsh., M.L. *LRFD Seismic Analysis and Design of Bridges No. 130093* Publication No. FHWA-NHI-15-004, U.S. Department of Transportation, 2014.
    4. Loehr., E.J. *Geotechnical Site Characterization Course No. 132031* Publication No. FHWA-NHI-16-072, U.S. Department of Transportation, 2017.
- Website
    1. Earthquake Hazards Mapping. *Design Ground Motions*. United States Geological Survey (USGS). earthquake.usgs.gov/hazards/designmaps. Accessed May 2018.
    2. New England Seismic Network. *Historical Events*. Weston Observatory Boston College. aki.bc.edu/quakes_historical.htm. Accessed May 2018.

Geotechnical Risks from Abandoned Coal Mines to Transportation Infrastructure and Mitigation - An Overview

David L Knott
WSP
51-55 Bolton St
Newcastle, New South Wales 2300 Australia
+61 2 4929 8345
David-L.Knott@WSP.com

Athena Livesey
WSP
The Victoria, 150-182 The Quays
Salford, M50 3SP United Kingdom
+44 161 200 5000
Athena.Livesey@wsp.com

Robert Kingsland
WSP
680 George Street
Sydney, New South Wales 2000 Australia
+61 2 9272 5380
Robert.Kingsland@wsp.com

Thomas Lefchik
Ohio Department of Transportation
650 Eastern Ave
Chillicothe, OH 45601 USA
+1 740 774 9059
Thomas.Lefchik@dot.ohio.gov

Elizabeth Dwyre
WSP
115 W Washington St #1270s
Indianapolis, IN 46204 USA
+1 317 287 3406
Elizabeth.Dwyre@wsp.com

Prepared for the 69th Highway Geology Symposium, September, 2018

Acknowledgements

The authors would like to thank the individuals/entities for their contributions in the work described:

Laura Rudling – WSP
Jennifer Whalan – WSP

Disclaimer

Statements and views presented in this paper are strictly those of the author(s), and do not necessarily reflect positions held by their affiliations, the Highway Geology Symposium (HGS), or others acknowledged above. The mention of trade names for commercial products does not imply the approval or endorsement by HGS.

ABSTRACT

Coal mining has been undertaken since prior to recorded history and is no longer performed in many areas; as such it is often forgotten and its risks misunderstood. The presentation will provide an overview of geotechnical risks associated with abandoned underground coal mining, such as subsidence, mine fires, and mine pools and subsidence risk mitigation.

INTRODUCTION

Commercial coal mining started in the United States in the Richmond basin in Virginia in 1748 (Wilkes, 1988) and progressed across the nation as coal was encountered as shown on Figure 1 and discussed in Eavenson (1942). Thus, mine subsidence can impact infrastructure in many regions. Its impact was recognized by the development of the Interstate Technical Group on Abandoned Underground Mines (ITGAUM), with 18 states and several federal agencies participating.

Impacts from abandoned mines result in project costs and closures. The annual cost to U.S. transportation agencies of mine subsidence and other impacts from abandoned underground coal mines is not known, but mitigation measures can cost millions of dollars for a major project. For example, the 8.5-mile long, $160M US 33 Nelsonville Bypass project in Ohio, completed in 2014, included $32M to inject approximately 50,000 cy of grout into abandoned mine workings. (Equipment World, 2013, and Construction Equipment Guide, 2014). The Centralia, Pennsylvania mine fire, which started in the early 1960's, expended over $5M in efforts to extinguish and mitigate the fire, followed by over $40M in relocation costs. In Centralia, a ¾-mile stretch of RT 61 was repaired and subsequently abandoned and relocated due to the fire having burnt and weakened the coal pillars supporting the mine roof. (PA DEP, Centralia Mine Fire Resources, Accessed May 27, 2018).

Closures for mitigation can last for months, for example I-70 in Ohio was closed for 4 months resulting in detours and cost to the public.

This paper provides an overview of potential geotechnical risks and mitigation options for abandoned underground mines as they relate to transportation projects.

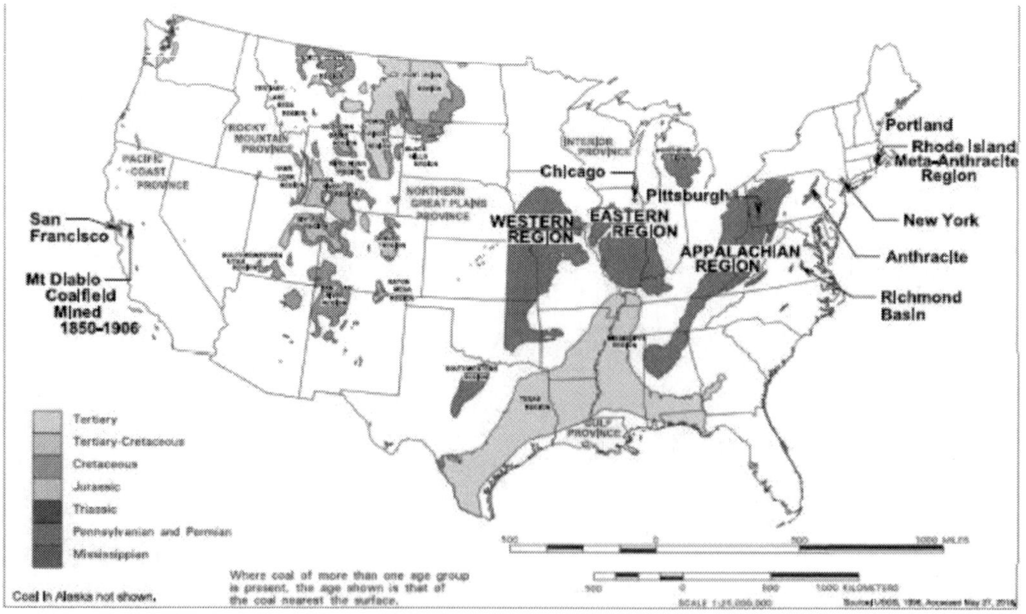

Figure 1 Coal Bearing Areas of The Lower 48 States Note Underground Coal Mining Has Also Occurred in Alaska. (Modified From USGS, 1996).

POTENTIAL GEOTECHNICAL RISKS

Potential geotechnical risks associated with abandoned underground mines include mine subsidence; various risks associated with mine entries; mine fires; unstable waste materials; flooded mine workings and associated water quality issues; impacts to cuts above alignments; and impacts to tunnels.

Mine Subsidence

Mine subsidence is defined by Bruhn et al., 1998 as follows:

"Mine subsidence is the downward movement of the ground surface due to gravity in response to a loss of support at mine level. The ground surface and whatever is constructed upon it is supported by a structural system that comprises the overburden (the soil-mantled sequence of rock strata situated between ground surface and mine level), the coal pillars, and mine floor. Excessive deformation or failure of one or more of these components over time can result in mine subsidence."

Mine subsidence typically occurs in the form of sinkholes from roof failure and troughs from pillar failure as shown on
Figure 2. These types are discussed as follows, including far field effects which can develop from pillar failures:

- Sinkholes – usually circular collapse depressions with sides that can be steep and abrupt or gently sloping towards the center. An example of a sinkhole which impacted an interstate in Ohio is shown in Figure 3. The sinkhole was triggered by dewatering of abandoned mine workings under the roadway due to auger mining that "broke into" the abandoned workings. Auger mining involves drilling horizontal holes for distances of 200ft or more into the coal at closely spaced centers. A photo showing sinkhole development initiating underground, is shown in Figure 4. Sinkholes rarely develop where the rock overburden is ≥ 10 times the mined height (Piggott and Eynon (1978).
- Troughs – broad, shallow depressions that can form over abandoned mines and are expected to occur over longwall mines and high extraction room and pillar mines. Trough formation depends on factors including mined width, mined height, thickness and hardness of floor and overburden rocks, and pillar stability. An example of the surface manifestation is shown on Figure 5 and Figure 6 and a view underground in Figure 7. Cover depth above the workings may not guarantee that subsidence will not impact site. Gray et al, (1996).
- Far field effects – Mine subsidence horizontal movements are driven by the release of energy when overburden strata subsides or with the release of horizontal stresses within the rock mass (Mills, 2011). Far field effects were first noticed at the Stanwell Park Viaduct south of Sydney, Australia, in 1985 (Hillerd, 1988), where significant distress to the structure occurred. It also occurred at Ryerson State Park Dam Pennsylvania (Hebblewhite and Gray, 2014). These far field movements can occur at substantial distances from the mining as Reid (1991) records movements of the order of 1.5km from active mining and Mills et. al. (2011) reports movements of 20mm at 1.6km from mining. An example is shown on Figure 8.

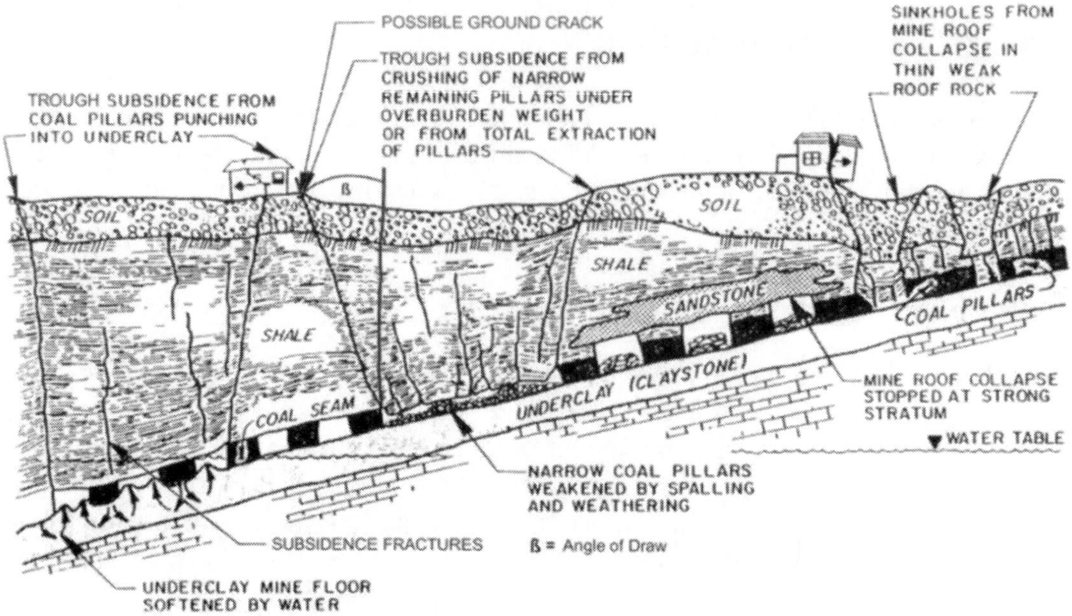

Figure 2 Types of Subsidence and Subsurface Conditions (Modified from Bruhn et al, 1978)

Figure 3 Over-Excavation of Sinkhole on I-70 in Ohio March 5, 1995

Figure 4 Examples of Sinkhole Development Underground Exposed in an Over-Excavation, Louth Park, New South Wales, Australia, (Courtesy Sam Mackenzie)

Figure 5 Failure of ~20ft to 30ft High Pillars at ~330ft to 500 ft. Depth With 4.6ft. of Subsidence in Ipswich, QLD, AUS, QLD Dept. of Mines and Energy Website

Figure 6 View Across Trough in Figure 5.

Figure 7 Squeeze (Creep) at Mine Level in Anthracite Region of Pennsylvania Bartoletti, 1996

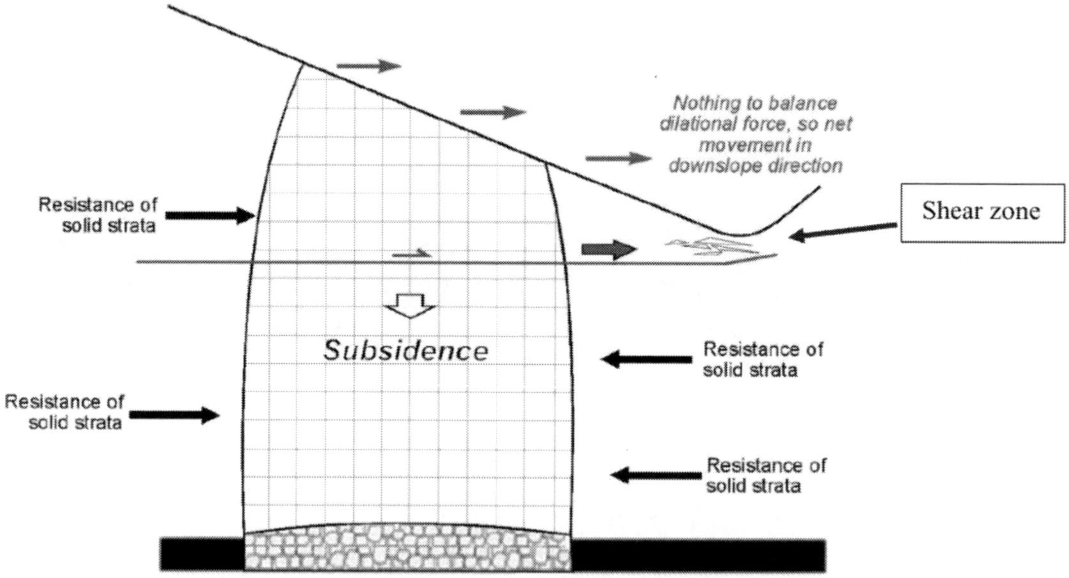

Figure 8 Lateral Dilation Mechanism for Horizontal Movement (Commonwealth of Australia, 2014)

Mine Entries

Mine entries can be drift entries, slopes, and / or shafts as shown on Figure 9. They provided access to the mine workings and ventilation. Drift entries and slopes may have shallow cover under the road and / or may be present in the adjacent cut slope. Both require mitigation to provide a stable road and cut slope. Potential issues in sealing / stabilization of entries include:

- Drainage - Entries may be located at a low point in the mine to allow drainage and facilitate haulage. Consequently, drainage flows toward the entry and out of the mine. If the entry is blocked and / or sealed, the mine may fill with water. If the blockage or seal fails or is breached, the water may catastrophically burst out of the entry. Water drainage from entries often needs to be treated to moderate acidity as discussed later.

- Gases – Potentially hazardous gases, should as methane, may emanate from entries that result in safety issues and this needs to be considered in closure.
- Wildlife - Access for protected bat populations may be needed.

Shafts present a significant hazard because of their depth, openness at mine level, degradation of shaft support / lining, and may have been improperly capped in the past and covered with soil, so they are no longer visible. A bare zone in a snow-covered area may indicate the location of a shaft as the warm air from the mine melts the snow. Figure 10 shows an example of a shaft with a road built over it. The potential for caving was recognized later and was mitigated by drilling angle holes into the sides of the shaft and injecting grout, with grout holes drilled trough the top and grout injected to complete the filling.

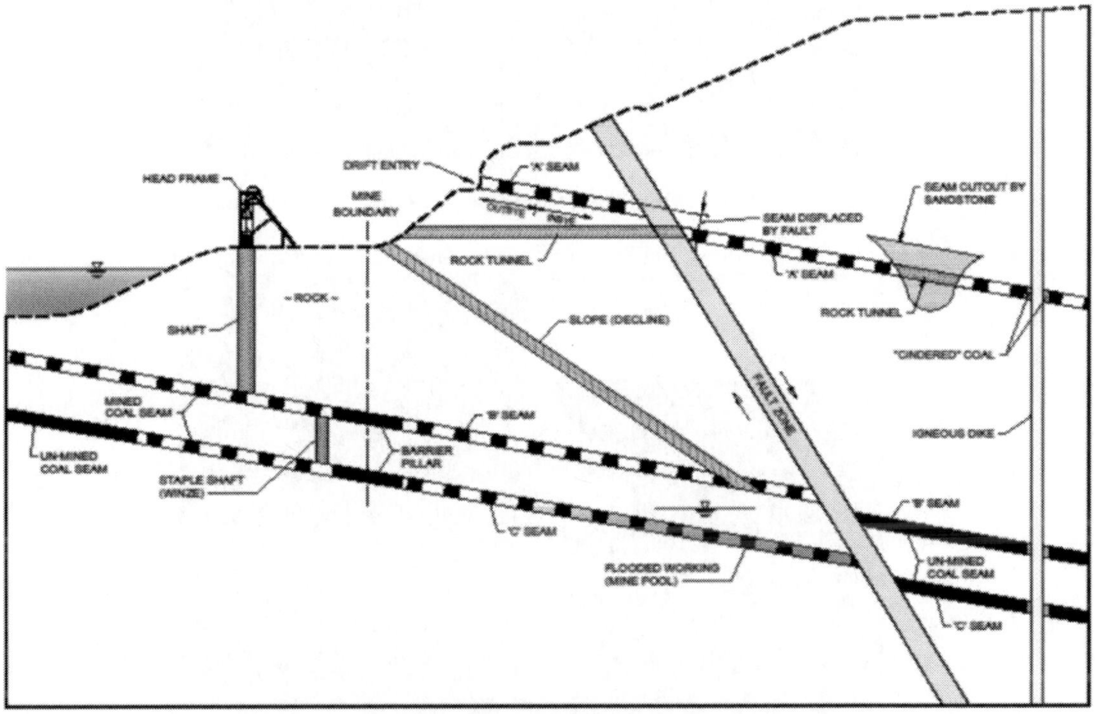

Figure 9 Mine Feature Terminology

Figure 10 Shaft Beneath Road Stabilized by Grouting After Hazard Was Recognized

Mine Fires

Fires in abandoned underground mines can also impact infrastructure. At Centralia, Pennsylvania, surface distress from abandoned anthracite mine workings burning under RT61 necessitated that the road be relocated. An example of the distress is provided on Figure 11 and the relocation is shown on Figure 12.

Figure 11 **Subsidence Feature with Steam from Underground Mine Fire Old RT61 in Centralia, PA**

Credit: Flickr/kaanahhttp://www.centraliapa.org/abandoned-centralia-old-route-61/ Accessed April 26 2018

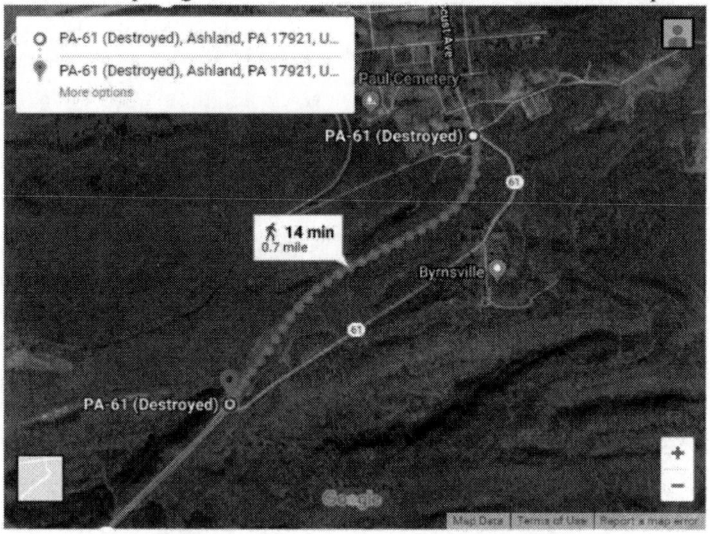

Figure 12 **Relocation of RT 61**

http://www.dangerousroads.org/north-america/usa/3843-abandoned-route-61-centralia.html accessed April 26 2018

Unstable Waste Materials

Coal was cleaned and sized as it came out of the mine. The waste material typically consisted of poor quality coal and carbonaceous rock types of various sizes and coal fines. In addition, the coal may have been washed and the wash water and fines disposed in slurry (sludge) ponds. The piles are uncompacted and may be unstable; subject to spontaneous combustion due to the presence of pyrite in the waste material and may be on fire; and present an acid drainage problem. Slurry ponds are difficult and costly to construct a roadway over as shown in Figure 13, where significant stabilization efforts were needed (see Bazán-Arias et al., 2004).

Figure 13 **Constructing a Road Over a Slurry Pond in Pennsylvania**

Flooded Mine Workings, Including Water Quality

As previously indicated, some entries were located near low points in the mine to allow drainage. It may also discharge where the workings or a seam are exposed by excavation for a infrastructure. Sometimes this water is acidic. An example of water discharging from mine workings in the Pittsburgh Coal Seam above road level is provided on Figure 14. This water can flow onto the road and cause icing problems in the winter, soften the material underlying the pavement, and corrode piping and structures that it passes through.

Figure 14 **Acid Mine Drainage Flowing from Abandoned Workings in the Pittsburgh Coal Seam, RT 837 West of Elrama, PA. The Water Was Treated by Applying Limestone to the Slope.**

The water can also impact mitigation activities as shown in Figure 15, where water was ejected from previously drilled grout holes due to the air pressure from drilling. In other cases, pumping of the water during grouting has been used to maintain the water level in the mine to reduce the potential for subsidence and control discharge as indicated in Millar and Holz (2011) and shown in Figure 16. The water may require treatment and may be reused for the grout as indicated in Millar and Holz (2011).

Figure 15 **Water Being Ejected During Drilling into Flooded Mine Workings**

Figure 16 **Ipswich Water Treatment Plant on Left and Grout Plant on Right Millar and Holz (2011)**

Impacts to Cuts Above the Alignment Due to Subsidence or Outcropping Coal Seams

Mine workings and coal seams may be encountered in cuts above the alignment. Sometimes the instability of the workings can destabilize a large area of the cut slope as shown on Figure 17 or localized areas as shown in Figure 18 and Figure 19. The pattern and types of the discontinuities may vary significantly and extensive mapping needed during excavation to assess support options and make changes (Livesey and Morgan, 2017). Mine openings may need to be stabilized to provide support for the slope. Grouting of the mine workings and the broken rock in the slope may also be needed.

Fires can also occur in outcropping seams due to grass fires and / or spontaneous combustion.

Figure 17 **Close-Up of Slope Failure**

Figure 18 **Example of Broken and Caved Rock and Void Exposed in Cut Above Mine Workings, Newcastle, UK**

Figure 19 **Mitigation of Cut Instability Using Mesh and Bolts, Newcastle, UK**

Mine Gases

Intersecting coal seams and / or workings by drilling or exposure of seams during earthwork can introduce hazards that should be recognized and planned for to mitigate risks which can include explosions and human health effects related to exposure. Gases of concern include methane (CH_4), hydrogen sulfide (H_2S), and carbon dioxide (CO_2). Exposure of the seam and workings to oxygen can lead to spontaneous combustion and the generation of CO, CO2, SO2, NOx, VOC, PaH, while burning may generate acid mists (Butcher, 2013).

Tunnels Impacted by Workings

Tunnels may encounter workings or pass above or below them. The Rook tunnel in Pittsburgh was completed in 1904. It passed through abandoned workings in the Pittsburgh Coal seam at an angle. Some measures were undertaken during construction to pass through the seam. However, in 1941 cracking and spalling of the concrete liner was observed in the portion of the tunnel passing about 15ft below the mine workings. The distress was found to be due to a roof fall in the overlying mine workings "rupturing" the floor of the mine. Temporary supports were installed until the area was supported by ARMCO heavy duty tunnel liner plates which were covered by shotcrete (Railway Age, 1942). The repair is shown in Figure 20.

In addition, a tunnel may pass through stress modified and / or broken ground due to mine workings above and below the tunnel.

Figure 20 **View of Rook Tunnel Repairs with View on Left of Liner Plates Covered with Mesh in Preparation for Shotcrete Placement and View on Right Showing Shotcrete Placement (Railway Age, 1942)**

SUBSIDENCE MITIGATION OPTIONS AND COMMENTS

Risk Assessment

Most transportation agencies planning new construction or major highway improvement projects over abandoned underground mines take a risk-based approach to the implementation of mitigation measures. In addition, many agencies, notably the Ohio Department of Transportation (ODOT, 1998) apply risk assessment methods to the management of existing transportation assets. Assessment of the likelihood of occurrence of subsidence, and the consequences of subsidence should it occur, are an essential part of the planning process for managing transportation assets over abandoned underground mines. Risk-informed decision making can aid agencies in allocating funds for various mitigation measures as described later in this paper, technology-based structural health monitoring, or a "watchful waiting" approach with repairs performed if needed.

The Indiana Department of Transportation (INDOT) used a risk-assessment based decision process for coal mining areas along the new-terrain Interstate I-69 project in southwestern Indiana. Risk-based approaches were used to define the limits of coal mineral rights to be acquired in an area permitted for future room-and-pillar coal mining. A risk-based approach was also used for a section of new roadway constructed over an area of a where highwall mining was performed in 1996. In the risk assessment, subsurface conditions were investigated and the likelihood of subsidence was evaluated, with a finding that the pillars appeared to have been adequately designed and were considered unlikely to fail. The consequences of failure, should it occur, were assessed as 4 inches of subsidence over 50 to 60 feet, which was deemed repairable, given that no structures were required in this section of highway. Therefore, no mine mitigation measures were required prior to construction of the new interstate, which opened to traffic in 2012.

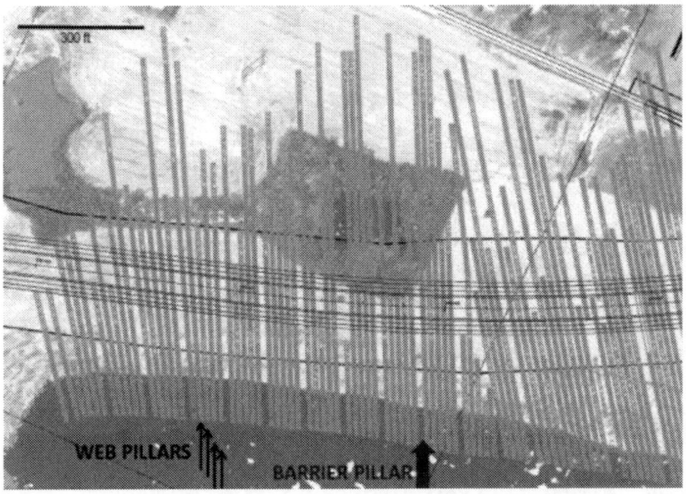

Figure 21 1996 Highwall Mine Under New I-69, Southern Indiana

Potential Land Ownership And/or Coal Sterilization and Rights

Mitigation approaches need to consider land ownership and / or coal sterilization rights as the approach may block off future access to the coal. Jones (2004) reported on a project where the grout was not allowed to flow beyond the property boundaries.

Stabilization of Workings by Cementitious Grout Injected into Boreholes Drilled into the Workings from the Surface.

Grouting Practice

Selection of appropriate mix flowability and strength depends on if the workings are open / caved, flooded or dry, seam dip, and sterilization of the coal. Appropriate mixes vary from those with specified flowability and strength values to those with specified mix ratios. Experience indicates that the flowabilities and strengths in Table 1 can achieve the desired results for sinkhole stabilization and possibly pillar support in many cases (Knott, 2004). In the UK, a common mix consist of 1 Part Ordinary Portland Cement (OPC):10 parts Pulverized Fly Ash (PFA) (Livesey and Morgan, 2017). For pillar support, higher strengths may be needed, such as 725 psi to 1160 psi (5 to 8 MPa) in 90 days for projects in Newcastle with pillars up to 16ft (5m) high. In some cases, the cement type may need to be specified to deal with the mine water corrosively.

Table 1 – Potential grouting conditions and grout properties

Condition	Material[1]	Flowability	Strength
Caved / broken conditions at mine level and non-continuous roof voids	Cementitious fly ash grout	25 to 35 second Flowcone	300psi (2.1MPa) @ 7 days
Open voids at mine and continuous roof voids >3ft (1m) high	Cementitious fly ash concrete	8 to 10-inch (200-250mm) slump	300psi (2.1MPa) @ 7 days
Barrier grout	Cementitious fly ash concrete	3 to 4-inch (75 to 100mm) slump	300psi (2.1MPa) @ 7 days

1 - Note additives may be needed, particularly where shrinkage is an issue

The amount of grout injected into a grout hole has varied over time and by region. Previous US practice typically involved injecting a given amount of grout into grout hole daily until it filled. Current US and Australian practice is to generally inject grout into a grout hole until it fills, if the grout flow can be tracked. The grout is usually injected under gravity or a slight pressure from the pump. In the UK, grout is injected under a nominal pressure, not exceeding 0.4psi per foot (10kPa per m) depth below ground level (i.e. not more than 50% of overburden pressure). Thus, avoiding blow out of the grout. If no more than a nominal/agreed tonnage of grout is introduced under that pressure over a 2-minute period, that area is considered to be stabilized. If more than the nominal/agreed tonnage is taken, additional grouting of the existing grout holes, or additional holes are undertaken until testing demonstrates that the ground will take no more grout.

Injection of grout into water filled workings should always be with a tremie pipe with the tip maintained in the grout and is raised as the grout level rises so that it does not become stuck. The use of the tremie pipe reduces the potential for dilution of the grout with water. Antidotal information from a site in Newcastle indicated that grout with a 2,175psi (15MPa) surface strength injected into flooded workings with the water level near the surface without a tremie pipe has a strength of about 145psi (1MPa) when cored samples were tested.

Grout Hole Pattern

The grout hole pattern can vary depending on the type of stabilization required and "tieing" of the mine map to the surface. Typically grout holes with closer spacing are required in areas where sinkhole mitigation is being performed as the grout needs to support the roof and / or the orientation of the mine working with respect to the surface is unknown. In the Pittsburgh region, a grid with a 15ft to 25ft (4.6m to 7.6m) grout hole spacing has been used, while a spacing of about 15ft (4.5m) was used on a UK project (Figure 22) (Livesey and Morgan, 2017). Bruhn et al. (1998) illustrate the savings associated with using a mine based grout hole pattern over a grid pattern. In areas where pillar support is being performed, the spacing of the grout holes is further apart.

Figure 22 Grouting for Proposed Roadway Widening in UK

In some areas, environmental constraints may control where the drill rig can setup and angled holes are needed to reach the target. An example of this is provided in Figure 23 and discussed in Kingsland (2013) where the mine workings for a bridge on the Hunter Expressway in an environmentally sensitive area were stabilized by drilling many holes at an angle from individual pads (platforms).

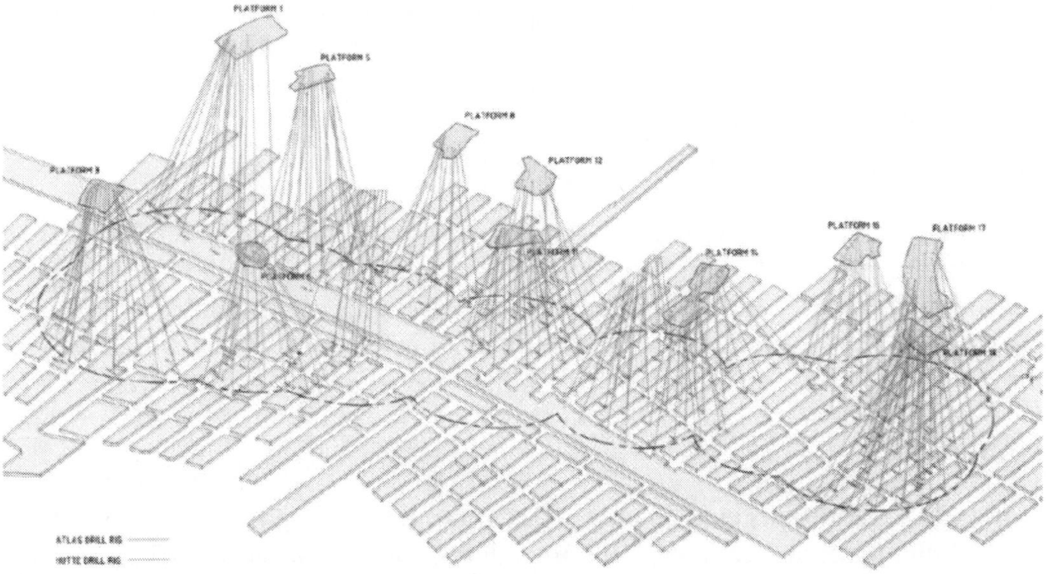

Figure 23 Example of Grouting Using Designated Drilling Pads (Kingsland, et al., 2012)

Drilling Conditions

The conditions encountered during drilling may vary from stable to unstable due to caving rock. In stable holes, casing may only be needed in the soil zone, while in unstable holes, the hole may need to be cased to mine level.

The importance of casing the soil zone is particularly important where sandy soils are present as they can flow down the borehole into voids, resulting in surface distress.

Mine Gas Issues

A case study will be presented to illustrate some of these risks and how they were mitigated based on Butcher (2013). The Hunter Expressway is a new four-lane motorway near Newcastle, Australia, with mining in two seams as discussed in (Kingsland et al., 2013). More than 1,600 grout holes were drilled and about 262,000cy (200,000 m^3) of grout was placed to mitigate subsidence impacts (Kingsland et al., 2012). With such a large amount of drilling and grouting required, a high level of potential interaction with gases both at mine level and within unmined coal seams that were intersected was present. Gas sampling performed prior to the start of grouting indicated gases of concern included: Methane - 76%, Oxygen - 0.39%, and Carbon Dioxide- 9.3%.

Special procedures were undertaken to reduce the potential for explosive conditions at the surface and underground as follows):

- Gas sampling and analysis – before and during work
- Administration – management plan and training
- Weather monitoring (especially barometer) - The atmospheric conditions will influence whether the mine voids will be "breathing in" (air sucked into the hole during high air pressure days) or "breathing out" (mine gas moving out of the hole on low air pressure days), which in turn influences the hazard location and controls adopted
- Hard controls – Examples include capping of boreholes to reduce "breathing', anti-static clothing, and foam drilling
- Demarcation of high risk zones.

In addition to the potential for explosive conditions at the surface, the injection of oxygen as part of the drilling process could lead to an explosive situation underground and / or increase the potential for spontaneous combustion. In addition to the procedures above, on the Ipswich motorway, nitrogen was injected into the workings to displace methane, which was flared off (Millar and Holz, 2011).

Drilling Monitoring

Monitoring of the conditions encountered in the grout hole is important to assess if the grout holes are hitting their intended targets and mine level conditions and if subsidence caving / has occurred. Figure 24 shows an example of a field marked up grout plan showing the conditions encountered and interconnection of the workings. The interconnection was the workings was based on air from drilling blowing out of other holes. The plan is also useful for grouting to help assess grout hole takes and where the grout is flowing.

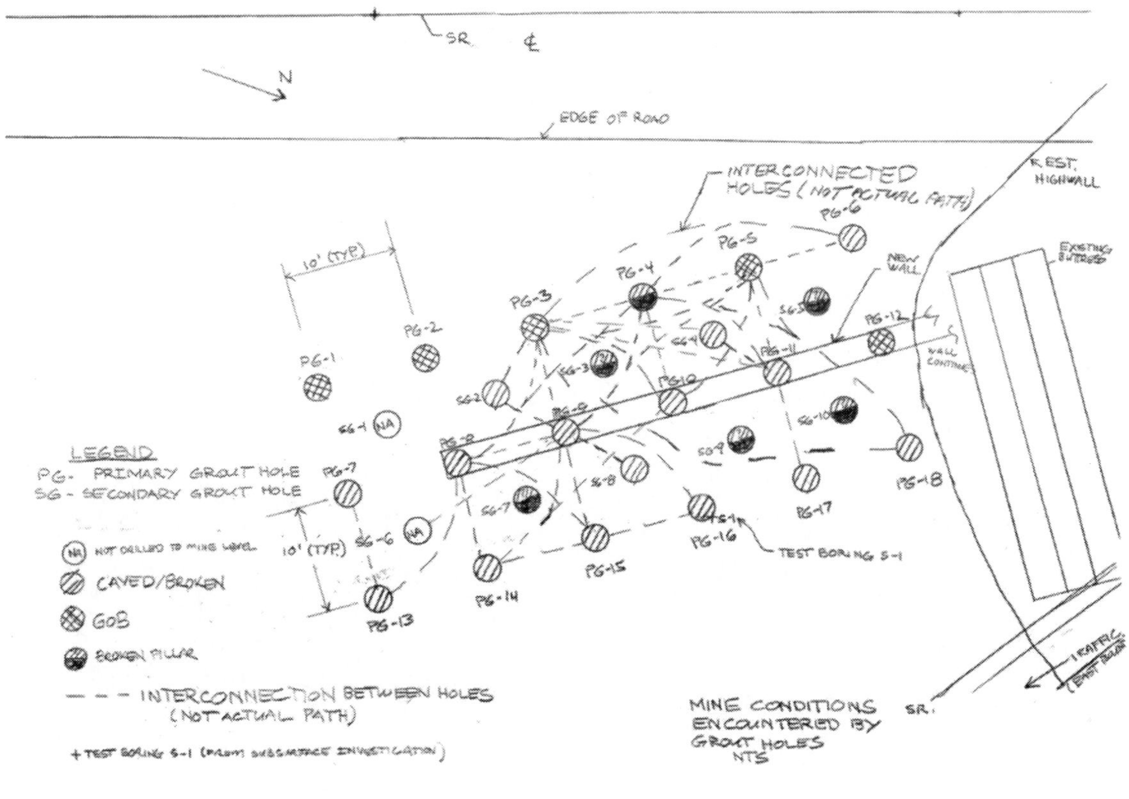

Figure 24 **Field Mark-Up Showing Conditions Encountered and Interconnection of Workings**

Assessment of Conditions After Drilling

Borehole camera, downhole geophysical logging, borehole deviation, laser mapping of unflooded workings, and sonar mapping of flooded workings are extremely beneficial in assessing borehole and mine level conditions. Additional information is provided in Knott and Streater (2017).

Grout Monitoring

In addition to the approaches presented in the previous section, checking grout levels with a weighted tape or a cable pushing cable in inclined holes is commonly performed. It is noted that "soft" materials at mine level can be displaced by the grout and move up into borehole. Checking hole conditions also reduces the need for re-drilling of holes due to potential caving since drilling since the cause of the change in depth if known.

Mine Water Impacts

As the grout is denser than water, it will displace water in the mine workings and flow out into the workings and up through old boreholes, shafts, subsidence features, and grout holes. In some cases, the grout may act as a "dam" and allow water on the updip side to "backup". The water may flow around the grouted zone and / or flow in the materials overlying the seam in discontinues too small for grout penetration.

Quality Control

Testing of grout flowability and strength are needed to assess if the production grout is satisfactory. Grout samples for strength testing may be cylinders or cubes. Testing can be at specific period or several periods to asses if the required strength is being obtained; as once the grout is in the ground, it cannot be taken out. Proper curing is essential, as samples that are not kept at the right temperature give low strength values.

Verification

Coring to assess the effectiveness of grouting is used on many projects. An example of good grout recovery is shown on Figure 25. Poor grout recovery can occur if the grout has insufficient strength to be cored and / or was impacted by flowing through water.

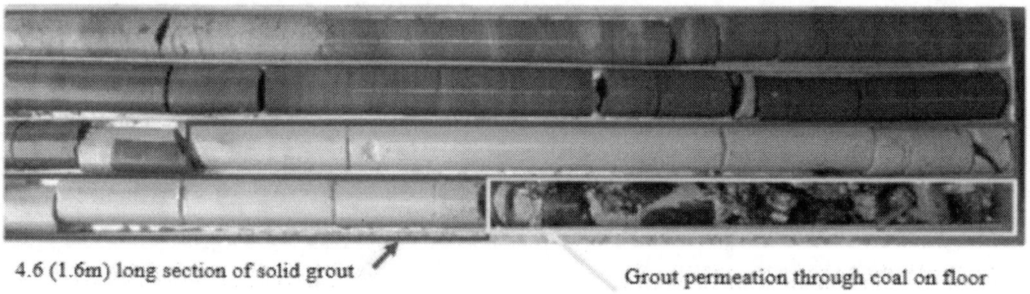

Figure 25 **View of Grout Recovered from Cored Verification Hole**

Over-Excavation of Workings

Over-excavation is the one method that where voids are addressed and will not present future problems. Considerations include mine pool level, acidic materials, gases, broken materials, economic depth of excavation, dealing with exposed entries, excavation backfilling and settlement, mineral rights, and traffic maintenance. However, it is noted that fills settle. An example is shown in Figure 26.

Figure 26 **Over-Excavation of Mine Workings**

Bridging / Spanning

Geosynthetics and other materials have been used to span opening. However, they need to account for deflection (sagging) and anchorage, which are influenced by material stiffness and strength. As geosynthetics achieve their strength through strain, a significant thickness of geosynthetics may be needed. Concrete slabs designed to span opening may also work.

An example of how far field effects were dealt with on one project is presented. The Hunter Expressway near Newcastle, NSW had potential far field impacts due to abandoned workings and potential future mining as discussed in Kingsland et al. (2011). Failure of the pillars in the abandoned workings could cause vertical subsidence in the range of 1.6ft to 2.6ft (0.5 m to 0.8 m) with associated horizontal movements estimated to be up to 1ft (0.3 m) in areas of steep topography.

The strategies adopted for managing the subsidence risk included both mine grouting and bridge design components with instrumentation. The bridge structures were designed to accommodate both the low levels of vertical subsidence that could not be prevented by grouting alone and the potentially much larger horizontal ground "valley closure" movements that were considered possible if adjacent areas of pillars were to collapse. An example of the grouting is shown on Figure 23. Double-sleeved drilled shafts were used for foundations intersecting the potential plane of horizontal shear movement. This plane was inferred to occur at a level close to the base of valley floor (nominally between 7ft (2 m) above and 30ft (10 m) below the base of the valley). The double-sleeving of piles provides an annulus of free space between the outer and inner sleeves to accommodate potential horizontal movements. Provision for upsidence beneath the pile cap was also made.

SUMMARY

Abandoned mine workings are "often out of sight and out of mind" and can result in large costs during construction and / or remediation for existing facilities. However, they can be investigated, assessed, and mitigated to achieve the desired results.

REFERENCES:

1. Anonymous. No Delays to War Traffic on this Tunnel Repair Job. Railway Age, December 5, 1942.
2. Bazán-Arias, N.C., Michalski, S.R., Glogowski, P.E. and Howard, F. In-Place Solidification of Coal Tailings for Expressway Subgrade" in Proceedings 2004 Conference of the American Society of Mining and Reclamation.
3. Butcher, B. Hazards from mine gases when conducting field investigations and subsurface works. Presentation to Newcastle Chapter – Australian Geomechanics Society, Newcastle, Australia, 7 March, 2013.
4. Bruhn, R. W., Knott, D. L., Sanger, D. B., and Kyper, T. N.; "Effectively Designing for Potential Mine Subsidence at a Disposal Facility in Northern Appalachia"; In *Land Subsidence Case Studies and Current Research: Proceedings of the Dr. Joseph F. Poland Symposium on Land Subsidence*; Edited by James W. Borchers, association

of Engineering Geologists Special Publication No. 8, Sacramento, California, published in 1998, pp 291–301. Note conf in 1995.

5. Commonwealth of Australia, Temperate Highland Peat Swamps on Sandstone: longwall mining engineering design—subsidence prediction, buffer distances and mine design options, Knowledge report, prepared by Coffey Geotechnics for the Department of the Environment, Commonwealth of Australia'. 2014
6. Construction Equipment Guide. *Nelsonville Bypass Project Nearly 50 Years in the Making*. March 28, 2014.
7. Eavenson, H.N. The First Century and a Quarter of American Coal Industry. Privately Printed, 1942.
8. Equipment World, *A glance at the $160 million Nelsonville Bypass project*. March 6, 2013.
9. Gray, R.E.; Bruhn, R.W.; and Knott, D.L., "Subsidence Misconceptions and Myths", 15th International Conference on Ground Control in Mining, Colorado School of Mines, Golden, Colo, Aug, 1996,
10. Hebblewhite, B. and Gray, R.E., "Non-conventional Surface Ground Behavior Induced by Underground Mining in Pennsylvania", The Appalachian Coalition for Geological Hazards in Transportation 2014 Technical Forum.
11. Hillerd, P.R.; "Damage to Stanwell Park Railway Viaduct", in Proceedings of Conference on Buildings and Structures Subject to Mine Subsidence, Newcastle Division of the Institution of Engineers Australia, Maitland, NSW August 28 to 30, 1988.
12. Jones, C.J.F.P., Personal communication, 2004
13. Kingsland, R.; Mills, K.; Stahlhut, O.; Huang, Y.; and Butcher, R. "Subsidence Hazards and Mitigation Strategies on Minmi to Buchanan Section of the Hunter Expressway" In 8th Triennial Conference Proceedings – Management of Subsidence: State of the Art, Mine Subsidence Technological Society, Pokolbin, New South Wales, Australia, 15-17 May, 2011. pp. 121-129.
14. Kingsland, R.; Mills, K.; Stahlhut, O.; and Huang, Y.; "Mine Subsidence Treatment and validation strategies on the on Minmi to Buchanan Section of the Hunter Expressway"; RMS Geotechnical Conference, Sydney, NSW, Australia, 2012
15. Kingsland, R. Stahlhut, O., Wheatley, D., Rees, P., Aryal, S. and Russell, G. Hunter Expressway, Australia: Dealing with poor ground and subsidence. Proceedings of the ICE - Civil Engineering. 166. 22-27. 10.1680/cien.12.00018. May, 2013.
16. Knott, D.L., "My Experience Grouting Abandoned Coal Mines", 5th Biennial Workshop of the Interstate Technical Group on Abandoned Underground Mines, April, 2004, Tucson, AZ, USA
17. Knott, D.L. and Streater, H.; "Downhole Techniques to Assess Mine Subsidence and Mitigation Effectiveness", 10 Triennial Conference proceedings of the Mine Subsidence Technological Society – Mine Subsidence: Adaptive Innovation for Managing Challenges, Pokolbin, NSW, Australia, November, 2017

18. Livesey, A. and Morgan, T., "A1 Coal House to Metro Centre Improvement Scheme" The Geological Society, London, May 2017

19. Millar, D. and Holz, B. "Ipswich Motorway Upgrade – Filling of Abandoned Coal Mines" Department of Transport and Main Roads Engineering and Technology Conference, Brisbane, Queensland, 2011.

20. Mills, K.W., Morphew, R.H. and Crooks, R.J. "Experience of Monitoring Subsidence at Ulan Coal Mine". In 8th Triennial Conference Proceedings – Management of Subsidence: State of the Art, Mine Subsidence Technological Society, Pokolbin, New South Wales, Australia, 15-17 May, 2011. pp. 89-99.

21. Piggott, R.J. and Eynon, P. "Ground Movements Arising from the Presence of Shallow Abandoned Mine Workings"; in Large Ground Movements and Structures, Ed by Geddes, J. D.; John Wiley & Sons, NY 1978, conf held in Cardiff, Wales 1977.

22. Reid, P. "Monitoring of Mine Subsidence Effects Adjacent to Cataract Reservoir". In 2nd Triennial Conference Proceedings, Mine Subsidence Technological Society, Maitland, New South Wales, Australia, 25-27 August, 1991. p. 46.

23. Wilkes, G.P. Mining history of the Richmond Coalfield of Virginia. Virginia Division of Mineral Resources Publication 85, Charlottesville, Virginia, 1988.

ICEFALL HAZARD PREDICTIVE INDICATORS & MITIGATION TECHNIQUES – RESULTS OF A 3-YR. RESEARCH STUDY IN ALASKA

David J. Scarpato, P.E.
Scarptec, Inc.
President & Principal Geological Engineer
P.O. Box 326
Monument Beach, MA 02553
(603) 361-0397
dave@scarptec.com

Matt Murphy
Alaska Department of Transportation and Public Facilities
Transportation Data Programs
Anchorage Field Office
Highway Data Manager (Acting)
(907) 269-0876
matt.murphy@alaska.gov

Prepared for the 69th Highway Geology Symposium, September, 2018

Acknowledgements

The authors would like to thank the individuals/entities for their contributions in the work described:

Alaska Dept. of Transportation & Public Facilities

Disclaimer

Statements and views presented in this paper are strictly those of the author(s), and do not necessarily reflect positions held by their affiliations, the Highway Geology Symposium (HGS), or others acknowledged above. The mention of trade names for commercial products does not imply the approval or endorsement by HGS.

ABSTRACT

Falling pieces of ice ("icefalls") may present an impact hazard along transportation corridors in U.S. northern-tier states subject to winter conditions. Hazards from falling ice include direct impact and secondary shatter events. Until recently, there have been very little research and design criteria relative to icefall hazard mitigation. The Alaska Department of Transportation & Public Facilities (DOT&PF) and Scarptec, Inc. completed icefall hazard research in February 2018. The 3-yr. research project, undertaken with cooperation from the FHWA, was broken down into two distinct phases – Phase No. 1 Literature Review (completed 28 February 2016) and Phase No. 2 Site-Specific Icefall Hazard Studies (completed 28 February 2018). Phase No. 2 entailed site visits in September 2016 and March 2017 to collect data on slope and ice conditions at seven sites along the Seward and Richardson Highways. The site-specific information was subsequently used for preliminary technical evaluations and an initial icefall impact risk assessment, both mitigated and unmitigated. This submission summarizes the results of both phases, with specific emphasis on potential mitigation solutions at Mile Post 113.2, which was the site of a large icefall in April of 2012. The results of this study may be beneficial to transportation planners and engineers responsible for highway design and maintenance.

KEYWORDS

Snow and ice control, ice phenomena, rock slopes, hazard evaluation, hazard mitigation, catchment, ditches, design practices, icefall, risk assessment, predictive indicators

INTRODUCTION

The term "icefall" generally describes the action of falling ice particles under the influence of gravity, similar to that seen with rockfalls. The term is both a verb and a noun, with the former referring to the action of falling ice and the latter referring to a thing, in this case a hazard. Icefall is a real hazard in northern tier states subject to winter conditions where ice can form and potentially fall from slopes, powerlines and structures. While icefall occurrence from towers, powerlines and structures is well-documented (1), falling ice emanating from slopes is not. One of the reasons for this lack of coverage in engineering and geohazards research is that ice is very much a transient hazard – its residence time is limited by temperature (i.e. air and substrate). Simply put, ice is "there one minute, then gone the next". Based on the above, it appears likely that icefall occurrence is more common than industry research would indicate; however, it is difficult to monitor, predict and mitigate due to its transient nature. Direct evidence of icefall impacts to passing motorists along transportation corridors also appears relatively limited, implying that icefall impact risk to motorists is generally very low; however, for those motorists who are impacted directly by falling ice, the results can be catastrophic.

Figure 1 – Post-impact photo from 6 April 2012 icefall event near MP 113.2 NB along the Seward Highway. (Photo adapted for use from KTUU article pub. 6 April 2012)

An April 2012 Seward Highway event demonstrated that icefall source zones nearest to highways can present significant impact risk. On the afternoon of 6 April 2012, a large slab of ice fell in the vicinity of Mile Post ("MP") 113.2 on the Seward Highway (Alaska Highway 1), severely injuring a motorist. The slab that fell was estimated to be on the order of 60 to 80 ft. in height by 20 ft. in width (Fig. 1). As a result of this direct icefall impact event, which was rare based on industry reporting of similar incidents, DOT&PF initiated site-specific icefall hazard evaluations in the winter of 2015 in order to better understand the hazard and develop initial strategies for icefall mitigation. The proposed course of work consisted of an initial literature review (Phase No. 1) followed by site-specific studies (Phase No. 2), both of which were completed over a 3-yr. period.

The civil/geotechnical engineering and natural/geologic hazards communities do not currently plan for icefall hazard mitigation during project scoping or design. This fact makes it challenging for DOTs nation-wide to track, manage, monitor, mitigate or even plan for icefall hazards because there is no industry-accepted technical basis for consideration of icefall hazards. This paper summarizes the results of the 3-yr. study, and it is hoped that the described studies help pave the way for subsequent evaluation and consideration of icefall hazards.

It should be noted that the authors have published previous articles on the subject of icefall hazards, many of which are cited in this paper and included in the References section.

LITERATURE REVIEW

In order to understand nationwide (and international) experience and documentation with icefall hazards, Phase No. 1 entailed a detailed literature review and summary report in 2015. We completed a public domain, open-source review by searching the internet, contacting various DOT entities and by communicating with geohazard specialists and mitigation system manufacturers from around the world. The report also contains links to specific media accounts of icefall events recorded on news websites, periodicals, technical journal articles, online newspapers and even YouTube videos. What we found was that there was only 3 documented cases in North America where falling ice had impacted a motorist resulting in injuries; two events in Terrace, British Columbia (B.C.) in 1988 and 2011; and, the April 2012 Seward Highway, Alaska event noted above. The B.C. events resulted in a fatality and minor injuries, respectively. There are however multiple documented instances (approx. 14 to 16 cases) where ice had entered the roadway but did not impact a vehicle. Furthermore, the incidence of icefall entry into roadways is likely underreported, as the ice could be completely melted in just a matter of hours. Interestingly, we also completed a general survey of 12 various DOT's in northern-tier states subject to cold weather conditions and found that many either "did not think icefall hazards were a problem" in their specific state, or "had not considered icefall" due to low event frequencies. The link to the entire report containing the results of the initial Phase No. 1 Literature Review is included within the References section (2).

Figure 2 – Google Earth image capture showing location of four study sites along Seward Highway south of Anchorage.

FIELD STUDIES

Phase No. 2 Site-Specific Studies were kicked-off in summer of 2016. The evaluations were completed at seven locations of documented ice development during September of 2016 and March of 2017, in order to observe the sites during periods of non-frozen and frozen conditions, respectively. The sites are indicated on the locus maps which are included as Fig. Nos. 2 and 3, and consisted of the following Mile Post (MP) locations along the Seward and Richardson Highways:

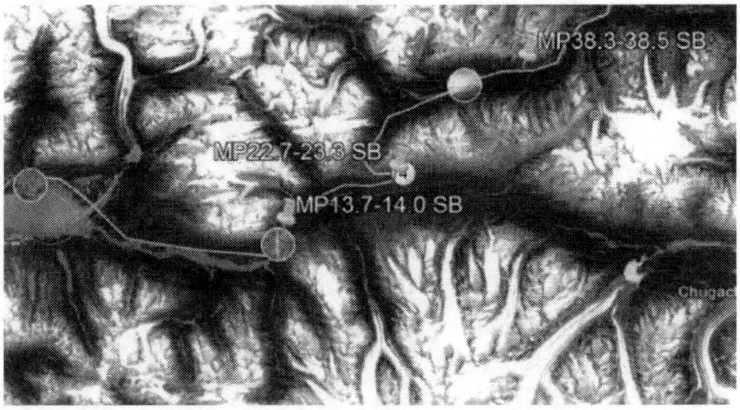

Figure 3 – Google Earth image capture showing location of three study sites along Richardson Highway east of the port of Valdez.

A. Seward Highway (Alaska Highway 1): MP 113.1 to 113.3 NB; MP 74.1 to 74.9 SB; MP 57.9 to 58.2 NB; MP 51.9 to 52.3 SB;

B. Richardson Highway (Alaska Highway 4): MP 13.7 to 14.0; MP 22.7 to 23.3; MP 38.3 to 38.5.

Data collected during the September 2016 visit consisted of information on existing conditions relative to the roadway, slope geometry, geotechnical observations, evidence of historic ice development and existing rockfall catchment conditions. These data were used to help understand substrate conditions at ice-bedrock contact surfaces, which is important in understanding sliding mechanics.

Information collected during the March 2017 site visits was specific to ice development conditions and included:

A. Ice Dimensions – We utilized a laser range finder and tape measure to collect data on ice slab shape, thickness, width (slope-parallel) and height;
B. Ice Strength – Qualitative estimate of ice strength based on direct observations, color, evidence of fractures, and hammer impact penetration, all of which can be used to use existing correlations with compressive strength;
C. Slab Support Mode – Documented mechanisms of ice slab support, including adhesion (i.e. rock-ice contact surface), toe-support (i.e. bearing), top support (i.e. hung, cantilevered);
D. Evidence of Icefall – Documented evidence of recent icefall occurrence, including horizontal run-out (shatter) distance;
E. Location of On-Site Ice Development;
F. Presence of Running Water;
G. Ice Color – Documented color of ice
H. Ice Quality – Presence or absence of entrained fines (e.g. silt, sand, mud);
I. Available Ditch Width – Including loss of width from ice development or wind rows from snow plows;
J. Available Ditch Height – Includes height of ditch infilling snow above roadway and/or from base invert of catchment ditch.

All of the data collected during both visits was included as an appendix to the Phase No. 2 report, which is included as a Reference at the end of this article (3).

ICE SLAB DEVELOPMENT

Ice development on slopes has been documented to cause short and long-term degradational effects on rock slopes, including effects like ice jacking, surcharging, accelerated weathering, ice-jammed discontinuities (which impedes drainage and allows destabilizing water pressures to increase), and increased frequency of rockfall and icefall (1). Ice will generally develop when two criteria are initially met:

1. The presence of "free" available liquid water; and,
2. Average daily temperatures fall below the freezing point (32 deg. F/0 deg. C).

Figure 4 – Stable ice slab feature near MP 74.8 SB on Seward Highway (Photo by Scarptec, Inc.)

Significant water discharge velocities and high winds will tend to impede ice development on slopes; however, both of these are relatively short-lived phenomena with respect to the duration of winter conditions. As the ice is subject to freezing, it will develop an adhesive interface strength that allows it to bond to the slope surface. The ice slab will grow outward and downward (vertically) and will generally mimic the flow path of the water seepage along the slope face (Fig. 4). Lateral ice cascade growth parallel to the slope face will occur significantly slower than the outward thickening and downward components of growth, resulting in smaller "icicle" or larger "slab" type formations.

Once bonded to the slope, the mass of ice will be supported either in suspension (e.g. localized free-hanging growth like an icicle), by interface "bonding" along the slope face (e.g. frozen waterfall feature) and to a lesser extent by direct bearing on the substrate. Ice slab stability is primarily derived from this interface bond strength, which when removed, results in slab displacement and subsequent deformation.

Sources of Water for Ice Slab Generation

Steady-state surface water overflow appears to be the primary causative factor in generation of large ice slab formations. Well-developed slabs of ice will be formed from the freezing of consistently flowing upslope water as it is intercepted or "captured" by the slope crest. Direct impact by precipitation (e.g. snow, rain) falling on the actual slope face does not appear to be a major source for significant ice development. Furthermore, presence of fracture-controlled seepage appears to be less significant in the development of ice than direct overflow of upslope surface water. Although joint-controlled seepage will add to total available water supply, it does not appear that fracture flow alone will supply the necessary volume of water needed to generate large ice slabs. Addition of water, either from existing up-gradient perennial streams or downslope migration of meltwater from snowpack, provided the consistent discharge needed to generate significant ice growth.

Temperature Controls on Ice Development

Temperature plays a very important role in both the formation and mass wasting of ice slabs. The temperature of the rock slope surface (i.e. "substrate") and the air both become important at the onset of initial freezing. The air temperature must be at or below 32 deg. F (0 deg. C) in order for fresh water ice to develop, assuming that the water seepage is at a relatively low discharge velocity. Once ice forms on the outer surface of the slope and penetrates the surface asperities, the mere presence of ice will help to further reduce the slope surface temperatures along the ice-rock interface, as bedrock will take longer to cool than the surrounding air. In order for significant ice slab development to occur, temperatures must be consistently cold enough to allow ice to grow but not so cold that all available upgradient water sources are completely frozen. The initial freezing period must be followed by prolonged periods of cold weather to generate continual ice growth. During periods where the average daily temperature falls below the freezing point, ice can be expected to develop. The longer the duration of below freezing average daily temperatures, the higher the likelihood that ice growth will be maintained and not be subject to melting or mass wasting. Depending on location with the State of Alaska, under most normal circumstances the onset of highway ice development would be expected to occur between October and November of any given year.

Solar Radiation Intensity

Incoming solar radiation will warm the air mass and bedrock, and initiate melting of the snow pack when sun intensity is high. In this regard, slope aspect played an important role in ice development. In general, south facing slopes were subject to increased duration of sunlight, which in turn increases the likelihood of snow pack melting on back-slopes above the ice-forming rock slopes. Incoming direct solar radiation intensity is high enough to induce partial melting of snow pack and subsequent slope overflow. As observed during site-specific studies, snow melt can still occur on days when the ambient air temperature is at or even 4 to 6 deg. F below freezing. This melting produced surface water that is ultimately captured by local topographic lows – in this case, the roadway limits. Along the drainage path, the water was captured by the slope crest allowing subsequent ice cascade growth at and near points of interception. Surfaces that absorb more incident energy (e.g. dark-colored bedrock) appeared to warm faster than those that were good reflectors (e.g. snow).

Air Temperature

The air is heated by incoming solar radiation. Ice will consistently develop and be retained on the slope during prolonged periods where the average daily temperature is less than 32 deg. F (0 deg. C). Ideal conditions for ice development are during periods of consistent daytime seepage from upslope water sources, followed by late day and night time re-freezing. Periods where the daily temperature never rises above the freezing point will limit upslope melting and downslope migration of water, both of which "feed" the ice and allow subsequent growth. Conversely, periods of excessive warmth will virtually eliminate the possibility of freezing due to elevated surface temperatures and rapid water flow velocities.

Bedrock Surface Temperatures

Bedrock surface temperatures appeared to play an important role in initial ice bonding (and melting). The outer surface of the rock needs to be at or below freezing temperature (32 deg. F) for initial crystallization of ice. The bedrock may be significantly warmer only a matter of inches into the surface. The initial ice crystallization acts as a "seeder" for subsequent adhesion and build-up of additional ice. High density bedrock surfaces with a higher proportion of dark minerals (or staining) will have lower albedos and will heat-up faster than light colored rocks. For example, under the similar conditions, basalt would heat faster than a typical granite due to the high content of dark-colored minerals. Conversely, the onset of initial freezing is likely also delayed on surfaces with low albedos. Bedrock surface heating also appeared to exert a critical control on slab de-bonding.

Initial Bonding

Ice will develop an "adhesion" bond strength between the substrate rock slope surface and the ice mass itself. This is important because once the slope surface warms sufficiently, the ice slab surface contact area will be reduced due to melting. Reductions in bonded surface area are directly translated into a net loss in the minimum adhesive contract strength required to support the weight of the ice structure. Adhesion is bond strength along a surface developed between two *dissimilar* materials (whereas cohesive bond strength is developed between two *similar* materials). For example, the bond between concrete and steel or concrete and rock would be an adhesive type bond. Similarly, the surface area along an ice to bedrock contact is an adhesive bond, characterized by its adhesive interface strength.

Large-scale slope "roughness" also aids in the development of ice growth. Irregularly-shaped slopes or those with benches provide additional surface area available for bonding. Vegetation can be considered a type of roughness perpendicular to the slope, which will assist in ice retention, especially with woody growth like saplings and small trees.

Slab Support Mechanisms

Figure 5 – Partial ice column direct bearing near MP 113.2 NB on the Seward Highway (Photo by Scarptec, Inc.)

Ice slabs may be locally, partially or temporarily supported on a slope over the residence period. For example, well-developed ice sheeting on near vertical slope faces could be supported at the toe by direct bearing on rock. In similar fashion, a mid-slope bench from previous blasting activities could serve as a local ice slab bearing feature. We found that ice slab stability is unlikely to be compromised during extended periods of sub-freezing temperatures due to contact adhesive strengths; however, slab loads will be distributed to contact points as the bond adhesion is lost due to melting.

Depending of slope angle and profile shape, ice structures may also garner a portion of their overall stability from external structural support mechanisms, including the following:

A. Direct bearing – Slab bears on bedrock (or soil) surfaces where a portion of the weight is transferred to the medium below as shown in Fig. 5;

B. Top-support – Ice is held up in part due to presence of a near horizontal slab on a bench or slope crest feature;

C. Frictional Interlocking – The "waviness" from asperities on the slope face will contribute to localized support as the slab grows outward.

In general, direct bearing and top support conditions would be most likely on steep slopes. The total contribution to overall slab stability consists of the adhesion, frictional and structural support components; however, we found that adhesion remains the single most important factor in overall stability during below-freezing conditions, as only a fraction of the load is supported by the other three components which become mobilized once surface temperatures rise and adhesion bonding is minimized.

ICE SLAB RELEASE

By definition, an initial triggering mechanism will induce failure in a slope system with the potential for instability. The system could be stable or even meta-stable until the onset of this trigger at the moment of incipient failure. For example, ground vibrations, like those imparted by earthquakes, can be a trigger for rockfalls or landslides. Falling ice hazards are also subject to triggering mechanisms, and understanding what these triggers are and how they develop will help shed light on the timing and frequency of the hazard. At any given site, conditions may already exist for instability, including unfavorable geology, steep slopes, and presence of water, to name a few; however, it is the triggering mechanism that frequently gets these icefall events to "go".

Although there may always be more than one trigger, the primary icefall triggering mechanism is a net positive fluctuation in ambient air and rock surface temperature (slope warming). As described by Graveline & Germain (4), increases in average daily temperature above a baseline (i.e. "degree-days") have been correlated with ice block releases. Short duration changes in temperature can occur in the winter; however, the likelihood of longer duration warming periods typically occurs in the late winter and early spring. Increases in temperature will affect icefall occurrence and will apply to the air, slope surface (i.e. "substrate") and to the ice mass itself. Heating of the fluid air mass occurs from convective heat transfer ("convection") and heating of the solid rock slope surface and ice mass will occur via convection and solar radiation (from direct sunlight).

Ice-Rock Interface Strength

Based on our interpretation of previous documented icefall events and our observations in the field, it is apparent that most icefall events originating from rock slopes will ultimately fail from a loss of interface adhesive strength due to slope warming. The loss of support exposes the slab to shear forces directed along the ice-rock contact surface. This is not to exclude localized tensile failures which could occur during rotation (i.e. toppling) of the slab, or local crushing of the slab toe due to loss of compressive strength. Given the extensive slope-parallel contact area that ice will have in comparison to other cross-sectional orientations, it's the contact shear strength that appears to play the greatest role in the stability of large ice slabs on highway rock slopes.

Ice-rock contact strength is primarily characterized by the adhesive and cohesive strength of ice along the interface. Attributes of the host bedrock joint surface strength become minimized as the triggering mechanism of temperature increase controls heating of the bedrock and air, which induces changes to internal ice strength and ice-rock contact strength. Ice slab interface strength is reduced to the strength of the ice adhesive bond and the ice's own internal cohesive strength component. The adhesive bond of ice may be greater than the cohesive strength in cases of significant surface area.

The cohesive strength component on the other hand, may result in nearly linear failure ("decohesion") primarily through the ice (5). The loss of adhesion ("delamination" or "debonding") would typically happen first, as the contact area of the adhesions is reduced through increases in temperature and subsequent melting from heating of the host rock and air mass as shown in Fig. 6. Ice failure through delamination does not typically leave much ice on the slope face, whereas decohesion failures will leave small pieces of ice remaining (adhered) to the face. This is significant because most failures observed (and post-failure photos reviewed) by us indicated very little remaining ice on the face, pointing toward delamination failures through loss of adhesive strength.

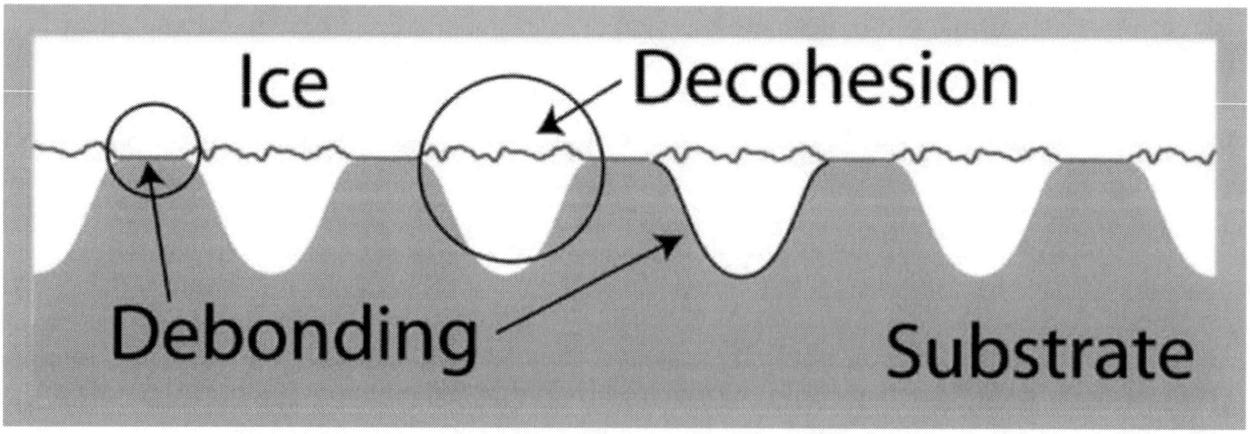

Figure 6 – Idealized graphic of cohesion vs. adhesion loss along an ice-rock interface. Image adapted from Fortin & Perron (5)

Based on published research, adhesion values of ice to various metal surface types can range between 10 and 300 psi (0.07 to 2 MPa) in the ice temperature range of 23 deg. F (-5 deg. C) to -4 deg. F. (-20 deg. C), respectively (5). Alternatively, research for aggregates used in pavement design shown an adhesive bond strength (under tension loads) of 15 to 36 psi (0.1 to 0.25 MPa) (Note: The studies cited this range were for ice adhesion to various surfaces including metal and gravel size pieces of granite and gabbro. It is expected that these adhesion values would be low for most rough bedrock surfaces. Intact cohesion values of fresh water ice can vary between 783 psi (5.4 MPa) and 1,363 psi (9.4 MPa) at ice temperatures between 32 deg. F (0 deg. C) and 3 deg. F (-16 deg. C), respectively (6).

Published literature (5, 7) indicates that at temperatures close to the freezing point, adhesion bonding will dictate tensile and shear interface strength. Loss of surface area due to melting effectively reduces the minimum bond strength needed for ice slab stability.

The frictional component of the interface shear strength comprises two components, the first being, base friction (small scale) afforded by grain roughness. The second component is large-scale roughness afforded by undulations in the slope surface. Natural or mechanical breaks, discontinuities and weathered surfaces all add to the large scale ("macro-scale") roughness, all of which would typically be visible from some distance away. Ice will tend to develop and prefer slopes with significant large-scale roughness profiles, including benches created by rock excavation and blasting activities.

Given the likely ice creep rate during melting along the ice-rock interface, base friction is expected to be overcome and will be very low. Furthermore, studies of ice-rock contact frictional strength appear to indicate significant spread in strength estimates; however, larger-scale roughness along the slope will increase shear strength due to dilation (overriding) of the slab over asperities during sliding. The contact surface of the sliding plane will break-down as sliding progresses. Macro-scale roughness may also result in the development of localized shear resistance pockets that can then be overcome in periods of rapid temperature change, resulting in a torrent of cascading ice material.

The base friction angle of fresh water ice has been shown by researchers to be temperature dependent and vary between 2 deg. at 32 deg. F (0 deg. C) to as much as 14 deg. at -40 deg. F (-40 deg. C) (8). The warmer temperatures and onset of melting associated with icefall would indicate that a very low base friction angle would apply, if used at all.

If we assume based on our field observations that icefall typically occurs during periods of warming, as noted above, then we can make a few simplifying assumptions about the shear strength of ice-rock contacts. Firstly, the values for base friction angle are expected to be very small at temperatures close to 32 deg. F (0 deg. C) (8, 9) such that we can neglect them in-lieu of larger-scale asperities. Secondly, delamination failures are most frequently observed in the icefall reviewed events, indicating a loss of adhesive shear strength. With these simplifications, the shear strength of an ice-rock contact surface can be written as:

$$\boldsymbol{\tau = A_v + \sigma_N \tan i}$$

Where τ is the ice-rock interface shear strength; A_v is ice adhesive strength; σ_n is the slab normal stress applied perpendicular to the surface; and "i" is the average asperity angle departure from the measured slope surface angle (in degrees).

Ice Slab Failure Mechanisms

The process of ice slab wasting on highway rock slopes is complex and degenerative, and is first characterized by loss of adhesive ice-rock contact shear strength, followed by periods of subsequent slab deformation as the structure tries to support its own self-weight. Ice mechanical strength is temperature dependent; however, in order for icefall to take place, general warming in the late winter or spring months is the typical case that would induce failures.

Slab Sliding at Source Zone

Failure commences from loss of shear strength along the ice-rock contact, as described above. Heating of bedrock substrate will speed up the rate of melting along contact even if air temperatures are close to freezing, and reduce adhesive bond shear strength along this interface. For slopes less than approx. 4V:1H (76 deg.), loss of interface shear strength and subsequent slab sliding appears to be the most likely failure scenario. A slab of ice failing by shear can be modeled as an equivalent block of weak rock sliding along an ice-rock contact surface, according to the equation shown in the preceding section. Once failure commences, frictional resistance helps to reduce ice slab velocity and energy; however, it can conservatively be assumed that micro-scale frictional contributions to shear strength along the rock-ice contact are effectively nil as they are likely very low and at the critical case, are reduced to the macro-scale roughness of the slope.

Slab Toppling

This type of rotational failure can occur if the slope face (or outside ice face angle) is between vertical and battered backwards into the slope (i.e. "overhung") such that the slab center of gravity falls outside the slab limits. As with shear failure case noted above, the ice-bedrock contact must be partially debonded in tension in order for the slab to release. Evaluations of this failure mode rely on adhesion bond tension strength and/or cohesive bond tensile strength. Toppling ice slabs are expected to be limited in overall thickness as ice strength is likely compromised at the onset of failure, resulting in crushing action as the ice rotates and falls. Likewise, the slab rotation arm length is expected to be reduced by the non-rigid nature of the ice as it rotates outward. In other words, low ice strength will reduce the horizontal extent of slab impact (the same could not be said for a rigid slab of rock, for example, which could rotate large distances before breaking-up). Large-scale topples

Figure 7 – Direct icefall impact & shatter at MP 13.9 Richardson Highway. Slab partially rotated outward during fall from overhung slope on 13 December 2017. Photo courtesy of AKDOT&PF.

would transition to pure block falls at relatively small distances away from the slope as shown in Fig. 7.

Slab Buckling

Slab buckling can occur in cases where the slab structural rigidity is low due to reduced cross-sectional area (i.e. from melting) and excessive slab heights. Reductions in ice strength due to melting, crushing and fracturing could also result in eventual buckling type failures. Much like other failures cases cited above, slab debonding must be present in order to allow for slab movement away from or down along the slope face. Ice strength is likely compromised at the onset of failure, resulting in crushing action as the ice is subject to deformation.

Slab Toe Failure in Weak Material (External Bearing Capacity Failure)

This failure mode would consist of loss of bearing support in soils, weak rock, or frozen ground underlying the ice slab. The slab would either "punch" through the soil and/or start sliding within the weaker substrate material. This mode of failure is expected to be infrequent.

The critical factor that ties all of the preceding scenarios together is that loss of ice-rock interface strength is required for initial displacement of the ice slab at the onset of failure. If strength of the ice-rock contact is not compromised, the release mechanisms that lead to global ice failure will not be initiated.

Icefall Impact Hazards

After the initial release, falling ice slabs will start to break-up either from "first impact" (direct strike on slope or other hard surface) or from internal slab deformation (e.g. rotating, crushing). The impacts discussed within this section are those within the "impact zone", which is the zone where falling ice blocks make contact with objects along their fall path after initial release and displacement.

Figure 8 – Direct icefall impact damage to bus near Terrace, British Columbia, courtesy of Terrace Daily Online, February 5, 2011.

As further described by Scarpato & Woodard (1), there are three icefall hazards that result after the ice slab failure and that are presented below, those being:

- Direct impact (primary);
- Impact shatter (secondary);
- Impact splatter (secondary)

Direct icefall impact hazards have the potential to be catastrophic depending on the location of initial impact. Direct impacts result from actual point-to-point contact with pavement, pipelines, utilities, or vehicles. Impact shatter results when an ice particle breaks-up upon initial contact with a substrate, like pavement, walls, rock outcrops, or a roadside ditch and casts material horizontally. Impact splatter results from initial ice block contact, where the substrate material (e.g. soil) yields and is sent travelling away from the point of impact.

Direct impact hazards for icefall can be similar to those for rockfall with respect to kinetic energy and collision damage. The 6 April 2012 icefall event near MP 113.2 NB along the Seward Highway was a direct impact event (Fig. 1). Another example of an icefall direct impact event is shown in Fig. 8, where a slab of falling ice impacted a tour bus outside of Terrace, British Columbia. Direct impact to the roadway was also observed more recently at approx. MP 13.9 SB along the Richardson Highway, although no vehicles were struck (Fig. 7).

PRELIMINARY ASSESSMENTS & FINDINGS

Our technical evaluations work consisted of a two-phased approach. The first phase was characterized by an initial "screening" evaluation to provide a way to differentiate impact risk at each of the seven sites, as each site has unique qualities. The initial risk screening was based on average daily traffic, site icefall history, catchment width and slope height. The screening evaluation results are summarized in Table 1 below and the full table entitled Preliminary Icefall Impact Risk Matrix (PIIRM) was included as Appendix 4 in the final report (3).

The "screening" table demonstrates that there are three specific sites with an elevated level of icefall risk to the roadway, those being MP 113.1 to 113.3 NB Seward Highway ("High"); MP 13.7 to 14.0 SB Richardson Highway ("Moderate to High"); and, MP 51.9 to 52.1 ("Low to Moderate"). The other four sites may develop icefall hazards and impart periodic (yet less frequent) icefall impact risks from year-to-year; however, these four sites tended to have significant catchment ditch widths with respect to slope height and unobstructed sight distances. So while the icefall hazard may exist, the actual impact risk to the highway is expected to be relatively low.

Site	Prelim. Risk Ranking
MP 113.1 – MP113.3 SEW	HIGH
MP 74.1 – MP 74.9 SEW	LOW
MP 57.9 – MP 58.2 SEW	LOW
MP 51.9 – MP 52.3 SEW	LOW-MOD
MP 38.3 – MP 38.5 RICH	LOW
MP 22.7 – MP 23.3 RICH	LOW
MP 13.7 – MP 14.0 RICH	MOD-HIGH

TABLE 1 - Abbreviated Prelim. Risk Ranking by Site

Based on the initial screening, more detailed site-specific technical evaluations were completed at locations of elevated impact risk. Technical evaluations at sites of "low risk" to the highway consisted of geometric evaluations only. Sites with an expected elevated level of risk (i.e. "Moderate" and "High") required technical evaluations which were more detailed, in addition to the geometric analyses completed for the other "Low" risk sites. The types of evaluations consisted of the following:

A. Simple Geometric Analyses – Assessment and comparison of observed (and probable) ice slab dimensions and roadway and catchment dimensions. Considered probable mechanisms of failure for falling, sliding or rotating slabs based on geometry. This includes geometric evaluation of known previous failure events;

B. Sliding Ice Block Analysis – Used equivalent rock block sliding approach using industry-available planar sliding analysis software called RocPlane© from Rocscience, Inc. The model runs allowed us to complete evaluations of ice-rock contact shear strength, including sensitivity analyses relative to adhesion strength and slope roughness.
C. Icefall Impact & Force Analysis – Included developing estimates of icefall weight, impact force and energy. Where possible, we attempted to "back-analyze" previous failure events to understand the magnitude of future icefall impacts.

Site-Specific Icefall Histories

Locations of known icefall events are included in Table 2 below along with ice height, direct impact and shatter horizontal distances. These data were used to develop an understanding of icefall frequency and impact distance by site location.

RECORDED EVENTS			ICE HEIGHT ABOVE ROAD(FT.)	PRIMARY IMPACTS			SECONDARY IMPACTS			
				DIRECT IMPACT(1) DISTANCE (FT.)			SHATTER DISTANCE (FT.)			SPLATTER (FT.)
MP	TYPE	DATE	MAX	MIN	MAX	AV.	MIN	MAX	AV.	MAX
113.2 SEW	SLIDE/FALL	4/16/2012	80	0	22	11	0	35	18	U
113.2 SEW	FALL	3/20/2017	85	4	6	5	6	55	31	N.O.
52.0 SEW	SLIDE	U	U	23	52	38	U	U	U	U
52.0 SEW	SLIDE	U	U	0	24	12	U	U	U	U
23.0 RICH	SLIDE/FALL	U	33	U	U	U	0	46	23	U
13.9 RICH	LOCAL TOPPLE/FALL	12/13/2017	20	4	22	13	22	54	38	N.O.

TABLE 2 – Icefall History & Estimated Impact Distances

Comparisons of Ice Development by Site

Based on our site observations, we can draw some preliminary comparisons relative to ice growth. Well-developed ice formations were observed at six out of the seven sites (86%). This included a total of 17 well-developed ice formations, as there were multiple discrete formations at certain sites. Ice coverage areas within the lateral extents described herein varied between <1% of the exposed slope surface (MP 14 Richardson Highway) to as much as 13.3% (MP 52 Seward Highway).

All of the well-developed slabs appeared to have been formed from the freezing of upslope water as it was intercepted by the slope face. Direct impact by precipitation (e.g. snow, rain) falling on the slope face did not appear to be a major source for significant ice development. Presence of fracture-controlled seepage appeared to be less significant in the development of ice than direct overflow of upslope surface water over the crest of the rock slopes, based on the following site observations:

- surface water discharge locations observed in September of 2016 were in close proximity to major ice features observed in March of 2017;

- Trickling water was heard behind (or within) major ice slabs observed in March 2017;
- Ice formation morphology (vertical columns or spires fused together) suggests steady-state source of water available for consistent cascading, freezing and outward growth of ice crystals;
- Re-freezing of upslope water generated from snow melt also appeared to be a causative factor in the addition of surface water to the slope system;
- Ice was found to have a blue hue at four of the seven sites (57%), which includes eight of the 17 (47%) specific ice slab features witnessed in March 2017. The blue hue indicates relatively thick ice growth as light is diffracted through the ice medium, and thickness is an indication of consistent addition of water.

MP 113.1 to 113.3 NB Seward Highway

To highlight how the technical approach was used as a basis for development of initial icefall mitigation options, we present an example of a site-specific evaluation at MP 113 northbound along the Seward Highway. This site was the location of the April 2012 impact event described above, and serves as a case study for this section of the paper. Remaining sites are described in further detail in the main report which is posted in its entirety to the TRB website (3).

Figure 9 *– Photo of ice conditions near MP 113.2 on 13 March 2017 (Photo by Scarptec, Inc.)*

The site is composed of a steeply cut south-southwest facing rock slope with slope heights ranging between 50 and 140 ft. Existing slope angles range between 70 and 86 deg. based on direct measurements of the slope; however based on available as-built plans, it appears that the slope is generally between 76 deg. (4V:1H) and 82 deg. (8V:1H). The slope appears to have been cut in "lifts", resulting in small lineations or remnant benches at approx. 30 ft. height intervals. These remnant benches provide efficient ice support ledges where ice can "hang-on" and support overlying upper slabs. Free water in the form of surface and fracture-controlled discharge is abundant at this site due primarily to upslope run-off from snow melt. Natural surface water drainage channels exist at the top of the slope, as witnessed in September 2016.

The available (i.e. "effective") catchment width between the toe of the slope and the paved shoulder is limited, and on the order of 8 to 14 ft. based on our measurements. Ditch depth ranges between 0 and 3 ft. depending on specific location. The sight distance available to motorists along this two-lane (12-ft. ea. lane) stretch of highway is on the order of 1,250 ft. and is generally sufficient to observe debris which is already in the roadway. There is also a 41-foot-wide viewing turn-out available for temporary lane closures or traffic pattern changes through a majority of the affected area, resulting in a total paved roadway width of approx. 90 feet. Ice development during the March 2017 visit is depicted on Fig. 9.

Geometric Analyses

We completed geometric evaluations based on slope height, ditch width and roadway width. Simple geometric analyses yielded the following findings:

1. Width of the catchment ditch from MP 113.1 to 113.3 was insufficient with respect to capture of potential icefall events. The existing catchment is between 8 and 14 ft. in overall width with slopes between 50 and 140 ft. in height.
2. The distance from slope toe to centerline of road (dividing line between lanes) is approx. 29 ft., resulting in a roadway that is too close to the slope. The closer the highway is to the slope, the higher the likelihood of direct impact events. Based on information from the 6 April 2012 icefall event, we estimated that the area of direct impact extended out approx. 22 ft. from toe of slope, which is within the northbound travel lane.
3. Based on the estimated volume ice of the 6 April 2012 icefall event, the available volume of the roadside ditch is inadequate for retention of fallen ice even if all the ice fell directly into the ditch.
4. This site is unique with respect to the other six sites, as the Average Annual Daily Traffic (AADT) is significantly larger (by upwards of three times) and the slope-parallel width of ice development can approach as much as 80 to 100 ft. The extensive ice development width can result in multiple ice shedding events throughout the effected interval.
5. In addition to the direct impact risk that exists at this site over the winter and early spring months, secondary impact shatter events could also periodically enter the roadway when icefalls do occur.

Figure 10 – Ice conditions along ditch near MP 113.2 on 13 March 2017 (Photo by Scarptec, Inc.)

Sliding Ice Block Analysis

We completed sensitivity analyses for ice sliding from loss of adhesive strength along ice-rock contact surfaces using RocPlane from Rocscience©. Although this is ideally intended for use with rock block sliding along an inclined plane, we assumed an equivalent rock block approach for sliding ice blocks, as the mechanics are similar. We used an interface adhesion shear strength value of 75 psi (0.5 MPa). What we found was that frictional properties, as expected, play a very small role in contributing to stability of sliding ice blocks. At temperatures approaching 32 deg. F (0 deg. C), the base ice friction angle (φ_b) is approx. 2 degrees. which is very low. Based on this, we chose to be "conservative" and assume that the frictional component of shear resistance is essentially zero. This is not an unreasonable approach, given that adhesion strength will be the dominant control on shear resistance at the moment of failure. This simplified the model and allowed us to focus on the major factor in ice slab stability – adhesion strength.

Factor of Safety (FS) is defined as the ratio of those forces resisting instability ("capacity") to those forces inducing ("driving") instability. This deterministic approach is frequently used with slope stability analyses, especially with preliminary studies. As described above, the shear strength along the ice-rock contact will be dominated by adhesive strength. If we use a relatively low end value cited in the literature for adhesive shear strength 75 psi (10,800 pounds per sq. ft.) and assume this is essentially a constant, then the intrinsic adhesion value per unit area during times of stability is the same during times of instability – meaning that failure will commence when the available adhesive contact area is minimized due to melting.

Loss of adhesive bond contact area due to melting will result in a reduction of ice slab shear resistance. This observation is unique to ice, as geotechnical slope stability analyses usually assume a relatively constant adhesive (or cohesive) contact area, but because ice is subject to melting, the contact area is actually reduced as the system tends toward instability. To assess the loss of net adhesive area at failure, we ran two separate analyses. We started with an initial adhesive shear strength of 10,800 psf, which resulted in safety factors that were very high as expected. We then "destabilized" the system by setting the FS equivalent to approx. 1.0, whereby the ice block is considered meta-stable and at the point of sliding. The required minimum shear strength along the slide plane at the onset of failure was found to be approx. 300 psf; however, the loss of contact area makes this an "apparent" adhesive strength. Because adhesion strength is assumed to be a constant property of ice at temperatures close to freezing, we back-calculated the loss of adhesive contact area and found that there was a reduction of approx. 96%.

The key takeaways from this ice sliding evaluation included the following:

1. Loss of apparent adhesive strength is actually due to loss of adhesive contact area, which is directly related to an increase in ice-rock interface temperature, with the bedrock surface heating up faster than the surrounding air mass due to solar radiation;
2. Even relatively small adhesive ice-rock contact "bridges" will support an ice slab given its relatively low density. The ice slab is also subject to melting internally, and so total weight is likely being reduced at the same time that the interface is melting; however, the interface is expected to melt faster than the overall ice mass given rock's higher coefficient of thermal conductivity;
3. Icefall events resulting from ice slab failures on the slope face do not happen instantaneously. Slab failure takes consistent input of direct solar radiation heating the ice-rock interface to temperatures at and above freezing.

Post-Failure Impact Energy & Force

Figure 11 – Damaged vehicle from 6 April 2012 icefall event near MP 113.2 on the Seward Highway (Photo by KTUU)

By our estimates based on photos, the falling slab of ice was approx. 60 to 80 ft. high by 20 ft. wide by 4 to 6 ft. thick. It appears that the direct impact zone extended approx. 22 ft. from toe of slope and maximum lateral extent of ice debris field was approx. 35 ft. from the slope toe (Fig. Nos. 1 & 11), just beyond the centerline of the road into southbound lane. It also appears that the this specific ice sheet failed in discrete sections, as there appeared to be other slabs that came down at different times. Based on our site observations, photographic evidence and recorded accounts from that day, we believe that the slab ultimately failed by loss of adhesive shear strength along the rock-ice contact in a period of increased average daily temperatures, resulting is a large cascading mass of falling ice. The failure event was likely a prolonged direct impact event lasting between 3 and 6 seconds as ice fell on itself and was subject to self-crushing at impact. In other words, this was not one discrete block but rather a series of slabs that delaminated vertically up the slope face. This resulted in significant damage to a small sports utility pickup truck (e.g. Ford® Ranger or equivalent), and it appears from photos that the truck was compressed vertically between 2.5 and 3.5 ft., as indicated in Fig. 11 above.

We estimated the impact force and energy of the 6 April 2012 event based on post-accident photos of the damaged vehicle and the ice blocks. An abbreviated version of the results is provided as Table 3 below. This analysis used kinematic equations for vertical fall of an equivalent ice block weighing 22.3 kips falling from an average distance (to center of mass) of 45 ft., and of equivalent plan dimension to the vehicle that was struck. Although ice also fell around and outside the limits of the truck, we can assume that the entire footprint area of the vehicle was impacted by falling ice. Based on vertical compression estimates indicated above, it is our opinion that this specific icefall event resulted in a vehicle impact force of between 175 and 225 kips with an impact energy (KE) of between 1,300 and 1,400 kJ.

To put the results in context, the KE associated with an automobile moving 60 mph is approx. 500 kJ, so this icefall impact event was certainly significant and on-par with what we could see with a large rockfall event. Given the magnitude of this icefall, and the frequency of ice recurrence at this site, the results were used to develop some of the mitigation strategies presented below.

			Fi = KE/S	
KINETIC ENERGY		STOP DIST.	IMPACT FORCE	
KE (KJ)	KE (FT-LBS)	S or δ (FT)	Fi (LBS)	Fi (KIPS)
1,361	1,003,990	1	1003990	1004
		2	501995	502
		3	334663	335
		4	250998	251
		5	200798	201
		6	167332	167
		7	143427	143

TABLE 3 - Estimate of Vehicle Impact Energy & Force

Mitigation Options

Given the site's icefall history, high traffic volume, low catchment capacity and high slopes, MP 113.1 to MP 113.3 has a high risk for direct impacts to the roadway. As such, we recommended consideration of the following measures for mitigation of the icefall impacts at the site:

1. Installation of Remote Active Onsite Monitoring – Although monitoring is not a specific mitigation method per se, it would allow the DOT&PF to observe the site and record conditions real-time. A weather station, slope sensor and camera type system could be installed adjacent to the slope so that ice development and behavior can be monitored. The data should be capable of being remotely-transmitted to decision makers so that further short- or long-term mitigation measures can be taken. The rock slope surface adjacent to the site's ice development location could also be instrumented with a pyranometer to measure incoming solar radiation and a temperature gauge to measure rock slope surface temperatures within 2 to 4 in. of the slope surface.

2. Icefall Mitigation Measures – Included consideration of the following icefall mitigation measures throughout the effected interval:

   A. Slope Excavation - Consider cutting the slope back by a minimum of 25 ft. This would allow for additional ditch catchment width (10) adjacent to the shoulder and would provide additional horizontal offset between the slope and the travelled roadway. The rock slope cut angle (and ditch width) would require design for rockfall and global rock slope stability, in addition to icefall. This solution could be implemented in conjunction with the drainage solution described below for maximum effect.

   B. Provide Upslope Drainage Diversion – This is an ideal long-term solution and would facilitate diversion of upslope drainage water (including meltwater) such that persistent upslope sources of surface water would not be captured by the slope crest. Diversion of drainage water will also help to reduce local incidence of rockfall.

C. Traffic Pattern Alteration – Would entail use of a modified traffic pattern that provides additional distance for potential failure of large ice slabs. This pattern would mitigate direct impacts of large-scale, relatively thick, well developed ice formations that constitute near-continuous vertical slabs covering large portions of the slope face. This proposed pattern is intended to account for the additional outward rotational component of slab failure due to interaction of ice blocks as they fall to the ground. We recommended that traffic be diverted further away from the slope and into an existing turn-out area and could result in a lane shift of approx. 24 ft. (two lanes).

3. Alternative Mitigation Measures – The methods shown below could be adapted at the site; however, they present some logistical challenges and are not used routinely as icefall mitigation measures:

A. Install Pre-Hung Drapery – The "ice drape" option is in concept an effective option, applicable to almost all field conditions. Wire rope anchorage points would be installed at the top of the slope in order to support the weight of the netting and assumed ice loads. Lower anchorage points and boundary cables would also be installed in order to prevent ice debris from exiting the drape and entering the roadway.

B. "Reinforced" Ice – Consists of installation of bars nearly perpendicular to the slope face in order to provide shear resistance along the ice-rock interface. The bars would need to be designed to resist shear and bending forces imparted by the ice, including minimum embedment. Additionally, the bars would need to be designed for maximum probable ice thickness at a given location.

MP 113.1 – MP113.3 SEW	
Unmitigated Risk Ranking	HIGH
Slope Excavation	LOW
Add Drainage Diversion	LOW
Add Prop. Traffic Pattern	LOW
Pre-Hung Drapery	LOW - MOD
Reinforced Ice	LOW - MOD

TABLE 4
Effect of Mitigation on Prelim. Risk

In order to show the effect of the mitigation options on icefall impact risk, we included a site-specific mitigated risk table within the final report. Table 4 below summarizes post-mitigation impact risk at MP 113.1 to 113.3, and the entire report table presents post-mitigated risk at the other sites referenced in this paper. This tool will allow planners to consider the effect that mitigation techniques will have with respect to the original unmitigated icefall hazard scenario.

Concluding Statements

Additional research, testing and field work is required to develop reliable icefall hazard mitigation techniques; however, upslope drainage diversion, slope excavation (rock slope and catchment ditch geometry) and partial traffic pattern or roadway geometry reconfigurations are expected to aid in significantly reducing the risk of direct icefall impacts. Slope monitoring in areas subject to persistent upslope drainage is important to predicting potential locations for ice development during winter and spring time periods.

REFERENCES:

1. Scarpato, D. & Woodard, M., Evaluation & Mitigation of Icefall Hazards for Civil Engineering Works, Proceedings from: *International Snow Science Workshop* (ISSW), Anchorage, Alaska. 2012.
2. Final Deliverable Report entitled: *Catastrophic Icefall Hazard Assessment, Avoidance Procedures & Mitigations Strategies-Phase I Literature Review*, February 2016. Doc. Link: http://www.dot.state.ak.us/stwddes/research/assets/pdf/4000-158.pdf
3. Final Deliverable Report entitled: *Catastrophic Icefall Hazard Assessment, Avoidance Procedures & Mitigations Strategies-Phase II Site-Specific Studies*, February 2018. Doc. Link: http://www.dot.state.ak.us/stwddes/research/assets/pdf/4000-168.pdf
4. Graveline, M., Germain, D., Ice-block fall and snow avalanche hazards in northern Gaspesie (eastern Canada): Triggering weather scenarios and process interactions, *Cold Regions Science and Technology*, December 2015.
5. Fortin, G. and Perron, J., Ice Adhesion Models To Predict Shear Stress At Shedding, *Journal of Adhesion Science and Technology*, March 2012.
6. Fish, A., Zaretsky, Y., CRREL Report, *Ice Strength as a Function of Hydrostatic Pressure and Temperature*, October 1997.
7. Saeki, H., Mechanical Properties Between Ice and Various Materials Used in Hydraulic Structures: The Jin S. Chung Award Lecture 2010, *International Journal of Offshore and Polar Engineering*, Vol. 21, No. 2, June 2011.
8. Ladanyi, B. and Archambault, G., Shear strength and creep of ice-filled indented rock joints, in: *Soil Rock America*, 2003.
9. Hamza, O., Laboratory measurement of the shear strength of ice-filled rock joints, *Annals of Glaciology*, January 2000.
10. Scarpato, D., Icefall Hazards Along U.S. Transportation Corridors – Are Rockfall Catchment Ditches Sufficient? Proceedings from: *66th Highway Geology Symposium*, September 2015, Sturbridge, Massachusetts.

Rock Slope Scaling Investigative Approach and Volume Estimation Method

John D Duffy
Yeh and Associates, Inc.
391 Front Street, Suite D
Grover Beach, CA 93433
805-440-9062 mobile
805-481-9590 office
jduffy@yeh-eng.com

Prepared for the 69th Highway Geology Symposium, September, 2018

Acknowledgements

The author would like to thank these individuals for their contributions in the work described:

Charlie Narwold, California Department of Transportation
Thomas Whitman, California Department of Transportation
Steve Balaban, California Department of Transportation
Bill Webster, California Department of Transportation

Disclaimer

Statements and views presented in this paper are strictly those of the author, and do not necessarily reflect positions held by their affiliations, the Highway Geology Symposium (HGS), or others acknowledged above. The mention of trade names for commercial products does not imply the approval or endorsement by HGS.

ABSTRACT

Rock scaling is generally defined as the removal of loose rock from a slope using hand tools, small explosive charges, pry bars, and other mechanical methods. Scaling work is typically performed by the contractor under force account or change order, by in-house maintenance forces, and/or specialty contractors. The demand for scaling rock slopes has been steadily increasing within the past few years. Rock cuts are a recognized asset that have the potential to generate rockfall that can impact traditional hard, constructed assets (e.g., pavements, drainage structures, walls). Scaling is an effective method to manage rockfall, reduce damage to roadways and structures and improve traveler safety.

While techniques and procedures used to conduct scaling operations have been steadily improving, methods to quantify rock scaling have lagged behind. More and more, designers are faced with the challenge of quantifying and preparing specifications for scaling projects. Most often designers rely on geotechnical professionals to recommend scaling and identify the areas to be scaled. These recommendations are commonly based on limited information or an approximate area to be scaled. As a result, there is often not enough information to provide a reasonable estimate of the volume of rock to be removed to bid the scaling work with any degree of certainty. Rock scaling projects need to be investigated like other geotechnical projects (i.e. rock excavations, foundation recommendations, landslide repairs) with analytical methods and recommendations that allow the designer to prepare plans, specifications and construction cost estimates that reasonably reflect the nature and scope of the scaling work. This paper presents a method that geo-professionals can use to assess and quantify rock scaling operations through a specific field investigation based on nationally recognized rock condition descriptors from Federal Highway Administration, the California Department of Transportation, the Rockfall Hazard Rating System, and the Geological Strength Index classification systems.

INTRODUCTION

Rock scaling or slope scaling is generally defined as the removal of loose rock from a rock slope face. This method is largely regarded as a temporary mitigation measure to reduce the potential and frequency of rockfall from the slope, and sometimes as a stabilization method. A rock slope that is scaled will typically need to eventually be re-scaled either as part of a planned or programmed activity or reacting to an emergency rockfall event. In either case, the scaling work is either done in-house by trained maintenance crews, under an emergency contract or is included as an item of work in a PS&E (plans, specifications and estimates) package that is competitively bid. The demand for scaling slopes has been steadily increasing within the past few years. Scaling techniques are improving as are the results of the scaling operations. What has not been keeping pace, are methods to reliably estimate the approximate quantity of material that will be scaled and the time it will take to scale a slope. The challenges associated with not having a known volume of scaled material and an idea of the duration of the scaling operations became quite evident during the 2018 Transportation Research Boards (TRB) "Rock Scaling" workshop sponsored by the Engineering Geology committee (*1*) and the Rockfall Management Subcommittee (*2*). This workshop, attended by practitioners from the construction business, consulting community, and government agencies, consistently asked the same, simple question: "How do we specify scaling projects for bid?"

CURRENT GEOTECHNICAL PRACTICE

In practice, most preliminary geotechnical investigations recommend options for mitigation that might include avoidance, stabilization, protection or management measures. Whatever solution is selected, often through a complex process involving various stakeholders, a final report is prepared detailing recommendations for the chosen solution. If the solution is a wall, then a structure investigation is required; if the solution is a slide removal then a landslide investigation is required and so forth. All too often when a scaling solution is recommended, no additional detailed rock scaling investigation is performed. The result is that the design team relies on anecdotal information and experiences of the geo-professional to render a guess at scaling volumes and duration.

Methods to quantify the scope and volume of rock that will be removed by scaling have remained subjective and almost exclusively rely on the experience (or lack thereof) of contractors, engineering geologists and maintenance engineers to guess about how much volume of rock will be removed during scaling operations and the duration of the scaling operations. And while for years this has proven to be adequate or folks were willing to pay the price for not having specific information, the increased demand for rock scaling has often made it unacceptable to have overruns in cost, schedule and extended road closures to perform scaling. Designers are unable to respond to the challenge of quantifying and preparing specifications for scaling projects that include a reasonable quantity of the time to complete the work and volumes for the work. These challenges arise because the "experience factor" that most often rock scaling recommendations are based on is backed by limited field data and not enough information to provide a reliable quantity to bid the scaling work with any degree of certainty.

CURRENT SCALING PRACTICE

Scaling operations can be classified into two categories: mechanical scaling and hand scaling (3). **Mechanical scaling** utilizes construction equipment such as cranes dragging heavy weighted materials across a slope, excavators crawling on a slope or suspended by cables onto a slope face, or long-reach excavators raking a slope face from the base of the slope (right side of Figure 1).

Figure 1: Mechanical Scaling and Hand Scaling. (Photos Courtesy of the Transportation Research Board)

Hand scaling is defined as the removal of loose rock using hand held equipment, which includes, scaling bars, air bags, hydraulic jacks, etc. (left side of Figure 1). Hand scaling operations are performed by specially trained personnel working in teams or crews. Scalers, as they are commonly called today, have special training and experience in slope access using ropes (*4*), slope site assessment procedures (*5, 6, and 7*), and an understanding of fundamental principles that govern the stability of a rock slope face.

The most common and widely used scaling method today is to use a team of specialized scaler personnel who are skilled in the use of rope access, hand tools and air bags. This method can be very effective and can significantly reduce unnecessary excavation and further destabilization on the slope that can occur when using more aggressive mechanical scaling methods involving heavy machinery and untrained non-scaler personnel.

RECOMMENDED ROCK/SLOPE SCALING INVESTIGATION

Rock scaling projects need to be investigated like other geotechnical projects (i.e. rock excavations, foundation recommendations, landslide repairs). Rockfall investigation methods are well defined in the TRB rockfall book Part 2, Chapter 6 "Site Characterization" (*3*). The rock scaling investigation is not necessarily an expensive and time-consuming process. The investigation does require a geotechnical study as would any other type of mitigation requiring geotechnical expertise. The main goals of the rock scaling investigation are as follows;

- Site Investigation
    - Identify the area or areas to be scaled
    - Describe the rock conditions of each area
    - Establish the scaling method required to scale each area
- Quantifying Volumes and Hours
    - Estimate the quantity of material to be scaled
    - Estimate the scaling hours required for each area

Rock scaling operations may focus on a specific unstable feature or may be applied broadly over a large slope area. The slope should be examined in enough detail to identify these areas. There might be several areas for each site and each area could have differing rock conditions that require different scaling methods. This investigation often requires rope access to carefully measure and evaluate each area. Radar imaging and photogrammetry are also useful tools for assessing a slope, but field verification of that type of information is still required. The areas that need to be scaled should be delineated on the plans to show the limits that need to be scaled both along and above the road.

Table 1: Comparison of Four Nationally Recognized Methods to Describe Rock Conditions

	FHWA	GSI	Caltrans	GSI	Caltrans	RHRS	RHRS	FHWA	Caltrans	RHRS	RHRS
1	Rock Material Strength	Strength	Hardness	Structure	Structure	Structural Condition Case 1	Rock Friction Case 1	Weathering	Weathering	Structural Condition Case 2	Difference in Erosion Rates Case 2
2	Extremely Strong Rock	Extremely Strong	Extremely Hard	Intact or Massive	Unfractured	Discontinious Joints, Favorable Orientation	Rough, Irregular	Fresh	Fresh	Few Differential Erosion Features	Small Difference
3	Very Strong Rock	Very Strong	Very Hard	Blocky	Very Slightly Fractured	"	"	Slightly	Slightly	"	"
4	Strong Rock	Strong	Hard	Very Blocky	Slightly Fractured	Discontinious Joints, Random Orientation	Undulating	Moderately	Moderately	Occasional Differential Erosion	Moderate Difference
5	Medium Strong Rock	Medium Strong	Moderately Hard	Blocky/ Disturbed/ Seamy	Moderately Fractured	Discontinious Joints, Adverse Orienattion Orientation	Planar	Highly	Moderately	Many Differential Erosion Features	Large Difference
6	Weak Rock	Weak	Moderately Soft	Disintegrated	Intensely Fractured	Continious Joints, Adverse Orienattion	Clay Infilling or Slicensided	Completely	Intensely	Major Differential Erosion Features	Extreme Difference
7	Very Weak Rock	Very Weak	Soft	Laminated / Sheared	Very Intensely Fractured			Residual Soil	Decomposed		
8	Extremely Weak Rock	Extremely Weak	Very Soft								

Steep rock slopes and certain types of rock are prone to rapid degradation through ongoing relaxation and weathering, exposure to severe climatic conditions, and gravity. To describe these effects on rock slope faces and develop a system to categorize those slopes as they relate to scaling, four nationally recognized methods to describe rock conditions were reviewed: the Federal Highway Administration evaluation of soil and rock properties (*8*), the California Department of Transportation soil and rock logging manual (*9*), the Rockfall Hazard Rating System slope condition descriptors (*10*), and the Geological Strength Index classification system (*11*). A comparison of the four methods is shown in Table 1. For this study the California

Department of Transportation soil and rock logging manual (*9*) rock condition method was used to describe the rock conditions as observed on slopes.

The field work should also include an assessment of the expected rock dimensions to be scaled. An average and maximum size rock should be identified and categorized in ½-foot to 1-foot increments. Specific outcrops or individual boulders that are to be scaled should be identified. These features requiring individual attention by the scalers should be located on the plans and should be carefully described and measured to determine an area and a volume.

It is equally important to make a determination and specify the type of scaling (hand or mechanical) that would be required or allowed for each area identified for scaling. Specifics are necessary for each area of the slope. One area may need standard hand scaling, another air bags, another a long reach excavator.

ESTIMATING ROCK SCALING VOLUMES AND SCALING HOURS

Twenty eight sites where scaling operations were performed for projects in California were evaluated to develop the methodology presented in this study (Figure 2). Scaling data were provided by the California Department of Transportation (*12, 13, and 14*). The data included the rock slope location, the volume of rock that was removed from the slope by scaling, and the number of scalers used to perform the scaling. Each site was located on Google maps (Figure 2). The Google maps tool was then used to measure the area of the slope that was scaled. Six locations were field verified by the author in 2018; most of the sites at one time or another were visited by the author over the past 20 years, and in many cases the author participated in the scaling operations at these sites. For these 28 project sites, approximately 1,214,700 square feet of slope were scaled, 3,143 cubic yards of rock were generated by the scaling, and 1,416 hours were spent by personnel to perform the scaling. The areas that were scaled ranged in size from 2,500 ft^2 to 130,000 ft^2 of slope face. Volumes scaled from the slopes ranged from 7 yds^3 to 536 yds^3.

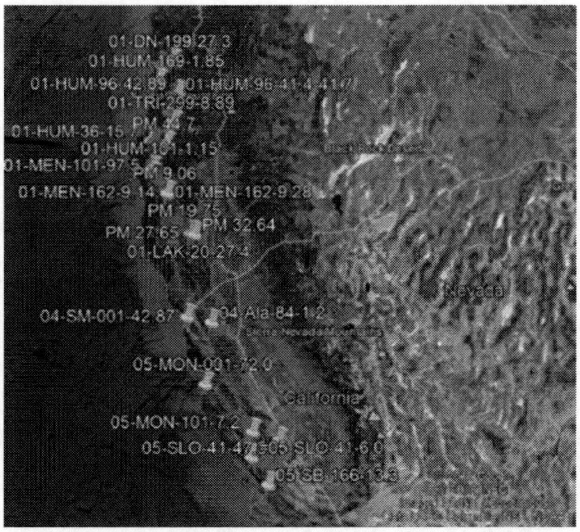

Figure 2: Location and Distribution of Rock Scaling Locations. (Google Maps)

The scaling at these sites was performed by state forces experienced and specially trained for scaling operations. **The scaling operations were performed using hand scaling techniques only. No air bags or any other methods were used.** The volumes were estimated based on the number of loader buckets to load a truck and the number of truck loads that were used to haul the material away. The scaling interval (time spent by the work force actually scaling material from the slope), while intended to be 15 minutes on and 15 minutes off, was varied depending on traffic volumes. However, an estimate of scaling effort of ½ cubic yard per hour per scaler is still considered reasonably accurate for these locations (*12, 13, and 14*).

The scaling operations at these sites took place in the Coast Ranges and Klamath Mountain Geomorphic Provinces (Figure 3). The Coast Ranges are generally composed of thick Mesozoic and Cenozoic meta-sedimentary strata (*15*). In several areas, Franciscan rocks are overlain by volcanic cones and flows (*15*). The Klamath Mountain province is considered a northern extension of the Sierra Nevada province composed of massive granites and metamorphic rocks (*15*).

Figure 3: Geomorphic Provinces. (6)

For each site an evaluation of the rock condition was made based on the California Department of Transportation soil and rock logging manual. The rock conditions in the project areas scaled generally consisted of slightly, moderately, to intensely fractured rock (Table 1). All the slopes had some rock condition variability. The condition that comprised the majority of the slope was used as the governing rock condition.

SCALING VOLUMES

The slope area for each site was measured to establish the square footage (ft^2) that was scaled. Then, assuming the average rock size removed during the "hand" scaling operation was 1-foot in dimension, a depth of 1 foot applied over the slope face was multiplied by the area square footage to establish a maximum potential rock volume available for removal by scaling. This value, referred to as the "potential scaling volume (PSV)", represents the volume (cubic yards) of rock that could be generated if the top one foot of material was removed across the entire slope surface. The actual amount of material scaled during each scaling operation was then divided by the PSV to establish a scaling-factor (k_s) for each slope. A rock condition was determined using the California Department of Transportation soil and rock logging manual

(Table 1) for each slope. Three trends within the data set were realized as shown in Table 2 which reflected the three rock conditions.

Table 2: Rock Condition Description and Associated Scaling Factor	
Rock Condition	**Average Scaling Factor (k_s)**
Slightly Fractured	0.05
Moderately Fractured	0.14
Intensely Fractured	0.26

For this data set the potential hand scaling volume (PSV) can be estimated by measuring the area of the slope to be scaled and then multiplying that area by a depth representative of the average rock size to be scaled. The slope rock condition is identified and the corresponding scaling factor (k_s) is selected. Multiplying the two results calculates an estimate of the volume of rock to be scaled (V_s).

Eq. 1: $$V_s = PSV * k_s$$

SCALING HOURS

The number of scalers used to perform a scaling operation was converted to scaling hours by assuming the work was performed in one 10-hour day with traffic control interruptions. A typical scaling day consists of approximately 1 hour of preparation time, 1 hour of demobilization, and 8 hours of hand scaling (*12, 13, and 14*). With a typical schedule of 15 minutes of active scaling and 15 minutes of a stoppage time to allow traffic to pass for each scaler: the actual time for scaling averages 4 hours per day. By dividing the total volume of rock scaled from each slope by the total hours used to perform the scaling, a volume expressed in cubic yards per hour per scaler (V_{hour}) was calculated for each slope. Hand scaling hours were generally similar for all rock conditions. The data indicate that a hand scaler, working with hand tools and pry bars, can scale an average of approximately ½ cubic yard of rock from a slope per hour per scaler. The average weight of the rock encountered in this study is 2.12 tons per cubic yard, which equates to a single scaler using pry bars and hand tools removing 2,190 pounds of rock per hour per scaler. A scaling crew of three is removing approximately 6,570 lbs. of rock each hour the approximate equivalent of 1 ½ cubic yards of rock an hour and generating a total of approximately 12 yards per day.

Once the volume of rock to be scaled (V_s) is estimated, the volume of rock should be divided by the average cubic yards per hour per scaler (V_{hour}) to estimate the time (T_s) it will take to scale the slope.

Eq. 2: $$T_s = \frac{V_s}{V_{hour}}$$

CONCLUSION

It's time to realize that rock scaling mitigation requires a disciplined investigative approach. With the information presented in the geotechnical rock scaling report, as outlined herein, a designer can scope an appropriate quantity, cost, time, support equipment and plan for disposal of removed material for the scaling operation. Equations 1 and 2 allow the designer to prepare area plans and specifications that include a reasonable quantity and cost for rock scaling. Bid items might include an item for mobilization, scaled rock per cubic yard, and disposal of scaled rock per cubic yard. The time for the project should be estimated considering a traffic control schedule. Equations 1 and 2 can also be used by construction managers to provide a basis by which to measure the reasonableness of a scaling operations progress or lack of progress and potentially head off potential change orders and overruns.

The methodology of estimating scaling hours and volumes was discussed at length with practitioners from construction, maintenance, and geotechnical that are experienced at all levels of rock scaling operations (*12, 13, 14, 16, 17 and 18*) and the feedback from those practitioners was used to verify the numbers made sense.

What is the average amount of material a hand scaler can actually scale per hour per scaler? There are in fact physical limitations of what a person can move in a certain amount of time. When averaged over a full day of scaling with traffic, lunch, and rest breaks by all accounts it was agreed that a ½ yd^3 per hour per scaler (2,190 lbs.) for hand scaling made sense and was by everyone's experience a realistic number.

What percentage of rock on the slope will actually be scaled? Estimating quantities for scaling rock slopes has been elusive and unreliable. The method presented provides a tool to estimate scaling quantities. Three typical categories of rock conditions that frequently require scaling are identified and three correlative scaling factors have been established to estimate scaling quantities. On a rock slope comprising slightly fractured rock, approximately 5 % of the potential scaling volume would be scaled. On a rock slope comprising moderately fractured rock, approximately 14 % of the potential scaling volume would be scaled. On a rock slope comprising intensely fractured rock, approximately 26 % of the potential scaling volume would be scaled. When compared against various scaling projects, the method made sense and seemed realistic. It must be understood that the estimates of scaling quantity and time made using these methods is a mean or average value and is for hand scaling only.

While these numbers are based on only 28 hand scaling projects, consider this: for a site 100 feet wide with a slope length 100 feet long, the area is 10,000 ft^2, and for an average rock size of 1-foot in dimension the PSV is 370 yds^3. Even if the mean scaling factor values represent a variance of +/- 20 %, the volume estimates for scaling would be approximately 10 yds^3 over or under, representing less than a day's work for a 3-person scaling crew. Those estimates should be reasonable and appropriate to provide the input needed to prepare and administer a construction contract: a bid item for the quantities to be scaled, a volume that can be used to help identify suitable disposal sites, and a duration for the work. This information should help to provide information to prepare a fair bid for scaling and allow the owner and contractor to reasonably track the progress of the work.

REFERENCES:

1. Ortiz, Ty, *Chairman Committee on Engineering Geology*, Transportation Research Board, https://sites.google.com/site/trbcommitteeafp10/Welcome, 2018.
2. Arndt, Ben, *Chairman Sub Committee on Rockfall Management*, Transportation Research Board. https://sites.google.com/site/trbcommitteeafp10/Welcome/rockfall-subcommittee, 2018.
3. Turner A.K., Schuster, R.L., *Rockfall Characterization and Control*, Transportation Research Board, National Academy of Sciences, Washington, D.C., www.TRB.org/ Rockfall, 2013, pp. 469, 580.
4. Professional Climbing Instructors Association, *Slope Access Technician Program,* Bangor, Maine, https://pcia.us/, 2018.
5. Single Rope Access, Association of Geohazard Professionals, *Rope Access Committee,* http://www.geohazardassociation.org/committees/, 2018.
6. California Department of Transportation, *Bank Scaling and Rock Climbing Manual,* Sacramento, California, http://www.dot.ca.gov/hq/esc/geotech/references/Rockfall_References/13_Caltrans_Bank_Scaling_and%20Rock_V9.1_02-14-14.pdf, 2017.
7. Duffy, J.D., Ingraham, P., *Single Rope Access, Association of Geohazard Professionals,* Highway Geology Symposium, Colorado Springs, Colorado, http://www.highwaygeologysymposium.org/wp-content/uploads/67_HGS-OPT.pdf, 2016.
8. FHWA, *Evaluation of Soil and Rock Properties FHWA-IF-02-034*, Geotechnical Engineering Circular No. 5, AFWL-TR-65-115, http://isddc.dot.gov/OLPFiles/FHWA/010549.pdf, 2002.
9. Office of Geotechnical Design, Caltrans Bank Scaling and Rock Climbing Manual, California Department of Transportation, http://www.dot.ca.gov/hq/esc/geotech/references/Rockfall_References/13_Caltrans_Bank_Scaling_and%20Rock_V9.1_02-14-14.pdf, Accessed April 8, 2016.
10. Pierson, L.A., Van Vickle, R., 1993, Rockfall Hazard Rating System, National Pooled Fund Study FHWA-SA-93-057, Oregon department of Transportation, Salem, Oregon, http://isddc.dot.gov/OLPFiles//FHWA/009767.pdf.
11. Marinos, P., and Hoek, E., *GSI: A Geologically Friendly Tool for Rock Mass Strength Estimation,* GeoEng 2000, International Conference on Geotechnical and Geological Engineering, Volume 1, Melbourne, Australia, 2000, http://lib.hpu.edu.cn/comp_meeting/ICGGE%B9%FA%BC%CA%B5%D8%D6%CA%B9%A4%B3%CC%BB%E1%D2%E9%C2%DB%CE%C4%CA%FD%BE%DD%BF%E2/PAPERS/INVITED/MARINOS.PDF.
12. Balaban, Steve, *Personnel Communication*, California Department of Transportation, San Luis Obsipo, California, 2018.
13. Narwold, C., *Personnel Communication*, California Department of Transportation, Eureka, California, 2018.

14. Whitman, Tom, *Personnel Communication*, California Department of Transportation, Oakland, California, 2018.

15. California Department of Conservation, *California Geologic Provinces - Note 36*, California Geological Survey, Sacramento, California, http://www.conservation.ca.gov/cgs/Documents/Note_36, 2002.

16. McNeal, Brian, *Personnel Communication,* Access Limited Construction, Oceano, California, 2018.

17. Ortiz, Ty, *Personnel Communication*, Colorado Department of Transportation, Denver, Colorado, 2018.

18. Webster, B., *Personnel Communication*, California Department of Transportation, Sacramento, California, 2018.

What Were We Thinking

A Case History of Extreme Slope Scaling in Washington

Marc Fish
Washington Department of Transportation
1655 South Second Avenue
Tumwater, WA 98512
360-709-5498
fishm@wsdot.wa.gov

James Struthers
Washington Department of Transportation
1655 South Second Avenue
Tumwater, WA 98512
360-709-5409
struthj@wsdot.wa.gov

Mike Mulhern
Washington Department of Transportation
1655 South Second Avenue
Tumwater, WA 98512
360-709-5583
mulherm@wsdot.wa.gov

Prepared for the 69th Highway Geology Symposium, September, 2018

Acknowledgements

The author(s) would like to thank the individuals/entities for their contributions in the work described:

Brad Schut – Washington Department of Transportation

Jerry Wood – Washington Department of Transportation

Bob Hooker – Washington Department of Transportation

Andrew Whitmore – Triptych Construction, LLC

Disclaimer

Statements and views presented in this paper are strictly those of the author(s), and do not necessarily reflect positions held by their affiliations, the Highway Geology Symposium (HGS), or others acknowledged above. The mention of trade names for commercial products does not imply the approval or endorsement by HGS.

ABSTRACT

US12 is one of three east–west highways that is open year round and crosses the Cascade Mountains. Near MP 164, the highway is situated between a 1000-foot high rock slope that sheds frequent rockfall and an environmentally sensitive lake. The slope is composed of andesite bedrock with blocky structure that bounds detached blocks up to 40 feet in size. The construction contract included 900-crew hours of slope scaling, 12,600 cubic yards (CY) of debris removal, an energy-absorbing blanket with concrete barrier, and a detour around the slope. The Washington Department of Transportation (WSDOT) anticipated that by project completion two-thirds of the slope would be scaled, the risk of rockfall would be reduced, and nearly 5,400 CY of rock would be in the lake. Nearing project completion, 900-crew hours of scaling was exhausted, twenty-five percent of the slope had been scaled, 5,200 CY of debris was on the highway, and around 250 CY of rock was in the lake. As winter was approaching, WSDOT had serious concerns over leaving potentially unstable rock on the slope, but reopened the highway after careful consideration. Since the completion of this project, rockfall has continued, the slope has required continuous monitoring during several winter storm events, and there has been one reported rockfall incident of a softball-sized rock striking a vehicle. This project has left WSDOT with several important questions to ponder before attempting another project of similar magnitude, most importantly, whether or not a slope of this size should be scaled at all.

INTRODUCTION

US12 is one of three highways that is open year round and crosses the Cascade Mountains in Washington. Between MP 164.55 and 164.86, approximately 14 miles east of White Pass, the highway is sandwiched between Rimrock Lake to the south and a near vertical rock slope to the north (Figure 1). The slope is about 1600 feet long and 1000 feet high, with an inadequate ditch that cannot contain most rockfall (Figure 2). Rockfall routinely originates from areas outside of the Washington Department of Transportation's (WSDOT) right-of-way (ROW). Maintenance indicated that this slope regularly produces small quantities of rockfall, typically consisting of

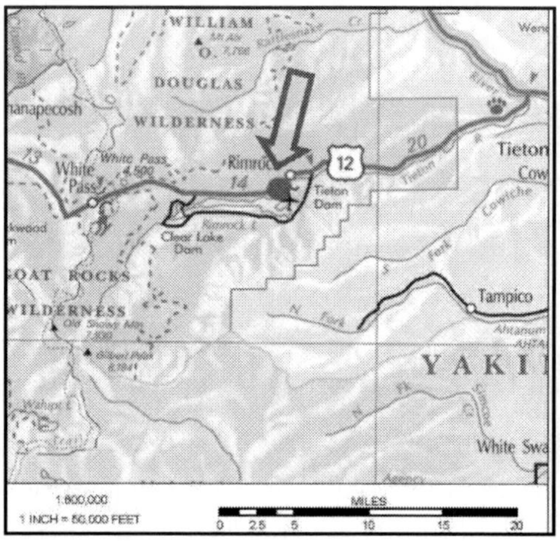

Figure 1: Site vicinity

Figure 2: Slope dimensions

one or a few rocks, 0.5 feet to 3 feet in size, that affect both travel lanes of the highway. Rocks typically hit the highway and often bounce over the eastbound guardrail and into the lake below. Observation of rock debris on the frozen lake surface in winter confirms that natural rockfall reaches the lake (Figure 3). Maintenance also indicated that 1 to 2 inches of ice, and occasionally up to 1 foot, accumulates on portions of this slope (Figure 4). One to two feet of snow has been

Figure 3: Rockfall on the frozen surface of Rimrock Lake.

observed on small sections of the steeper portions of the slope, and several feet of snow usually accumulates on the upper flatter portions of the slope after large snow events. Snow avalanches that originate on the upper portions of the slope have previously covered both travel lanes of the highway with 10 to 20 feet of snow (Figure 5). Previous rockfall events have covered both travel lanes of the highway with up to 1 foot of debris for approximately 100 feet laterally and several larger rockfall events containing rock, snow, and ice have deposited several hundred cubic yards of debris per event onto the highway (Figure 6). Maintenance conducts daily rock patrols beneath this slope and they clean the ditch twice per year, removing blocks up to 3 feet in diameter.

Figure 4: Ice build-up on slope.

Figure 5: Avalanches along the highway below chute 3.

Figure 6: Simular rockfall to previous events. This location is just east of the project location.

The original 2007 conceptual design for this slope included intensive scaling and the installation of rock bolts and dowels. The conceptual design, however, had access, environmental, traffic control, and utility constraints that were not considered at the time of the 2007 project scoping. The project was delayed until these constraints could be addressed, and the available funding (~$1.8 million construction) indicated that the original scope of the project needed to be reduced to remain within the programmed funding level.

GEOLOGY AND SLOPE MORPHOLOGY

This near vertical, unstable rock slope consists of intrusive andesitic bedrock that appears to be three separate flows that dip in an easterly direction (Figure 7). The lower and middle flows exhibit crude columnar structure and the top flow exhibits blocky structure with blocks sizes up to 40 feet in size (Figures 8 and 9). The lower and middle flows are only visible on the western half of the slope. An irregular slope provides for many rockfall launching features. Adversely oriented fractures and joints throughout the slope form planar, wedge, and toppling-type failures. We have identified three waterfall areas, or chutes, on the lower half of the slope that flow water most of the year (Figures 9 and 10). This slope emits rockfall, ranging in size between 0.5 and 3 feet, that originates high on slope and affects both travel lanes of US12. The existing ditch is up to 14 feet wide, at times is nearly 2 to 3 feet deep, and it provides limited rockfall catchment (Figure 11). The ditch is unable to contain the larger rockfall, which sometimes travel over the highway and into the lake below.

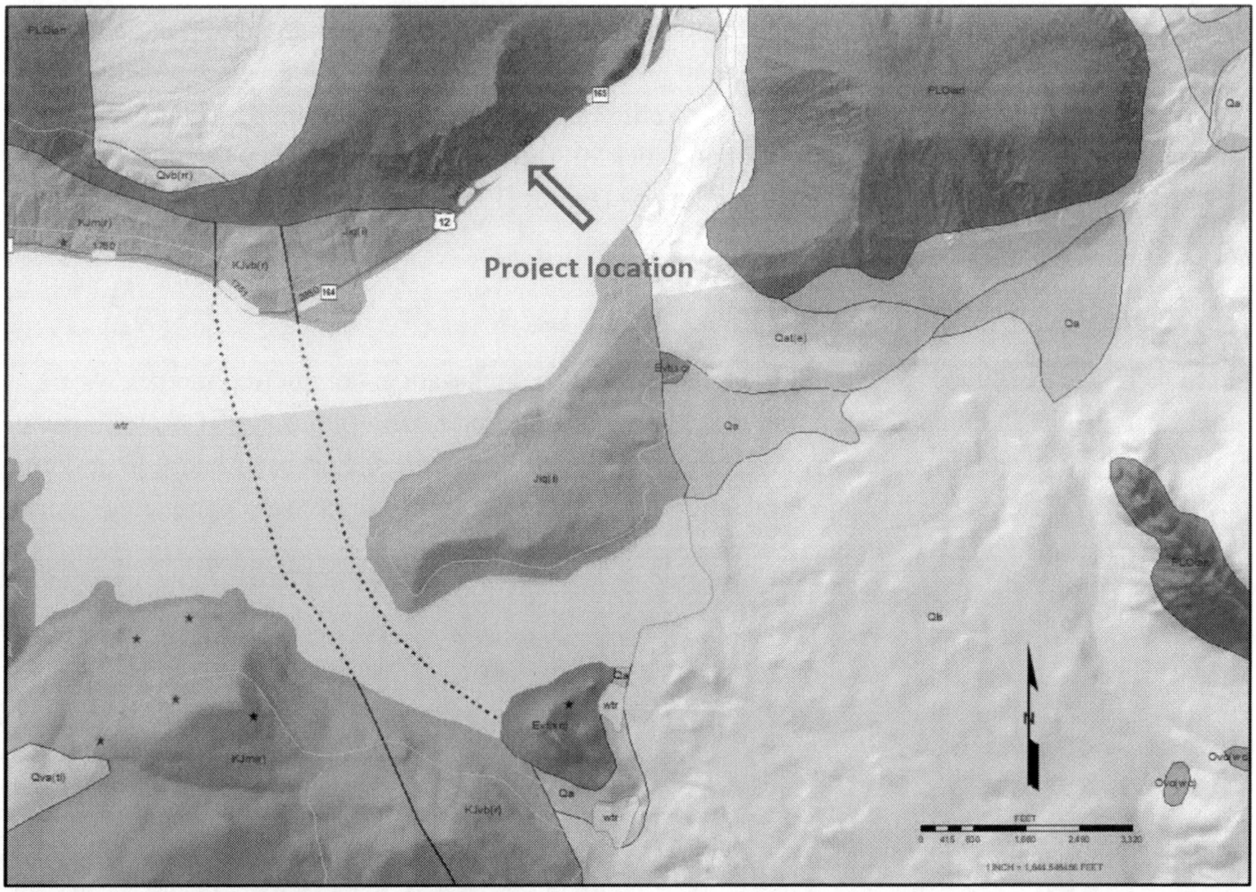

Figure 7: Geology map of project area.

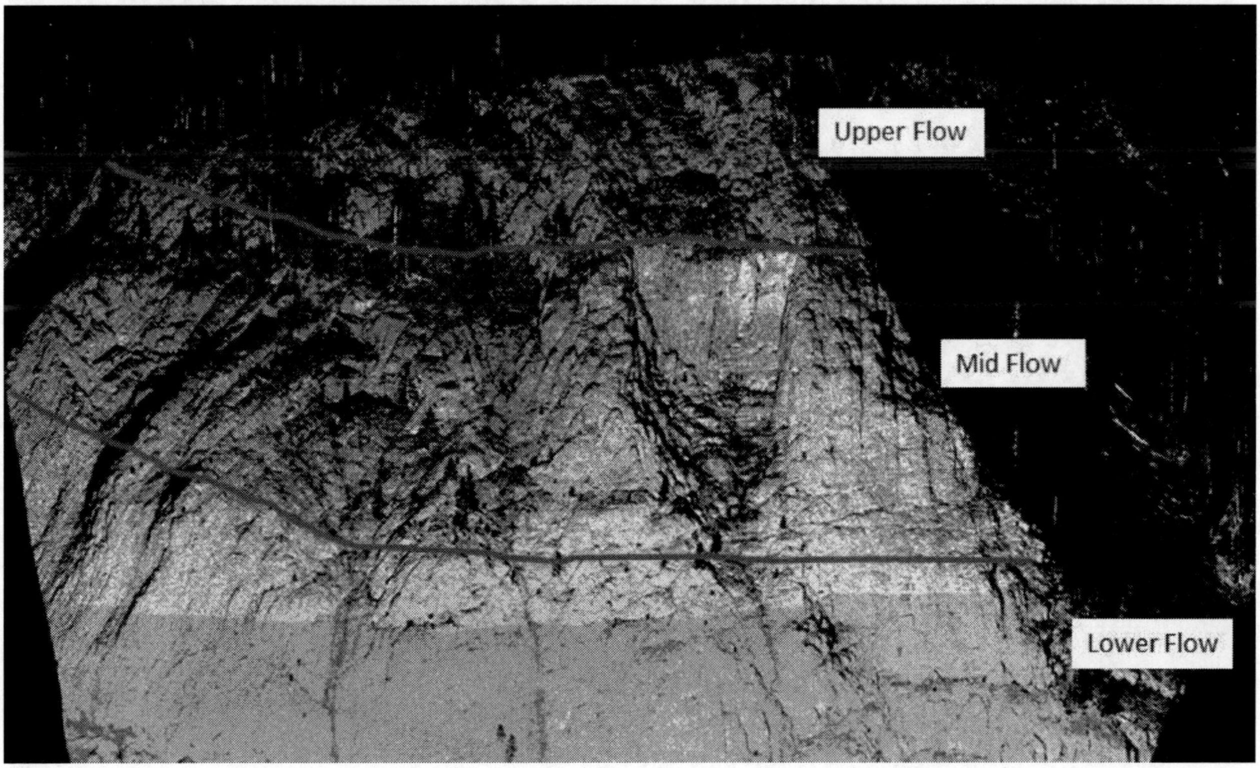

Figure 8: Approximate locations of lower, mid, and upper flows showing crude rock structure.

The lower quarter of the slope is steep, is oriented from 60 degrees to near vertical, and the upper portion has a somewhat flatter orientation that typically ranges between 35 and 50 degrees. The slope consists of a large bench area, located near one-half the height of the slope, which divides the upper and lower portions of the slope. On the upper portion of the slope, we observed numerous potentially unstable blocks, many rockfall origination points appearing as freshly exposed faces, and several chute and bowl areas (Figure 12). The irregular slope has many rockfall launching features (Figures 8 and 9).

GEOTECHNICAL INVESTIGATION

WSDOT's geotechnical investigation included site reconnaissance, conducting interviews, measuring slope heights, lengths, and block orientations, and taking photographs from across the lake, the crest of the slope, and from a helicopter. Due to difficult site access issues, an on-slope inspection was not performed. Several site reconnaissances to this slope were conducted. During

Figure 9: The critical cross section is down chute 3.

Figure 10: Surface water runoff traveling down chute 3.

Figure 11: A view of the limited ditch.

Figure 12: Upper slope area displaying recent rockfall origination points.

each site reconnaissance, recent rockfall was observed in the ditch, in conjunction with recent pavement divots below the slope, and rocks along the eastbound shoulder of the highway (Figures 13 and 14). From the highway, only the lower half of the slope is observable, but from across the lake the entire slope is visible. From the other side of the lake, the WSDOT GeoMetrix Office conducted a terrestrial lidar scan of the slope. Data from this scan was used to characterize slope geometry, generate cross-sections, and to estimate material quantities. An experienced contractor that specializes in rock slope mitigation also provided recommendations on how to mitigate this slope.

Figures 13: Rockfall in the ditch and pavement divots below the slope.

Figure 14: Rocks along eastbound shoulder of the highway.

MITIGATION OPTIONS CONSIDERED AND ROCKFALL CONTAINMENT

Combinations of slope scaling, block reinforcement, type 1 or 2 cable net slope protection, a special hybrid netting system, and rockfall fences were all considered to mitigate this slope. Concerns regarding most of these options included high costs, potential heavy ice accumulations, or damage from snow avalanches, long periods for construction, and difficulties reinforcing large unstable blocks high on the slope led to the selection of intensive slope scaling over the eastern two-thirds of the slope (Priority Area #1) as the preferred mitigation option. This option provided the highest safety improvements for the amount of available funding ($1.8 million). This provided a significant, laterally continuous segment of corridor safety improvement, which also included the reportedly most problematic rockfall source area, chute 3 (Figure 15).

We initially recommend the use of an 8-foot wide by 8.5-foot tall conex barrier system along the east bound lane in an attempt to reduce the amount of scaled rock debris entering Rimrock Lake. The vision was to move this system on and off the highway at the beginning and end of each week, to allow traffic to pass through the project site each weekend. Due to the increased costs of traffic control (~$1 million), an alternative method was developed that included a full time detour, the placement of concrete barrier between the guardrail and the eastbound lane of US12, and a 12-inch-thick energy-absorbing blanket over the highway. The energy-absorbing blanket was composed of loosely placed base course material. Our analysis indicated that this alternative method would provide a comparable reduction in the amount of scaled rock debris entering Rimrock Lake, when compared to the conex barrier system alone. This reduction was

Figure 15: Priority areas of slope to scale.

accomplished by increasing the width of the catchment area (~6 feet) at the expense of losing some catchment height (~5 feet), and by reducing the bounce height of the scaled rock debris that is striking the highway.

GEOTECHNICAL ANALYSIS

Since block reinforcement, netting, and rockfall fencing were not included as the preferred mitigation alternative, the geotechnical analysis consisted solely of rockfall simulation.

A critical cross section was developed from the terrestrial lidar. It was used in conjunction with field-gathered data on surface roughness, rockfall initiation locations, and slope hardness to estimate tangential and normal coefficients. Using this information, the potential/likely rockfall trajectories and the anticipated output values could be determined (Tables 1 & 2). The program RocFall by RocScienc was used to simulate rockfall originating from the slope (Figure 16).

A 2.6-foot diameter (average of large size) rhombus-shaped block of andesite bedrock was modeled with the rigid body analysis method and an energy-absorbing blanket over the highway. The following input values were used:

	Surface Rough ness	Surface Roughness Amplitude	Tang	Norm	Dynamic Friction	Rolling resistance
Slope	6	1	0.85	0.35	0.576	1.31
Energy Absorbing blanket	0	0	0.2	0	0.58	0.39

Table 1: RocFall (ridged body analysis) input values.

The simulation produced the following general results:

Analysis Point Location	Mean Velocity (ft./sec)	Mean Bounce Height (ft.)	Mean Kinetic Energy (ft-lbs)	Rocks reaching Analysis Point
Concrete Barrier	~138	~110	~900,000	50%

Table 2: RocFall (ridged body analysis) output values.

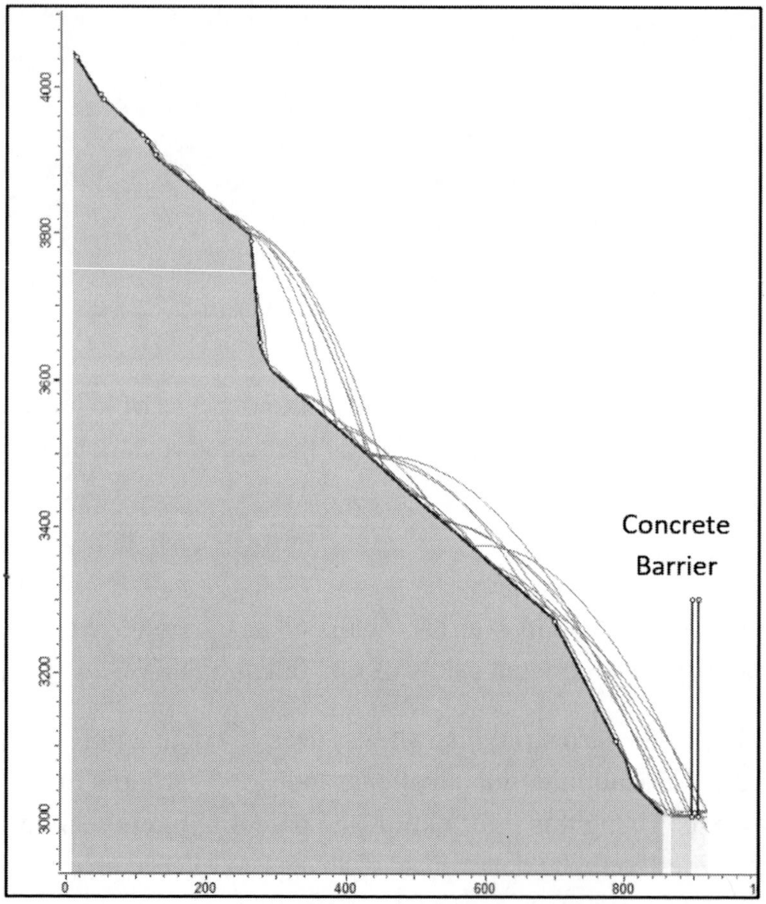

Figure 16: RocFall trajectories down the critical cross section (profile K).

The results of the rockfall simulation were used to assist in determining the percentage of debris that would likely enter Rimrock Lake, with or without the different types of protective systems.

Also, located at the base of the slope were overhead power lines and a buried fiber optic line. These were located within WSDOT right-of-way under an easement agreement. WSDOT was under no requirement to protect these utilities. The utility company wanted to bury the power lines in the ditch and use recommendations from their consultant to protect the buried fiber optic line. Their consultant indicated that blocks might penetrate as deep as 3.7 feet into the substrate, with an impact stress of 110 psi. Their recommendation was to:

1. Place a series of 12-inch thick by 12-foot long crane mats perpendicular to and centered on the cable. The purpose of these mats were to act as a safety measure to prevent rocks from penetrating the energy-absorbing blanket and damaging the cable below.
2. Place 3 feet of energy-absorbing material on the top of the mats to absorb the rockfall energy and to mitigate the potential for rock shatter projectiles.

Since WSDOT was under no requirements to protect the utilities and the recommendations by the utility company's consultant appeared costly and time consuming, WSDOT decided to use a 12-inch energy-absorbing blanket and steel plates to reduce bounce heights and to protect the buried utilities.

Slope Access

A 5-mile long, US Forest Service road provided access to the top of the slope. Unfortunately, this road had recently experienced a couple of large washouts and the Forest Service was in no hurry to initiate repairs (Figures 17 and 18). This limited the access to the top of the slope to helicopters, hikers, and ATVs. The contract made slope access the responsibility of the contractor.

Figure 17: The washed out US Forest Service road that was needed to access the top of the slope.

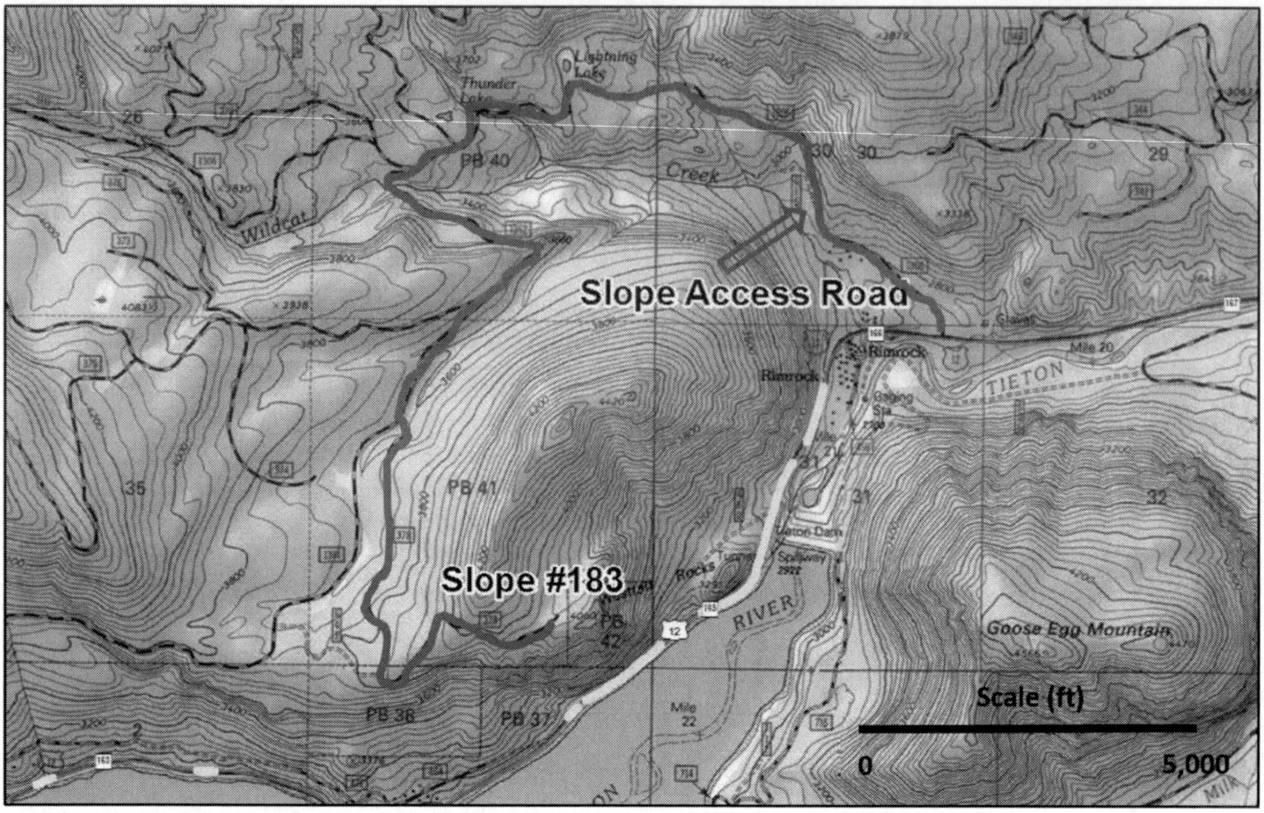

Figure 18: The 5-mile long US Forest Service road that was needed to access the slope.

Estimated scaling productivity and working days

By considering all other ancillary projects costs, the Geotechnical Office estimated that 900-crew hours of slope scaling would be available, and that the contractor would be able to scale the eastern two-thirds (Priority Area #1 in Figure 15) of the slope, and possibly more (Priority Area #2 in Figure 15). We anticipated that the contractor would use four scaling crews and work 10-hour days. We estimated approximately 23 working days to complete the scaling operations.

CONSTRUCTION

The project needed to adhere to several environmental, weather, and traffic related constraints. These constraints included bull trout spawning, water quality standards, spotted owl nesting, heavy summer traffic through Labor Day weekend, and the arrival of early winter weather in November. For these reasons, the construction on this project was scheduled for the months of September, October, and early November.

The contract called for 55 working days after September 6, 2017. The project consisted of 900-crew hours of slope scaling, 12,600 CYs of debris removal, 1,200 linear feet of energy absorbing blanket, 1,200 linear feet of ecology block wall, paving, and an approximate 16-mile detour around Rimrock Lake. To ensure that the project was not getting silt into the lake, environmental water sampling was required on a regular basis. The sampling was done in the lake from a boat, while the scaling activities were not in operation.

Since the top of the slope was difficult to access, time consuming, and provided no visibility of the work being conducted on the slope, project inspection was primarily accomplished from the other side of Rimrock lake (~3000 feet away) utilizing binoculars and radios. During the course of the project, two on slope inspections were done and on one occasion, a drone was used to observe the scaling operations up close. Scaling debris was cleared from the highway about halfway through the project, and again at the conclusion of the slope scaling (Figure 19).

Figure 19: Scaling debris captured along the highway before it was cleared.

The detour added about a half hour of additional travel time for through travelers on US12. It allowed for a complete closure of the highway during construction (Figure 20). Minor upgrades, including signs and widened shoulders were needed at several locations along the detour route.

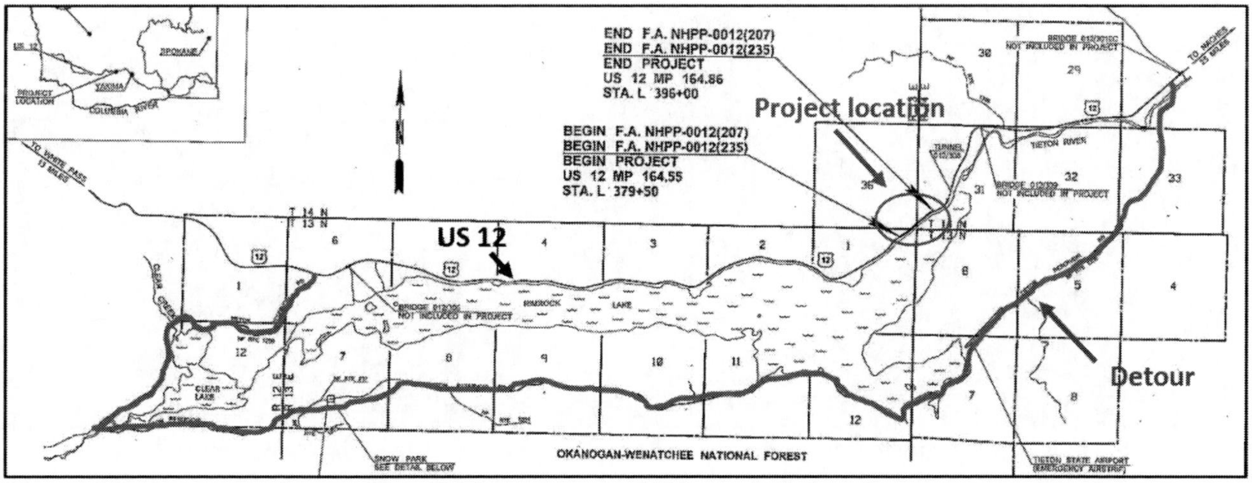

Figure 20: An approximate 16-mile detour around Rimrock Lake along a county road.

COMMENTS AND OBSERVATIONS

Direct observations of rocks falling down the slope varied significantly from our earlier rockfall simulations. Rocks appeared to stop on the slope with much greater frequencies than anticipated. They had much lower bounce heights and velocities, rollout distances, and fewer rocks made it into the lake than our simulations had shown. At many locations, the scalers encountered an abundance of loose rock sitting on the slope (Figure 21). Some of this rock was from the scaling operations, and some of it was from accumulated rockfall. If this rock was removed, it was done by hand, and repeated several times, before the rock would make its way down to the highway below. Since most of these rocks did not appear they would be migrating downslope anytime in the near future, WSDOT and the contractor decided that it would be better to leave this material on the slope and concentrate the scaling efforts on the loose detached blocks in the steeper areas.

Bypassing the scaling in some areas, and leaving loose rock on the slope led to some contractor prompted safety concerns. While the scalers were on the slope, they were observing rockfall from locations directly above (Figure 22). They had difficulties determining where the rockfall was originating from, but they were concerned that rocks falling from above could strike them. In addition, there were areas that were in direct line of this rockfall. If the scalers were to enter these areas, there would be no safe egress (i.e. chute 3). For these reasons, the scalers would not go into the chutes, and WSDOT was not comfortable with pressing this issue with the contractor. What made matters worse, was that scaled rocks from the crest and sides of the chutes were not always making their way down to the highway. They were collecting at locations where the scalers would not go (i.e. chute 3), due to the above mentioned safety concerns.

Figure 21: Many of the scaled rocks did not make it down to the highway below.

The immense overall size of this slope, and the quantity of large detached blocks certainly impeded the logistics of this slope mitigation project. On a daily basis, it took over an hour to get the scalers to and from the crest of the slope, via the US Forest Service road. Once the scaling operations got further down on the slope, a considerable amount of time was spent rappelling from and climbing back up to the crest of the slope, setting ropes, and maneuvering equipment on the slope. Some of the scaled rocks were also moved multiple times to get them further downslope and to the highway below. In addition to the safety concerns mentioned above, these logistical issues led to reduced scaling efficiencies on an hour-by-hour basis.

By late October, winter was closing in and the weather had changed to the worse. Rain and daily freeze thaw cycles were generating regular rockfall on the slope. The contractor was not comfortable working in these poor weather conditions and only worked when the weather was favorable (dry and above freezing). As the days in the contract and the scaling hours were ending, WSDOT was concerned about leaving potentially loose rock on the slope. Especially rock that was not previously at some of these locations (i.e. chute 3). Before the project was over, one worker that was clearing rock from beneath the slope was struck by debris, one falling rock had struck an excavator, and another rock had struck a vehicle. In addition, only about one-third of Priority Area #1 had been scaled (Figure 23). This was far less than the anticipated entirety of Priority Area #1 and possibly a portion of Priority Area #2.

To address these concerns, WSDOT considered several different alternatives on how to proceed:

1. Add additional days, scaling hours, and debris removal into the contract. With the arrival of winter, only a limited number of days remained where work could safely be accomplished.
2. Use helicopters with high capacity water baskets to dump water onto the slope to dislodge and wash down any loose material that remains on the slope.
3. Keep the highway closed through the next winter storm event (another 3 to 5 days), monitor how much debris stays on, and comes off the slope.

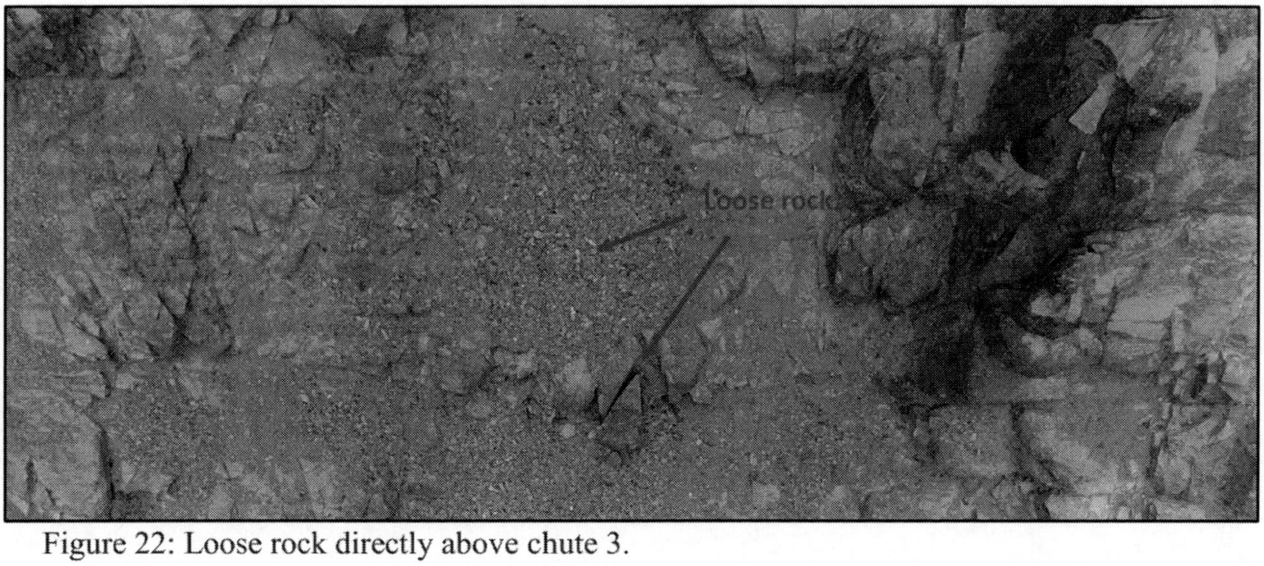

Figure 22: Loose rock directly above chute 3.

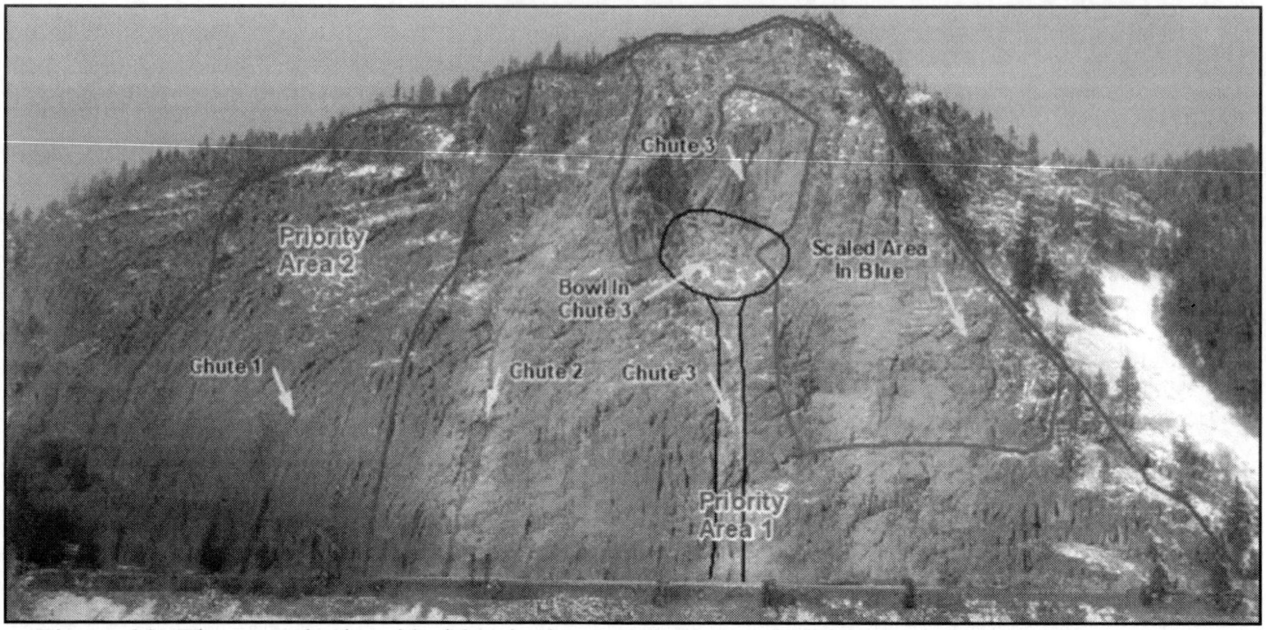

Figure 23: The area depicted in blue is the actual area scaled during the project.

With the amount of work that could still be done, option number 1 was not a favorable option. Everyone was ready for the project to end. Option number 2 was considered risky because WSDOT had never attempted this type of work before, and could not project how successful it would be. Option number 3 was the preferred option, because if little to no debris came off the slope during the next storm event, the remaining debris would probably stay on the slope over the foreseeable future.

One final comment is that all aspects of the project should have been considered at the time of project scoping. This includes a scheduled marathon beneath the slope and grazing cattle on adjacent US Forest land that will hinder traffic along the detour route (Figures 24 and 25). Unfortunately, WSDOT needed to scramble at the last minute to get the marathon re-routed (a Boston Marathon qualifier) and compensation for the ranchers for not using the adjacent US Forest grazing land.

Figure 24: The scheduled marathon beneath a nearby slope.

Figure 25: Cattle grazing in the vicinity of the detour route.

CONCLUSION

This project was the first slope of this size and magnitude that has ever been scaled within the state of Washington. The project did not go as planned, but it did stay within budget and within the allotted timeframe. As compared to the period before this project, WSDOT believes the slope is in a safer condition now. Maintenance reports that they had fewer reported rockfall events over the following winter and spring. The utilities that were buried within the ditch line and beneath the westbound travel lane were not damaged. The project did scale rock that was far outside of WSDOT right-of-way, but the Department will carefully consider whether or not they should ever try something this grand or aggressive again.

REFERENCES

Campbell, N. P.; Gusey, Daryl, 1992, Geology of the Naches Ranger District, Wenatchee National Forest, Kittitas and Yakima Counties, Washington: Washington Division of Geology and Earth Resources Open File Report 92-3, 12 p., 2 plates. Geologic map: Plate 1, scale 1:62,500.

Miller, R. B., 1985, The pre-tertiary Rimrock Lake inlier, southern Cascades Washington: Washington Division of Geology and Earth Resources Open File Report 85-2, 16p. 1 plate. Geologic map: Plate 1, Geologic map of pre-Tertiary rocks in the Mount Rainier 1:100,000 quadrangle, scale 1:100,000.

Schasse, H. W., compiler, 1986, Geologic map of the Mount Rainier quadrangle, Washington: Washington Division of Geology and Earth Resources Open File Report 87-6, 46 p., 1 plate. Geologic map: Plate, scale 1:100,000.

Swanson, D. A., 1978, Geologic map of the Tieton River area, Yakima County, south-central Washington: U.S. Geological Survey Miscellaneous Field Studies Map MF-968, 1 sheet, scale 1:48,000.

Vance, J. A., Clayton, G. A., Mattinson, J. M., and Naeser, C. W., 1987, Early and middle Cenozoic stratigraphy of the Mount Rainier-Tieton River area, southern Washington Cascades, in Schuster, J. E., ed., Selected papers on the geology of Washington: Washington Division of Geology and Earth Resources Bulletin 77, p. 269-290.

Dog Lake Marathon. Pictures. https://www.doglakemarathon.com/pictures. Accessed May 31, 2018.

Tieton Washington Cattle Drive. Images. https://www.google.com/search?q=tieton+WA+cattle+drive&source=lnms&tbm=isch&sa=X&ved=0ahUKEwie5ZaDyrDbAhViCDQIHe61AYoQ_AUICigB&biw=1344&bih=699#imgrc=1qMdO_-uACg1HM:&spf=1527791307551. Accessed May 31, 2018.

Slope Access Safety Evaluation (SASE) Form

William C. B. Gates, PhD, PE, PG
McMillen Jacobs Associates
1109 1st Ave., Suite 501
Seattle, WA 98101-2988
206-496-4829
gates@mcmjac.com

Prepared for the 69th Annual Highway Geology Symposium, September 10-13, 2018

Disclaimer

Statements and views presented in this paper are strictly those of the author(s), and do not necessarily reflect positions held by their affiliations, the Highway Geology Symposium (HGS), or others acknowledged above. The mention of trade names for commercial products does not imply the approval or endorsement by HGS.

ABSTRACT

The Slope Access Committee of the Association of Geohazard Professionals (AGHP) has been developing a practical approach to evaluate slope access safety, the following is one suggested approach. Before accessing a slope to conduct an inspection or geologic investigation, it is important for the inspector to evaluate, characterize and identify the potential geologic hazards and safety issues first. The Slope Access Safety Evaluation (SASE) form was drafted to facilitate an initial Job Safety Assessment (JSA) of the slope. The SASE was originally patterned after the CALTRANS Slope Scaling assessment form. However, it has been shortened to one page and simplified. The purpose of the SASE form is to produce an initial record of the slope condition by a competent individual. A competent individual is an engineering geologist or geotechnical engineer that can identify the inherent geologic hazards and safety issues on the slope and has been trained through a recognized rope access / climbing program. The SASE identifies the geologic hazards and describes the slope access / retreat and the support / anchor conditions. The SASE takes about 15 minutes to complete before accessing the slope. Also, the SASE includes an equipment check list to ensure that the competent individual has the appropriate safety and climbing equipment. The goal of the SASE is to insure the competent individual thinks about the potential safety hazards and plans for them before climbing onto the slope.

INTRODUCTION

Background of the SASE

The Slope Access Committee of the Association of Geohazard Professionals (AGHP) has been developing a practical approach to evaluate slope access safety (Duffy, et al, 2018). There are many jobsite safety programs used by various professionals. However, the programs typically don't address the site-specific safety concerns faced by the geotechnical engineers and engineering geologists working on rock slopes. John Duffy and colleagues with California Department of Transportation (Caltrans) saw the need for a detailed slope safety assessment about 25 years ago and developed the multi-paged Slope Scaling Assessment form (Caltrans, 2014) for geotechnical assessment and rock scaling along their highway system. This form has been modified to three pages and posted on the Association of Geohazard Professionals under Rope Access Committee as the Rope Access Assessment Form (AGHP, 2018). The Slope Access Safety Evaluation (SASE) form was drafted to facilitate an initial Job Safety Assessment (JSA) of the slope. The SASE was originally patterned after the Caltrans Slope Scaling Assessment form (Caltrans, 2014). However, it has been shortened to one page and simplified and may be used for rock slopes along highways, railways, quarries, and other slopes. The SASE has been adopted by McMillen Jacobs Associates (McMillen Jacobs Associates, 2018) and is included in the company Safety and Health Management Program. Moreover, Washington Department of Transportation (WSDOT) and other consulting professionals have adopted similar versions of the SASE and modified it to meet their organization or company specific field safety requirements (Fish, 2017; Shanahan & Marshal, 2018).

Purpose of the SASE

The purpose of the SASE form is to produce an initial record of the slope conditions by a competent individual. The completed SASE becomes part of the health and safety plan and is maintained as a permanent record. Moreover, the SASE is to compliment the Job Safety Assessment (JSA) of the slope. When conducting the slope access safety evaluation, the individual may not see all the aspects listed on the SASE, such as access to the top of the slope or anchorage conditions. However, the goal of the SASE is to insure the competent individual thinks about the potential safety hazards and plans for them before climbing onto the slope.

The SASE is qualitative rather than quantitative in that it rates the slope but doesn't provide a numerical value of safety severity. The SASE is to be completed at the site before accessing the slope (Figure 1). The form takes about 15 minutes to complete. Also, the SASE includes an equipment check list to ensure that the competent individual has the appropriate safety and climbing equipment. It is assumed the trained team with the competent individual will be accessing the slope to conduct the investigation as soon as they complete the SASE.

Figure 1: Slope Access Safety Evaluation at Deception Crags, Exit 38, I-90, North Bend, WA and BNSF Railroad, Stampede Pass, WA.

ELEMENTS OF THE SASE

Figure 2 is an example of the SASE and lists the elements. In general, the elements are self-explanatory. The following sections provide a more detailed discussion of each element.

Slope Location, Date and Weather

Location of the slope is important. At the top of the form, indicate location of the soil or rock slope, including state, city or county and federal or state highway route number and milepost. For the railway, indicate railroad owner and milepost. For other slopes, use some geographic locator to identify location. In addition, include date of the safety evaluation and weather conditions.

Competent Person & Training programs

Include the name of the competent person or individual completing the SASE. A competent person is a mature engineering geologist or geotechnical engineer or a technician with similar training that can identify the inherent geologic hazards and safety issues on the slope. In addition, the individual has been trained through a recognized slope access / climbing program. On the form, include last training date and any refresher training within the last year. It is important that all climbers on the slope, go through a refresher course on a routine basis to maintain their skills.

Slope Access Safety Evaluation (SASE) Form

Slope Location (HWY/RR & MP): ____________________
Competent Person: ____________ **Date:** ____________ **Weather:** ____________
Last Training Date: ____________
Any Training/Practice Within 1 year: Yes ○ No ○

Slope Description
Cut Slope ○ Natural Slope ○ Recent/Old Slide ○

Slope Materials
Soil ○ Soil and Rock ○ Rock ○

Slope Condition
Numerous detached blocks ○ Some detached blocks ○ Few detached blocks ○ NA ○

Slope Height
<50' ○ <100' ○ <150' ○ <200' ○ >200' ○

Slope Width W = ____________

Slope Angle
35° to 45° ○ >45° to 70° ○ >70° to 90° ○ Overhanging ○ If yes, <5' ○ >5' ○

Rock block Size S (<1') ○ M (1'–3') ○ L (3'-6') ○ XL (>6') ○ NA ○
Average ________ Maximum ______

Is there loose rock on the slope brow? Yes ○ No ○
Danger of rockfall from above during assessment Low ○ Medium ○ High ○ NA ○
Danger of rockfall onto highway below during assessment Low ○ Medium ○ High ○ NA ○
Are other personnel at toe of slope? Yes No ○ ○

Equipment Check List
- Level D, Safety ○
- Helmet, Climbing ○
- Harness- Body ○
- Ascenders: 2 ○
- Descender & Safety ○
- Safety Carabiner: 10 ○
- Daisy Chains: 3 ○
- Rope, Static: 150' ○
- Prussic Slings ○
- Assorted Slings ○

Access & Anchor Conditions

Access/Retreat	Easy ○	Difficult ○	
Angle	Flat ○	Moderate ○	Steep ○
Vegetation	Trees ○	Shrubs ○	None ○
Rock Outcrops	No ○	Yes ○	If yes, Many ○ Few ○

Other Anchors Type ____________
Mechanical Anchors Needed No ○ Yes ○ Type ____________

SLOPE DIAGRAMS

Cross Section Sketch	Front View Sketch (entry, exit, emergency access, chutes, etc. locations)

Comments / Additional Considerations: ____________________

Figure 2: The Slope Access Safety Evaluation (SASE) form.

Some companies such as McMillen Jacobs Associates have developed internal safety and training programs attendant to slope access and safety (McMillen Jacobs, 2018) (Figure 3). Caltrans developed their own in-house training program to train their slope technicians, engineering geologists and geotechnical engineers circa 1990 and developed the Caltrans Bank Scaling and Rock Climbing Manual (Caltrans, 2014); it is available online. Washington Department of Transportation (WSDOT) have an in-house training program. Duffy, et al, (2018) describe third party organizations that provide similar training and certification programs such as SPRAT (Society of Professional Rope Access Technicians), IRATA (Industrial Rope Access Trade Association) and PCIA (Professional Climbing Instructors Association Slope Access Technician (SAT) program).

Figure 3: Limited Rope Access Training Program conducted for WSDOT and other consultants by McMillen Jacobs Associates. Climbers in the left photos are working on Spire Training Rock in Tacoma, WA. Middle and right photos are at Deception Crags at Exit 38, I-90 near North Bend, WA.

Slope Description

On the form, record the general type of slope that your team will be assessing, such as; apparent cut slope, natural slope or some type of recent or old slide (Figure 4). Typically, a cut slope may have the cleanest face with fewest loose rocks and good access to the top. A natural slope may have good anchors such as trees, but access to the top may be difficult because of no trail or road necessitating either cross-country travel or rope access. A recent or old slide may be unstable with loose rock, tension fractures, chutes for rockfall and present an extreme safety challenge. Again, access and retreat may be difficult.

Slope Materials

Describe the primary geologic material that makes up the slope (Figure 5). On slopes consisting of soil, one would expect primarily rotational slides and slumps with possibly rills and gullies down the face from erosion. If the slope is composed of soil and rock, you may see rotational slides, with tension fractures near the head scarp and maybe rockfall from the rock outcroppings and springs seeping out of the face. Alternatively, the slope may be strong competent granite such as Half Dome in Yosemite, where structure controls the primary stability and natural anchors for rappels may be sparse requiring mechanical anchors such as pitons or rock bolts.

Figure 4: Slope description: Left photo is an example of a cut slope; note rock chutes between cores stones. Middle photo displays two natural slopes consisting of subvertical columns of basalt; primary stability issue is rockfall from toppling. Right photo is of recent rockslide in very weathered weak disintegrated granite (DG).

Figure 5: Example of slope materials. Left photo displays a rotational slide in residual soil. Center photo is an example of a composite rock slide in basalt rock overlain by the Latah formation consisting of lacustrine geo-intermediate material. Note seepage right side of photo. Right photo displays Half Dome at Yosemite National Park. The rock slope consists of massive strong granitic rock, where the rock structure controls stability.

Slope Condition

It is important to understand the slope conditions and estimate what kind of rockfall the team may expect (Figure 6). On the form, record the relative number of detached blocks and key blocks that one can observe on the brow and face of the slope that could present an unstable slope condition and rockfall. Tension fractures behind the block are a good indication of pending failure. Show location of blocks and/or tension fractures on a sketched cross-section of the slope at the bottom of the form.

Slope Dimensions

The evaluator should estimate the height and width of the slope before accessing the slope (Figure 7). Include the estimated dimensions on the sketch at the bottom of the form. Knowing the height and geometry of the slope will help to estimate rockfall runout. This information will also facilitate estimating the length of rope to use for vertical scanlines. The width of the slope measured along the toe will provide an estimate of how many vertical scanlines or future scaling lines are required. In general, estimate every 40 feet for a scanline. This assumes the engineering geologist or scaler on rappel can swing 20 feet each side of the drop line from the anchor. Provide a sketch of the slope with the chutes, seep areas and other important items on the form.

Figure 6: Example of slope conditions. Photo on left displays numerous detached blocks and tension fractures and a key block. Center photos displays one large detached block with several tension factures. Photo on right shows one rock tower with a tension fracture behind the block.

Figure 7: Example of width and height of slope. In this case the slope width measured at the toe is about 160 feet. The height measured from the brow to the toe is about 200 feet

Slope Angle

It is important to estimate the slope angle of the rock exposure (Figure 8). The geometry of the slope and the kinematic relationship of the discontinuities will control the stability of the slope. In addition, the geometry of the slope will establish the appropriate techniques used to assess or scale the slope. Slopes dipping at 35 to 45 degrees are easier to access to assess and scale than vertical or overhanging slopes. Include with the assessment a sketch of the cross-section of the outcrop displaying the steps and overhangs at the bottom of the form.

Figure 8: Example of slope angle. In the left photo, the slope dips about 35 degrees towards the road with a sub-vertical release joints as steps; slope failures typically occur as planar slides. Photo on the right displays a complex rockslope with an overall slope dip at about 55 degrees to the road with areas that are subvertical and overhanging, slope failures are typically from wedge sliding.

Rock Block Size

On the form, estimate the size of blocks observed on brow, face and at the toe of the slope (Figure 9). Provide and average size and a maximum size. This information will facilitate analysis of the rockfall. In addition, indicate if there are key blocks on the slope and show their location on cross-section sketch at the base of the form. Removal of a key block could exacerbate the stability of the slope.

Describe Rockfall Danger

On the SASE form describe the rockfall danger in general terms (Figure 10). The individual assessing the slope should indicate if he or she can observe apparent loose rock on the brow or face of the slope. Rate the danger of rockfall to the team from above during the assessment of the slope. In addition, note if there are or will be additional personnel on the slope that may be in the line of rockfall while the team is working on the slope. Also, rate the probability of rockfall reaching the road during the assessment. Check for rock chutes that tend to concentrate rockfall, which typically are the most dangerous avenues of rockfall to personnel on the slope (Figure 10).

Indicate on the slope diagram if there is a catchment ditch at the toe of the slope and establish if the ditch appears adequate to catch falling rock before it rolls onto the highway or railway (Figure 10).

Figure 9: Estimate average block size and maximum block size one can expect on the slope. Blocks on left average about 2 ft diameter. Pyramidal block on right next to the geologist measures about 12 x 13 ft.

Describe Slope Access, Retreat and Anchor Conditions

Near the bottom of the form, there is a section for the evaluator to evaluate the access to and retreat from the slope, anchorage area and anchor conditions of the slope brow. The evaluator should indicate if the brow of the slope and anchorage area is easy or difficult to gain access to or retreat from. Assess if there is a road or trail to the brow of the slope, or is cross-country travel or rope access required. The evaluator should keep in mind, access and retreat to the slope is critical if there is some type of emergency.

The evaluator should judge if the terrain behind the brow is flat, moderately inclined or sub-vertical. The backslope this will impact access to the top of the slope as well as locating and installing the anchors.

If ropes are to be used, record type of anchors available, such as; natural anchors including bushes, trees, large boulders or rock outcrops; or other anchors such as fence posts, guardrails, utility poles or heavy equipment (Figure 11). If natural anchors are sparse, indicate if mechanical anchors will be needed such as pickets, cams, pitons, or bolts.

Figure 10: In the left photo, the investigator is assessing the slope, checking for rockfall conditions and the adequacy of the catchment ditch along the railway. In the right photo, the climbers are standing at the base of a large rock chute which is source of major rockfall.

Figure 11: Evaluate slope for available anchors for safety and rappel lines. In the left photo the technician is using 1-inch webbing around a tree for a rappel anchor. Right photo, technician is demonstrating a 3-point anchor using mechanical camming devices.

Slope Diagram

At the bottom of the form, the SASE provides a block for the evaluator to sketch a general cross-section and front view of the rockslope. On the cross-section include location of overhangs, loose rock, potential anchors such as trees and catchment ditch with general dimensions. In the front view sketch include entry and exit areas, and emergency access areas. One of the important elements of the sketch is indication of the location of the rock chutes (if any).

Equipment Check List

For convenience and to jog the evaluators memory while assessing the slope conditions, an equipment check list has been provided to ensure that the team has the appropriate safety and climbing equipment for the project.

Comments/Additional Considerations

At the bottom of the form, the evaluator may record any additional comments of considerations attendant to the slope assessment, safety, equipment, or other pertinent information.

CONCLUSIONS

The SASE form is a working document and can be modified to meet the needs of the user. The form should be quick and simple to complete. It is not designed to cover every safety aspect of the slope. However, it is a tool to get the evaluator or team to think of the present and unforeseen safety issues that may be on the slope and prepare accordingly.

REFERENCES

The Association of GeoHazard Professionals (AGHP), Rope Access Assessment Form, 2018, http://www.geohazardassociation.org/committees/rope-access-committee/

Duffy, John, Marc Fish, and Colby Barrett, 2018, Safe Work on Dangerous Slopes; Geotechnical News, Vol. 36, No. 1, March 2018, www.geotechnicalnews.com

California Transportation Department (Caltrans), 2014, Caltrans Bank Scaling and Rock Climbing V9.0, Chapter 3, Site Evaluation, Slope Scaling Assessment, Version 4.1, PP. 23-30.

Fish, Marc, 2017, Washington Department of Transportation (WSDOT), Personal communication.

McMillen Jacobs Associates, Safety and Health Management Program, 2017, Appendix E, Fall Protection and Rope Access Work Safety Plans, 2018.

Shanahan, Matt and Mike Marshal, 2018, GRI, Personal communication.

Rockfall in New Jersey:

A Proactive and Collaborative Approach

Amber B. Granger, P.G.
Haley & Aldrich, Inc.
299 Cherry Hill Road, Suite 303
Parsippany, NJ 07054
(973)-658-3939
AGranger@haleyaldrich.com

Scott J. Deeck, P.E.
New Jersey Department of Transportation
P.O. Box 600
Trenton, NJ 08625
(609)-530-4579
Scott.Deeck@dot.nj.gov

John P. Jamerson
New Jersey Department of Transportation
P.O. Box 600
Trenton, NJ 08625
(609)-789-7798
John.Jamerson@dot.nj.gov

Edward M. Zamiskie, Jr., P.E.
Haley & Aldrich, Inc.
299 Cherry Hill Road, Suite 303
Parsippany, NJ 07054
(973)-658-3909
EZamiskie@haleyaldrich.com

Prepared for the 69th Highway Geology Symposium, September, 2018

Disclaimer

Copyright Notice

ABSTRACT

In heavily traveled New Jersey, a single rockfall event has the potential to cause serious injuries and costly damages. Within a landscape of budget constraints and competing project priorities, the New Jersey Department of Transportation (NJDOT), working with the consultant community, has been successful in developing and implementing a proactive rockfall mitigation approach. NJDOT's Engineering Geology unit and Project Management group have partnered in the development of a multi-year Rockfall Mitigation Program that advances projects for remediation that have been prioritized in NJDOT's Rockfall Hazard Management System. New Jersey has a wide variety of bedrock geologic conditions, ranging from rift basin sedimentary and igneous formations to the folded and faulted formations of the Highlands and Valley and Ridge physiographic provinces. Other factors that impact rockfall projects are equally diverse, including urban settings with tight spatial constraints and heavy traffic volumes, as well as rural settings with environmental, cultural and historical considerations.

Every rock slope and situation is unique, and there are numerous possible mitigation techniques. In the Concept Development phase, the NJDOT needs a succinct and detailed description of viable alternatives to help them choose an effective and economical option. Toward that end, NJDOT subject matter experts work closely with consulting engineering geology experts to evaluate the rockfall hazards and project constraints, and then build a consensus with the stakeholders regarding the preferred mitigation alternative. This collaborative approach helps streamline the subsequent project phases by reducing or eliminating re-work and facilitating an efficient path to final design and construction to stretch available project dollars.

INTRODUCTION

As the most densely populated of the 50 States, New Jersey relies heavily upon its widespread system of toll roads, interstate and state highways to connect communities that range from urban to rural. From the famous Jersey Shore to the Delaware Water Gap, from Cape May to High Point, from Washington Crossing to the Statue of Liberty, New Jersey's highways provide access to the state's wide variety of cultural, historical and natural attractions. In addition to its own cities, the state is 'bookended' by two major metropolitan areas, New York City and Philadelphia, and most of New Jersey's roads see heavy daily commuter traffic on a year-round basis. Safety of the traveling public is a primary concern, of which highway rockfall is one aspect. Factors that may impact rockfall projects include urban settings with tight spatial constraints and heavy traffic volumes, as well as rural settings with environmental, cultural and historical considerations.

In addition to its geographical and cultural variety, New Jersey has a wide variety of bedrock geologic conditions. The southern portion of the state is located within the Coastal Plain physiographic province, which consists mainly of unconsolidated sand, silt and clay deposits overlying deep 'basement' bedrock. However, in the northern half of the state, bedrock is either exposed or near ground surface. A full range of igneous, sedimentary and metamorphic rock types are present that contain numerous contacts, fractures, faults and other discontinuities. In the Piedmont physiographic province, rift basin sedimentary and igneous rock formations frequently outcrop, like the Palisades Sill along the Hudson River or the tall basalt slopes along Interstate 280. The folded and faulted formations of the Highlands and Valley and Ridge physiographic provinces are also often exposed, typically along roadways. The Valley and Ridge province and approximately half of both the Highlands and Piedmont provinces were affected by the most recent glaciation, while the remaining areas were not. When combined with a high degree of man-made impacts (both recent and historical), this wide range of geologic settings results in a multifaceted transportation environment, requiring an adaptable approach to engineering geology in general, and rockfall mitigation in particular.

Transportation agencies have long contended with rockfall events along roadways and rail lines wherever they cross, cut through or skirt mountainous terrain. Routes can be blocked for days depending on the severity of the incident and impact emergency services, evacuation routes, or other vital community interactions. The resulting costs of clean up and repair can far exceed the cost of timely stabilization measures. Therefore, a proactive approach should be taken to safeguard people and property. In the past, transportation agencies responded reactively by cleaning up debris to clear the route as soon as possible, until the next event. Over the last few decades, more and more states have adopted proactive approaches to rockfall events.

This paper discusses the approach to rockfall mitigation projects in New Jersey, including the evaluation and ranking of potential project sites, the funding and processes of project execution, and some details regarding the Concept Development phase, which includes collaboration between entities to streamline and optimize rockfall mitigation projects. These processes can reduce or eliminate re-work and facilitate an efficient and cost-effective path to final design and construction.

NEW JERSEY ROCKFALL HAZARD MANAGEMENT SYSTEM

The New Jersey Department of Transportation (NJDOT) has jurisdiction of all non-toll Interstate and State highways contained within New Jersey. As such, the NJDOT has the responsibility to address the rockfall hazards and risks on those highways containing rock cut slopes. *Rockfall Characterization and Control* (*2*) defines a rockfall hazard as a natural occurrence that creates a danger or threat and can be described by its geometry, failure mechanism, or other characteristics; and defines a rockfall risk as the consequences realized when the hazard fails and is measured in terms of adverse effects to people or property. A rockfall event can be composed of large masses of rock blocks or small, discreet blocks. Some of the possible triggers for these events include heavy or sustained rainfall, snowmelt, channel runoff, groundwater seepage, ice jacking caused by freeze-thaw cycles, differential erosion, and root jacking caused by vegetation growth (*3*).

There are currently 444 highway cut slopes on NJDOT-maintained roads. In 1994, NJDOT adopted and began using the Rockfall Hazard Rating System (RHRS) for evaluating and ranking highway rock cut slopes. The RHRS was originally developed by the Oregon Department of Transportation, sponsored by the Federal Highway Administration (FHWA), in order to address the need for a proactive rockfall methodology to uniformly evaluate rock slopes and numerically differentiate apparent risk at rockfall sites (*4*). The RHRS methodology forms the basis for the NJDOT's Rockfall Hazard Management System (RHMS), which is used to evaluate, prioritize and program rockfall mitigation projects for implementation on NJDOT-maintained roads. The RHMS is administered and maintained by the Engineering Geology unit within NJDOT's Division of Bridge Engineering & Infrastructure Management.

When evaluating rock cut slopes, NJDOT applies the RHRS' original two-phase approach, which consists of a Preliminary and Detailed rating phase. The Preliminary rating phase addresses the likelihood of rockfall events to occur (rockfall hazards) as well as the likelihood of material from such an event reaching the roadway surface (rockfall risks). There are three Preliminary rating values: 'A' (high), 'B' (moderate), and 'C' (low). The Detailed rating phase develops a numerical rating for each slope, utilizing 10 site-specific categories. However, while the original RHRS methodology stipulates that Preliminary ratings be performed on all cut slopes and that Detailed ratings are to be initially performed only on 'A' rated cut slopes, the NJDOT has implemented Detailed ratings on all cut slopes within the inventory, regardless of the Preliminary rating determination. This practice allows for equivalent evaluation of all slopes, and subsequent decision-making on whether to 'bundle' any or all 'B' slopes incurring higher detailed ratings along with high-priority 'A' slopes for mitigation. The 10 categories of rating criteria used for the Detailed ratings are:

- Slope height
- Catchment ditch effectiveness
- Average vehicle risk
- Sight distance
- Roadway width
- Rock structural character (case 1, discontinuity condition)
- Rock structural character (case 2, erosional condition)

- Average block size/average volume per event
- Presence of water on slope
- Rockfall history

The RHRS uses an exponential scoring system of 1-100 points, depicted by a graph of the function $y = 3^X$. For ease of ratings, four distinct conditional 'break points' within the range were established (3 points, 9 points, 27 points and 81 points) to assist with standard, easily-distinguished changes within the respective category. However, NJDOT utilizes the entire 1 to 100-point range graph for scoring, with the break points used only as a preliminary scoring guide, which maximizes the benefits of the exponential system. This practice allows for discrete adjustments in each of the individual scoring categories, thereby resulting in unique, differentiable final scores for nearly all slopes within the inventory. Rarely does a final rating score occur using all four of the break points as final category scores. The resulting numerical ranking can be easily used to develop and justify programming and funding requests for rockfall mitigation projects.

The RHMS is used to prioritize and program rockfall mitigation projects for implementation within NJDOT's Capital Project Delivery Process, described below. The highest ranked cut slopes are targeted as the main initiative, while adjacent or geographically nearby lower-ranked slopes are screened and evaluated for inclusion as a group, where appropriate for benefits of cost-efficiency or other factors, such as traffic impacts or public input.

TRANSPORTATION CAPITAL PROGRAM

The FHWA defines Asset Management as a "systematic cost-effective process of maintaining, upgrading and operating physical assets." In January 2008, the NJDOT adopted Asset Management as the official, institutional approach to managing its infrastructure assets and making capital investment decisions. With the current economy, and the need to spend public dollars wisely, Asset Management policy and practice are a high priority at the NJDOT. This approach supports and complements the NJDOT's federal and state-mandated investment planning documents. Focusing on the department's Core Mission—safety, infrastructure preservation, mass transit, mobility and congestion relief, and operations and maintenance—this Capital Program outlines projects and programs that rebuild New Jersey's bridges and roads, provide mass transit services, and reduce congestion.

Capital Investment Strategy

The NJDOT allocates funds to projects and programs through two main capital program documents: The Transportation Capital Program and the Statewide Transportation Improvement Program. The Transportation Capital Program is a document required by New Jersey State law. This program allocates state and federal transportation funding for the period of one state fiscal year (July 1 through June 30) for the NJDOT, New Jersey Transit (NJT), counties and municipalities. The Statewide Transportation Improvement Program (STIP) is required by federal law. Like the Transportation Capital Program, the STIP includes both state and federal funding and includes projects and programs of the NJDOT, NJT, and the counties and municipalities. The current STIP for New Jersey is for fiscal years 2018 through 2027.

A companion document, the Statewide Capital Investment Strategy (SCIS) provides transportation investment recommendations for transportation program categories based upon goals, objectives, and performance measures. The SCIS is a requirement of the Transportation Trust Fund Authority Act of 2000. The SCIS also represents an Asset Management approach to addressing New Jersey's transportation needs, and is a systematic, comprehensive process for maintaining, upgrading and operating physical assets cost-effectively.

Among many others, Safety Management is one of the asset categories. An annual investment amount seeks to maintain the current performance indicators for reducing fatality and injury severity rates, in addition to promoting strategies and partnerships to continue to achieve that reduction. Rockfall Mitigation is listed as one of the safety management programs.

Capital Project Delivery

The FHWA requires use of a formal project delivery process to obtain approval and access to federal funding. The NJDOT's Project Delivery Process aligns with FHWA's regulations. It controls and simplifies the process by which federal approval and funding are obtained. The Project Delivery Process consists of the Problem Screening Phase, Concept Development Phase, Preliminary Engineering Phase, Final Design Phase and Construction Phase, see Figure 1.

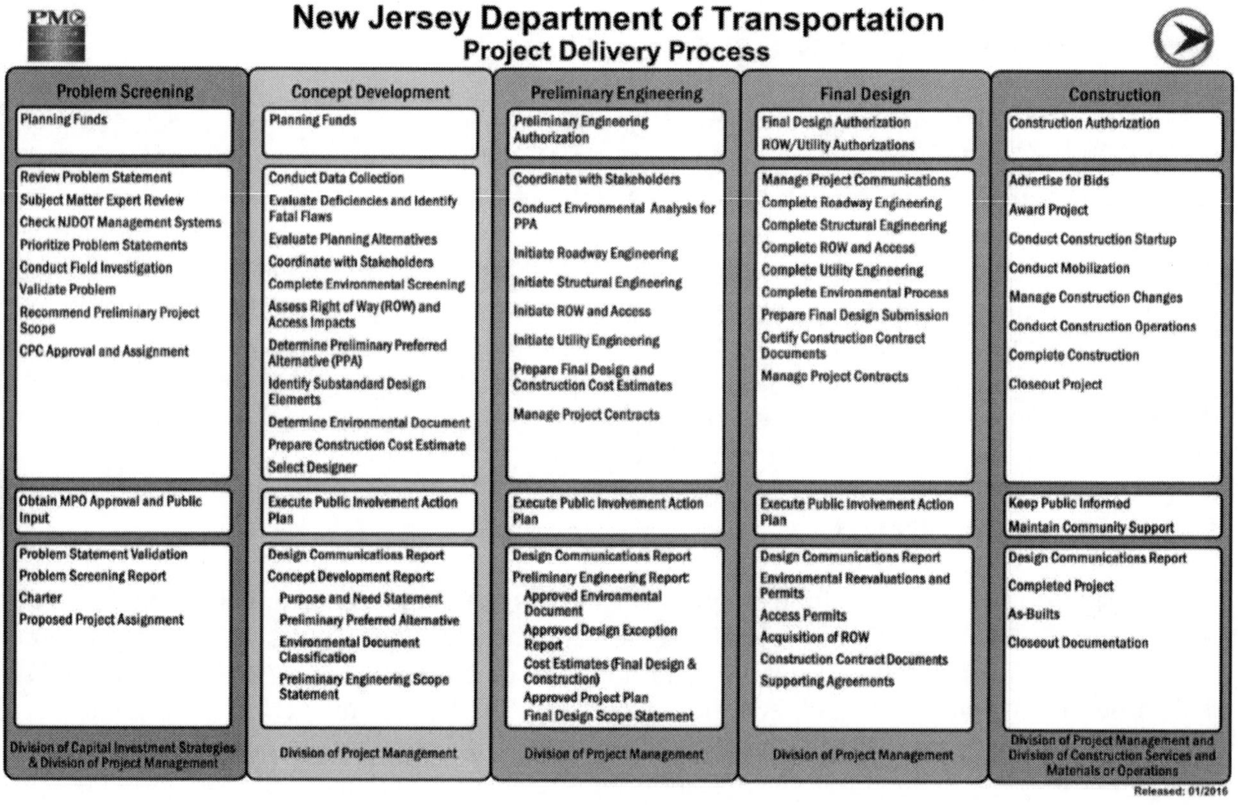

Figure 1. NJDOT Project Delivery Process.

NJDOT's Project Delivery Process begins with an evaluation of potential transportation problems in the Problem Screening phase. During evaluation, NJDOT researches the problem statement to have a clear understanding of the problem and its impact. The Problem Screening phase determines how important that problem is relative to other transportation problems. The RHMS is integral to this phase for rockfall mitigation projects.

Project planning occurs during the Concept Development (CD) Phase and considers the problems associated with the project and evaluates alternative solutions. An alternative is selected based on environmental impacts, constructability, cost effectiveness, how effectively the alternative addresses the project need, and if the project can be constructed in a timely manner. The selected alternative becomes the Preliminary Preferred Alternative (PPA). Once NJDOT approves the PPA, it is further developed using industry standards and practices.

In the full-scope Standard Delivery Approach, the Preliminary Engineering (PE) phase includes an environmental analysis of the PPA, initiates project design work in support of the environmental document, and initiates the Right-of-Way (ROW) acquisition process for temporary or permanent construction easements. Then during the Final Design (FD) phase, a set of detailed construction plans and specifications are developed for construction of the project. In this phase, NJDOT will secure the necessary permits to begin construction. Finally, during Construction the project team ensures that the contractor is building the project according to the contract documents while minimizing impacts to the existing infrastructure and the traveling public.

The NJDOT has developed a Limited Scope Project Delivery Approach to effectively administer the planning and design of transportation-related problems with minimal impact to the project surroundings and no need for ROW acquisition. Limited Scope project types are typically pavement resurfacing, bridge deck or superstructure replacement, Intelligent Transportation Systems (ITS) installation, simple intersection improvement, guiderail replacement and rockfall mitigation projects. The main difference between the Limited Scope Approach and the Standard Delivery Approach is that the Limited Scope process does not have a formal PE phase. Eliminating the formal PE phase for this approach is possible because the project scope should not change once the PPA is selected at the end of the CD phase. The Department can realize significant administrative and engineering cost savings and time savings by eliminating this phase.

CONCEPT DEVELOPMENT FOR ROCKFALL PROJECTS

By considering the Department's fiscal goals and objectives together with the established project delivery process, the NJDOT's Engineering Geology Subject Matter Experts (SMEs) in collaboration with its Division of Project Management have developed a multi-year program of rockfall mitigation projects which have been prioritized from the RHMS for design and construction. As such, the Rockfall Mitigation Program (RMP) fits within NJDOT's SCIS. The RMP uses the prioritized rankings and generates mitigation projects targeting either a single high priority cut slope, or a 'bundled' group of slopes, which typically incorporate a main high priority slope with one or more moderately prioritized slopes for the purposes of cost-efficiency, geographic nearness, similarity of mitigation measures, or minimization of the recurrence of traffic impacts in a local community. The RMP serves as a programing guideline for rockfall projects from Concept Development, into Preliminary and/or Final Design, and through Construction. In this manner, long-term funding needs can be evaluated, adjusted and requested for all costs incurred in multiple projects within different phases of the Project Delivery Process. The initial step in this pathway is a screening of the RHMS to propose mitigation project groupings based upon priorities within the RHMS, while also addressing other factors like cost-efficiency. Once a project grouping is screened, it is presented for approval into Concept Development within the Capital Project Delivery process.

The CD phase is the foundation for the remaining project phases. It builds on the rating determined through the RHRS and presents an overarching view of the project. One of the essential purposes of this phase is to establish the "Purpose and Need" for the project. It is vital to explain why the project is necessary to stakeholders and the public because the need for rockfall projects is not as intuitive as pavement or bridge repair projects, for example. Rockfall events are generally not covered in local news unless egregious. Minor rockfall events are typically removed from roadways by maintenance personnel in a timely manner and considered to be 'debris.' Therefore, the precise range and frequency of rockfall can be unclear to NJDOT geology personnel as well as misunderstood by the public.

The rockfall hazards and risks are evaluated by qualified engineering geologists through geologic mapping and characterization, in general accordance with the recommendations of FHWA's *Rock Slopes Reference Manual* (*5*). The results of the evaluation are presented in the CD report through annotated photographs and detailed descriptions. Some of the other details included in the CD phase are roadway and rock slope geometries, topographic maps, existing Right-Of-Way (ROW) boundaries, identification of stakeholders, as well as potential environmental or other constraints.

Approach to Identifying Rockfall Mitigation Alternatives

In addition to establishing the Purpose and Need for the project and presenting the overall project parameters, the CD phase is an alternatives analysis. The first alternative is "No Build". Typically, this alternative does not address the Purpose and Need but should be included for comparison purposes and may apply to certain low-risk slopes if the project contains more than one rock slope.

In general, viable alternatives are evaluated following the rockfall hazard mitigation approach hierarchy of (1) removal: get rid of it; (2) stabilization: don't let it fall; and (3) protection: let it fall safely. The removal approach physically eliminates the rockfall source zones, which makes it the most effective method in terms of long-term remediation of the hazards, future maintenance costs, aesthetics and other impacts. Strategies include mass rock removal through blasting, excavation or reshaping using methods such as trim blasting, hoe-ramming, boulder busting, scaling with prybars, and other mechanical means of removal, see Figure 2.

Figure 2. Photographs of rock scaling (left) and a trim-blasted slope (right).

Stabilization consists of securing and/or reinforcing the rockfall zone to prevent rocks from moving. Available methods include targeted rock dowels or anchors, cable lashing, anchored mesh, polyurethane resin 'grouting', shotcrete and buttressing. The photographs in Figure 3 illustrate two types of stabilization techniques: rock dowels and anchored mesh systems. The use of shotcrete and buttressing for the stabilization of shear zones along rock slopes is also a common method. Installation of these designs generally requires specialty contractors with experience executing the methodologies and working on steep slopes where access can be challenging. Permanent mechanical systems require periodic monitoring and maintenance, which should be considered in overall project costs.

Figure 3. Photographs of rock dowels (left) and anchored mesh (right).

Protection involves intercepting and retaining rockfall before it reaches the roadway once the event has occurred. Techniques include the construction of catchment ditches, rockfall barriers and fences, rockfall sheds, earthen or engineered rockfall embankments, hybrid fence and draped mesh systems that allow controlled rockfall descent. These remedial methods typically involve the most maintenance of the three approaches, as well as more significant visual impacts. Two of the most common protection measures are rockfall barriers and catchment ditches, as shown in Figure 4.

Figure 4. Photographs of rockfall barrier fencing (left) and a rockfall catchment ditch (right).

In most cases, a combination of all three mitigation approaches are used to design the most effective and feasible mitigation system. In practice, three to five viable alternatives are typically considered for each slope evaluated. Each alternative, except for the "No Build" alternative, are presented in a figure that includes an existing conditions photograph, the topography of the slope that shows the proposed alternative in plan view, and a sketch illustrating the proposed alternative in cross-section view (Figure 5). Presenting the most pertinent information in one relatively simple figure helps the project team focus on the most important aspects of each alternative.

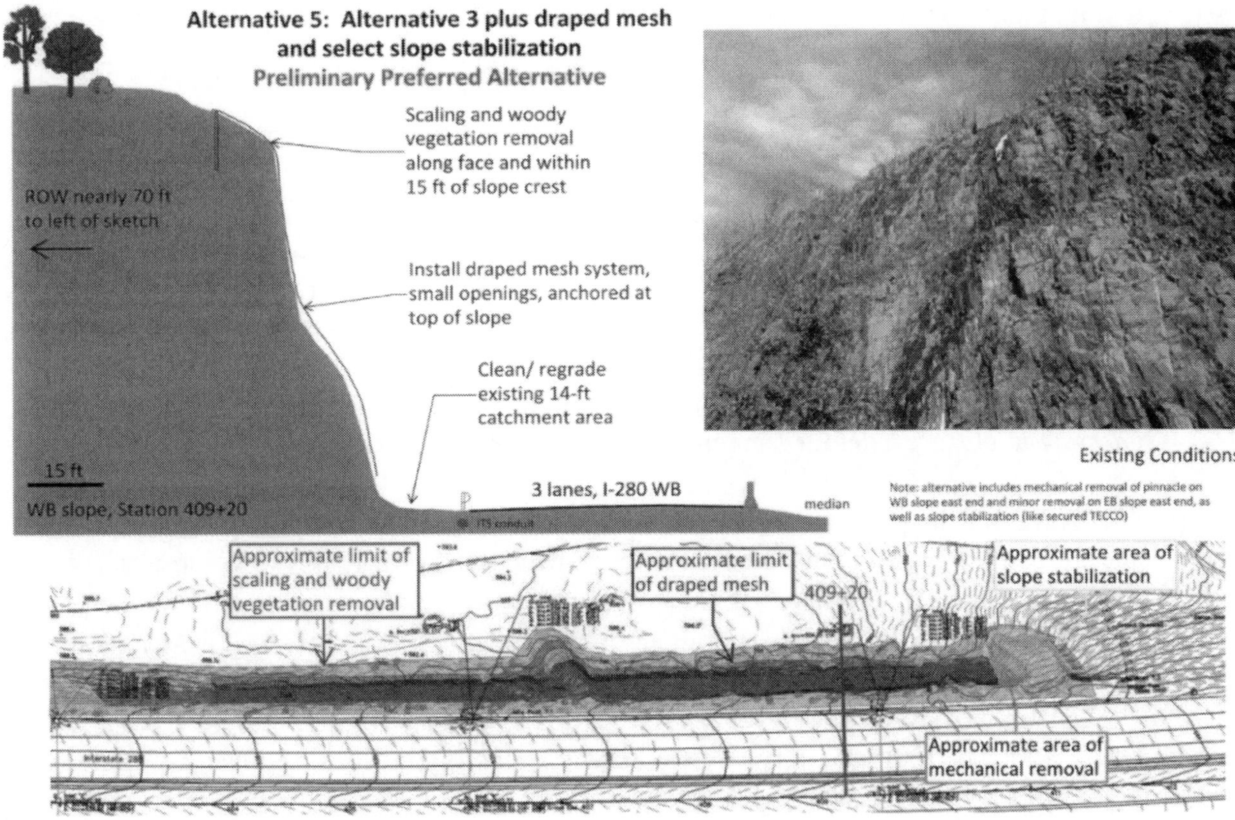

Figure 5. An example of a figure within a CD-phase report illustrating a proposed rockfall mitigation alternative.

Evaluating Alternatives with Simple Matrices

Evaluating the various alternatives is a process that needs to include more than simply considering the remediation of rockfall hazards and risks or technical performance. For example, constructability and safety during construction are paramount considerations; providing sufficient working space while maintaining and protecting the traveling public needs to be incorporated into the design. In addition, aesthetics play an important role in all cases but is elevated near or within designated and protected natural, historic or culturally significant areas. Strategies for improving aesthetics include colorized elements (like mesh, shotcrete or fencing), hydro-seeding or boulder-scaping. Beyond attaining the required performance and risk avoidance, the best mitigation designs are characterized by simple, practical components with an emphasis on constructing within the existing easements, durability, longevity, and no or low maintenance requirements.

A matrix of potential alternatives compares key elements or categories of concern for a rockfall mitigation project. The following is a list of typical categories within the matrix that are used by NJDOT project teams:

- Risk Reduction: Subjective measure of the effectiveness of the alternative in reducing rockfall risk or slope instabilities to the roadway/traveling public.

- Right-of-Way Impacts: Relative measure of whether the alternative can be constructed and maintained within the NJDOT ROW or whether ROW construction and maintenance easements will be necessary.

- Long-term Maintenance: Subjective measure of need for maintenance of mitigation systems, or of resultant need for maintenance on the roadway by NJDOT from rockfall or slope failures.

- Service Life: Estimated lifespan of the mitigation alternative (time period within which it performs effectively without needing replacement, major repairs or upgrades).

- Construction Impact: Subjective estimate of impacts to disturbed area and roadway traffic (from road closures, detours, inconvenience, and visual effects).

- Difficulty of Construction: Relative estimate of construction complexity, regarding access, availability of qualified contractors, specialized procedures or equipment, etc.

- Aesthetic Impacts: Subjective judgment on the negative visual effects during and after construction to roadway users and the general public.

- Utility Impacts: Estimate of the extent to which existing utilities, if present, will be impacted by the construction activity.

- Length of Construction: Estimate on time duration of active construction.

- Range of Costs: Estimated range of construction costs.

The categories can then be color coded or numerically rated with respect to desirability (green is desirable, yellow is neutral, and red is undesirable). An example of an alternative comparison matrix is shown in Figure 6.

Alternative	Description	Risk Reduction	Beyond ROW	Long Term Maintenance	Service Life	Construction Impact	Difficulty of Construction	Aesthetic Impacts	Utility Impacts	Length of Construction (days) Low	Length of Construction (days) High	Range of Costs (x $1,000) Low	Range of Costs (x $1,000) High
EB and WB													
1	No build	None	No	High	N/A	N/A	N/A	N/A	N/A	0	0	0	0
2	Rockfall Fencing	High	No	Moderate	Moderate to High	Low to Moderate	Low to Moderate	High	Yes	130	160	1,026	1,924
3	Scaling, woody vegetation removal, clean and regrade ditch	Low	No	High	Low	Low	Moderate	Low	No	140	170	557	1,043
4	Alternative 3 plus shotcrete expansion and repair, reinforced slope stabilization netting	High	No	Moderate	Moderate	Moderate	Moderate to High	Low to Moderate	No	210	250	1,173	2,199
5	Alternative 3 plus draped mesh and reinforced slope stabilization netting	High	No	Low to Moderate	Moderate to High	Moderate	Moderate to High	Moderate to High	No	220	260	1,462	2,742
6	Trim Blasting	High	No/Maybe	Low	High	High	High	Moderate	No	[illegible]	[illegible]	[illegible]	[illegible]

Color key:	Desirable	Neutral	Undesirable

Figure 6. A typical comparison matrix of rockfall mitigation alternatives.

Team Collaboration – Identifying the Preliminary Preferred Alternative

One of the processes that has been developed and is highly effective for NJDOT rockfall mitigation projects is the selection of the preliminary preferred alternative, or PPA. Part of that process is an alternatives workshop held with the designer and NJDOT personnel. During the workshop, the project team works together to complete the comparison matrices described above and come to a consensus as to the best rockfall mitigation strategy for the project. This consensus is the basis for the remaining steps in the PPA selection process, which include consideration of the National Environmental Policy Act of 1969 (NEPA) requirements and public outreach processes.

As the recipient of federal transportation funds, NJDOT must comply with FHWA's implementing NEPA regulations (codified at 23 CFR 771 *Environmental Impact and Related Procedures*). NEPA provides a planning and decision-making framework for selecting the most feasible and prudent project alternative that avoids or reduces negative social, economic, and environmental impacts. In practical application during NJDOT's Project Delivery Process, the principle "avoid, minimize, and mitigate" steers a project toward selecting the most feasible and prudent alternative that results in the least environmental harm (*6*), in balance with other engineering and transportation considerations (e.g., design standards, costs, ROW, utilities), to best address the Purpose and Need for the project. Many of NJDOT's rockfall mitigation projects comply with permitting for publicly owned lands, historic and cultural resources, and threatened and endangered species, among other issues.

The NJDOT also implements a Public Involvement Action Plan (PIAP) for every project. The PIAP is established during the CD phase and continues throughout the NJDOT Project Delivery Process. The plan will typically include meetings with local officials (Public Officials Briefings) as well as one or several Public Information Centers held in the project area. Other meetings may be held with identified special interest groups, such as neighborhood associations and community activist groups, watershed associations, local or regional bus service providers, emergency services providers, business groups, etc. As safety projects, rockfall mitigation projects are vital for traveling motorists, but they are often misunderstood. Therefore, the PIAPs for these projects often need to include an overview of rockfall hazards in general for educational purposes.

After coming to a design-related consensus, evaluating the environmental components of the project and soliciting public feedback, the NJDOT can then advance the project with the selected PPA to the PE phase (or the Final Design phase if the project qualifies for the Limited Scope Approach).

CONCLUSION

Historically, the NJDOT only developed rock engineering work through incorporation within other projects; for example, if a new roadway alignment was proposed and designed, and it would include a cut in rock areas, rock engineering aspects would naturally be incorporated into the project scope. However, in the case of rockfall mitigation on existing roadways, such work would normally only be achieved through 'piggybacking' onto other projects (whether structurally-related or otherwise) within the same geographic area. Unfortunately, since these sorts of partnerings were driven by other engineering goals (such as bridge replacements), this process was 'hit or miss' in terms of targeting the high-priority cut slopes for rockfall mitigation.

Now, to better implement the Asset Management approach, NJDOT's Engineering Geology SME's have succeeded in establishing internal partnerings with NJDOT's Project Management unit to develop a pathway to program, design and implement rockfall mitigation projects as 'stand-alone' projects of their own. Through this partnering, and the subsequent acceptance by NJDOT Senior Management, NJDOT's Engineering Geology SMEs have been able to put forward their RHMS as the primary element in screening high-priority cut slopes for programming into the NJDOT's Project Delivery Process. The result of this effort has been the establishment of a Rockfall Mitigation Program (RMP) that proactively initiates and furthers rockfall mitigation projects that are solely targeting the highest priority cut slopes within the RHMS inventory, thereby fulfilling the prioritization intent of the RHRS methodology.

Collaboration of the project team in the CD phase is essential to the success of the RMP. The systematic evaluation of mitigation techniques using simple comparative tools helps to minimize project schedules and soft costs by reducing waste and revisions. Projects that qualify for the Limited Scope Approach can be significantly streamlined by eliminating the PE phase. Working from an agreed-upon PPA should eliminate the need for further alternatives analyses and associated engineering costs in the PE and/or FD phases. The consensus-building workshops and 'on-the-board' review processes provide a solid foundation for future project stages.

To date, the NJDOT has successfully 'graduated' seven rockfall mitigation projects from the CD phase into the design phase of project development under this methodology, with another four projects expected to advance by the end of 2018. This programmatic approach has provided both NJDOT Senior Management and FHWA with a high level of confidence in the advancement of these unique projects. With this record of success, NJDOT is poised to continue moving ahead as a leader in rockfall hazard management.

REFERENCES

1. Brawner, C.O. and Wyllie, D.C. Rock Slope Stability on Railway Projects, *Proceedings of the American Railway Engineering Association Regional Meeting,* Vancouver, Canada, 1976, pp. 449-474.

2. Turner, A.K. and Jayaprakash, G.P. *Rockfall Characterization and Control*, A.K. Turner and R.L Schuster, Editors. Transportation Research Board (TRB), National Academy of Sciences, Washington, D.C., 2012.

3. McCauley, M. L., Works, B.W. and Naramore, S. A. *Rockfall Mitigation*, Report No. FHWA/CA/TL-85/12, U.S. Department of Transportation, 1985.

4. Pierson, L.A. and Vickle, R.V. *Rockfall Hazard Rating System, Participant's Manual.* Publication No. FHWA-SA-93-057. Federal Highway Administration (FHWA) and National Highway Institute (NHI), 1993.

5. Wyllie, D.C and Mah, C.W. *Rock Slopes Reference Manual*, Publication No. FHWA-HI-99-007. Federal Highway Administration (FHWA) and National Highway Institute (NHI), 1998.

6. *Environmental Overview of Regulations and Permits*, State of New Jersey Department of Transportation, 2016.

Development and Testing of a Modular Rockfall Protection Wall to Mitigate Earthquake-Induced Slope Hazards

Rori Green, PE
North Canterbury Transport Infrastructure Recovery
205 Annex Road, Christchurch 8024, New Zealand
+64 21 289 9688
greengeo71@gmail.com

Cédric Lambert
North Canterbury Transport Infrastructure Recovery
cedric.lambert@outlook.co.nz

Charlie Watts
North Canterbury Transport Infrastructure Recovery
charlie.watts@jacobs.com

Daniel Kennett
Stahlton Engineered Concrete
33 Waterloo Road, Christchurch 8042, New Zealand
Daniel.Kenett@fultonhogan.com

Emerson Ryder
Holmes Solutions
7 Canterbury St, Christchurch 8042, New Zealand
EmersonR@holmessolutions.com

Prepared for the 69th Highway Geology Symposium, September, 2018

Acknowledgements

The authors would like to thank the following individuals/entities for their contributions to the design and physical testing of the modular rockfall protection wall, and for the additional work that contributed to its approval for use on the NCTIR project:

Stuart Finlan – New Zealand Transport Agency
Eric Ewe – Geofabrics New Zealand Ltd.
Michal Tutko – Eliot Sinclair
Clive Anderson – NCTIR
Greg Saul – NCTIR

Disclaimer

Statements and views presented in this paper are strictly those of the author(s), and do not necessarily reflect positions held by their affiliations, the Highway Geology Symposium (HGS), or others acknowledged above. The mention of trade names for commercial products does not imply the approval or endorsement by HGS.

Copyright Notice

ABSTRACT

The November 2016 M7.8 Kaikoura earthquake resulted in excess of 40 landslides that directly impacted the key road and rail corridor on New Zealand's South Island. Within two months, the New Zealand Government formed the North Canterbury Transport Infrastructure Recovery (NCTIR) alliance, a team of more than 1700 workers who were tasked with restoring road and rail service by the end of 2017.

The work has involved a wide variety of landslide hazard mitigation measures that have included source treatment, installation of passive rockfall protection measures and relocation of sections of road further away from the base of the slope onto new seawalls. One of many challenges facing the geotechnical design team is space limitations along the narrow coastal corridor.

A modular rockfall protection wall has been developed to add to the suite of permanent rockfall protection structures in use on the project. The wall comprises interconnected concrete blocks with an upslope energy-absorbing layer of sand-filled and rock-filled gabions. The key advantages of the wall are a narrow footprint and a relatively fast installation time.

It was necessary to demonstrate the performance and capacity of the wall before it could be approved for use on site. Full-scale physical testing was performed at a vehicle impact testing facility. Six tests were undertaken to investigate sliding and overturning failure modes; impact energies were 250 and 750 kJ. Data collected during testing includes multiple high-speed videos and pre- and post-test laser scans.

The wall performed successfully, and it has been approved for use on site. The first installation is anticipated by mid-to-late 2018.

INTRODUCTION

The November 2016 M7.8 Kaikoura earthquake caused significant damage to transportation infrastructure located in the northeast of New Zealand's South Island. Part of the damage was due to nearly 1 million cubic metres of rock falling onto the Main North Line (MNL) railway and State Highway 1 (SH1) from more than 40 primary landslides, cutting off a major transportation corridor and isolating the town of Kaikoura and surrounding rural communities.

By the end of 2016, the New Zealand Government made the decision to form an alliance to undertake work to restore the coastal transportation corridor. NCTIR, the North Canterbury Transport Infrastructure Recovery, is an alliance partnership between the New Zealand Transport Agency (NZTA), Kiwirail and four major construction contractors *(1)*. The alliance team has consisted of up to 1700 people from more than 100 organisations. They were given the challenge of re-opening the corridor by the end of 2017.

One of two NCTIR geotechnical design teams was tasked with works related to characterization and mitigation of slope hazards along 28 km of coastal corridor affected by landslides. The work involved design and construction of landslide hazard mitigation works, from mapping and characterization of landslides to design and construction of protection structures.

A key part of the work involved finding robust rockfall protection solutions for fragile, earthquake-damaged slopes that could be constructed relatively quickly within a narrow corridor. To this end, a modular rockfall protection (MRP) wall has been developed and tested for use on the NCTIR project.

Overview of Earthquake and Damage

The 14 November 2016 M7.8 Kaikoura earthquake was a complex event that involved rupture along multiple faults. Figure 1 shows the area most affected where significant ground shaking occurred. The event was felt throughout most of New Zealand. Fault rupture propagated northwest from the epicenter; surface ground rupture was observed along at least 20 faults spanning a distance of about 100 km.

Due to the significant ground shaking, more than 10,000 landslides were generated over an area of about 10,000 km^2 *(3)*. The area affected by landslides is shaded red in Figure 1.

More than 80 landslides either directly affected or occurred upslope of the transportation corridor. The main part of the transportation corridor affected by landslides extends for a distance of about 7 km south

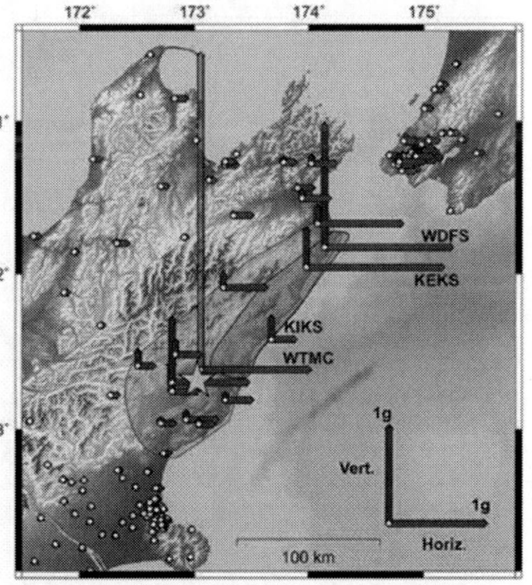

This graphic shows peak ground accelerations in the horizontal and vertical directions recorded by GeoNet strong motion instruments during the magnitude 7.8 Kaikoura earthquake of November 2016. The epicentre of the earthquake is shown as a yellow star. The vertical value of 3g recorded at Waiau in North Canterbury has unusual characteristics which scientists are still investigating. The approximate extent of mapped landslides is shown as pink shaded areas, with the majority occurring in the darker shaded region.

Figure 1 – Area affected by November 2016 M7.8 Kaikoura earthquake (*2*)

and 21km north of Kaikoura; the location of Kaikoura is shown by the KIKS in Figure 1. Figures 2 and 3 show the transportation corridor and provide some general context for the project setting and scope.

Figure 2 – Panoramic views of SH1 / MNL north (upper) and south (lower) of Kaikoura; bare areas along the lower slopes are where landslides have occurred

Figure 3 – Two of the largest landslides affecting the corridor; for scale, the landslide on the right is up to about 250 m high.

MODULAR ROCKFALL PROTECTION WALL

An additional rockfall protection solution with a relatively narrow footprint and low deformation under impact loading is needed along several areas of the corridor where there are space constraints between the slope toe and road/rail alignment. The slopes in many of these areas contain varying quantities of potentially unstable material and are expected to generate multiple rockfalls over time. Flexible barriers such as rockfall fences are not considered a practical option in some areas given the anticipated barrier deformation and the amount and frequency of rockfalls.

Stacked mass concrete blocks have been used in these areas as temporary rockfall protection walls, however the energy capacity, deformation and damage response of these of structures is unknown. The question arose as to whether a permanent rockfall protection wall could be developed using stacked concrete blocks that could be quickly erected at locations where the space requirements did not suit existing protection systems. Given the anticipated range of rockfall energies, the rockfall protection solution would need to be able to withstand moderate impact energies in the range of 300 to 400 kJ (or more, if possible).

Development

Work undertaken by others has been considered in the development of the MRP wall configuration. The concepts of particular interest are the use of gabion baskets as an energy-dissipating layer and the performance of concrete in rockfall protection structures. The work considered includes two separate PhD research projects involving cellular (gabion) structures (*4,5*) and concrete roadside barriers (*6*), and their use as rockfall protection structures. Both studies included physical testing programmes; the findings related to the behaviour and performance of structures during physical testing were considered in the development of the MRP wall configuration and its testing.

Cellular Gabion Wall

Researchers in France have undertaken work to evaluate the use of gabion baskets in rockfall protection structures. This work has included an evaluation of the deformation of individual rock and sand-filled gabion baskets, and small-scale and full-scale testing of cellular gabion sandwich structures composed of layers of rock-filled and sand-filled gabions (*4, 5, 7, 8, 9*).

Of particular interest for the development of the MRP wall was the performance of the rock and sand gabion layers in terms of deformation and dissipation of impact energy. The researchers undertook full-scale testing of a 2m-thick, 4m-high cellular wall backed by an earthen embankment; and a 3m-thick, 4m-high cellular wall. The cellular walls were formed by 1m-thick rock gabion and sand gabion layers. The test energies ranged from 200 kJ to 2200 kJ (*8*). This work was the basis for selection of rock gabion and sand gabion layers as a composite energy-absorbing layer, and it also helped to guide the selection of impact energies used in testing the MRP wall.

Concrete Barriers

The Ohio Department of Transportation (ODOT) recently sponsored research aimed at better defining the performance and energy capacity of concrete roadside barriers used as rockfall protection (*6*). Part of work involved physical testing of precast and cast-in-place concrete barriers. The barrier designs were modified to investigate the effects of various energy-absorbing features on the energy capacity and resulting damage to the barriers. This included varying the reinforcing steel type, size and spacing; as well as using different types of fibre-reinforced concrete. Test energies were up to 160 kJ. ODOT used the results of the work to modify the design of concrete roadside barriers where they are used as rockfall protection.

Of particular interest for the development of the MRP wall is the improvement in energy capacity and performance with the use of steel fibre-reinforced concrete. The addition of steel fibres significantly reduced concrete spalling and increased the energy absorption capacity by 30 to 100 percent, depending on the barrier type and test impact location.

MRP Wall Configuration

The modular rockfall protection wall configuration selected for testing (Figure 4) utilises a modified configuration of sea-wall blocks developed for the NCTIR project together with an upslope energy-dissipating layer consisting of sand-filled and rock-filled gabion baskets. The blocks are 2 m x 1m by 1m (L x W x H); each weighs about 5000 kg; they are chamfered on the upslope side to allow for easier installation around curves. The gabion baskets are 2 m x 0.5 m x 0.5 m (L x W x H). The rock fill is as per the gabion manufacturer's specification. The sand fill is concrete sand with a maximum grain size of 5 mm, lightly compacted within a geotextile-lined gabion basket.

The concrete blocks are installed in an interlocking arrangement and are joined together using vertical steel shear bars. The 32 mm diameter steel bars are installed within a 100 mm diameter open duct, affixed with a plate and nut in the top block. The system is able to dissipate energy on large-scale impact via deformation of the rock and sand gabion layers and sliding and rotation of the individual rigid blocks, while still remaining joined as a coherent barrier to further rockfall.

Development of the wall configuration was a collaborative effort amongst the NCTIR design team, Stahlton Engineered Concrete and Geofabrics New Zealand. Stahlton provided the modified sea-wall block design, including the steel shear bar connections within the concrete blocks. Geofabrics provided general information and advice on the gabion basket layers; this advice included input from Maccaferri who have expertise in rockfall protection solutions.

The motivation for using the sea-wall blocks was two-fold. First, they could be fabricated using the concrete molds developed for the sea-wall blocks, saving both cost and time. Second, with over 7000 sea-wall blocks being planned for use on the project, considerable experience will have been developed in their fabrication and installation.

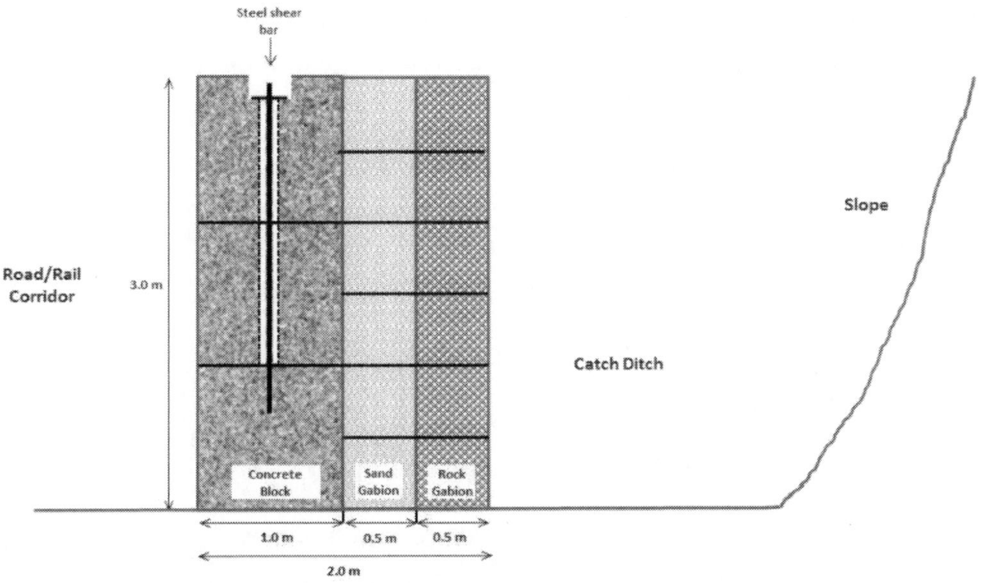

Figure 4 – Cross-Section through Modular Rockfall Protection Wall

Test Design

A testing standard specifically applicable to rockfall protection walls formed with rigid elements does not exist. Instead, researchers and manufacturers have undertaken numerical modelling, physical testing and back-analysis of actual rockfall impacts to evaluate the performance and energy capacity of these types of structures (*10*). Rockfall protection walls are typically designed on the basis of allowable or acceptable deformation, considering an impact by a design boulder with a specified impact velocity.

Testing Programme

In order to demonstrate the performance of the wall sufficiently, so that it could be approved for use on the project, it was decided to use a European testing guideline developed for dynamic rockfall fences as a basis for developing the testing programme. This was discussed and decided upon by the NCTIR design team, Stahlton and Holmes. The document used is the ETAG 027 Guideline for European technical approval of falling rock protection kits, published by the European Organisation for Technical Approvals (*11*).

The key aspect of the ETAG 027 considered in developing the MRP wall testing programme was the selection of two impact energy levels representing Serviceability and Maximum energy levels. Under ETAG 027, these broadly are:

- Serviceability Energy Level (SEL): The barrier should be able to withstand two impacts with no repairs after the first impact. SEL is typically used as a design criteria where multiple or frequent rockfall impacts are anticipated.

- Maximum Energy Level (MEL): The barrier should be able to stop a single impact. It is expected that the barrier will need substantial repair or replacement after an MEL impact. MEL is typically used as a design criteria where infrequent rockfall impacts are anticipated. MEL is defined as 3 x SEL energy level.

The SEL and MEL designations used in ETAG 027 are for dynamic rockfall barriers and are not terms that are typically used to designate energy levels for rigid-type barriers, such as an embankments or this wall. The terms have been used here to indicate the likely "frequent" (SEL) and "infrequent" (MEL) rockfall impact energies.

In addition to impact energies, the other key factor considered in developing the testing programme was failure mode, either sliding or overturning. These failure modes were tested by varying the impact height. Sliding was evaluated by impacting the wall at mid-height; overturning was evaluated by impacting the wall in the upper third of the wall height.

A total of 6 tests were planned to evaluate the energy capacity and performance of the MRP wall; the testing programme is summarized in Table 1 and illustrated in Figure 5.

Table 1 – Testing Programme

Test No.	Failure Mode	Multiple Impacts (SEL)*	Single Impact (MEL)
1 to 3	Sliding	2 x 250 kJ	1 x 750 kJ
4 to 6	Overturning	2 x 250 kJ	1 x 750 kJ

*no repairs to wall after first test

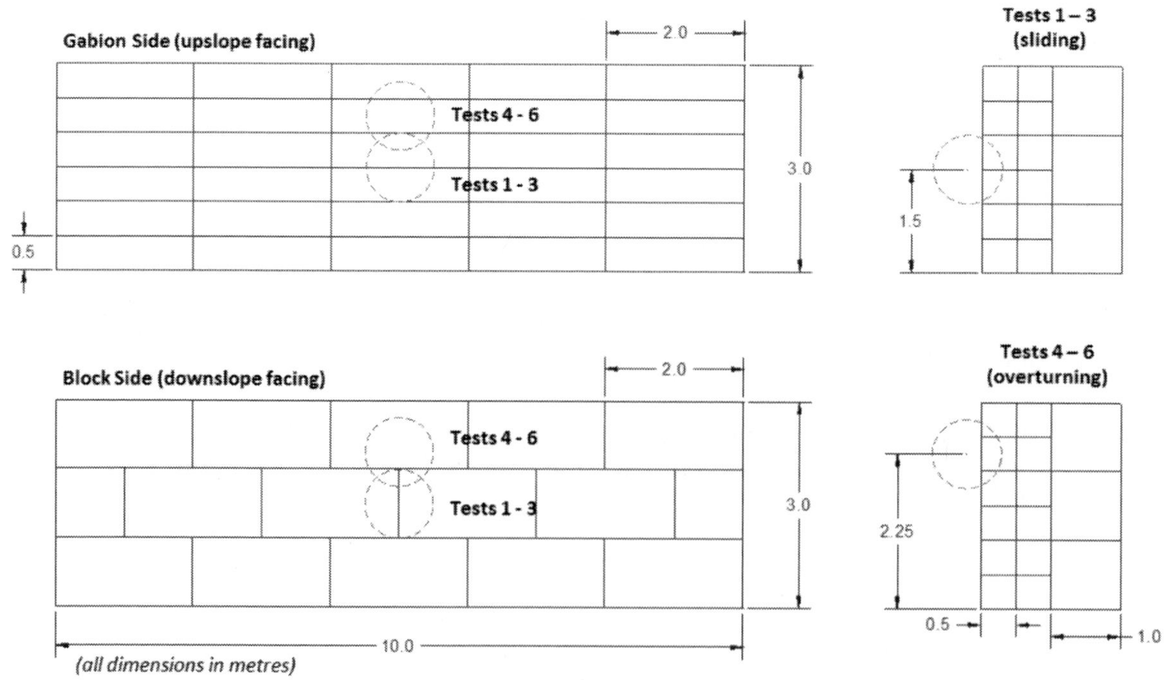

Figure 5 – Test impact locations for sliding and overturning failure modes

Test Walls

Two 3m-high by 10m-long walls were constructed for testing (Figure 6). The walls were constructed within a ditch in order to achieve the impact heights; the ends of the walls were not constrained. The two test walls were designated A and B. Test Wall A was used to investigate the sliding failure mode using a 1.5m impact height (Tests 1 to 3). Test Wall B was used to investigate the overturning failure mode using a 2.25m impact height (Tests 4 to 6). The test walls were substantially re-built following the 2 x 250 kJ impacts.

The composition of the concrete blocks differed in Test walls A and B. Test Wall A was constructed using 50 MPa (28-day strength) plain concrete; Test Wall B was constructed using 50 MPa concrete reinforced with steel fibres at a dosage rate of 20 kg/m^3. It was anticipated that spalling of the concrete blocks would potentially be an issue, and the option to add steel fibres to the concrete blocks for Test Wall B was included as part of the testing program.

Figure 6 – Overhead and side view of test wall

Test Set-up

Impact energies were delivered to the MRP wall via a rolling bogey fitted with a spherical impacting head. The impacting head consisted of a 1-m diameter concrete-filled, steel-reinforced spherical steel dome (Figure 7). The bogey was fitted with steel ballast to scale the weight up and down to achieve the target impact energies. The bogey travelled along a guide rail and was propelled using a tow rope attached to a drop-weight system. The drop-weight system was composed of a known mass of concrete blocks that were lifted with a crane and attached to the tow rope via a system of pulleys (Figure 8). The mass was lifted to a specified height and dropped such that the bogey reached a target velocity on impact with the MRP wall; target impact velocities were in the range of 20 m/s, which is within the mid-to-upper range of possible rockfall velocities. The tow cable dropped off of the bogey immediately prior to impact so that it was travelling freely on impact.

Testing was conducted at the Holmes Solutions testing facility in Christchurch, New Zealand. Holmes Solutions are an ISO 17025 Accredited Testing Laboratory under the International Laboratory Accreditation Cooperation (ILAC) scheme audited by International Accreditation New Zealand (IANZ). Holmes has substantial experience with full-scale dynamic impact testing of roadside safety barriers to US, European and Australia/New Zealand testing standards.

Figure 7 – Test bogey and guide rail system

Figure 8 – Test set-up

Data Collection

Data collected during the tests consisted of high speed video (up to 300 frames per second) from multiple cameras positioned at the sides, front and above the wall. An accelerometer was installed on the test bogey to record velocity before and during impact. Horizontal displacement measurements were taken manually at discrete locations for all tests. Additional displacement data was acquired via laser scanning to provide a comprehensive survey of the wall before and after each test. Laser scanning was undertaken by Eliot Sinclair surveyors for 4 of 6 tests.

Test Results

A total of 6 tests were conducted for the planned testing programme. The wall successfully stopped the bogey in all tests.

Energy dissipation in the MRP wall was observed through the following mechanisms:

- Deformation of rock gabion layer
- Deformation of sand gabion layer
- Displacement of impacted concrete block(s), both translational and rotational
- Displacement of adjacent concrete blocks; engaged through block-to-block contact and through steel connections
- Deformation of steel connections, both rebar and steel plates
- Deformation of foundation, including slight embedment of the toe of wall and rotation of the wall about the toe

A summary of the actual impact conditions, horizontal wall displacements and rotational displacements are presented in Table 2. The horizontal displacements are measured at the front face of the concrete blocks; the rotation is measured about the front toe of the wall.

Table 2 – Summary of Test Results

Test #	Impact Height	Impact Velocity	Impact Energy	Maximum Displacement		Maximum Rotational Displacement
				Base	Top	
1	1.5 m	17.4 m/s	246 kJ	95 mm	124 mm	1 deg
2	1.5 m	17.9 m/s	268 kJ	80 mm	61 mm	1 deg
3	1.5 m	24.0 m/s	769 kJ	352 mm	484 mm	3 deg
4	2.25 m	17.1 m/s	268 kJ	44 mm	200 mm	3 deg
5	2.25 m	17.1 m/s	267 kJ	55 mm	104 mm	2 deg
6	2.25 m	22.9 m/s	755 kJ	186 mm	538 mm	9 deg

Figure 9 shows a comparison of the horizontal displacements for selected tests; laser scan results are shown where available. Of note is the displacement pattern and greater number of blocks engaged for the higher energy impacts. Figure 10 shows the bogey penetration and gabion basket deformation for the same set of tests as in Figure 9. Damage to the gabions was generally confined to the impact zone, however there was increased deformation of the gabions above the impact point.

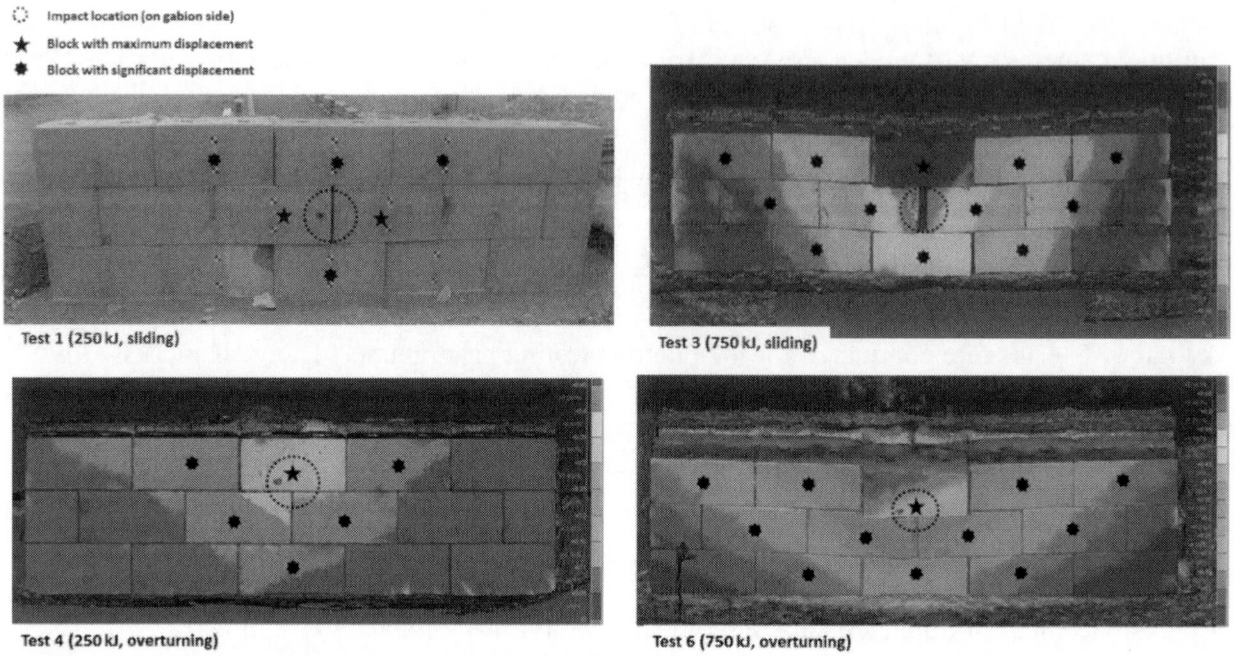

Figure 9 – Comparison of horizontal displacements for selected tests (colour scales differ)

Figure 10 – Comparison of bogey penetration and gabion deformation for selected tests

Figures 11 and 12 show the front face of the wall following the 750 kJ tests for sliding and overturning (Tests 3 and 6). Of particular note is the difference in damage to the plain and steel fibre-reinforced concrete blocks. Spalling occurred in the plain concrete blocks, with relatively large pieces of concrete being lost off the blocks due to contact between the blocks as they displaced during impact. The damage to the fibre-reinforce blocks consisted of cracking and crushing; no spalling occurred. The paint marks in both photos indicate damage that occurred following the 2 x 250 kJ tests; the blocks were re-arranged when each of the walls was repaired following the 250 kJ tests. Minimising spalling is an important road safety consideration if the MRP wall is to be located adjacent to a roadway.

Figure 11 – Test Wall A following Test 3 (sliding, 750 kJ); plain concrete blocks

Figure 12 – Test Wall B following Test 6 (overturning, 750 kJ); steel fibre-reinforced concrete blocks

Additional damage sustained by the wall consisted of bending of the steel bars and deformation of the steel plates (Figure 13).

Figure 13 – Damage to steel bars and plates following Test 5 (2 x 250 kJ, overturning)

Test to Destruction

A 7th test was undertaken immediately after Test 6 in order to further investigate the failure mode of the MRP wall. No repairs were made to the wall and the test was undertaken using the same target impact energy and height as for Test 6.

The gabion layers were substantially damaged during Test 6 (Figure 14). The rock and sand gabions deformed and had a reduced thickness; additionally the sand gabion would have undergone some compaction.

Figure 14 – Test Wall B following Test 6

The impact velocity for Test 7 was 23.1 m/s and the impact energy was 771 kJ. The failure mode was detachment of the upper central block with punching of the steel anchor plates and bending of the steel bars (Figure 14). The 5000 kg block came to rest about 1.6m from the front of the wall with its top face resting on the ground.

Figure 15 – Test Wall B following Test 7

FUTURE USE

Based on the results of the physical testing programme, the MRP wall has been approved for use on the NCTIR project. The recommended energy capacity limits that have been adopted are presented in Table 3.

Table 3 – Recommended Energy Capacity Limits

Case	Energy Level	Expected Displacement
Multiple impacts	250 kJ	< 100 mm*
Single impact	750 kJ	< 400 mm

*for first impact

Additional work that was undertaken in the order to gain approval includes consideration of seismic and debris loading conditions, design life, and inspection and maintenance.

Some of the advantages of the MRP wall for the NCTIR project are:

- Re-use of concrete molds developed for the sea-wall blocks.
- Leveraging of site experience with fabricating and installing sea-wall blocks.
- Reduced footprint width in comparison with a similar embankment-type structure.

- Anticipated reduced construction time in comparison with similar embankment-type structure; this has the advantage of minimizing the time workers spend in potentially hazardous areas, and it reduces the road closure time.
- Potential for staged installation of MRP wall, with concrete blocks installed in advance of gabion baskets; this may allow for use of concrete blocks as temporary rockfall protection.
- Potential for further reduction of construction time if gabions are pre-filled and lifted into place.

REFERENCES:

1. New Zealand Transport Agency. *Kaikoura Earthquake Response*. https://www.nzta.govt.nz/projects/kaikoura-earthquake-response/. Accessed May 2018.
2. GNS Science. *Kaikoura quake produced strongest ground shaking in NZ, new research shows*. https://www.gns.cri.nz/Home/News-and-Events/Media-Releases/strongest-ground-shaking-in-NZ. Accessed April 2018.
3. C. Massey, D. Townsend, E. Rathje, K. E. Allstadt, B. Lukovic, Y. Kaneko, B. Bradley, J. Wartman, R. W. Jibson, D. N. Petley, N. Horspool, I. Hamling, J. Carey, S. Cox, J. Davidson, S. Dellow, J. W. Godt, C. Holden, K. Jones, A. Kaiser, M. Little, B. Lyndsell, S. McColl, R. Morgenstern, F. K. Rengers, D. Rhoades, B. Rosser, D. Strong, C. Singeisen, M. Villeneuve; Landslides Triggered by the 14 November 2016 MwMw 7.8 Kaikōura Earthquake, New Zealand. *Bulletin of the Seismological Society of America*. doi: https://doi.org/10.1785/0120170305. 2018.
4. Bertrand, D. Modélisation du comportement mécanique d'une structure cellulaire soumise à une sollicitation dynamique localisée - Application aux structures de protection contre les éboulements rocheux. 2006.
5. Heymann, A. Approche expérimentale du comportement mécanique des géos-ouvrages à technologie cellulaire. Application aux ouvrages pare-blocks. 2012.
6. Musa, A. *Evaluation of Concrete Barrier as Rockfall Protection*. PhD dissertation, University of Akron, Ohio, 2015.
7. Lambert, S., P. Gotteland and F. Nicot. Experimental study of the impact response of geocells as components of rockfall protection embankments. *Natural Hazards and Earth System Sciences*, European Geosciences Union, 9, pp. 459-467. https://hal.archives-ouvertes.fr/hal-00455337. 2009.
8. Lambert, S., A. Heymann, P. Gotteland and F. Nicot. Real-scale investigation of the kinematic response of a rockfall protection embankment. *Natural Hazards and Earth System Sciences*, European Geosciences Union, 14, pp. 1269-1281. www.nat-hazards-earth-syst-sci.net/14/1269/214. 2014.
9. Bourrier, F., S. Lambert, A. Heymann, P. Gotteland and F. Nicot. How multi-scale approaches can benefit the design of cellular rockfall protection structures. *Can. Geotech. J.* 48 : 1803-1816, 2011.
10. A. Volkwein, K. Schellenberg, V. Labiouse, F. Agliardi, F. Berger, et al. Rockfall characterisation and structural protection - a review. *Natural Hazards and Earth System Sciences,* European Geosciences Union, 2011, 11, p. 2617 - p. 2651.
11. European Organisation for Technical Approvals. *ETAG 027 – Guideline for European technical approval of falling rock protection kits*. 2013.
12. NCTIR. *Detailed Design Report – Modular Rockfall Protection Wall*. Document number 100001-DD-GT-RP-0001. Internal report. 2018.

Rock Slope Remediation at the Penobscot Narrows Bridge

Bryan C. Steinert, P.E.
Haley & Aldrich, Inc.
75 Washington Avenue, Suite 1A
Portland, Maine 04101-4617
(207)-482-4607
BSteinert@HaleyAldrich.com

Laura Krusinski, P.E.
Maine Department of Transportation
16 State House Station
Augusta, Maine 04333-0016
(207)-624-3441
Laura.Krusinski@maine.gov

Amber B. Granger, P.G.
Haley & Aldrich, Inc.
299 Cherry Hill Road, Suite 103
Parsippany, New Jersey 07054-1124
(973)-658-3939
AGranger@HaleyAldrich.com

Wayne A. Chadbourne, P.E.
Haley & Aldrich, Inc.
75 Washington Avenue, Suite 1A
Portland, Maine 04101-4617
(207)-482-4607
WChadbourne@HaleyAldrich.com

Prepared for the 69th Highway Geology Symposium, September, 2018

Acknowledgements

The authors would like to thank the following individuals/entities for their contributions to this project:

Mike Wight – Maine Department of Transportation
Andrew Lathe – Maine Department of Transportation
Brad Miller – Haley & Aldrich, Inc.

Disclaimer

Statements and views presented in this paper are strictly those of the author(s), and do not necessarily reflect positions held by their affiliations, the Highway Geology Symposium (HGS), or others acknowledged above. The mention of trade names for commercial products does not imply the approval or endorsement by HGS.

ABSTRACT

In 2003, Haley & Aldrich, Inc. (Haley & Aldrich) was retained by Figg Bridge Engineers, Inc. and the Maine Department of Transportation (MaineDOT) to provide geotechnical consulting services for the replacement of the Waldo-Hancock Bridge and associated approach roadways, which included the design of a concave, semi-circular rock slope that varied in height up to approximately 100 ft along the Prospect (westerly) approach to the Penobscot Narrows Bridge (replacement bridge).

During construction of the Prospect approach and excavation of the rock slope in 2005 several areas of the rock slope were identified by Haley & Aldrich as needing remediation and in 2012, a long-term rock slope maintenance and monitoring (M&M) program was implemented. Between 2005 and 2016 a total of 23 areas along the rock slope were judged to pose potential safety and long-term maintenance issues of varying degree. In 2016, MaineDOT approved and secured funding to remediate nine different areas of the rock slope judged to be "most critical" and "moderately critical" as it related to public safety and annual maintenance.

In the Fall of 2016, Contract Documents (CDs; plans and specifications) were prepared and "most critical" and "moderately critical" areas of the rock slope were remediated using a combination of rock scaling and vegetation removal, rock dowels, wire rope cable lashing and anchored wire mesh netting. Less critical areas will continue to be monitored and could be remediated during future phases of the project, if additional funding is available.

INTRODUCTION

Beginning in 1931, all traffic heading up U.S. Route 1 (Route 1) along the coast of Maine crossed the historic Waldo-Hancock suspension bridge to access the colorful Down East Maine communities of Bar Harbor, Blue Hill, Castine, and Eastport. The narrow, two-lane, steel bridge soared over the Penobscot River, providing views of the Civil War-era Fort Knox and the town of Bucksport to the north, and Penobscot Bay to the south.

During the spring of 2003, the idyllic scene from the bridge was interrupted by engineers and contractors checking the condition of the main-span suspension cables. They found that the 75-year-old cables were far more severely deteriorated than believed, jeopardizing the integrity and safety of the bridge. Subsequently, the bridge was posted, and access was denied for vehicles weighing over 24,000 pounds until viable stabilization and/or remedial repair options could be provided. An immediate decision was made by the Maine Department of Transportation (Maine DOT) to replace the bridge with a new, modern structure and approach roadways while a stabilization contract was undertaken to strengthen the main-span cables until the new bridge and approaches could be completed.

The location of the replacement bridge (Penobscot Narrows Bridge) is parallel to and immediately downstream of the existing bridge as shown on **Figure 1**.

Figure 1 – Project Locus

PROSPECT APPROACH

Locating the Penobscot Narrows Bridge immediately downstream of the Waldo-Hancock Bridge required realignment of an approximately 775-ft long section of the Prospect approach to the west and into a bedrock-controlled hillside to provide access to the new bridge, as illustrated by the blue dotted line in Figure 1.

The Prospect Approach roadway varies between 40 and 60-ft wide and (shoulder-to-shoulder) and generally consists of two, 14-ft wide travel lanes and two, 8-ft wide outside shoulders. A portion of the roadway has an approximate 14-ft wide curbed median. Ground surface elevations along the concave, semi-circular approach to the new bridge ranged from approximately El. 135 to El. 140 in the vicinity of Route 1 to as high as about El. 250. The proposed grade for the new (i.e., current) roadway ranged from approximately El. 141 to El. 144. As a result, a rock cut up to approximately 100 ft was required to construct the Prospect Approach to the Penobscot Narrows Bridge.

GEOLOGIC SETTING

Based on the Maine Geological Survey surficial geology map of the Bucksport Quadrangle, the near surface soil conditions along the proposed roadway consists of thin drift, which is a glacial till deposit that is generally less than 10 feet thick and overlies bedrock.

The Maine Geological Survey bedrock geology map of the area indicates that the bedrock at the site consists of sulfidic schist that contains graded beds (1/32 to 2-in. thick) of quartz-chlorite-muscovite-plagioclase siltstone and pelite of the Penobscot Formation. Andalusite, corderite, and biotite are present in contact metamorphic aureoles adjacent to granitic rocks. Immediately to the west of the site there is a mapped contact between the Penobscot Formation and the Granite of Mount Waldo. The Mount Waldo rock is a light-gray, medium grained, equi-granular biotite granite with no apparent foliation. An excerpt from the Maine Geological Survey bedrock geology map of the area is shown on **Figure 2.**

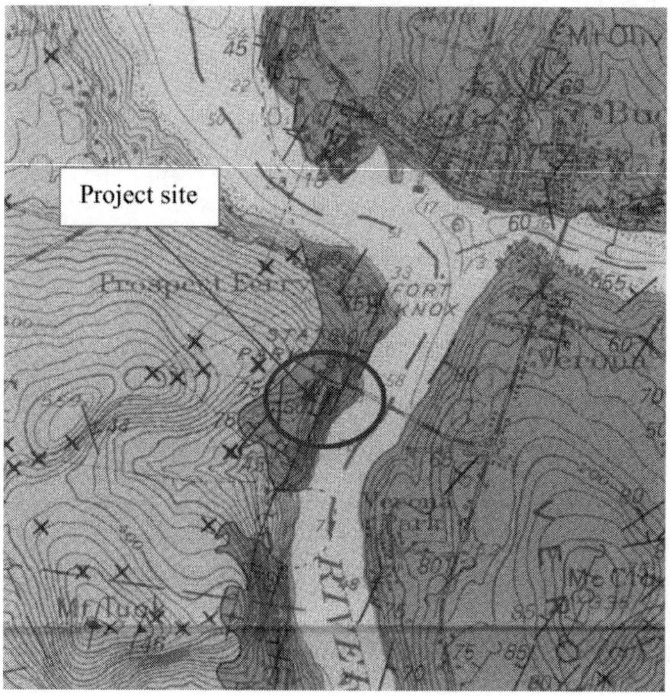

Figure 2 – Bedrock Geology

ORIGINAL FIELD INVESTIGATIONS AND CONDITIONS

Test Borings and Bedrock Sample Descriptions

Haley & Aldrich completed a design phase subsurface exploration program at the site in September and October 2003. A total of four test borings, designated PRCB1-03 through PRCB4-04, were drilled along the proposed approach roadway as shown on **Figure 3**.

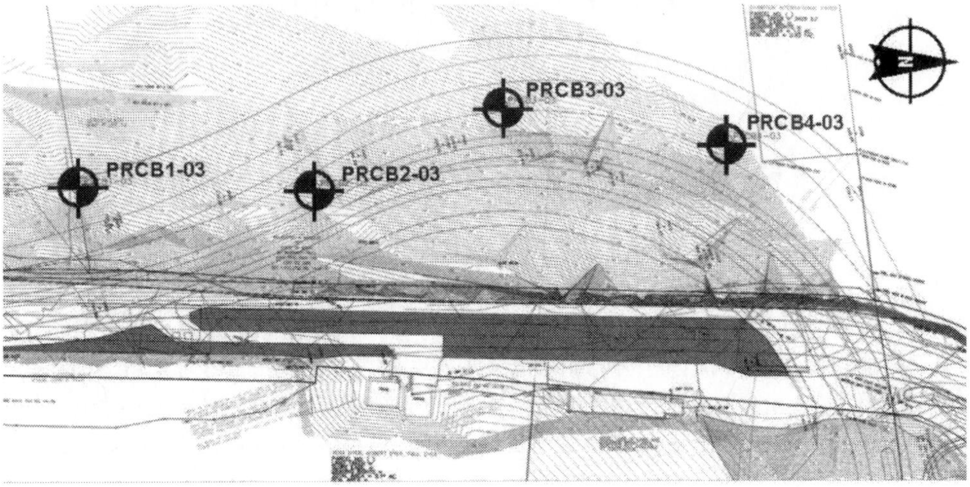

Figure 3 – Test Boring Locations

The test borings were terminated at depths ranging from approximately 28 to 83 ft below the top of bedrock surface.

The bedrock sampled in the test borings generally consist of gray, fine-grained, metamorphic, hard, fresh, metaquartzite. Joints in the rock are typically low angle with steep to vertical foliation joints. The joints are generally tight and discolored, some with heavy oxidation. Veins of gray, medium to coarse grained, igneous intrusive granite were encountered in several of the test borings. Rock quality designation (RQD), a common parameter used to help assess the competency of sampled bedrock ranged from 85 to 100 percent, indicating very good to excellent rock mass quality. Highly fractured bedrock was encountered in localized zones with RQD values as low as 15 percent.

Bedrock Outcrop Observations

In addition to drilling test borings, the geologic conditions at the site were investigated by collecting rock mass data on exposed bedrock outcrops along the existing roadway alignment as shown on **Figure 4**.

Figure 4 – Existing Rock Slope

A geologic reconnaissance was conducted by a Haley & Aldrich geologist in August 2003. While onsite, Haley & Aldrich collected data on structural geologic properties (e.g., strike, discontinuity dip and dip direction, infilling, visible seepage, persistence, aperture) and general rock mass properties (e.g. weathering/alteration, intact rock compressive strength).

The observed bedrock consists of hard, gray, slightly weathered, fine-grained to aphanitic quartzite with occasional pyrite mineralization and a few calcite veins up to 2-in. thick. The rock mass contains three main joint sets. One set is parallel to foliation and dips steeply to the northwest. Another set dips steeply to the northeast, and the third set is low angle to nearly horizontal. The combined orientation of the joint sets results in a blocky structure. Typical block sizes range from about 2 to 5 feet.

DESIGN ROCK SLOPE GEOMETRY

Based on rock engineering analyses of the data collected and the conditions present along the proposed roadway, Haley & Aldrich recommended that the proposed rock cut be sloped at a nominal 4 vertical to 1 horizontal (4V:1H). Haley & Aldrich also noted the potential for localized geologic features with adverse orientations that may not become apparent until rock slope excavation and that may require stabilization. As a result, Haley & Aldrich recommended that stability assessments be made during construction if fractured or jointed rock was exposed.

Rockfall analyses were completed to determine catchment area geometry at the toe of the rock slope. A catchment area is intended to retain rock blocks that may become detached from the rock slope and would otherwise enter the roadway, creating a hazard.

Haley & Aldrich evaluated the catchment area using the computer program RocFall (Rocscience Inc., 2001). The program simulates falling rocks on the slope to determine percentage of rockfall retained by a catchment area. The program allows for variation of the geometry of the rock slope and catchment area to optimize design. The analyses assumed that the rockfall was generated from a 20-ft tall zone at the top of the rock slope. Rock blocks were assumed to have a mean weight of 1,500 pounds with a standard deviation of 500 pounds, which corresponds to a 2 ft x 2 ft x 2 ft rock block with a volume of 0.3 cubic yards. Considering that the results of the rockfall analyses are highly dependent on irregularities on the rock face that act as launch points for a falling block, launch points were given a 10-degree inclination toward the roadway (based on the observation of a secondary joint set at the site) and four variations to the geometry of the rock slope were analyzed to simulate likely configurations resulting from bench blasting of the slope. An irregular rock face can result from less-than-ideal perimeter control blasting that often occurs in a blocky rock mass.

In addition to the rockfall analysis, a rock slope up to 80-ft high was evaluated using the design criteria presented in the Oregon Department of Transportation (Oregon DOT) Rockfall Catchment Area Design Guide, dated February 2002. The design guide relates the height and slope of the rock face with the width of the catchment area (horizontal distance from the toe of the slope to the edge of the pavement) and the backslope of the ditch.

Based on the results of the rockfall analyses and the guidance provided by the Oregon DOT Rockfall Catchment Area Design Guide, Haley & Aldrich recommended a 22-ft wide catchment area including a 14-ft wide unpaved foreslope (4H:1V) and an 8-ft wide paved shoulder. Haley & Aldrich estimated that the recommended catchment area would contain between 80 and 90 percent of rockfall depending on the quality of perimeter control blasting and other factors. Haley & Aldrich also recommended that a clearing limit of 25 feet be established at the top of the rock slope and all soil within 10 feet of the top of the slope be removed and that a rockfill toe buttress be provided to prevent soil from encroaching on the top of the slope. The recommended rock slope geometry and catchment area are shown on **Figure 5**.

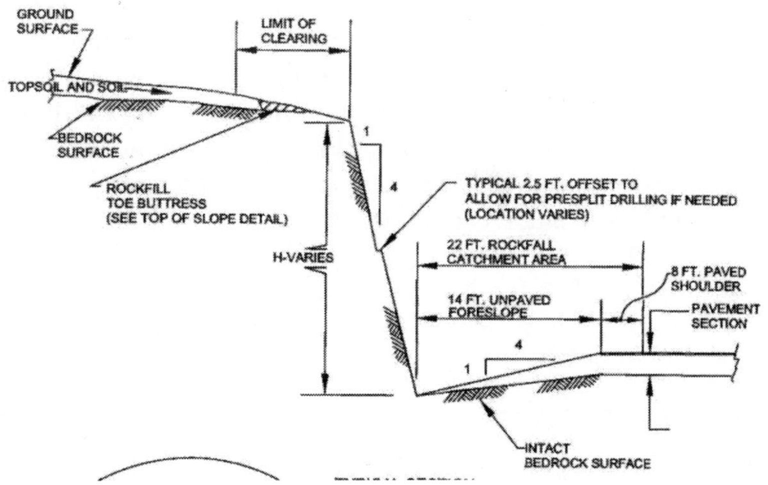

Figure 5 – Design Rock Slope and Catchment Area Geometry

ROCK SLOPE CONSTRUCTION AND INSPECTION

Blasting and excavation of the rock slope began in late 2004/early 2005 and was substantially complete by June 2005. Construction progress photographs are shown below on **Figure 6**. The completed rock slope is shown on **Figure 7**.

Figure 6 – Blasting and Rock Slope Excavation

During construction and excavation of the rock slope multiple site visits were made by Haley & Aldrich staff during which portions (Areas) of the rock slope were identified as needing remediation (stabilization). Draft sketches, details and/or specifications for remedial measures were prepared during construction and again in 2005/2006 in an effort to stabilize the identified Areas. MaineDOT elected not to perform the recommended rock slope remedial work during the original bridge and approach roadway construction due to project-specific constraints at the time of the work.

In 2009 and again in 2012, Haley & Aldrich was re-engaged by MaineDOT to further evaluate the condition of the rock slope, design new and/or refine previous stabilization measures and prepare bid documents in an effort to stabilize identified Areas along the rock slope during demolition of the Waldo-Hancock Bridge. MaineDOT elected to temporarily delay proposed rock slope remedial work until after the completion of the bridge demolition. As a result, the 2012 Haley & Aldrich work plan was modified to include recommendations for a long-term rock slope maintenance and monitoring (M&M) program. The condition of the rock slope was monitored and documented by MaineDOT in 2014.

Figure 7 – Completed Rock Slope

2015 FIELD INVESTIGATIONS AND ROCK SLOPE AREA ASSESSMENT

July and October 2015 Site Inspections

In 2015, MaineDOT approved and secured funding to remediate (stabilize) portions of the rock slope judged to be "most critical" and "moderately critical" as it relates to public safety and annual maintenance. As a result, Haley & Aldrich was re-engaged and conducted a site visit with MaineDOT geotechnical engineers in July 2015 in conjunction with MaineDOTs annual M&M inspection. The primary purpose of the site inspection was to:

- Observe and document rock slope conditions in the Areas where remedial measures were previously (i.e., between 2004 and 2012) recommended and compare the previous and current rock slope conditions to assess whether the recommended remediation measures were still appropriate and what additional remedial measures, if any, may be needed;

- Observe and document rock slope conditions in Areas where remedial measures were not previously recommended and compare the previous and current conditions to assess whether remedial measures may be needed;

- Identify Areas where additional inspection (e.g., rope access inspection) would be needed to collect additional structural information to determine the final priority/ratings for more critical Areas; and,

- Assign preliminary ratings to each of the identified Areas, ranging from "least critical" to "most critical" in an effort to further refine work scope to be completed during subsequent phases of the project.

Several Areas of the rock slope were assigned preliminary ratings of "most critical" and "moderately critical" as a result of the July 2015 site inspection and were judged by Haley & Aldrich as needing follow-up investigation so that they could be accessed from the top of the rock slope using rope access techniques, and observations made in Areas and from perspectives that are not visible from the base of the slope. In addition, the supplemental field investigation provided an opportunity to collect sufficient information to determine vegetation/tree removal requirements both on top of the rock slope and on the rock slope face itself. In general, the primary purpose of the site inspection was to:

- Observe rock slope Areas initially ranked "most critical" and "moderately critical" that were previously judged to pose the highest potential risk for rockfall and where remedial measures were previously recommended.
- Observe, measure and document dimensions of specific rock slope Areas, key rock block attributes, discontinuity location/orientation/condition and identify zones of loose rock to support determination of the final priority/ratings and preliminary and final design of remedial measures.
- Determine the final priority/ratings for each Area based on the additional data collected.

A site inspection was completed by a two-person team of Haley & Aldrich engineering geologists in October 2015 that allowed for a detailed examination of several rock slope Areas previously ranked as "most critical" and "moderately critical". Rope access techniques were used, as shown in **Figure 7**, to descend the rock slope face from the area above the top of the slope.

Figure 7 – Site Inspection using Rope Access Techniques

The rope access approach allowed Haley & Aldrich personnel to directly observe rock structure and the spatial relationships between rock blocks that were not visible or discernable by routine observations made from the base (roadway level) or from the top of the rock slope. Observed attributes included near-vertical separation joint orientations controlling potential block release, zones of weakened and sheared rock, sliding plane conditions where controlling joints dip out of the rock slope face towards the roadway, and measurements of rock block and other critical area dimensions. Observation and documentation of the rock slope conditions also included determining the structural geologic properties of the bedrock (e.g., discontinuity dip and dip direction, frequency, infilling, visible seepage, persistence, aperture) and rock mass properties (e.g. weathering/ alteration, estimation of intact rock compressive strength).

Rock Slope Area Assessment

Between 2005 and 2015 Haley & Aldrich identified a total of 23 Areas along the rock slope, designated Area 1 through Area 19 (including 1A, 4A through 4C and 5A), that were judged to pose potential safety and long-term maintenance issues of varying degree. Final ratings were assigned to each Area after completion of the October 2015 site inspection and are summarized in **Figure 8**.

Area No.[1]	Approximate Location (Sta.)	Year Identified	Original Stabilization Recommendations	2016 Final Rating[2]	Preliminary Stabilization Recommendations[3]	Remediation in Initial Phase of Work[4]
1	360+50	2005-2009	Type B Rock Dowels		2 Rock Dowels	Yes
1A		2015	NA		NA	No
2	361+50	2005-2009	Type B Rock Dowels, Shotcrete, Shotcrete Rock Drains		NA	No
3	362+00	2005-2009	Type B Rock Dowels, Shotcrete, Shotcrete Rock Drains		NA	No
4A	364+00	2005-2009	Type B Rock Dowels, Rock Drains		2 Rock Dowels	Yes
4B			Type B Rock Dowels		2 Rock Dowels	Yes
4C			Type B Rock Dowels		5 Rock Dowels	No
5	365+00	2005-2009	Type A and B Rock Dowels, Wire Mesh		Anchored Mesh, 32 rock dowels	Yes
5A			Type B Rock Dowels		Check Scaling	Yes
6	365+50	2005-2009	Type B Rock Dowels		Cable Lashing, 4 rock dowels	Yes
7	366+00	2005-2009	Type A Rock Dowels, Shotcrete, Shotcrete Rock Drains		NA	No
8	366+50	2005-2009	Shotcrete, Shotcrete Rock Drains		NA	No
9	366+75	2005-2009	Type B Rock Dowels, Shotcrete, Shotcrete Rock Drains		Scaling, 8 Rock Dowels	Yes
10	359+50 to 360+00	2005-2009	Wire Rope		1 Vertical Rock Dowel	Yes
11	358+75	2005-2009	NA		NA	No
12	358+75	2012	NA		NA	No
13	359+75	2012	NA		NA	No
14	364+50	2012	NA		Check Scaling	Yes
15	364+50	2012	NA		Check Scaling	Yes
16	364+50	2012	NA		2 Vertical Rock Dowels, 4 Sub-Horizontal Rock Dowels	Yes
17	359+75	2015	NA		NA	No
18	362+50	2015	NA		NA	No
19	360+50	2015	NA		NA	No

Legend:
"MOST CRITICAL"
"MODERATELY CRITICAL"
"MARGINALLY CRITICAL"
"SLIGHTLY CRITICAL"
"LEAST CRITICAL"

Figure 8 – Final Rock Slope Remediation Area Assessment

Considering that available funding for rock slope remediation was limited, MaineDOT requested that Haley & Aldrich develop rock remediation design recommendations for Areas rated as "Most Critical" and "Moderately Critical". Less critical Areas (i.e., "Marginally Critical" to "Least Critical" Areas) will continue to be monitored during future M&M inspections and potentially remediated during future phases of the project, if additional funding is secured and made available.

ROCK SLOPE REMEDIATION DESIGN AND CONSTRUCTION

In general, rock slope remedial measures included the use of passive rock reinforcement elements (i.e. dowels), anchored ("pinned") wire mesh netting, wire rope cable lashing and rock scaling and vegetation removal.

Haley & Aldrich was responsible for rock remediation design and full-time field engineering and construction oversight that was provided by a combination of experienced engineering geologists and geotechnical engineers during the period 5 October to 22 November 2016. The rock remediation work was completed by Apex Rockfall Mitigation, LLC. (Apex) who was the specialty rock remediation subcontractor to the Lane Construction Corporation (Lane). In general, Apex was responsible for rock slope scaling and vegetation removal, installation of rock dowels, and the installation of anchored wire mesh and cable lashing systems.

Rock Slope Scaling and Vegetation Removal

In general, scaling was completed along the entire rock slope, from the top to the bottom, to remove loose rock fragments/blocks, soil and vegetation that posed a falling hazard both during and after construction. All scaling activities were completed at night and while vehicular traffic was stopped during 25-minute (maximum) intervals as shown on **Figure 9**. All traffic was allowed to clear prior to the next 25-minute stoppage in accordance with project requirements.

The majority of scaling was completed using hand tools consisting of pry-bars, picks and/or shovels. Areas that contained heavily fractured rock and soil were scaled using pressurized air. Large rock blocks were scaled using a combination of hand tools and inflatable air bags.

Figure 9 – Rock Slope Scaling at/near Area 9

Rock Dowels, Wire Mesh Netting and Wire Rope Cable Lashing

A total of 67 rock dowels were installed (61 were included on the contract drawings) within seven different areas of the rock slope, which included six additional rock dowels that were installed in Areas 1, 5 and 10 as summarized below in **Table 1**.

Table 1 – Rock Dowel Summary

Area	Total Number of Rock Dowels		Notes
	Designed	Constructed	
1	2	3	rock dowel A1-3 (15-ft long) added
4	7	7	
5	32	36	four boundary cable anchors (4-ft long) added in the upper left and right and lower right and left corners of the wire mesh
6	4	4	
9	6	6	
10	2	3	rock dowel A10-3 (5-ft long) added
16	8	8	

All rock dowel locations were marked by Haley & Aldrich prior to drilling. The rock dowel holes were drilled using either a specialty "wagon" drill rig suspended from ropes or a "plugger" drill mounted to a manlift as shown in **Figure 10**. The 2-½ to 3-½ -in. diameter holes (minimum 2-3/8 in. required) were generally drilled in close proximity to the marked location with the exception of two, which were relocated by Haley & Aldrich after scaling activities were completed. The holes were drilled to the depths specified unless fractures with significant soil infilling were encountered in the holes during drilling, based on contractor-estimated drill action, like variable drilling rates and loss of air pressure.

Figure 10 – Rock Dowel Drilling with Wagon Drill (left) and Manlift (right)

In accordance with project requirements, a minimum of one pull test was completed in each area of the rock slope where rock dowels were installed (12 total). After receipt of acceptable grout compressive strength laboratory test results and hydraulic jack calibration information, Haley & Aldrich selected the rock dowels to be tested either at random or based on drilling or installation conditions, like the presence of soil seams or lower grout strengths. Each rock dowel was loaded incrementally up to 125 percent of the design load (i.e., 84 kips) and displacement/deformation was measured via two dial gauges that were setup on opposite sides of the dowel bar. Total displacement/deformation for each rock dowel tested was less than the maximum allowed.

Upon successful completion of pull testing, rock dowels were outfitted with the appropriate hardware (bearing plate, washers, nuts) as shown in **Figure 11**. Prior to installation, any voids observed beneath the dowel bearing plate following grouting were backfilled with dry-packed grout. Per specification, the setting force was applied by tightening the nut against the washer and plate to remove loose float from the washers and plate using a torque wrench. The nuts were tightened with a minimum applied torque of 150 ft-lbs.

Figure 11 – Wire Mesh Netting and Wire Rope Cable Lashing in Areas 5 and 5A

CONCLUSIONS

Beginning in 2005, MaineDOT has effectively managed one of their greatest assets in the rock slope along the westerly approach to the Penobscot Narrows Bridge as shown in **Figure 12**. Through their continued persistent efforts to secure funding and in implementing an annual M&M program MaineDOT has reduced risk, kept the infrastructure in good condition while maximizing the available funding. High risk rock slope scaling was completed at night to reduce the potential for impact to the traveling public and aesthetic features (i.e., color selection of wire mesh netting and cable lashing powder coating) were used to blend the remedial measures seamlessly into the natural surroundings.

Figure 12 – Final Rock Slope Condition

Advantages of Using a Downhole Optical Televiewer for Rock Cut Slope Design—An Example in Central Pennsylvania

Jeremy Robinson, P.G.
Gannett Fleming, Inc.
207 Senate Ave
Camp Hill, PA 17112
jsrobinson@gfnet.com
Ph: (717)763-7211

Andrew Smithmyer, P.G.
Gannett Fleming, Inc.
207 Senate Ave
Camp Hill, PA 17112
asmithmyer@gfnet.com
Ph: (717)763-7211

Prepared for the 69th Highway Geology Symposium, September 2018

Disclaimer

Statements and views presented in this paper are strictly those of the author(s), and do not necessarily reflect positions held by their affiliations, the Highway Geology Symposium (HGS), or others acknowledged above. The mention of trade names for commercial products does not imply the approval or endorsement by HGS.

Copyright

ABSTRACT

The Central Susquehanna Valley Transportation (CSVT) Southern Section, is a proposed four-lane, 5-mile long, limited-access highway located in Snyder County, Pennsylvania. The construction of the CSVT Southern Section crosses multiple geologic formations characterized by limited bedrock exposure. Multiple rock cut slopes are proposed with slope heights ranging from 25 to 150 feet. To meet the project needs, exploration of subsurface conditions was performed using a downhole optical televiewer (OTV) sonde to collect borehole data during the subsurface exploration program. Several roadway borings were located in the high rock cut areas for collection of downhole OTV imagery to provide a more robust bedrock discontinuity set for rock slope stability analyses. OTV imagery provides a 360-degree view of the borehole sidewalls that can be used to identify and analyze rock discontinuities in a well log software package, such as WellCAD. Rock discontinuities were identified in the software as joints/fractures, bedding joints/fractures, and bedding features (no fracture), and a well log was created for each surveyed borehole. Discontinuity measurements were analyzed using the computer program DIPS to analyze the potential for failures within the proposed rock cuts.

Collection of OTV data in boreholes on this project provided the ability to collect and analyze a significant amount of bedrock discontinuity data in an area with limited rock exposure. The use of the downhole OTV allowed for collection of a robust set of localized bedrock discontinuity data, which enhanced the analyses for rock cut slope design.

INTRODUCTION

The CSVT Southern Section is a proposed limited-access highway section along U.S. Highway 15 in Snyder County, Pennsylvania that includes multiple rock cut slopes with heights ranging from 25 feet to 150 feet (Figures 1 and 2). Due to the limited rock exposure in the project area, a downhole OTV sonde was used to collect rock discontinuity data. The rock discontinuity data was required to perform the rock slope stability analyses in order to develop cut slope recommendations. This paper summarizes the site geology, describes the process used to obtain the discontinuity data using the OTV, and discusses some of the advantages of using the OTV discontinuity data for rock cut slope design.

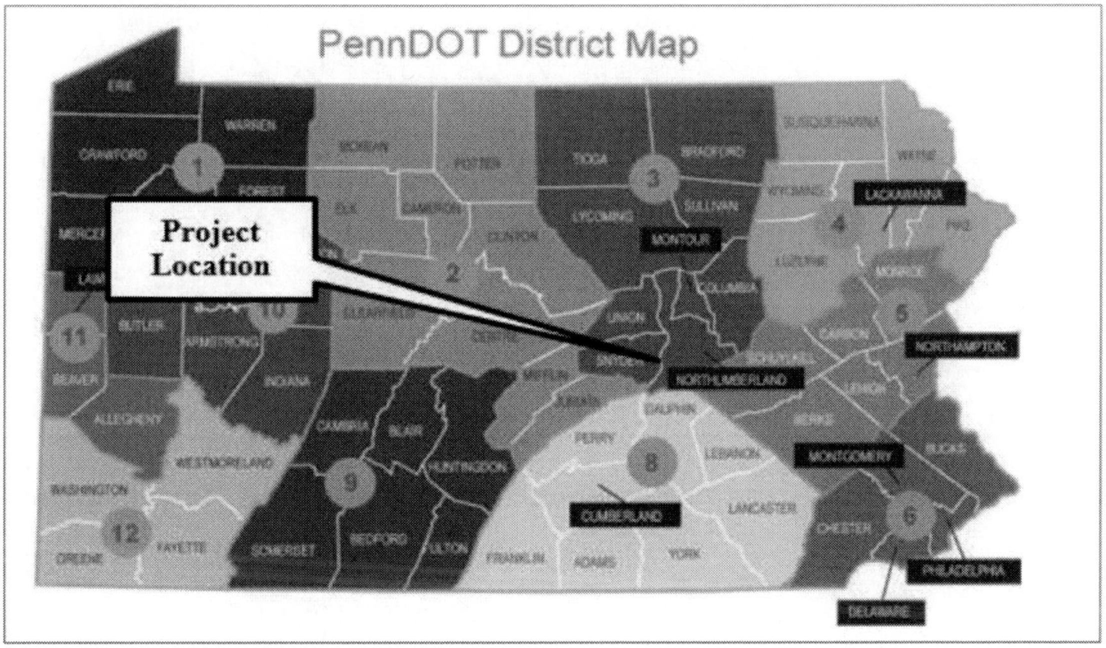

Figure 1 Project Location Map

PROJECT SETTING

Regional Physiography and Topography

The project is situated in the Appalachian Mountain Section of the Ridge and Valley physiographic province. The Appalachian Mountain Section is characterized by numerous, long, narrow mountain ridges separated by narrow to wide valleys. The section is generally comprised of very tough sandstones at the crests of the ridges, relatively soft shales and siltstones in most of the valleys, and limestone and dolomite in some of the valleys (*1*). The project area lies on the west side of the Susquehanna River and is located just south of the confluence of the main and west branches. The relief in this area is moderate to high. Ground surface elevations within the project area range from 426 to 755 feet. The project study area is inclusive of several types of land uses including agricultural, forest, field, and developed.

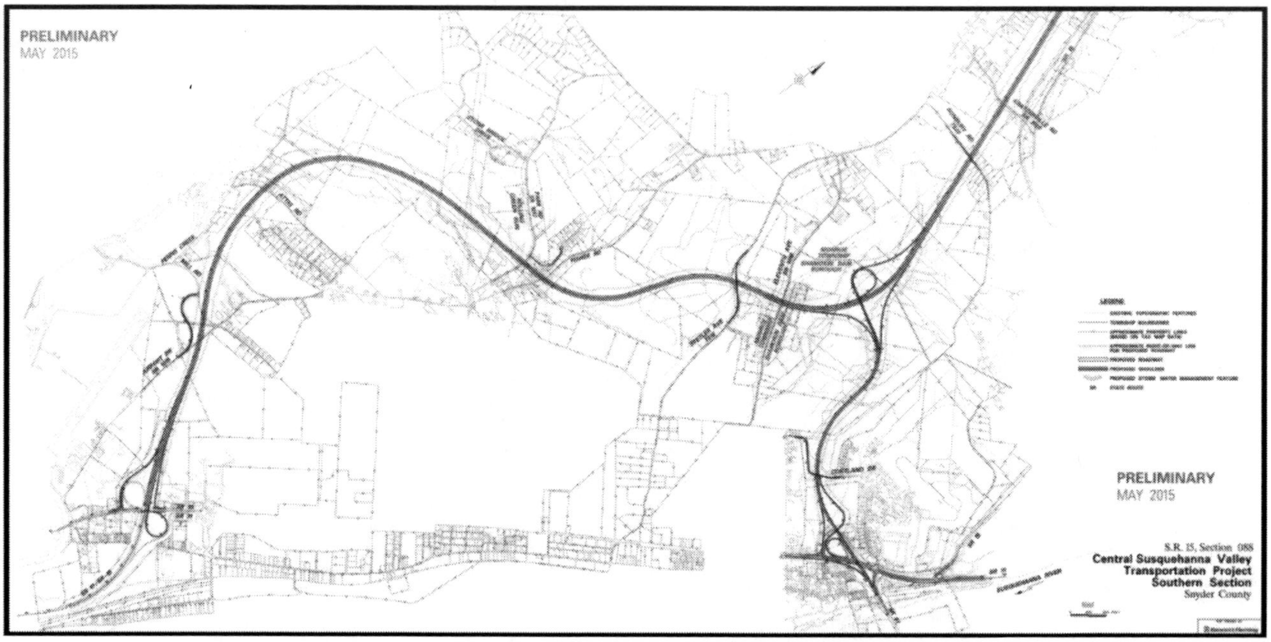

Figure 2 Project location showing proposed alignment of the CSVT Southern Section

GEOLOGY

From south to north, the proposed CSVT alignment passes through areas underlain by Silurian and Devonian age bedrock that is younger to the north and represents an overall transition from shallow marine to deltaic depositional conditions (Figure 3) (*2*). The geologic formations underlying the project area include the Keyser and Tonoloway Formations, undivided, Onondaga and Old Port Formations, undivided, Hamilton Group, Trimmers Rock Formation, and the Catskill Formation as shown in the generalized stratigraphic section in Figure 4 (*3*).

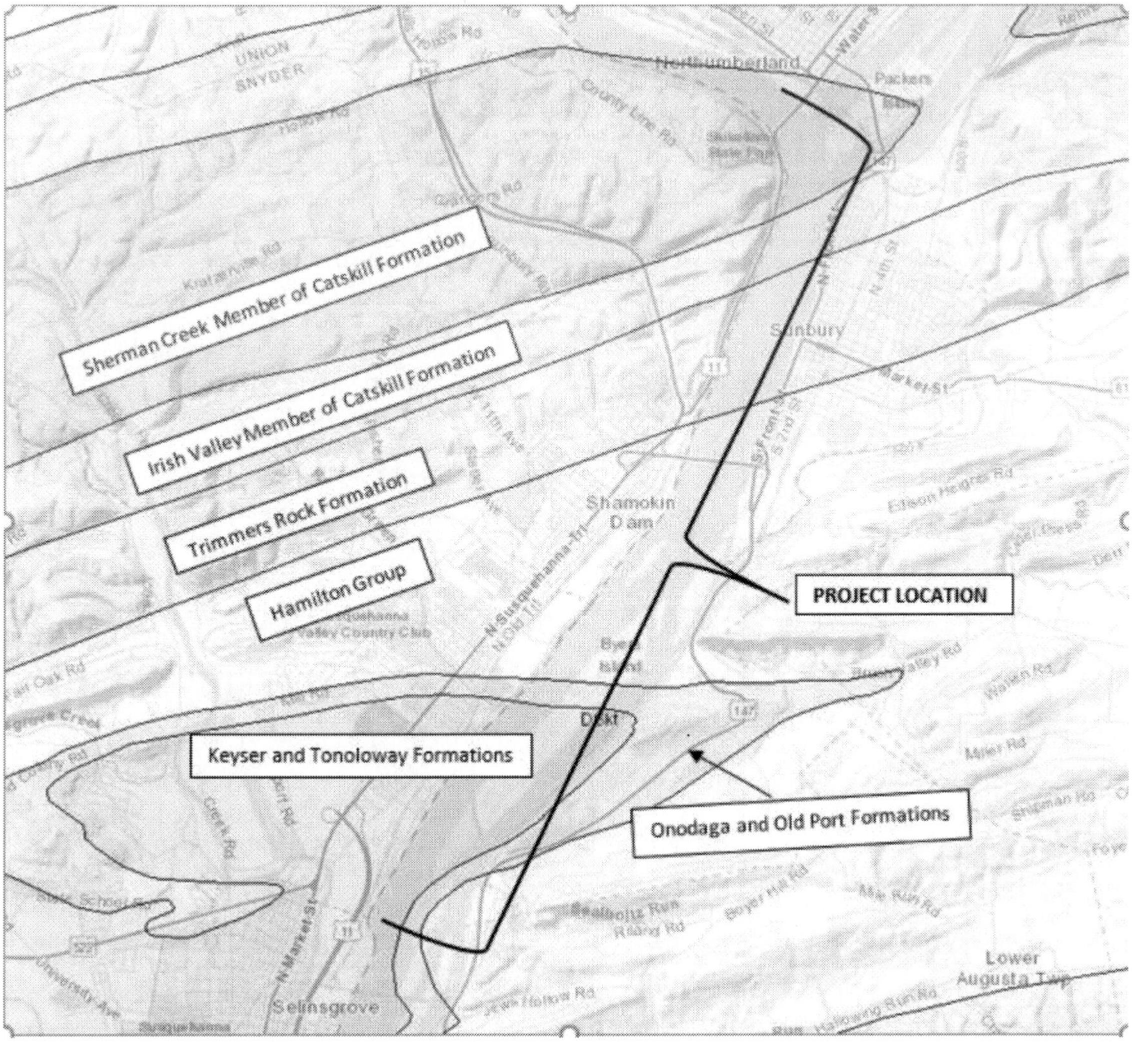

Figure 3 Project Location Shown on Geologic Mapping obtained using GEODE (*2*)

Geologic Discontinuity Measurements

Due to the very limited bedrock exposures present within the project area, only a small dataset of bedrock discontinuity measurements was available from other studies or could be surficially measured. A total of 20 discontinuity measurements of bedding and jointing orientations were acquired from several sources that included: the map of the adjacent Freeburg quadrangle (*3*), Geotechnical Engineering Report (*4*), and manual field measurements collected from outcrops by Gannett Fleming. To develop a more robust bedrock discontinuity set for rock slope stability analyses, 14 geotechnical borings comprising approximately 1,245 lineal feet were scanned with a downhole OTV. A total of 1,440 discontinuity measurements were obtained for use in rock slope stability analyses at proposed rock cut locations along the proposed alignment.

System	Series	Formation		Member	Rock Type
Devonian	Upper	Catskill 7,100'		Sherman Creek-2,000'	Interbedded grayish red claystone and fine grained, cross-bedded sandstone
				Irish Valley-2,100'	Interbedded, grayish red and olive gray sandstone, siltstone, and shale
		Trimmers Rock 2,000'			Medium gray siltstone and shale, with fine grained sandstone in upper part
		Harrell 200'			Olive and medium light gray shale
	Middle	Hamilton Group	Mahantango 1,200'	Sherman Ridge-500'	Olive gray, fossiliferous, claystone with interbedded fine sandstones
				Montebello-250'	Olive gray, medium grained, locally conglomeratic, fossiliferous sandstone, interbedded with siltstone and claystone
				Fisher Ridge-450'	Laminated gray silty shale
			Marcellus 300'		Highly fissile, dark gray to black shale
		Onondaga 200'		Selinsgrove-70'	Medium to dark gray, fossiliferous argillaceous fine grained limestone
	Lower			Needmore-130'	Medium gray, fissile shale
		Old Port 170'			Dark gray, whitish weathering chert, underlain by shale limestone beds, and locally overlain medium to coarse grained sandstone
Sil.	Upper	Keyser 170'			Medium gray, fossiliferous, lumpy, fine to coarse grained limestone
		Tonoloway 600'			Medium gray, laminated thin-bedded, fine grained limestone

Figure 4 Generalized Stratigraphic Column of Central Pennsylvania Geologic Units (*2*)

SUBSURFACE EXPLORATION PROGRAM

A subsurface investigation began in December 2015 to determine and evaluate the geotechnical engineering properties of the soils, bedrock, and groundwater conditions. Approximately 146 roadway borings were drilled during this exploration program. The construction of the CSVT Southern Section includes nine separate roadway sections that require rock cut slopes reaching depths of up to approximately 150 feet.

Fourteen boring locations were identified to provide subsurface information at the proposed rock cut slope areas. These borings were sited at locations of the proposed rock cut slopes and were drilled and constructed to allow collection of the downhole OTV surveys (Figure 5). Temporary casing was installed in the soil overburden and HQ-sized (3.78-inch diameter borehole (96 mm)) rock coring was performed to selected elevations.

Figure 5 Downhole Logging Equipment Setup at a Boring Location

OPTICAL TELEVIEWER SURVEYS

Early use of OTV logging was developed in the late 1960's in the petroleum industry. More recent advances since the 1990's included development of the instruments to be used with common geophysical logging systems and better data processing software. These improvements have allowed for more widespread use of the equipment. Current use of downhole OTV systems can be used in typical subsurface exploration borings with diameters between 1.6 inches to 7.8 inches (40 mm to 200 mm) (*5*). These systems allow for the collection and interpretation of in-situ conditions of the subsurface rock.

The OTV instrument provides oriented downhole optical images that are captured using a Mount Sopris OTV model OBI40-1G. The OTV operates with a fisheye type lens that continuously captures a borehole wall image as it is lowered deeper in the borehole. The OTV utilizes a three-axis magnetometer and accelerometers to orient the instrument in space. The recorded bearings and inclinations are used to orient the captured images in reference to magnetic north and angle from vertical.

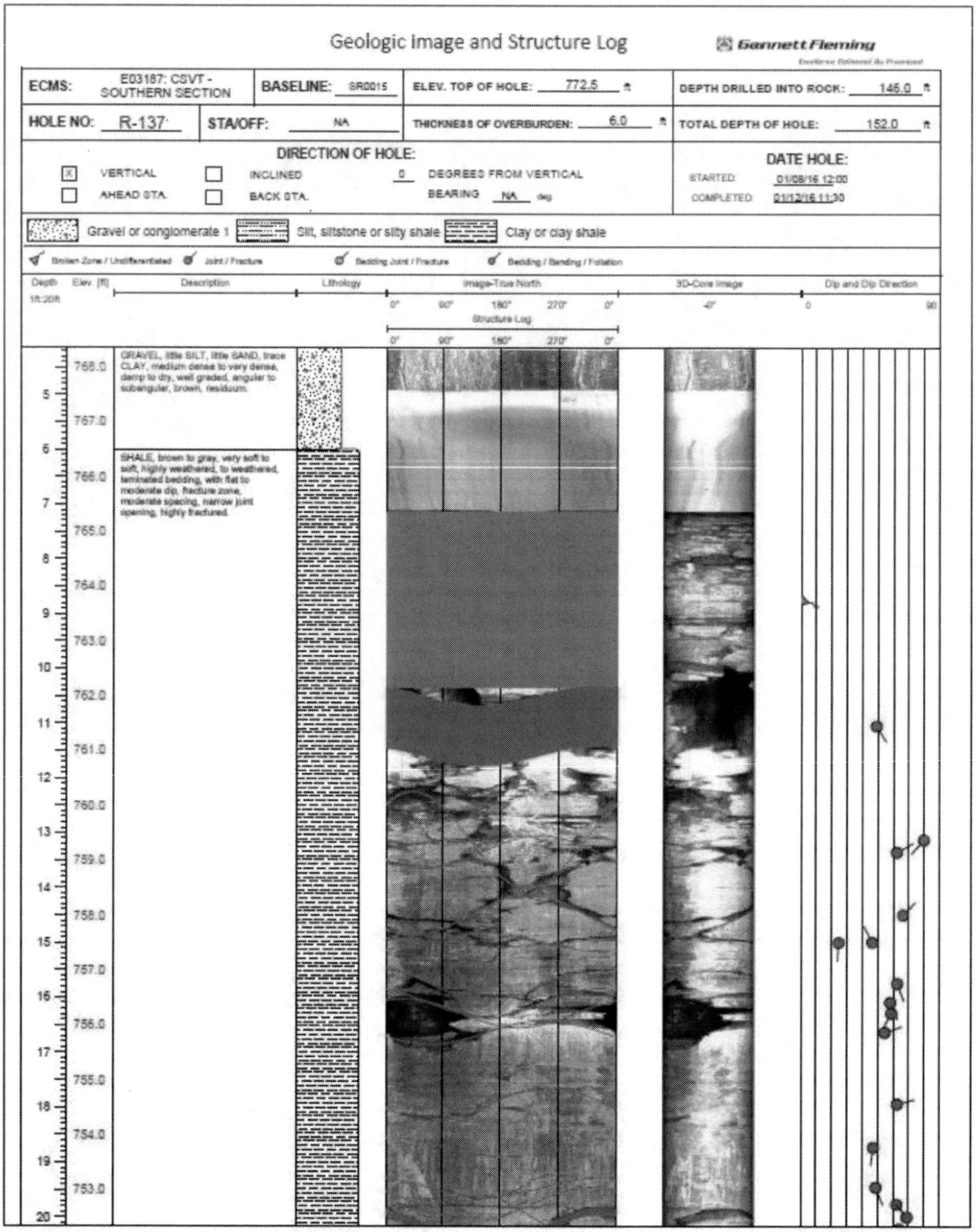

Figure 6 Example of Geologic Image and Structure Log

Image processing software (WellCAD) was then utilized to process the recorded images and orientation data to define geologic structural features (i.e., joint, fracture, bedding). All fractures and bedding features identified were used for a conservative stereographic and kinematic analyses. The software calculates the true orientation of these structural features by knowing the borehole diameter and orientation. All structural data measured and oriented by the WellCAD software was corrected to true north. Outputs generated using WellCAD consisted of Image and Structure Logs (Figure 6).

The Image and Structure log is used to display the OTV image and other downhole information. The log can be customized to meet specific project needs. For this project, the Image and Structure Log includes the general lithology and descriptions recorded in boring logs, the structure log, the three-dimensional core image, and discontinuity data tadpole plots. The general lithology and descriptions were recorded in boring logs and imported into the log. The structure log includes the OTV image and discontinuity pics that were manually assigned a sinusoidal curve to match discontinuities identified in the unwrapped OTV image. WellCAD calculates the orientation of the discontinuity based on the orientation of the borehole (i.e. bearing and angle) and the diameter of the hole. The structural discontinuity data are also plotted as tadpole plots, which show the dip magnitude and dip direction. The dip direction is shown by the direction of the tail of the tadpole.

The structure data was imported into and analyzed using the computer program Dips (version 7.006 by Rocscience, Inc.) for analysis of rock discontinuities used in rock slope analyses. Structure data exported from WellCAD included the depth of discontinuity, the Azimuth bearing of the dip direction of the discontinuities, the dip magnitude, the aperture of the discontinuity, and the discontinuity type. The discontinuities in the analysis for this project included four categories: Broken Zone, Joint, Bedding Joint, and Bedding (for measurement, no joint/fracture).

ROCK DISCONTINUITY CHARACTERIZATION

A total of 343 discontinuities collected from two borings located at proposed rock cut Slope No. 2 were plotted and contoured using Dips. Three principle discontinuities were identified: Bedding $36^{o}/357^{o}$ (dip/dip direction), a steeply dipping joint set (Set B) 80° /235° and a moderately dipping joint set (Set A) $46^{o}/154^{o}$. The large dataset shown in the contour plot illustrates the degree of variability present within each discontinuity set. In many discontinuity investigations, this variability is often unidentified when relying on a limited number of manual measurements. Rock slope practitioners are often compelled to conservatively design new rock cut slopes due to uncertainties related to an inadequate characterization of the variability of the discontinuities present within the rock mass. Results of the borehole discontinuity data collection and subsequent contouring suggests in most cases at least 100 to 150 discontinuity measurements adequately characterize the rock mass for rock slope stability evaluations (*6*).

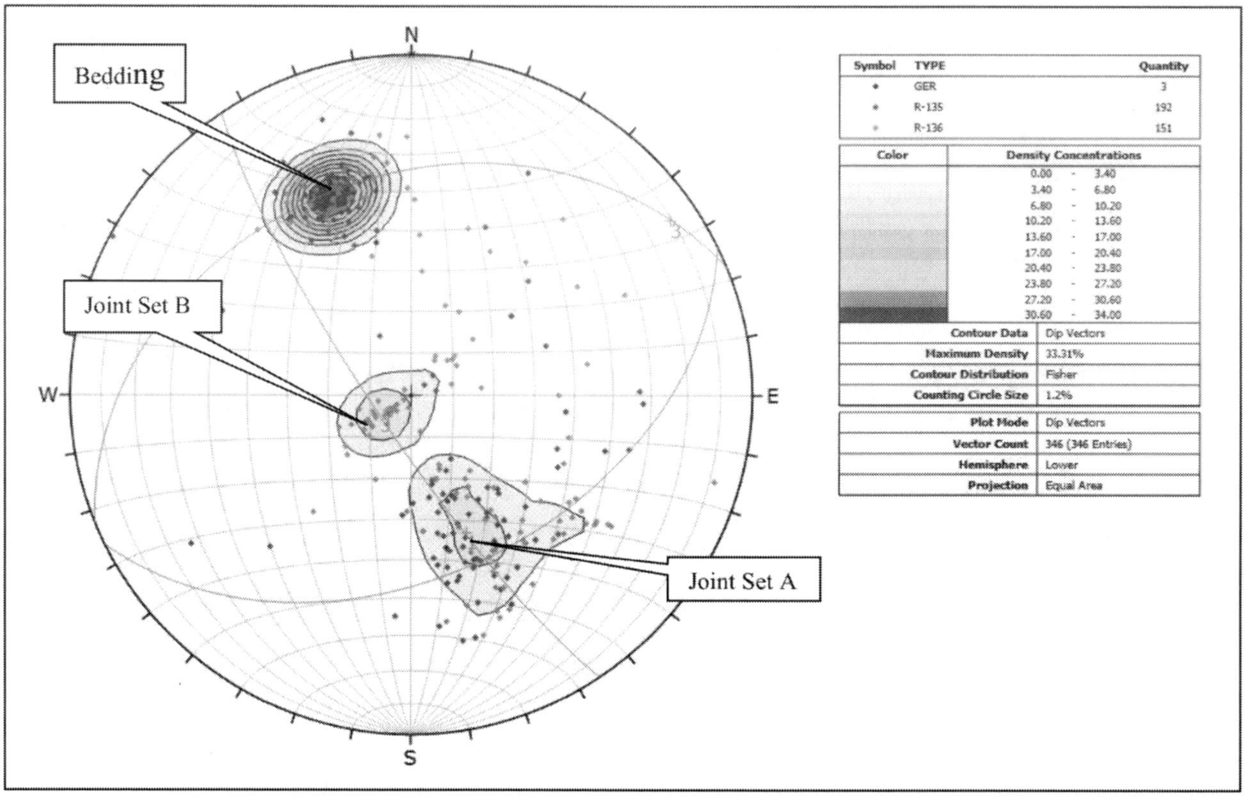

Figure 7 Rock Cut Slope No. 2 – Discontinuity Contour Plot

KINEMATIC ANALYSES

The preliminary roadway alignment resulted in nine proposed rock cut slopes ranging from 40 to nearly 150 feet in height. The stereographic analyses were completed for each proposed cut slope using the Dips computer program. Stereographic projections of the discontinuity data collected from two borings located at proposed rock cut slope No. 2 were used to complete kinematic analyses for planar, toppling, and wedge failure modes and are presented in Figures 8 through 10. An assumed friction angle of 30 degrees was selected based on a published range of (25°-35°) by Hoek and Bray (*6*) and by performing a triple core tilt test on rock cores obtained from the borings.

The kinematic analyses resulted in four critical bedding discontinuities with respect to planar failure and no critical discontinuities with respect to toppling failure. The wedge failure analyses identified nearly 60,000 intersecting discontinuities with approximately four percent of these intersections located within the critical zone. The kinematic analyses were completed considering a preliminary 1.5(H) to 1.0(V) cut slope.

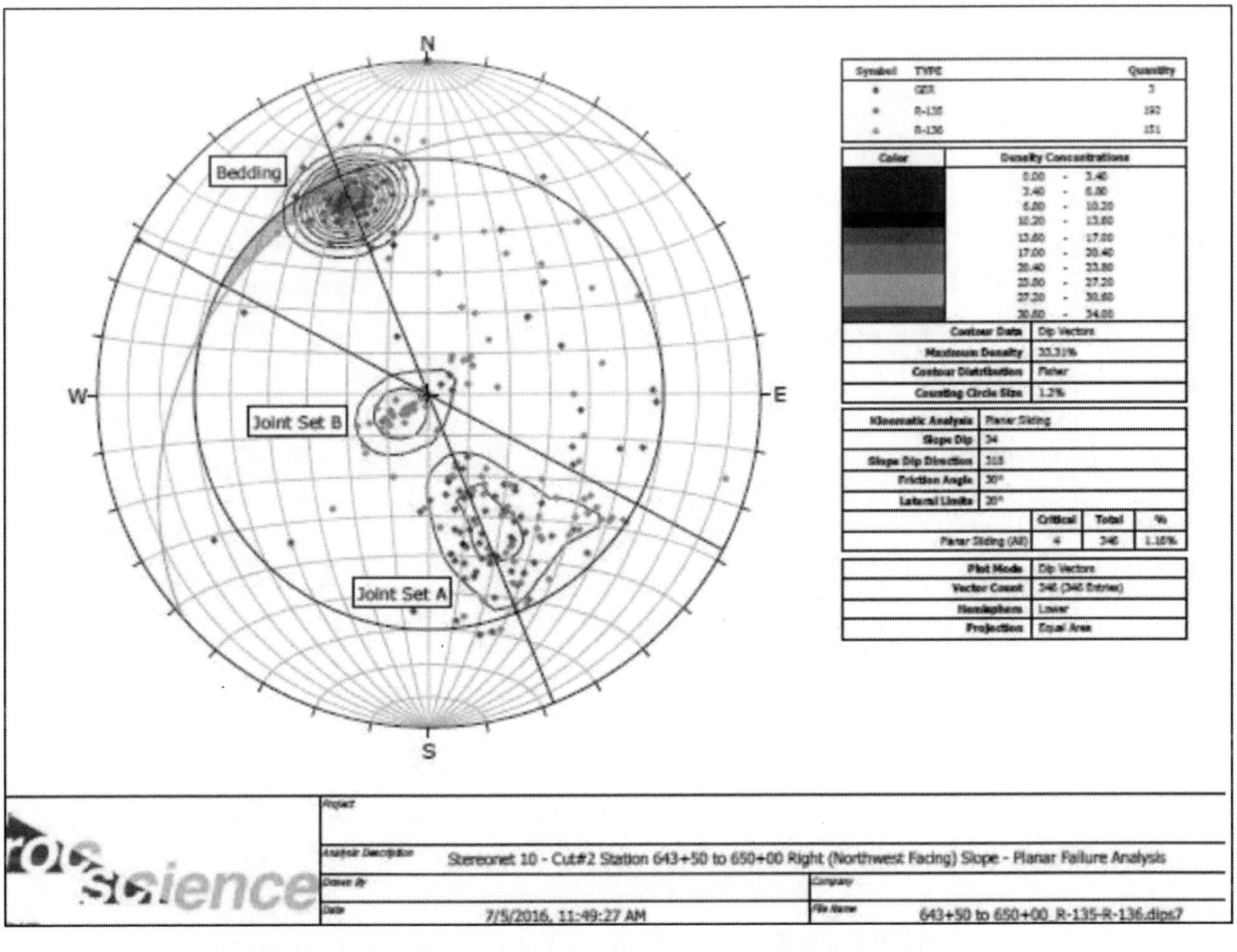

Figure 8 Rock Cut Slope No. 2 – Planar Failure Analysis

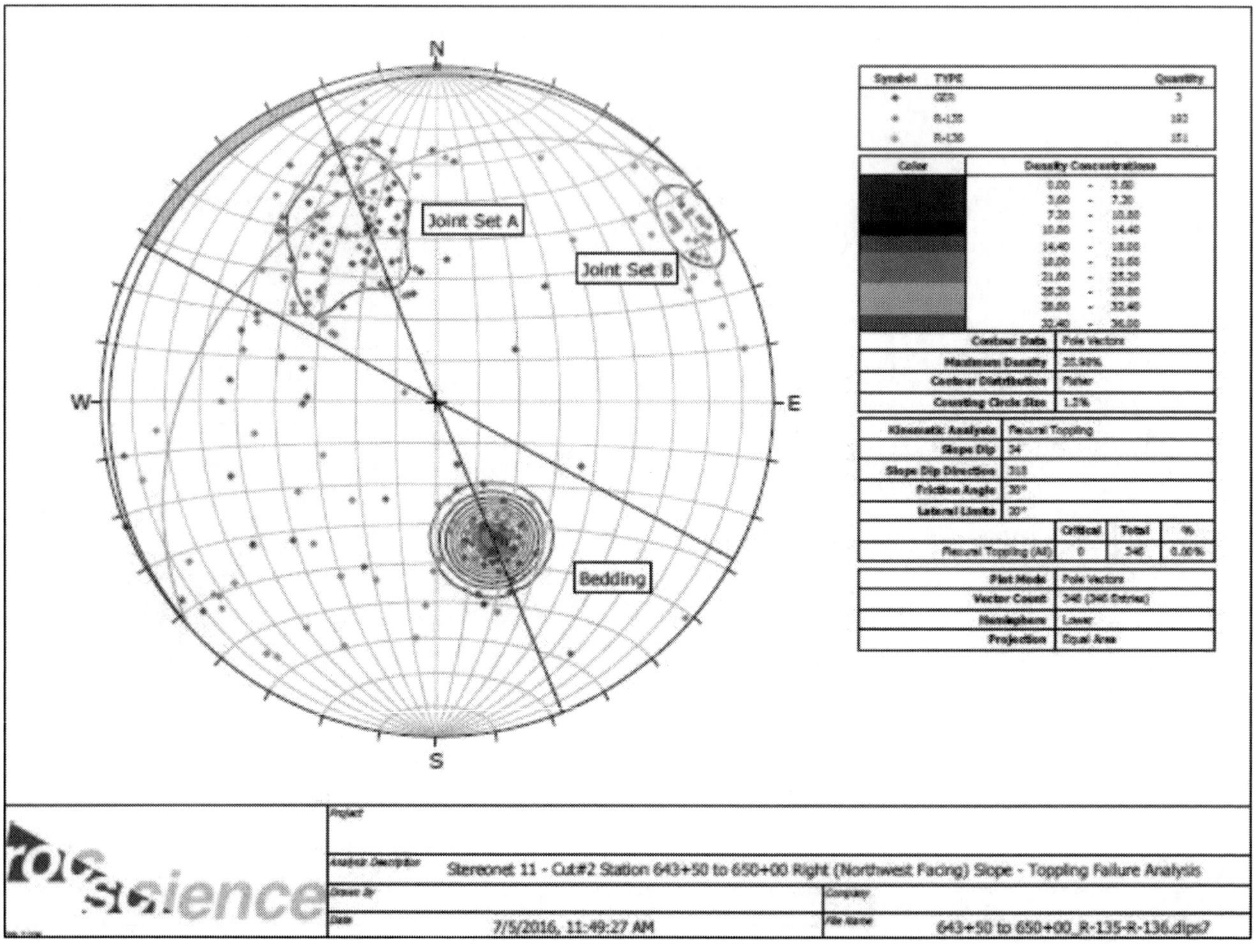

Figure 9 Rock Cut Slope No.2 - Toppling Failure Analysis

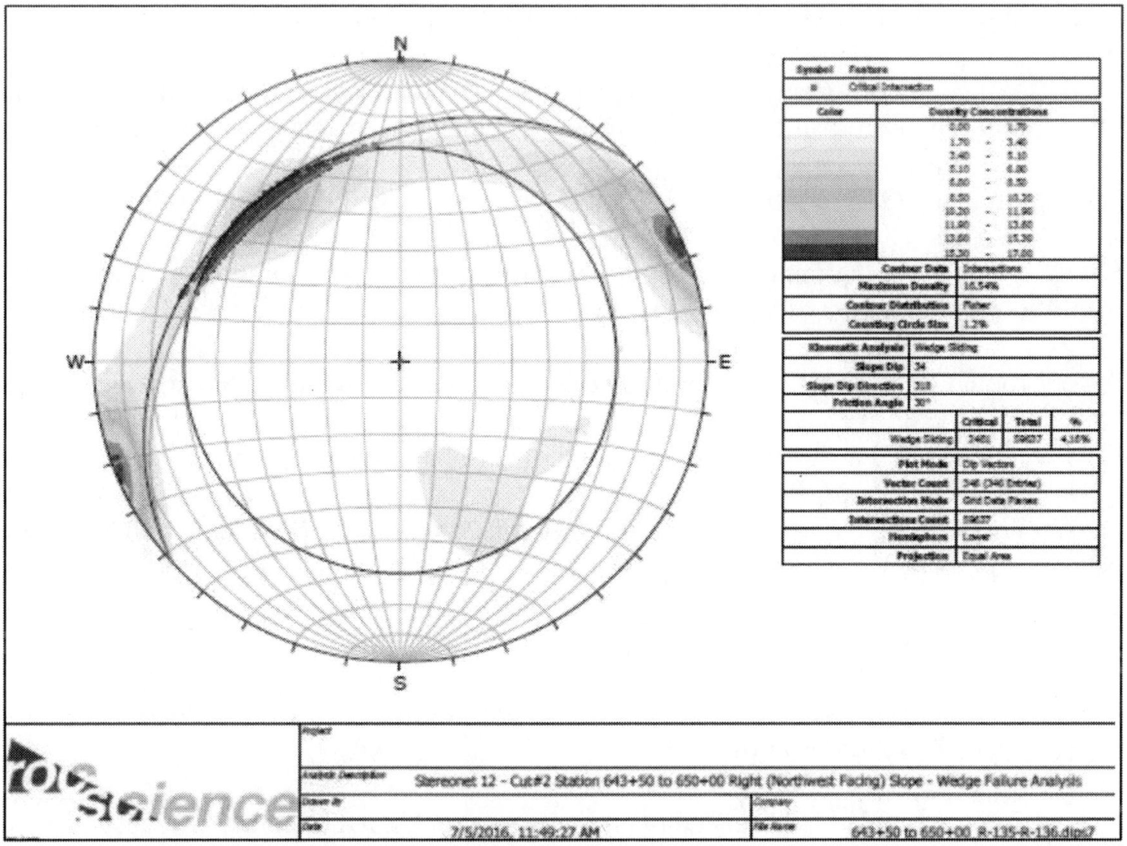

Figure 10 Rock Cut Slope No. 2 – Wedge Failure Analysis

ADVANTAGES OF USING OPTICAL TELEVIEWER IMAGERY

The collection of discontinuity data using an OTV and analysis in WellCAD provided an effective tool to obtain rock discontinuity measurements used in analyses of rock cut slopes.

Advantages in using OTV borehole imagery for this project included the following:

1. Relatively portable equipment that operates using laptop computer and provides a high-resolution digital image of borehole wall.
2. A very robust set of measurements was collected in an area with limited surface exposure.
3. Data were collected at specific borings at the locations of the proposed cut areas.
4. Data collection was incorporated into the subsurface program. OTV data collection was performed independent of a drill rig.
5. A robust set of measurements was collected in a short period of time.
6. The image processing software provided the discontinuity type, strike, dip and dip direction of each identified discontinuity.
7. Robust dataset enabled more efficient and accurate rock cut slope analysis.

REFERENCES

1. Sevon, W. D., *Physiographic Provinces of Pennsylvania*, 4th ed. Map 13. Pennsylvania Geological Survey, 4th ser., scale 1:2,000,000. http://www.dcnr.state.pa.us/topogeo/maps/map13.pdf, 2000.

2. MacLachlan, D. B., Hoskins, D. M., and Payne, D. F., *Bedrock geology of the Freeburg 7-1/2' Quadrangle: Pennsylvania Geological Survey*, 4th ser., Open File Report 95-04. http://www.dcnr.state.pa.us/topogeo/publications/pgspub/openfile/index.htm. 1995.

3. Pennsylvania Geological Survey. Interactive Web-mapping application for Pennsylvania GEOlogic Data Exploration (*GEODE*). https://www.gis.dcnr.state.pa.us/geology/index.html. Accessed May 23, 2018.

4. GEO-Technical Services, Inc. (d/b/a GTS Technologies, Inc.), *Geotechnical Engineering Report in Preliminary Design, S.R. 0015, Section 088, Central Susquehanna Valley Transportation Project, Snyder, Northumberland, and Union Counties, Pennsylvania*, prepared for Pennsylvania Department of Transportation District 3-0, January 2004.

5. Williams, J.H., and Johnson, C.D. Acoustic and Optical Borehole-Wall Imaging For Fractured-Rock Aquifer Studies. *Journal of Applied Geophysics* vol. 55, Issue 1-2, p. 151-159, 2004.

6. Hoek, E., and Bray, J.D, *Rock Slope Engineering*, 3rd ed., Institution of Mining and Metallurgy, London, 1981, 358 p.

Design of Pinned Drapery Systems for Rockfall Protection

Mike Koutsourais, PE
Rockfall Business Unit Manager
Maccaferri, Inc.
10303 Governor Lane Blvd.
Williamsport, MD 21795
301-223-6910
mkoutsourais@maccaferri-usa.com

Marco Deana
Corporate Rockfall Sector Technical Specialist
Officine Maccaferri S.p.A.
Via Kennedy 10-40069
Zola Predosa Italy
m.deana@maccaferri.com

ABSTRACT

Pinned, or secured, drapery systems are composed of steel wire mesh and anchors. The purpose of the pinned drapery system is to improve the surficial rock face stability and maintain the debris/rock in place.

Currently, no standard design procedure exists for determining the improvement provided by steel wire mesh in pinned drape applications, and designers are forced to adopt one of the proprietary design methodologies and software programs available from the few manufacturers of steel wire mesh for rockfall mitigation in the United States. Each methodology has its own unique assumptions and design approaches to analyzing the surficial stability of steel wire mesh in pinned drape applications. Yet from a purely technical design standpoint, the geomechanics of the problem are the same irrespective of the product.

This paper will discuss the analytical differences between the various design approaches for the steel wire mesh for pinned draperies currently available in the market with the intent of opening dialogue toward the development of an industry standard design methodology for this critical and growing rockfall mitigation application.

Introduction

Over-steepened soil slopes and certain rock slopes with an unstable surface layer often require additional surface stabilization measures. The use of high strength steel wire mesh pinned to the slope surface with soil nails or rock bolts has proven to be a viable, cost-effective and aesthetically pleasing alternative to conventional techniques. These systems are often referred to as secured, or pinned, drapery systems. The goal of this solution is to stabilize the surficial portion of the slope with the nails and keep in place the unstable material that can move between the anchors with the mesh. Figure 1 illustrates the components of a pinned drapery system, consisting of a combination of anchors and high strength steel wire mesh.

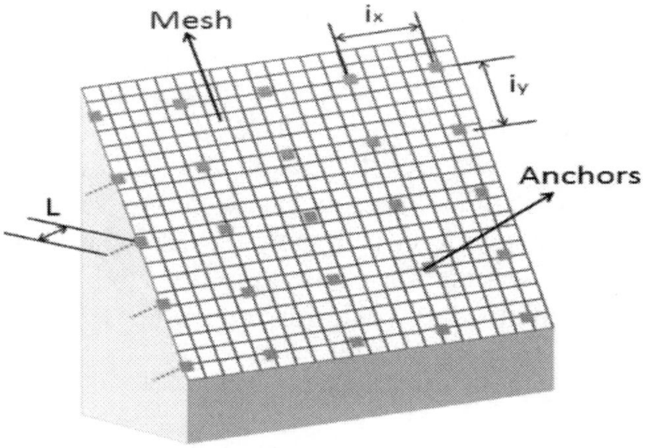

Figure 1 - Pinned drapery system: the intervention is composed of a steel wire mesh and a pattern of anchors (L = length of the anchors; i_x and i_y = distance between the anchors, respectively horizontally and vertically).

Pinned drapery systems are used to retain unstable rock and soil slopes and are more focused on the stability of the facing or shallow depth features of the material. Given the high degree of variability between rock and soil, many different analysis methods and outcomes are possible. For rock slopes, kinematic analysis that addresses the planar, wedge, and toppling modes of rock failure is the general method used. This method considers the global failures of the rock mass but is typically not used to evaluate and analyze the localized facing aspects. For soil slopes, a limit equilibrium analysis method is typically used to evaluate the deeper seated, global stability issues and in many instances is also used for evaluating shallow, surficial features [5].

For each of the material types ranging from rock to soil, the analysis generally consists of limit equilibrium methods, but there are limitations to using this approach depending on what the fundamental failure mode concepts and assumptions for the rock or soil are. One main question involves whether the pinned drapery is providing confinement to a rock or soil mass. Confinement in a soil mass would be analogous to an "at-rest" earth pressure. If deformation is reduced or assumed to be minimal, then a limit equilibrium approach may be justified; however, if the main design concept is that the system will be allowed to deform and bulge outward, this approach is better analyzed using more rigorous finite element type modeling. The difficulty for a design engineer is that limit equilibrium methods are generally straightforward and provide a relatively short design timeline, whereas finite element methods take longer and are much more complex and, in many instances, cost prohibitive for an owner/agency to fund for what can be a relatively small section of rock slope [5].

The analysis of a pinned drapery system on an unstable slope can be performed by either: (i) decoupling the slope-parallel stability contribution given by the nail from the stability contribution given by the mesh for the retention of the inter-nail soil/rock volume, or (ii) taking into account all the contributions to stability from both the nail and mesh.

For this second approach, proprietary design programs often require the designer to choose which mesh to consider, then analyze the stability of the prescribed mesh based on a certain set of design conditions, anchor types and patterns. They are dependent upon the reliability of the calculation model and on the designer's familiarity with the performance characteristics of the various wire mesh products available. In essence, the wire mesh characteristics then dictate the maximum anchor spacing and thus the overall cost-effectiveness of the design. Figure 2a illustrates a conceptual design iteration for these proprietary approaches.

A standardized design approach would allow the designer greater flexibility in determining the most cost-effective nail type and pattern for the client while being empowered to specify the minimum wire mesh performance characteristics required to achieve stability without necessarily being intimately familiar with the numerous products offered by various manufacturers. Figure 2b illustrates a conceptual design iteration of a hypothetical standardized approach. At this very last level of the design process, there are some basic questions to be addressed: what are the technical features of the mesh involved in the design process?

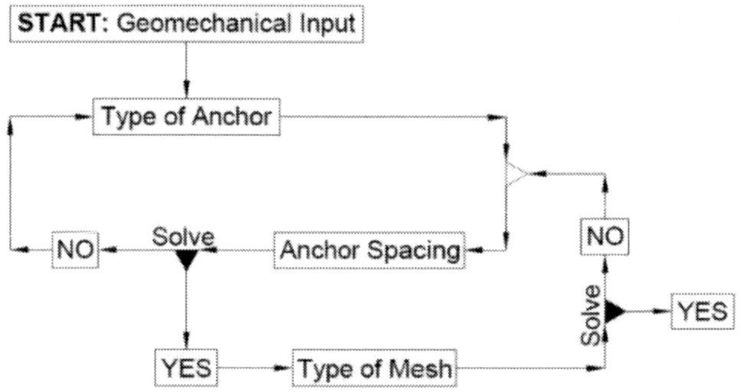

Figure 2a – Conceptual Design Iteration for Proprietary Systems

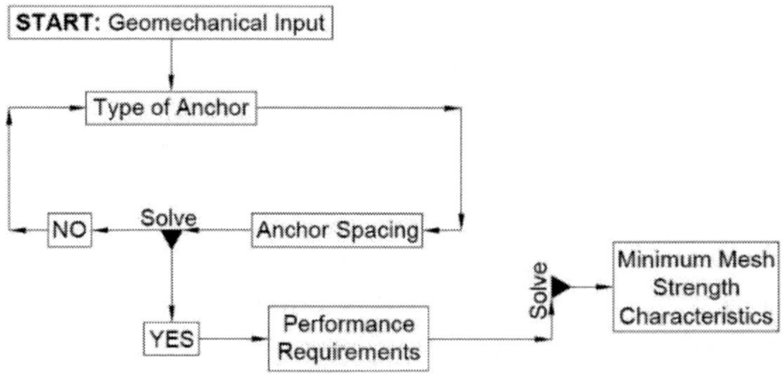

Figure 2b – Conceptual Design Iteration for Hypothetical Standardized Approach

Comparisons of Various Design Approaches

Slope stability is most commonly determined through either a Limit Equilibrium or Finite Element Analysis. Each of these methods has its own peculiarities, which should be known when selecting the one to use. For instance, the Limit Equilibrium Approach is simpler and less time-consuming than the Finite Element Method, both for the definition of the model and for the calculation time.

The Limit Equilibrium Method (LEM), developed in the middle of twentieth century, studies the Factor of Safety of the slope. As a first step, a failure surface is defined, then the equilibrium of the forces along the direction of this surface is studied, with the result being:

$$Fs = \frac{R}{F}$$

Where:

- Fs is the Safety Factor
- F are the driving forces
- R are the resisting forces

The required value of Fs is greater than 1.0, typically 1.3 to 1.5, depending on how R and F were calculated. A major assumption of the LEM is that the soil is a rigid, perfectly plastic material which is analysed to prevent collapse of the slope without considering deformation. This approach considers the equilibrium of the slope before any movement takes place [9].

The Finite Element Method (FEM) was also developed during the twentieth century, but it has become popular with the development of computer science, which has made the often iterative numerical analyses affordable.

The FEM studies the behaviour of the entire model, without any assumption regarding the sliding surface. The only assumption made is with respect to the material model used to simulate the behaviour of the materials making up the model. Although the FEM analysis gives good results, it has high computational time, and the reliability of the results is highly dependent on definition of the model parameters.

Although no standard design procedure exists, commonly used design approaches include Maccaferri's MacRO 1, GeoBrugg's Ruvolum®, Rocscience's SLIDE, and Plaxis® finite element analysis. The purpose of this paper is not to critique or support one procedure versus the other, but to offer an objective 'birds-eye' view of these procedures with the intent of moving the industry toward standardizing a methodology that may, and probably should, consider elements of each of these existing approaches.

While the MacRO 1, Ruvolum®, and SLIDE design programs provide a limit equilibrium analysis on 'slope-parallel instabilities' associated with the stability mechanisms related to the anchors, only the MacRO 1 and Ruvolum® programs include an analysis of the stability mechanisms related to the steel wire mesh facing. These two proprietary design approaches include a limit equilibrium analysis of both 'slope-parallel instabilities' and 'between-nail' instabilities [1, 4, 8], as shown in Figure 3a and 3b, respectively. Superficial slope-parallel instabilities concern the potential for the cover layer to slide off the stable subsoil, as the nailing is intended to stabilize this unstable cover layer. Local instabilities between the nails are also addressed to ensure the surface stabilization system, nailing in combination with a mesh cover, retains any potential surface mass that may become unstable. Both design procedures also calculate and check the minimum length and pattern for the anchors in order to improve the equilibrium condition of the slope face. They also require the designer to pre-select the steel wire mesh facing to be analyzed. The MacRO 1 design further considers a 'serviceability' limit state criterion to ensure that excessive facing deformation will not occur, as excessive deformation could cause stripping on the anchors, increased force on the anchors, or could interfere with close infrastructure or vehicles [1].

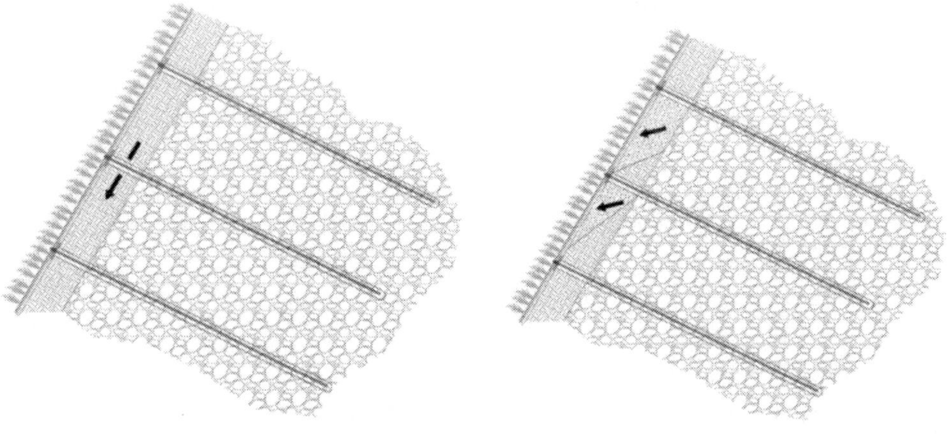

Figure 3 – (a) Slope-Parallel Instabilities and (b) Between-Nail Instabilities

While the failure mechanisms analyzed by these two design procedures are consistent, the underlying assumption of the type of restraint is quite different. The theory behind the MacRO 1 design approach considers that under the weight of the debris, the mesh deflects and generates a pocket of debris. Since the steel wire mesh is assumed to deflect under load, the mesh cannot therefore be modeled as a beam which is able to transmit pressures uniformly distributed on a surface by means of the nails [1] [3]. Based on the MacRO 1 design principles, the performance of the system is given in terms of resistance and deformation. The characteristics used for the steel wire mesh include the tensile strength, the load bearing (punching) resistance, and punching deformation. Figure 4 illustrates the basic theoretical model for the MacRO 1 design approach.

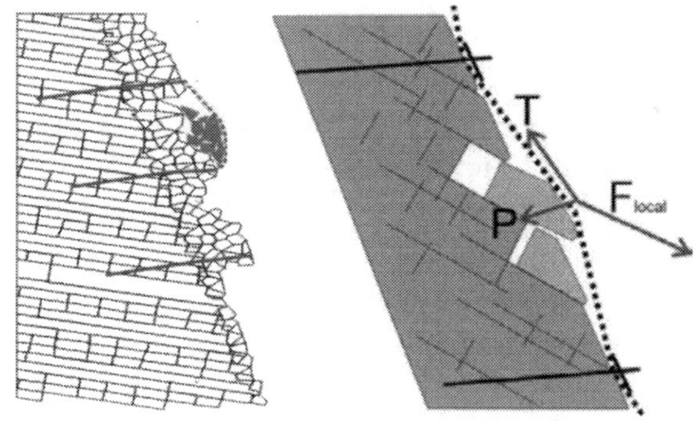

Figure 4 – Theoretical Model for the MacRO 1 Design Approach

The MacRO 1 calculation procedure allows for determining both the ***ultimate limit state*** (verification of breaking loads of the system components), ***and serviceability limit state*** (maximum permissible deformation of the facing). The MacRO 1 theory assumes that debris may slide outward from the slope face and cause the mesh to deform. Since the load pushing is asymmetric and the mesh deforms unevenly, the forces acting on the facing are represented as shown in Figure 5 [1]. The MacRO 1 software requires the input of geological parameters and allows the application of external loads. It is not based upon the Mohr-Coulomb criterion, but on rock mechanics.

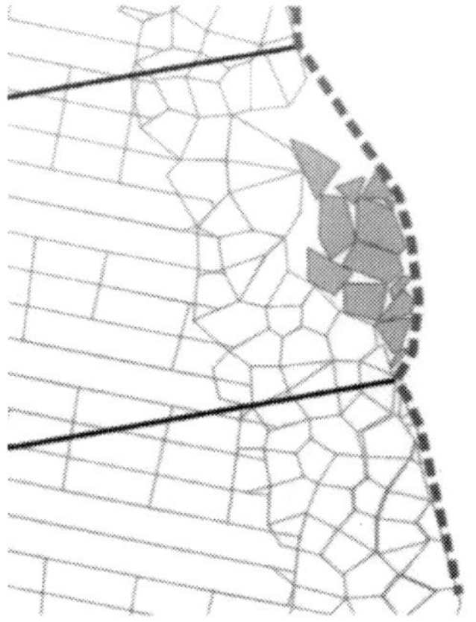

Figure 5 – Deformed Mesh with Forces Characterized by In-Isolation Punch Test

In this simplified scheme, the force of the debris acting on the mesh is resisted by the mesh's tensile strength and in-isolation punch resistance. Figure 6 illustrates the Punch Resistance Test per ISO 17745:2016 and ISO 17746:2016. The punch test is carried out on a sample having a size of 3.0 x 3.0 m ± 20%, restrained into a large steel frame and loaded by means of a punching device with a diameter of 1.0 m.

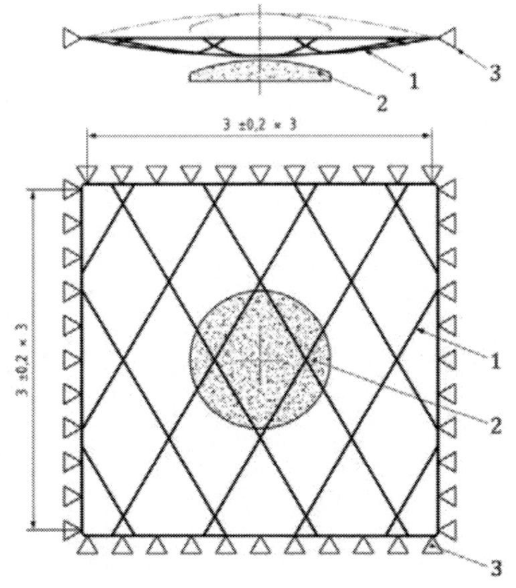

Legend:

1) Tested mesh sample
2) Hemispherical shaped load sharing device
3) Perimeter constraint

Figure 6 – Illustration of the ISO Test to Determine the In-Isolation Punch Resistance and Deformation

The serviceability limit state analysis ensures that the mesh deflection, per the in-isolation punch test, does not exceed the maximum allowable design displacement, as illustrated in Figure 7, where M is the design punch force acting on the mesh, Z_{bulg} is the punch displacement under load per the punch test, B_{ulg} is the allowable design displacement calculated by dividing the maximum design displacement per the project requirements, D_{mbulg}, by a factor-of-safety, F_s.

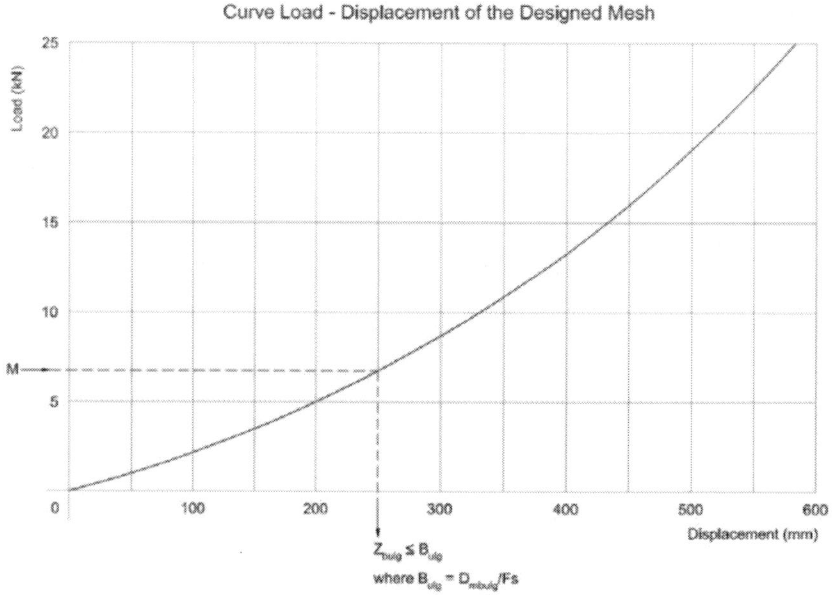

Figure 7 - Example of a load-displacement curve used for the design of the mesh at the Serviceability Limit State

The Ruvolum® design procedure is based on the theory that by tightly pressing and if possible slightly impressing the spike plates in the ground to be stabilized, the mesh is tensioned in the best possible manner [4, 8]. As such, the mesh is assumed to 'actively' apply pressure to the slope face, thereby preventing any surface movement and mesh deformation.

By this assumption, the mesh is prestressed against the nail head and tightening of the nut causes the spike plate to be pressed firmly onto or even slightly into the ground [4]. The resistance to these forces in the direction of the nail is determined by a punch test where the load pushes the mesh into the substratum as shown in Figure 8.

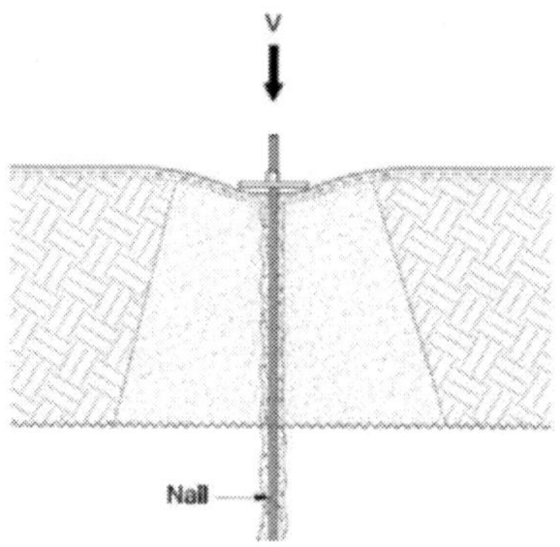

Figure 8 – Nail-Direction Load Characterized by In-Soil Punch Test

The Ruvolum® program is based on a Mohr-Coulomb failure criterion (c', tanø) and analyzes the equilibrium of the infinite slope, considering also the pre-tensioning of the mesh prescribed by the manufacturer (force V) [8]. However, this "active" stabilization assumption has been a point of dispute since to tension the mesh and impart an active stabilizing force to the slope, a mesh must first be in contact with the slope and be uniformly stretched for the design loading condition to impart an active force [3]. If these two conditions are not met, slope movement can occur, and little to no slope stabilization is achieved [3]. It has been reported [3] that inspections of secured drapery installations in California and Washington State found, despite concerted construction efforts, that the mesh commonly was not in contact with the slope or could be easily lifted from the slope, large portions of the installations were providing only passive restraint for rockfalls [3].

SLIDE is a limit equilibrium analysis that uses the method of slices in various forms. The method of slices implements the division of the slope in slices and calculates the equilibrium of the forces along given sliding surfaces.

The SLIDE program is easy to use and very versatile in its global analysis of anchored soil and rock slopes. The program model shown in Figures 9 and 10 also efficiently analyzes 'slope-parallel' instabilities, but while it is possible to analyze 'between-nail' instabilities, the program does not allow the designer to include the stabilizing effect of a steel wire mesh facing system.

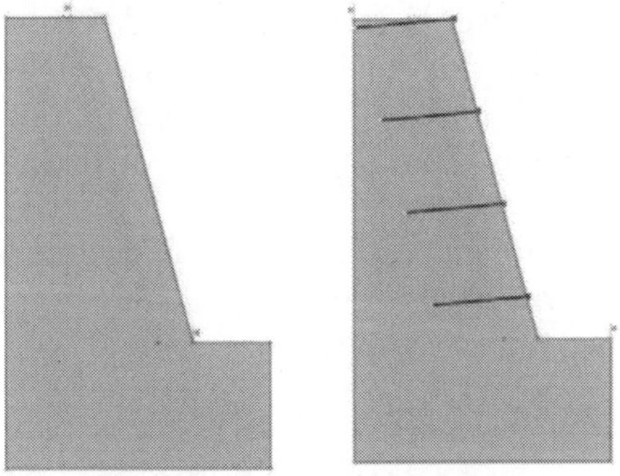

Figure 9 – SLIDE Program Model

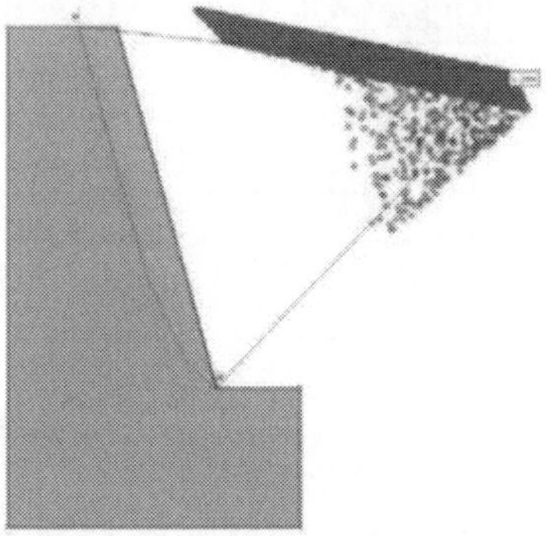

Figure 10 – SLIDE Program Analysis

The SLIDE program easily allows the user to analyse different scenarios of nail spacing and investigate the forces acting on the nails, for instance, all based on the same starting model.

The Plaxis® model, as illustrated in Figure 11, is a finite element analysis which allows the user to define different soil layers and different interacting structures, such as piles, anchors geotextiles, beams, etc., and considers the deformation of the slope face and the soil-nail behavior. The program analyzes 'slope-parallel' and 'between-nail' instabilities and can be used to determine the stabilizing effect of a steel wire mesh facing system. However, the program requires significant user input, and the analytical time can be very slow and inefficient. As with any finite element software, it is mandatory that the input properties be well defined.

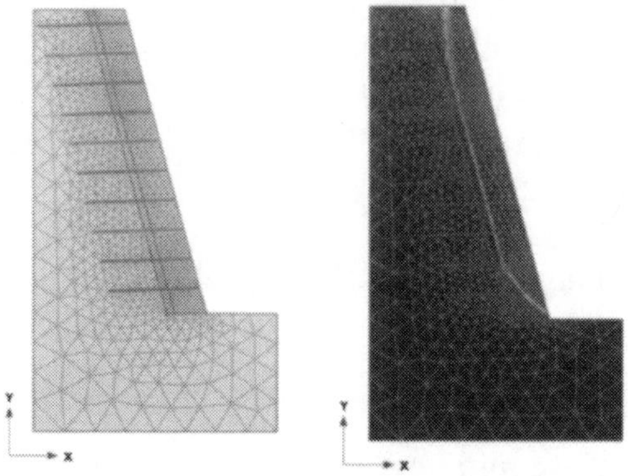

Figure 11 – Plaxis Program Model

All of the programs provide information about the stress acting in the nails, but only Plaxis® can perform a complete analysis concerning all the components (i.e. tension, shear, bending moment). Plaxis® also allows the user to investigate the deformation at each stage, in every single point of interest, and the deformation of structures (such as the deformation of the anchors). Only FEM allows the user to evaluate the behaviour of the soil-structure interaction, since the LEM analyzes the slope without any deformation taking place.

Table 1 summarizes the similarities and differences between the MacRO 1, Ruvolum®, SLIDE and Plaxis® design theories from a 'birds-eye' view. As previously mentioned, a standardized design methodology for pinned drapery applications using flexible high strength steel wire mesh facing may adopt elements of each of these existing approaches.

	MacRO 1[1]	**Ruvolum**®[4]	**SLIDE**	**Plaxis**®
Design Approach	Limit Equilibrium w/Serviceability	Limit Equilibrium	Limit Equilibrium	Finite Element
Assumed Mesh Restraint	Passive	Active	N/A	Passive
Slope-Parallel Instabilities	Yes	Yes	Yes	Yes
Between-Nail Instabilities	Yes	Yes	Possible	Yes
Mesh Tensile Strength Analyzed	Yes	No	No	Possible
In-Soil Mesh Puncture Strength Analyzed	No	Yes	No	Possible
In-Isolation Mesh Puncture Strength Analyzed	Yes	No	No	Possible
In-Isolation Mesh Puncture Deformation Analyzed	Yes	No	No	Possible
Parallel-to-Slope Mesh Puncture Strength Analyzed	No	Yes	No	Possible
Soil and Nail Deformation	No	No	No	Yes

Table 1 – Summary of MacRO 1, Ruvolum®, SLIDE, and Plaxis® Design Approaches for Pinned Drapery

Design Example and Results

A main component of this paper is to compare the analytical results of a slope stability example using the limit equilibrium and finite element methods presented. A comparison of the different results was performed, with a focus on the relation the results have with the real phenomenon and the model used.

For this work, the following methods were used:

1. Software Slide (by Rocscience)
2. Software MacRO 1 (by Maccaferri spa)
3. Sofware Ruvolum® (by Geobrugg AG)
4. Plaxis® (Finite Element)

For this study, an "ideal" soil was chosen to compare the results of these different approaches. The reference material is defined with Mohr-Coulomb strength parameters as follows:

- Internal Friction Angle $\acute{\phi} = 38°$
- Cohesion c = 400 psf (20 kPa)

The scope of the paper being the evaluation of the information resulting from the use of software, it was decided to investigate only the stability of the slope-parallel instabilities, being aware that the stability between the nails is provided by the mesh.

The geometry of the problem was chosen to be compared to the solution that can be analysed with Ruvolum® and MacRO 1, so the model consists of a 75° infinite slope with a 1.0 m (3 ft) surficial layer of the defined characteristics underlain by a stable layer of infinite strength.

Setting the parameters (pre-tensioning, load diffusion angle, partial safety factor) to analyse only the infinite slope problem, allows the user to compare its results to the other software. The MacRO 1 program, starting from equilibrium, evaluates the safety factor reached with the use of the mesh. Even in this case, the program was used to evaluate the contribution of the nails to the overall stability, not considering the mesh.

Using the parameters mentioned above, analysis of the slope was performed without anchors, and then with anchors varying from a 1 m to 4 m apart.

These analyses allowed for comparisons between the different solutions and the influence of different approaches on the solution.

Theoretical Approach

The theoretical approach has been used to solve the problem as a common benchmark. It is useful to have a closed analytical solution as a comparison for the numerical analysis: it allows the capability of the software analysis and the error it implies. The used model is the unstable block upon an inclined slope as illustrated in Figure 12.

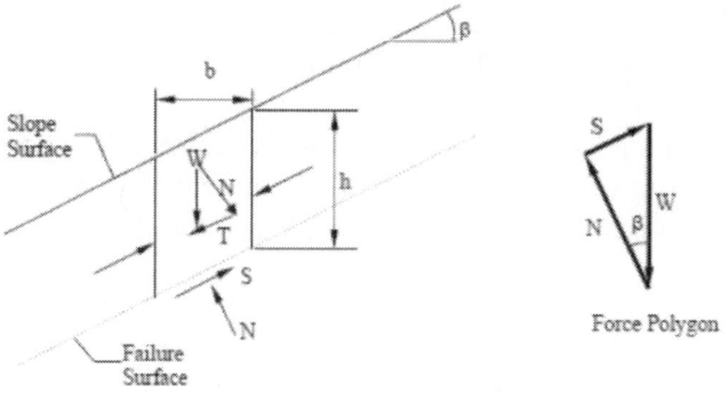

Figure 12: Reference Theoretical Model

Where:

- W is the weight
- N is the component of the weight perpendicular to the failure surface
- T is the component of the weight tangential to the failure surface
- S is the friction resistance plus the nail shear resistance, if present

With this model there is the assumption that the forces between each slice is equal since the slope is infinite and there is no difference between slices [6, 9]. It is possible to use this model because the failure of the slope on a planar surface parallel to its face, with nails as stabilizing elements, by breaking the problem down to its simplest module - the equilibrium of the soil element relative to a single nail.

In the reference model, the only stabilizing action, if the driving forces are bigger than the resisting forces, is the shear strength of the nail. It should be remembered that 'equilibrium' is the main assumption, so no deformation is allowed. This is coherent with the scenario, since the shear resistance of the nail is mobilized for minimal deformation along the sliding plane.

To allow a simple comparison among programs, only the slope-parallel instabilities of Ruvolum® will be analyzed, since this model is the same as the theoretical model. The reference is the theoretical model of a block on an inclined plane where:

- β = inclination of the slope,
- W = weight of the block,
- c = soil cohesion,
- α = angle of shear resistance on the sliding surface,

It is possible to define the Factor of Safety of the slope along the sliding plane as:

$$Fs = \frac{R}{F}$$

Where:

- R is the sum of the resisting forces
- F is the sum of the driving forces

Since the only driving force is the weight (its component along the sliding surface), it is possible to define:

$$F = W \cdot sin(\alpha)$$

At the same time, it is possible to define the resisting force, since the material is assumed to have a Mohr-Coulomb failure criteria, as follows:

$$R = W \cdot cos(\alpha) \cdot tan(\varphi') + c'$$

If *Fs* is smaller than 1, it means that the slope (with the assigned value for the materials) is unstable. The difference between R and F will give the needed value for additional resistance to bring the slope to a stable condition.

Using the nail shear strength of a single nail, T_{nail}, will result in the following equation for the resistance:

$$R = W \cdot cos(\alpha) \cdot tan(\varphi') + c' + T_{nail}$$

The SLIDE program is free to look for the critical surface, using different methods: Bishop simplified, Janbu simplified, Janbu corrected, Fellenius, and Spencer. This has been done to have a benchmark with respect to a global stability analysis. The program performs the analysis with nails (passive anchors), and the above-mentioned anchor patterns.

The Finite Element Software Plaxis® (2D) was chosen, since it is one of the most used geotechnical finite element software programs available and is a good benchmark. It is often used for complex and important projects, but for this work it has been used for a simple task to evaluate the safety factor of a uniform slope.

The underlying stable rock layer is represented as a *Linear Elastic* material with a high *E* modulus to model a rigid subsoil. For the surficial weathered rock layer, a Mohr-Coulomb material was chosen with the already mentioned internal friction angle and cohesion.

The Plaxis® program calculates the Factor of Safety by reducing the value of the materials' parameters until the model fails. This procedure allows us to study the slope at failure and can be done for all nail patterns studied.

Results

The main result of the slope stability analyses using the various methods is the Factor of Safety, expressed as a function of the nail spacing. All the analyses have been made with nail spacings of 1 m x 1 m, 2 m x 2 m, 3 m x 3 m, and 4 m x 4 m; while the scenario without nails has been analysed only with the theoretical approach and with SLIDE and Plaxis®. All of the analytical methods considered the case without wire mesh facing; only Plaxis® was used to determine the stability effect that might be realized with the wire mesh facing. With Plaxis® an analysis was made trying to simulate a steel wire mesh acting as a membrane, set on the slope and tied to the nails with hinges. The membrane model was elasto-plastic, with a tension resistance of 170 kN and negligible bending stiffness. Since the connection with nails was modelled with hinges, the nails and mesh cannot exchange bending moments, only tension. The scenario analysed was the same slope as before, but with the use of steel wire mesh at the face.

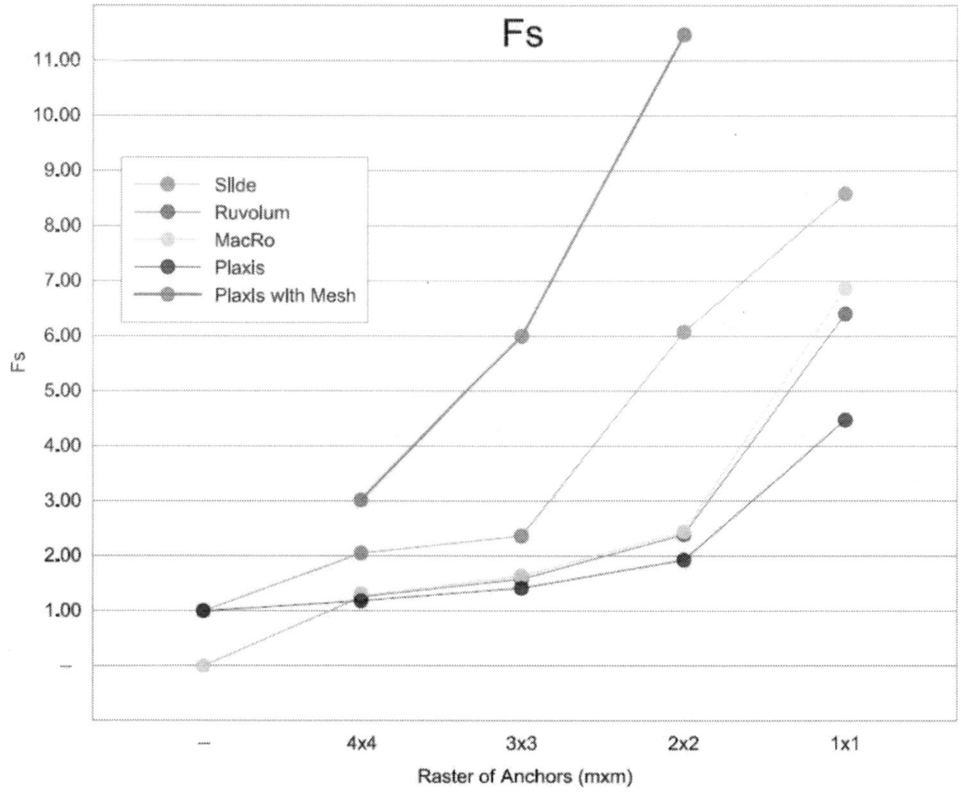

Figure 12 – Results Showing the Variation of Fs Versus Nail Spacing as a Function of Method of Analysis

Figure 12 shows the factor of safety results versus nail spacing for each method of analysis. These results show that without mesh, and for tight nail patterns less than 3 m x 3 m, a great variation of the safety factor can be observed, while for nail patterns greater than 3 m x 3 m the difference is not as great. However, the nail spacing at which Fs divergence occurs between the methods of analyses may vary with different soil and slope parameters.

Further, the SLIDE program generally gives the highest Fs, Plaxis® the lowest, while MacRO 1 and Ruvolum® provide similar results.

Finally, the Plaxis® model with wire mesh facing shows the significant stability improvement that may be gained by incorporating the steel wire mesh at the face. It is important to underline that the construction of the model of the mesh can be quite difficult in FEM, especially regarding the soil-mesh interaction. The FEM analysis considers the stiffness of the solution by taking into account the behaviour of the materials. The results show that the Fs significantly increases for closer spacing of the nails, where the confinement effect of the mesh on the entire slope and the stiffness of the solution is more evident.

Conclusions

Pinned drapery systems using a high strength steel wire mesh facing have proven to be a viable, cost-effective and aesthetically pleasing solution to stabilize the surficial portion of a rock or soil slope. The most common design approaches developed during the last century include the Limit

Equilibrium Method and the Finite Element Method. The LEM gives information only about the slope stability before any movement takes place. While the FEM allows the user to gain important information about the condition of the slope, considering the deformation and the soil-structure interaction. By FEM analysis, it has been shown that the steel wire mesh facing could provide significant stability improvement to the surficial portion of an unstable slope.

The specific method and software used should be done with a sound knowledge of how it works, assumptions that are made, and how the parameters are used. For instance, disregarding deformations using LEM, including the resistance of the steel wire mesh is an approach affected by a conceptual error: on one hand, we are analyzing the equilibrium of the slope before any displacement takes place, while on the other hand, we are considering as a stabilizing force the tensile resistance of the mesh, which needs deformations to be developed in order to mobilize this tensile resistance. Analyzing the slope deformation further allows the user to consider the soil-structure interaction, an important factor in the evolution of the slope, and the potential impact of the system deformation on adjacent infrastructure.

For the above-mentioned reasons, it would be beneficial for the engineering community if a standardized design approach could be developed which would take into consideration the evolution of the stress-strain condition of the slope with a pinned drapery system, evaluating the soil-structure interaction, to fully capitalize on the potential stability impact of high strength steel wire mesh facing systems. Figure 13 provides a proposed flowchart for the design of pinned drapery systems in an effort to move this conversation forward for the benefit of the profession.

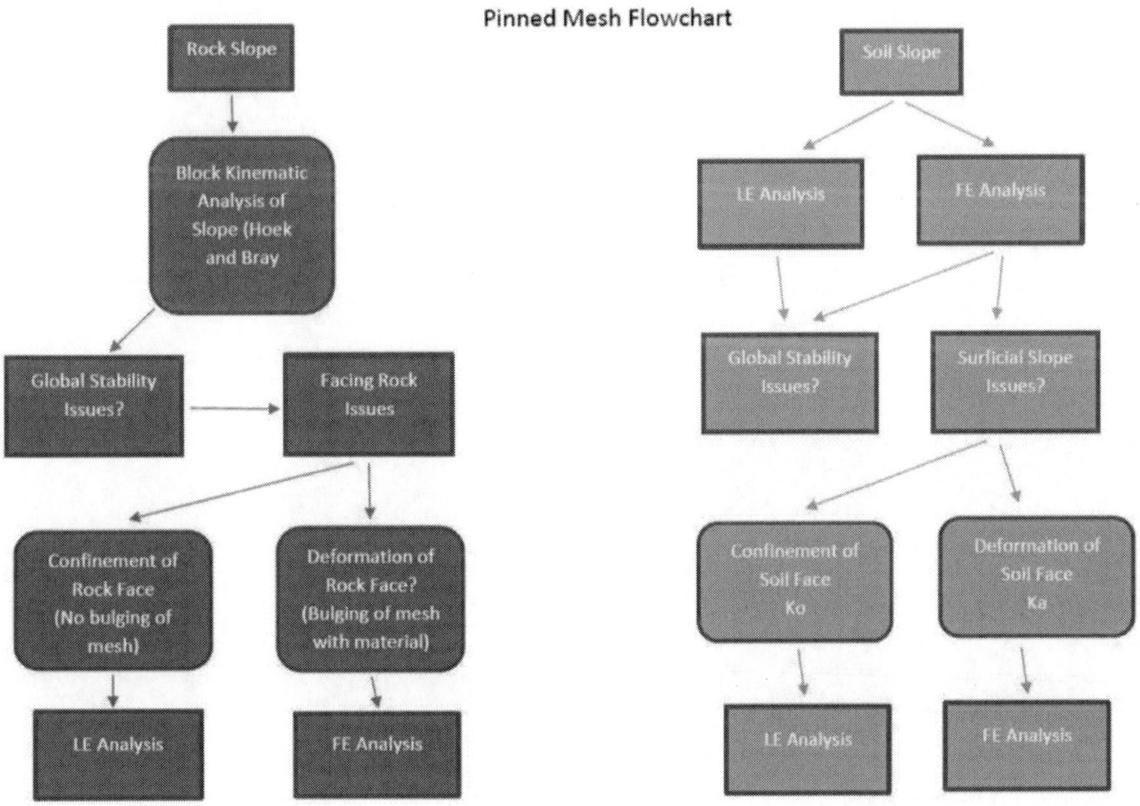

Figure 13 – Proposed Flowchart for the Design of Pinned Drapery Systems [5]

References

[1] Grimod A., Giacchetti G. (2013): A new design approach to design pin drapery systems. GeoMontreal 2013, 29th September – 3rd October 2013, Montreal (QC).
[2] Bertolo P., Giacchetti G., (2008): An approach to the design of nets and nails for surficial rock slope revetment – Interdisciplinary Workshop on Rockfall Protection, June 23 - 25 2008, Morshach, Switzerland.
[3] Keith Turner A. and Schuster Robert L., Editors (2013) – Rockfall: Characterization and Control – TRB Transportation Research Board - The National Academies Keck Centre 500 Fifth Street, NW, Washington, DC 20001, pgs. 568 - 570.
[4] Daniel, Flum and Rudolf, Ruegger, (2002) – Tecco® Slope Stabilization System, Technical Academy Esslingen, January 22 – 23, 2002.
[5] Arndt, Ben P. (2018) - Communication regarding design concepts and methodologies for designing rock vs soil slopes with pinned drapery systems.
[6] R. Lancellotta – Geotecnica – Zanichelli (2007)
[7] R. Nova – Soil Mechanics, McGraw Hill – McGraw Hill (2002)
[8] Cala, Flum, Roduner,Ruegger, Wartmann – Tecco® Slope Stabilization System and Ruvolum® Dimensioning Method – Geobrugg (2012)
[9] R.F. Craig – Soil Mechanics – E & FN Spon 1997
[10] David Muir Wood – Geotechnical Modelling – CRC Press (2004)
[11] D.C. Wyllie, C.W. Mah – Rock Slope Engineering and Mining – Spon Press (2001)

Formatting Issue - Best Available Paper/Graphic

Hard Bedrock, Marine mud, and High Tidal Range: Highway Geology and Engineering Challenges along Maine's Mid-coast

Edited by:

Robert G. Marvinney

Maine Geological Survey

69th Annual Field Trip of the Highway Geology Symposium

Portland, Maine, September 10-13, 2018

Highway Geology Symposium Guidebooks

September 2018

Formatting Issue - Best Available Paper/Graphic

OFFICERS OF THE HIGHWAY GEOLOGY SYMPOSIUM
2018

Chairman ----------------------Ken Ashton
West Virginia Geological Survey
P.O. Box 879
Morgantown, WV 26507

Vice Chairman---------------- Krystle Pelham
New Hampshire DOT
P.O. Box 483
5 Hazen Drive
Concord, NH 03302

Secretary ---------------------- Bill Webster
CalTrans
5900 Folsom Blvd.
Sacramento, CA 95819

Treasurer --------------------- John Pilipchuk
NCDOT Geotechnical Engineering Unit
1589 Mail Service Center
Raleigh, NC 27699

2018 FIELD TRIP

Field Trip Organizer and Editor:

Robert G. Marvinney
Maine Geological Survey

Field Trip Leaders:

Charles Hebson, MaineDOT
Lindsay J. Spigel, Maine Geological Survey
Stephen M. Dickson, Maine Geological Survey
Marleigh L. Snow, Haley & Aldrich, Inc.

Special Acknowledgements:

Tammara Roberts, Maine Geological Survey

Formatting Issue - Best Available Paper/Graphic

HARD BEDROCK, MARINE MUD, AND HIGH TIDAL RANGE: HIGHWAY GEOLOGY AND ENGINEERING CHALLENGES ALONG MAINE'S MID-COAST

EDITED BY:

ROBERT G. MARVINNEY

MAINE GEOLOGICAL SURVEY

69TH ANNUAL FIELD TRIP OF THE HIGHWAY GEOLOGY SYMPOSIUM

PORTLAND, MAINE, SEPTEMBER 10-13, 2018

HIGHWAY GEOLOGY SYMPOSIUM GUIDEBOOKS

SEPTEMBER 2018

CONTENTS

FIELD TRIP OVERVIEW

Field Trip Logistics

Welcome to the 69th Highway Geology Symposium in Portland, ME. Perched above scenic Casco Bay, Portland is Maine's largest city with a population of about 67,000. The Greater Portland metropolitan area is home to over half a million people, more than one-third of Maine's total population. Portland's economy is heavily dependent on tourism and has arisen as a popular "foodie" destination, with the Old Port district figuring prominently in that regard. With the Portland International Jetport nearby and easy access via Interstate highways, Portland is the jumping off point for millions of tourists each year who come from other regions of New England, the nation, and the world to explore Maine's coastal wonders, rugged mountains, and thousands of pristine lakes. Also anchoring the economy is the Port of Portland, the largest tonnage seaport in New England and a major center for Maine's fisheries, which in 2016 landed over 130 million tons of the iconic American lobster statewide.

We will explore sites of geologic and historical interest along Maine's rock-bound south-central coast from Portland north to Bucksport. Our trip will begin from the Holiday Inn by the Bay and travel north, mostly on U.S. Route 1, passing by Bath Iron Works enroute. Ships have been built there since 1884 and today BIW is an important manufacturer of the most advanced U.S. Navy vessels, including the *USS Daniel Inouye* (DDG 118) currently under construction. Our first stop is at Sherman Marsh just east of the quaint coastal village of Wiscasset. First settled in 1663, this seafaring village is fondly known as the prettiest village in Maine, with numerous Federal-style and Victorian-style mansions. In the 1930s a causeway dam was built at Sherman Marsh to carry Route 1 across a tidal inlet, impounding a 200-acre lake (Figure 1). When the causeway dam was washed out by a major storm in 2005, the Maine Department of Transportation elected to restore the salt marsh to its original state. This stop will explore the engineering and social challenges of salt marsh restoration, a common issue along the entire Maine coast, and one that is exacerbated by accelerating sea-level rise.

From Sherman Marsh, we will continue north on Route 1 to Rockland, first settled in 1769 when it was known for shipbuilding and production of lime from Precambrian carbonates unique to this region. Our second field trip stop will be at a location on the north

Figure 1. Sherman Marsh after restoration. The former lake occupied the flat marsh area shown in this image.

shore of Rockland Harbor where in April of 1996 a significant landslide threatened lives and destroyed two homes. Along much of coastal Maine, the hard Paleozoic bedrock ledge is overlain with a veneer of Pleistocene glacial-marine mud (silt and clay) known as the Presumpscot Formation (Figure 2). Immediately following deglaciation, which along the central coast occurred around 16,000 years ago, the sea transgressed inland over the crust that had been depressed under the immense weight of glacial ice that was at least

Figure 2. Layered mud of the Presumpscot Formation at a construction site in Augusta, Maine.

a mile thick. Before the crust could rebound significantly, a veneer of glacial-marine mud of varying thickness was deposited. Consequently, the Presumpscot can be found well inland along the major drainages of the Penobscot and Kennebec rivers (Figure 3).

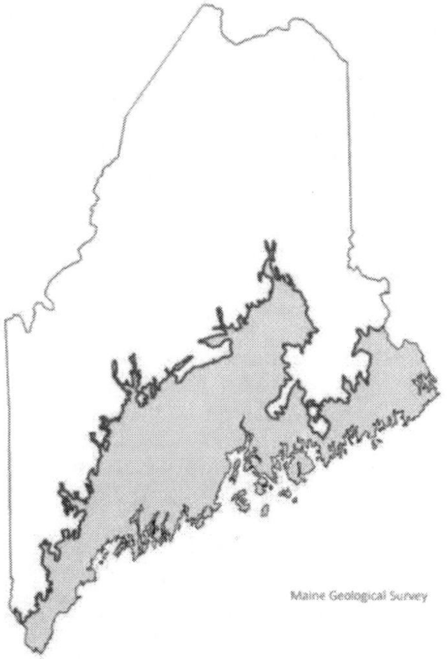

Figure 3. Maximum landward extent of the sea following deglaciation.

While long known to be susceptible to landslides, the full extent of this hazard in the Presumpscot was unknown until lidar elevation data were collected across southern Maine, revealing scores of previously uncatalogued features. The Presumpscot presents particularly challenging geotechnical problems for bridge abutments and slope stability.

No trip to Maine would be complete without enjoying a full lobster dinner! Lunch will be at Young's Lobster Pound (Figure 4) in Belfast where you will be treated to the uniquely Maine experience of eating lobsters on the shores of scenic Penobscot Bay, your favorite libation in hand.

Figure 4. Young's Lobster Pound, Belfast.

From Belfast, the trip will continue north on Route 1 to our last stop at the Penobscot Narrows Bridge (Figure 5) and Fort Knox State Park (Figure 6). This was the site of the historic Waldo-Hancock suspension bridge that served for 75 years until inspections indicated that the main-span suspension cables were corroded to the point that the bridge would need to be replaced. Construction for the new cable-stay bridge began in 2004 with a significant road cut on the western approach that will be the primary focus of our visit. Modelled after the Washington Monument, the towers reach 420 above sea level. Opened in 2007, the south tower of Penobscot Narrows Bridge hosts the first bridge observatory in the United States and the tallest in the world! From the Observatory, one has a spectacular 360-degree view of the Maine coast that on a clear day spans south to the Camden Hills, east to Mount Desert Island and northwest to Mount Katahdin. Further innovations in construction include a cradle system that carries the strands within the stays from bridge deck to bridge deck, as a continuous element, eliminating anchorages in the pylons. Each epoxy-coated steel strand is carried inside the cradle in a one-inch steel tube. Each strand acts independently, allowing for removal, inspection and replacement of individual strands. The cable-stay design uses a system of pressurized nitrogen gas to defend against corrosion.

Figure 5. Penobscot Narrows Bridge, Prospect, Maine.

Figure 6. Historic Fort Knox, Prospect, Maine.

Field trip itinerary

Wednesday, September 12, 2018

Figure 7 is a map of the route.

6:30 – 7:00 am	Buses arrive at Holiday Inn for boarding
7:00 am	Buses depart Holiday Inn
8:00 – 8:40 am	Stop 1: Sherman Marsh
9:30 – 10:15 am	Stop 2: Rockland Harbor landslide site
11:30 am – 1:00 pm	Lunch at Young's Lobster Pond
1:30 – 3:15 pm	Stop 3: Penobscot Narrows Bridge and Observatory
5:30 pm	Return to Holiday Inn

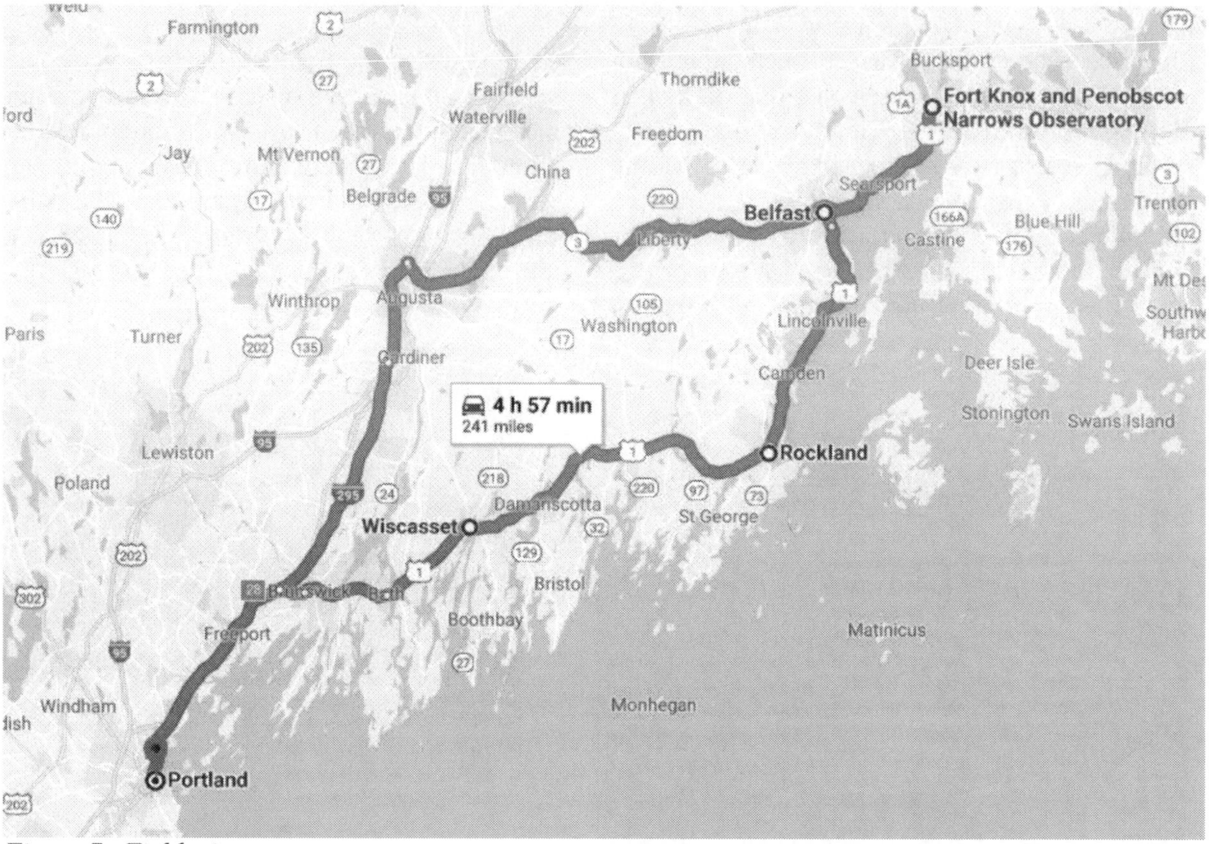

Figure 7. Field trip route.

Overview of Maine's geology

Bedrock

Maine's bedrock records more than half a billion years of geologic history. Over this period, geologic processes such as erosion and sedimentation, mountain-building, deformation, metamorphism, and igneous activity produced the complex pattern of bedrock geology that we see today. Over centuries of mapping, geologists have identified hundreds of bedrock formations and igneous intrusions distinguished on the basis of age and rock type (Osberg, et al., 1985). On the simplified geologic map in the back pocket of this guide, these rocks have been grouped into units of similar geologic age. With the advent of plate tectonics, we now know that the crust of Maine is composed of multiple, small plate fragments, both continental and oceanic in composition, which are distinctive and have had separate histories. While ongoing research continues to refine the nature and exact boundaries, it is generally accepted that the geology of Maine is composed of a mosaic of such terranes (e.g. Berry and Osberg, 1989; Robinson et al., 1998; Tucker et al., 2001; Hibbard et al., 2010). These were once widely scattered microplates in Iapetus, an ocean which preceded the modern Atlantic Ocean. The geologic history recorded in Maine's bedrock spans several major cycles of deposition, deformation, and igneous activity related to plate tectonic movements. The simplified chart (back side of simplified map in pocket) recounts the histories of the various terranes that were later to become Maine's bedrock. In the chart, while the terranes have separate histories, they are shown in separate blocks. Laurentia refers to the ancient eastern margin of North America. The Iapetus terranes comprise a composite island arc, formed in Iapetus, that collided with Laurentia during the Ordovician Period. Avalon is a microplate which collided with early North America in the Devonian to form eastern North America as we know it today.

The landscape we will drive across on the field trip is underlain with medium- to high-grade metamorphic rocks that were once marine sediments and volcanic rocks developed in former ocean basins marginal to Laurentia. While deposited horizontally originally, the beds throughout the region are now vertical, as you will see in most of the road cuts along our route, due to the immense compressive forces of plate collisions. These metamorphic rocks are punctured by large intrusive rock bodies, mostly granite, but also of intermediate and mafic compositions, mostly of Silurian or Devonian age. These plutons represent the roots of the once-formidable Appalachian Mountains, in this region now reduced to gently rolling hills by 400 million years of erosion capped by the last glacial episode.

Glacial geology

Continental glaciers similar to today's Antarctic Ice Sheet probably extended across Maine several times during the Pleistocene Epoch, which lasted from about 2.5 million to 10,000 years ago. The slow-moving glacial ice changed the landscape as it scraped across mountains and valleys, eroding rock debris and carrying it for miles. The sand, gravel, marine mud, and other unconsolidated sediments that cover much of Maine are largely the products of glaciation. Some of these materials were deposited directly from glacial ice as an uneven blanket of stony till; others washed into the sea or accumulated in meltwater streams and glacial lakes as the ice receded. Glaciation also disrupted earlier drainage patterns and helped create the thousands of ponds and lakes that make the state so attractive to vacationers.

The most recent glacial episode in Maine began about 35,000 years ago, when the Laurentide Ice Sheet overspread southern Quebec and New England. During its peak development, this ice sheet was centered over eastern Canada and flowed east to southeast across Maine. It became several thousand feet thick and covered the highest mountains in the state. The weight of the glacier pushed the land downward several hundred feet. Climatic warming forced the Laurentide Ice Sheet to start receding as early as 21,000 years ago, soon after it reached its terminal position on Long Island, New York (Sirkin, 1986). The ice margin withdrew from the Gulf of Maine to the present position of the Maine coast by 17,000 to 16,000 years ago (Borns et al., 2004). The Earth's crust was still depressed by the weight of the ice sheet, causing the sea to flood southern Maine as the glacier retreated to the northwest.

The landforms and sedimentary deposits resulting from marine submergence are among the most distinctive glacial features in the state. The sea extended far inland, reaching present elevations to at least 420 feet in central Maine. Great quantities of sediment washed out of the melting ice and into the sea, which was in contact with the receding glacier margin. Sand and gravel accumulated as deltas and submarine fans where streams discharged along the ice front, while the finer silt and clay dispersed across the ocean floor. The marine environment favored accumulation of till and washed sediments in moraines along the bottom edge of the glacier margin, recessional sequences of which have been dramatically revealed by lidar (Figure 8).

History of Fort Knox

Located on the west bank of the Penobscot River in Prospect, Maine, in an area known as the Penobscot Narrows, Fort Knox is one of the best-preserved fortifications on the New England seacoast. The Fort has many architectural features present only to itself, as well as a rich history behind its cannon batteries. Maine was repeatedly involved in northeast border disputes with British Canada, and the area between Castine and the rich lumber city of Bangor was invaded and occupied by the British during the American Revolution and the War of 1812. Despite the Webster-Ashburton Treaty of 1842, Fort Knox was established in 1844 to protect the Penobscot River valley against a possible future British naval incursion.

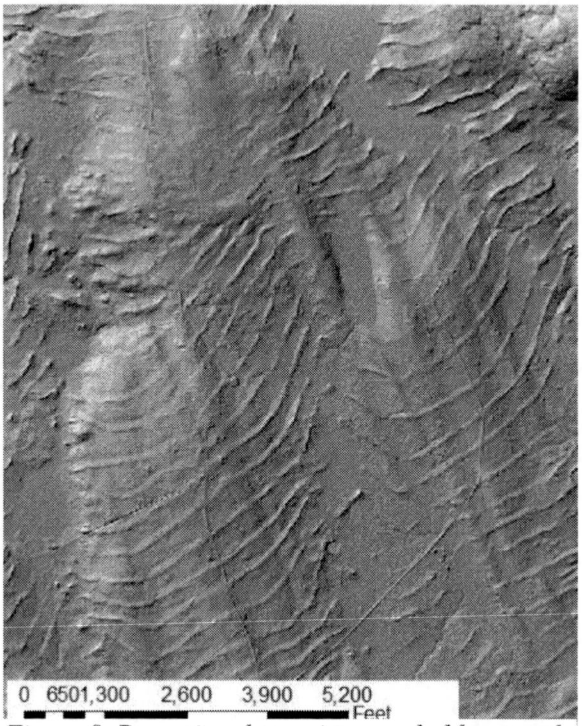

Figure 8. Recessional moraines, probably annual, revealed by lidar west of Belfast.

The Fort was designed by Chief Engineer Joseph Totten and constructed from 1844 – 1869, the first such fort in Maine to be constructed entirely of granite. The Mount Waldo granite, quarried just 5 miles away, served as the source. An important note is that this same quarry provided granite for construction of the Washington Monument in Washington, DC. The Fort was named for Major General Henry Knox, America's first Secretary of War, who was born in Boston but retired to Thomaston, Maine in 1796. The Fort garrisoned its first troops from 1863 to 1866. These troops were mostly volunteers undergoing training before being sent to their active posts and included members of the celebrated 20th Maine. Troops were also briefly stationed at the Fort during the Spanish American war in 1898, but never saw military action. As a virtually intact example of a mid-19th century granite coastal fortification, it was added to the National Register of Historic Places in 1969 and declared a National Historic Landmark on December 30, 1970.

References

Berry, Henry Newhall, IV, and Osberg, Philip H., 1989, A stratigraphic synthesis of eastern Maine and western New Brunswick: in Tucker, Robert D., and Marvinney, Robert G. (editors), Studies in Maine Geology: Volume 2 - structure and stratigraphy: Maine Geological Survey, p. 1-32, 32 p., 5 figs., abstract, references.

Borns, H. W., Jr., Doner, L. A., Dorion, C. C., Jacobson, G. L., Jr., Kaplan, M. R., Kreutz, K. J., Lowell, T. V., Thompson, W. B., and Weddle, T. K., 2004, The deglaciation of Maine, U.S.A., in Ehlers, J., and Gibbard, P. L. (editors), Quaternary glaciations - extent and chronology, Part II: North America: Elsevier, Amsterdam, p. 89-109.

Hibbard, James P., van Staal, Cees R., Rankin, Douglas W., 2007, A comparative analysis of pre-Silurian crustal building blocks of the northern and the southern Appalachian Orogen: American Journal of Science, v.307, no.1, p.23-45.

Robinson, P., Tucker, R. D., Bradley, D., Berry IV, H. N., and Osberg, P. H., 1998, Paleozoic orogens in New England, USA: GFF, v. 120 (Pt. 2, June), p. 119-148, Stockholm.

Sirkin, L., 1986, Pleistocene stratigraphy of Long Island, New York, *in* Cadwell, D. H. (editor), The Wisconsinan stage of the first geological district, eastern New York: New York State Museum, Bulletin 455, p. 6-21.

Tucker, R. D., Osberg, P. H., and Berry, H. N., IV., 2001, The geology of a part of Acadia and the nature of the Acadian orogeny across central and eastern Maine: American Journal of Science, v. 301, no. 3, p. 205-260.

Other sites utilized as references:
https://en.wikipedia.org/wiki/Fort_Knox_(Maine)
http://fortknox.maineguide.com/
https://www.maine.gov/dacf/mgs/explore/surficial/facts/surficial.htm

FIELD TRIP SPONSORS

MAINE GEOLOGICAL SURVEY

Formatting Issue - Best Available Paper/Graphic

69th Highway Geology Symposium, pages 1-18, 2018

SHERMAN MARSH IN NEWCASTLE SALT MARSH RESTORATION PROJECT BY THE MAINE DOT

CHARLES HEBSON AND MARIA GUERRA
Manager, Surface Water Resources Division
MaineDOT
16 State House Station
Augusta ME 04333-0116

INTRODUCTION

In the 1930s, a 200-acre salt marsh at the headwaters of the Marsh River in Newcastle, Maine, was impounded by a causeway dam that carried US-1, creating Sherman Lake. Prior to this, the Marsh River was crossed by a timber swing bridge (Figure 1) open to tidal exchange and boat traffic. The area around the marsh was used for agriculture and several brick yards were located along the marsh shoreline. When the dam was built, it was a simple earthen structure with a stone block spillway just above high tide elevation (Figure 2). The resulting lake was generally quite shallow (<= 4' deep, except over the old tidal river channel). Figure 3 shows the lake as created by the 1930's causeway. In the early 1960s, the "new" US-1 was built over the old causeway; this was a span on steel pilings driven through the old causeway.

Figure 1. Swing Bridge prior to causeway and impoundment

Figure 2. Old US-1 Causeway, downstream face.

DAM DESTRUCTION

On Columbus Day Weekend of 2005, a 50-year storm washed out the dam under US-1, draining the lake and recreating the salt marsh (now informally referred to as Sherman Marsh). The thought that the 2005 breach was preceded by mats of marsh peat and vegetation clogging the spillway, allowing water to rise and flow over the unarmored earthen dam (Figures 4 and 5).

Figure 3. Sherman Lake prior to breach.

Figure 4. Suggested failure mode – the clogged spillway facing upstream in 1995 (a) and 2004 (b).

Figure 5. Immediate aftermath of Columbus Day 2005 breach; US-1 overhead.

SITE REMEDIATION

Historically, similar scenarios were addressed by the DOT with intentions to quickly repair the dam; emergency reconstruction, such as this, does not require a permit. However, in 2005, the choice was made to leave the breach open, and there were several reasons for this:

1) The dam no longer served a transportation purpose.
2) Dam inspection and maintenance are not typical duties for MaineDOT and outside department expertise.
3) A manmade, shallow, freshwater pond did not belong in what was formerly a salt marsh.

The breach would require follow-up stabilization (Figure 6), however, this was not a long-term solution, for two reasons:

1) The breach (as stabilized) was not large enough to allow for the needed tidal exchange. The openings were still too small and the inverts were too high, preventing adequate drainage.
2) The steel pilings for the "new" US-1 span overhead were exposed to tidal salt water and air.

Over the course of just a couple years, significant rust and steel loss was observed and without corrective action, US-1 was at risk. The choice to allow for reversion to salt marsh was celebrated by environmental agencies and nonprofits as an environmentally responsible choice to restore wetland function. At the time of the breach there was talk of partnering to finish the restoration that nature had started and the MaineDOT had chosen to facilitate. However, it became apparent that institutional arrangements, timing and available grant opportunities would not work in the context of a DOT project. In the meantime, something had to be done to protect the steel pilings. Furthermore, the post-breach temporary channel stabilization beneath the bridge was starting to fail. Ultimately it was decided that the MaineDOT would proceed at its own expense to enlarge and stabilize the opening by encasing the steel pilings in grout-filled fiberglass sleeves (similar to sonotubes). The final cost would be approximately $850,000. Salt marsh conditions before and after restoration are shown in Figure 8.

Figure 6. Failure of temporary post-breach stabilization.

Figure 7. Start of construction for permanent stabilization

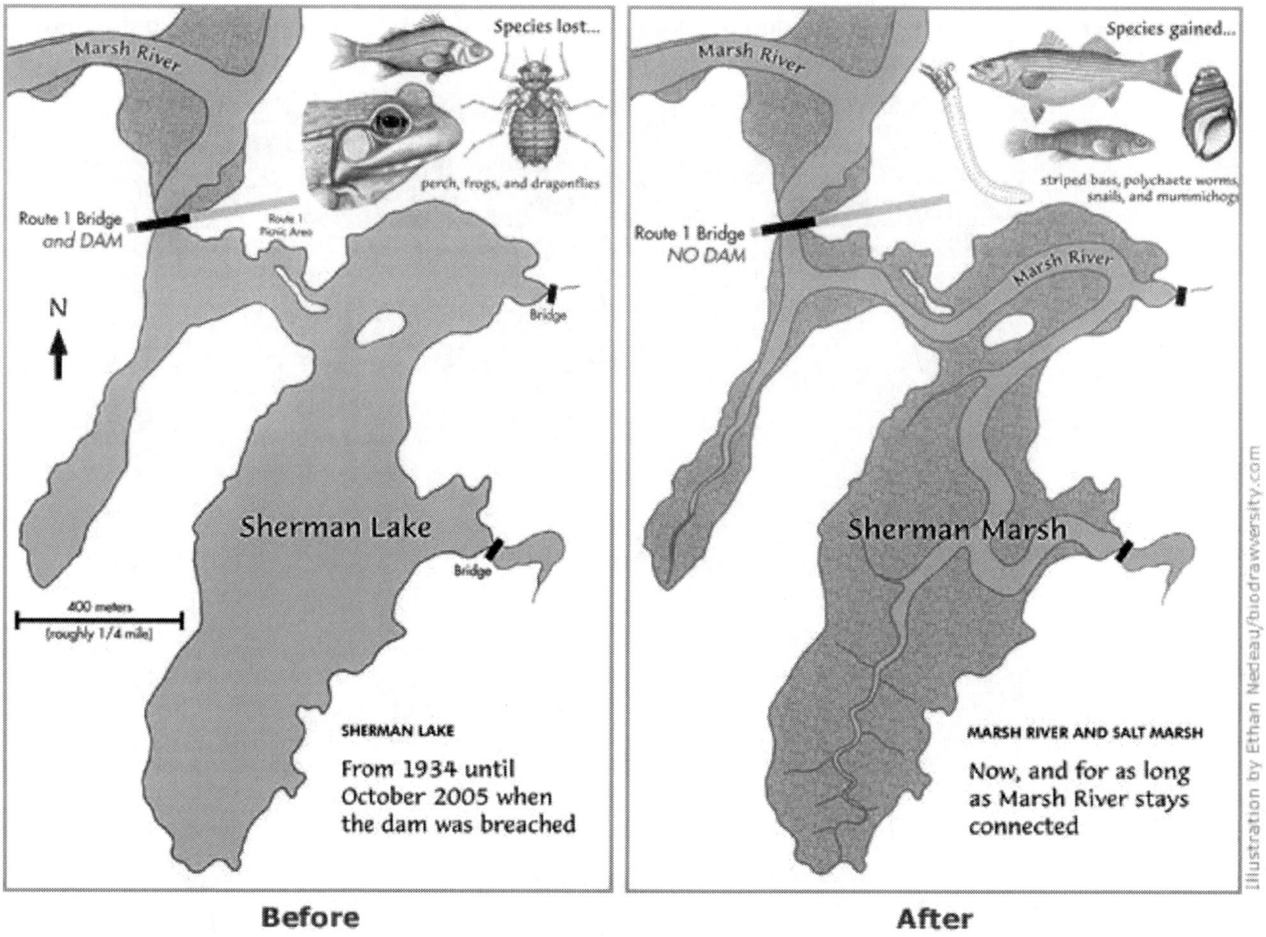

Figure 8. Configuration of Sherman Lake and Sherman Marsh before and after the dam breach and restoration.

SITE MONITORING

Over time, monitoring done by the University of Southern Maine revealed that *Phragmites australis* (Figure 9) was spreading throughout the newly restored marsh. This plant can crowd out native species, creating monocultures and increasing the vulnerability of the marsh to dramatic changes. It was anticipated that improving tidal exchange would help control *phragmites* by increasing salinities on the marsh. However, the naturally increased salinity by itself would not be enough to eliminate the established *Phragmites*. Freshwater inputs from the Sheepscot River maintains relatively low salinities and maintains the water in a brackish state, at best. Therefore, salinities can only reach maximums within the high 20's (mg/L), and can often drop into single digits or low teens, particularly in Spring. (See Addendum 1.)

MaineDOT contracted for a major phragmites spray program with 2 years of follow-up maintenance spraying. Following that, MaineDOT assumed responsibility for annual treatment. The *phragmites* elimination program has been remarkably successful. The marsh is essentially *phragmites*-free with just a few days required each year for "maintenance" treatment. Seeking to get something back in return for this elective restoration effort, MaineDOT attempted to deposit this site into a wetlands mitigation bank. After facing repeated setbacks and a lack of enthusiasm on the part of the Federal agencies involved, this effort was abandoned in the spring of 2017.

Figure 9. Example of phragmites stand prior to the beginning of treatment.

Restoration of Sherman Marsh has been a resounding success. The flow restriction at the bridge has almost been eliminated (see Addendum 2: hydrology) with very little head loss across the outlet. The marsh surface is inundated several times a month as required and drainage on the outgoing tide has been greatly enhanced by lowering the outlet channel inverts. *Phragmites* is essentially eliminated; it occurs as random individual plants as opposed to dense stands; chemical usage is down to a *de minimis* level. The time actually spent applying chemical is a small fraction of the time spent walking the 200-acre marsh. This stands as one of the largest salt marsh restorations in northern New England.

Special thanks to Ryan Tarte for sharing his paper that served as a basis of this guide:
Ensuring Stakeholder Engagement in Mitigation Banking: A Case Study of Sherman Marsh, March, 2018.

69th Highway Geology Symposium, pages 1-18, 2018

ADDENDUM 1: SALT MARSH SALINITY

Sherman Marsh Salinity Analysis

Salinity (specific conductance) data were collected at various times in 2008, 2009, and 2010 using the Solinst LTC data logger. Data collection was somewhat problematic as the equipment was not entirely reliable. However, enough data were collected as to allow for a general understanding of salinity behavior in the marsh.

The location of Sherman Marsh with respect to the open sea and the Sheepscot River estuary is critical to understanding the marsh salinity regime. The marsh is effectively the headwater of the Marsh River, which subsequently discharges into the Sheepscot River far upstream from the mouth in the Gulf of Maine. The source of saline water for the marsh is at the junction of the Marsh and Sheepscot Rivers. Salinity at this point is greatly influenced by freshwater discharge of the Sheepscot. Maximum salinity in seawater is on the order of 35 ppt; maximum salinity at the Sheepscot-Marsh junction can be significantly less.

Loggers were installed at various times between early spring and late fall in 2008, 2009 and 2010. Salinities were generally low in April and May as a result of the spring runoff in the Sheepscot River. Salinities then gradually increased over the summer, peaking at about 25 ppt in late summer. However, these periods of higher salinities could be dramatically impacted by large storm runoff events in the Sheepscot. No doubt runoff from the Marsh River watershed could also have an impact, but likely much less due to much smaller size as compared to the Sheepscot.

Selections of data that illustrate general patterns are offered below. Due to the mass of data obtained, only representative examples can be given in a written report. Furthermore, the natural variability of the system, due to variability in Sheepscot River and Marsh River runoff, means that it is difficult to characterize the salinity regime except in the broadest of terms.

The single most important observation is that the Sheepscot River discharge and the location of the marsh high up in the system and away from direct seawater inputs effectively cap the maximum attainable salinity in the Marsh River and Sherman Marsh. Two subsidiary observations are

1) freshwater runoff in spring severely depress salinity until from early spring through early summer;
2) even after salinity approaches maximum attainable levels in mid to late summer, runoff events can quickly and sharply dilute salinity; recovery is a lengthy process.

The highest observed salinities were about 30 ppt, but these only persisted for short periods of time only to be diluted by runoff events. Average values in the range of 15 ppt to 25 ppt were more likely for extended periods of time.

Salinities during Spring Runoff

Figure 1 shows the Sheepscot River long term average monthly flows as well as the 2010 monthly flows. Flow values are from the US Geological Survey gage at North Whitefield, Maine. Figure 2 shows mid-marsh salinity and Sheepscot daily runoff in April and May 2010. There is clearly less freshwater to dilute incoming seawater as the season progresses from spring into summer. By June, daily peak salinity is up to about 20 ppt. This peak generally continues to increase deep into summer and early fall, subject to dilution by individual runoff events.

Salinity in Response to Runoff Events

Salinity shows a strong response to individual runoff events. Figure 3 shows monthly flows for 2008 as context to consideration of runoff events in summer 2008.

May, August and September were all somewhat wetter than average while June and July were somewhat drier. Figure 4 shows a data trace for August through October 2008. The Sheepscot flow data are at 15-minute intervals; salinity data are at 5-minute intervals. The salinity logger was located just downstream of the US 1 bridge over the Marsh River (Sherman Marsh outlet). There are three (3) major flow events; each event triggers a significant dilution of salinity. The recovery takes much longer than the initial dilution. Similar graphs are shown for 2009 in Figures 5 and 6.

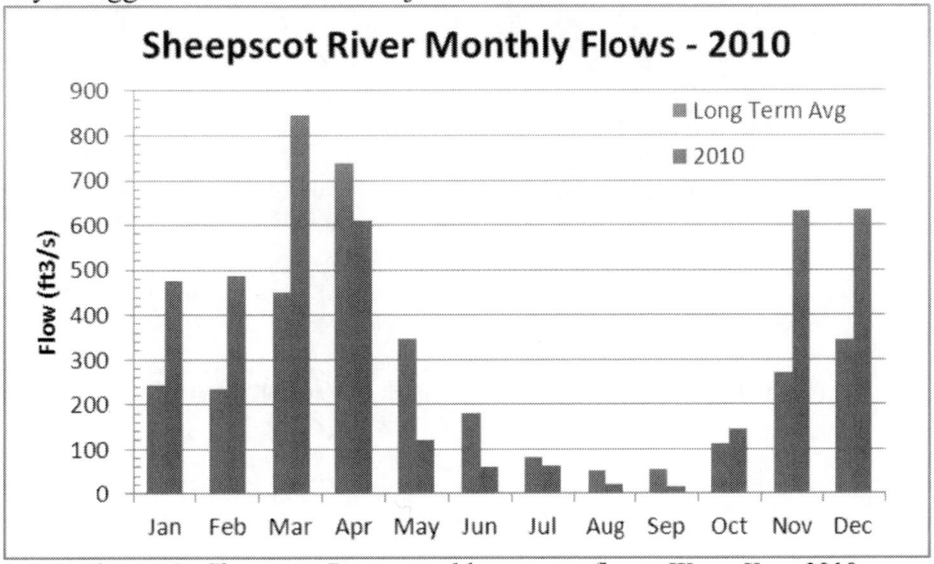

Figure 1. Sheepscot River monthly average flow – Water Year 2010.

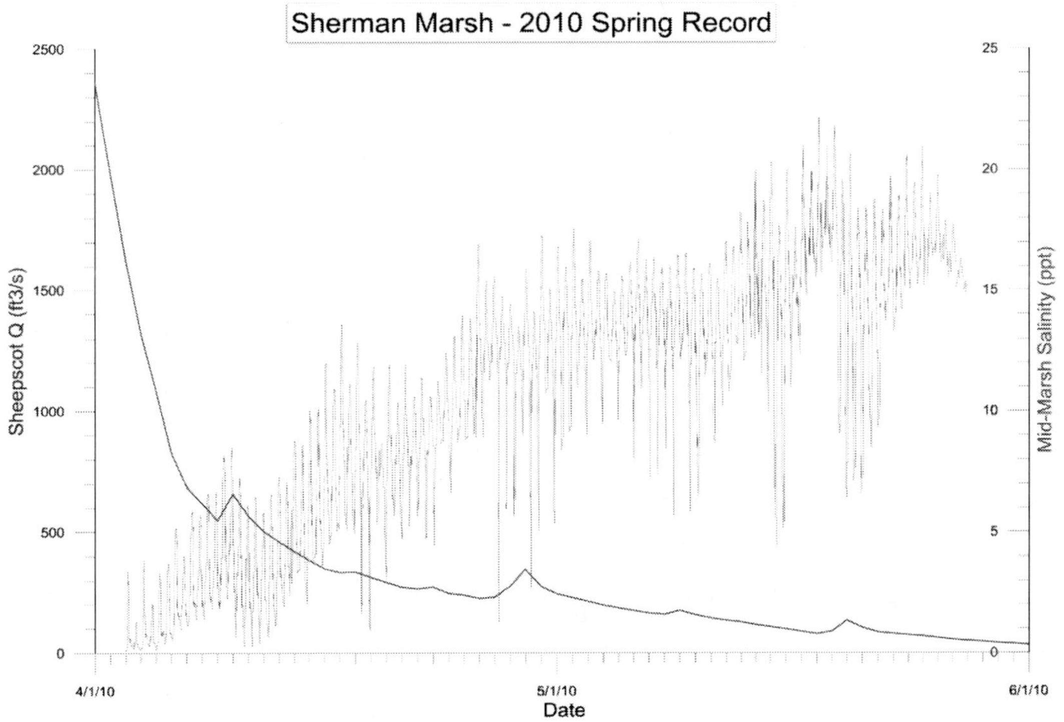

Figure 2. Spring runoff mid-marsh salinity – 2010

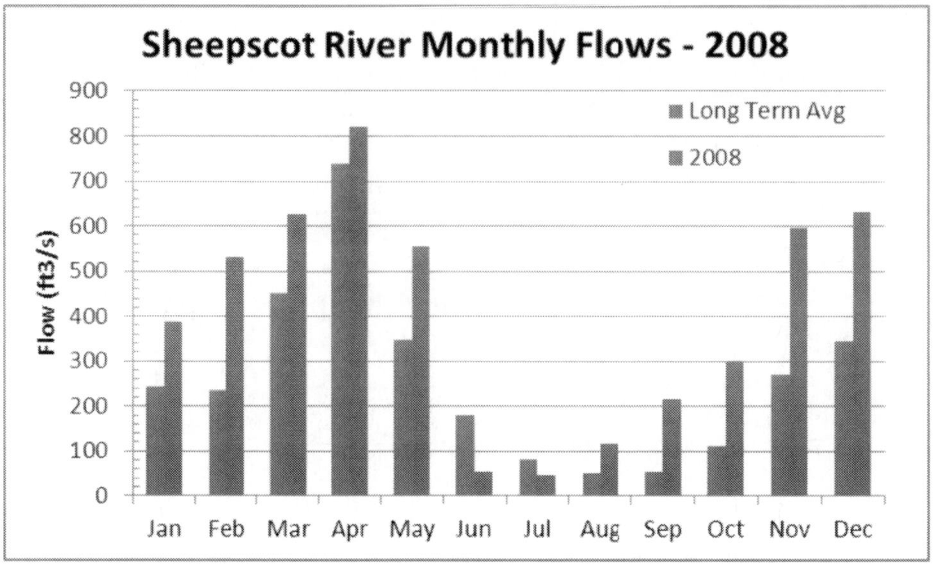

Figure 3. Sheepscot River monthly average flow – 2008.

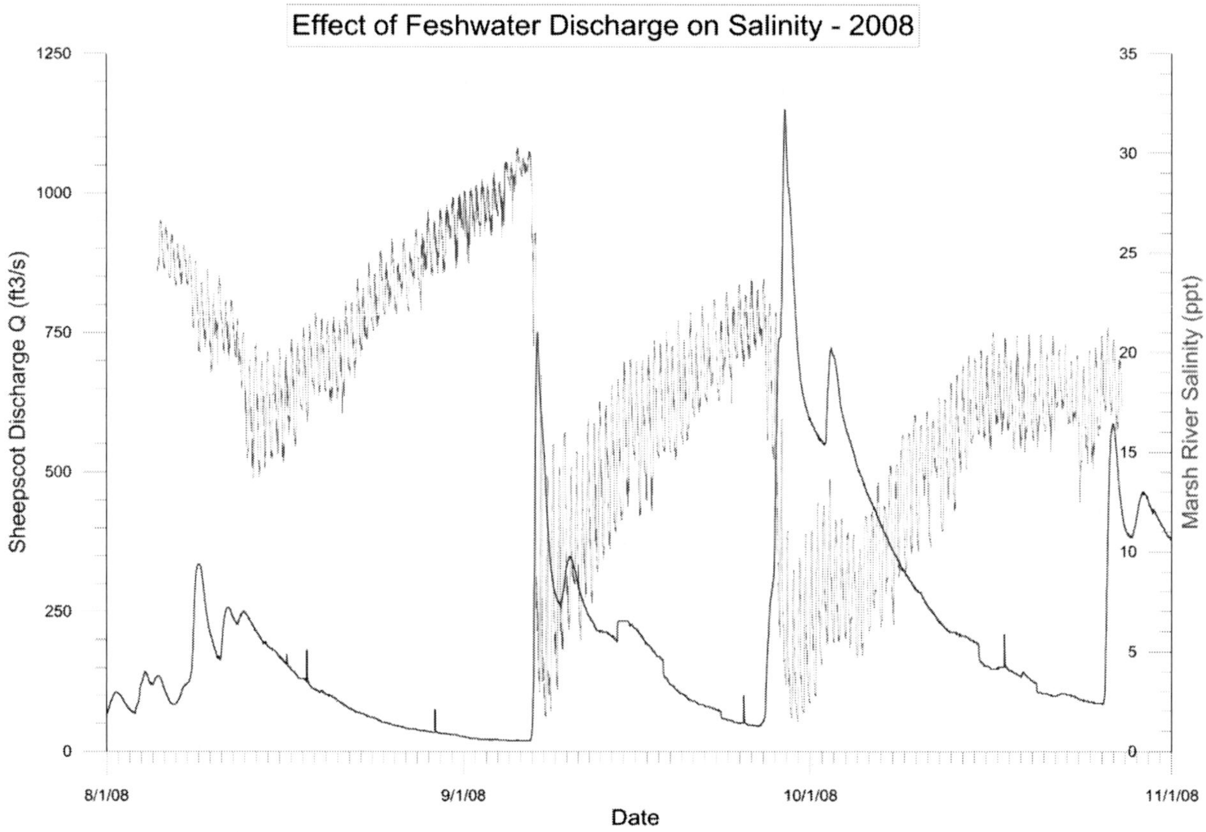

Figure 4. Effect of freshwater dilution on salinity – 2008.

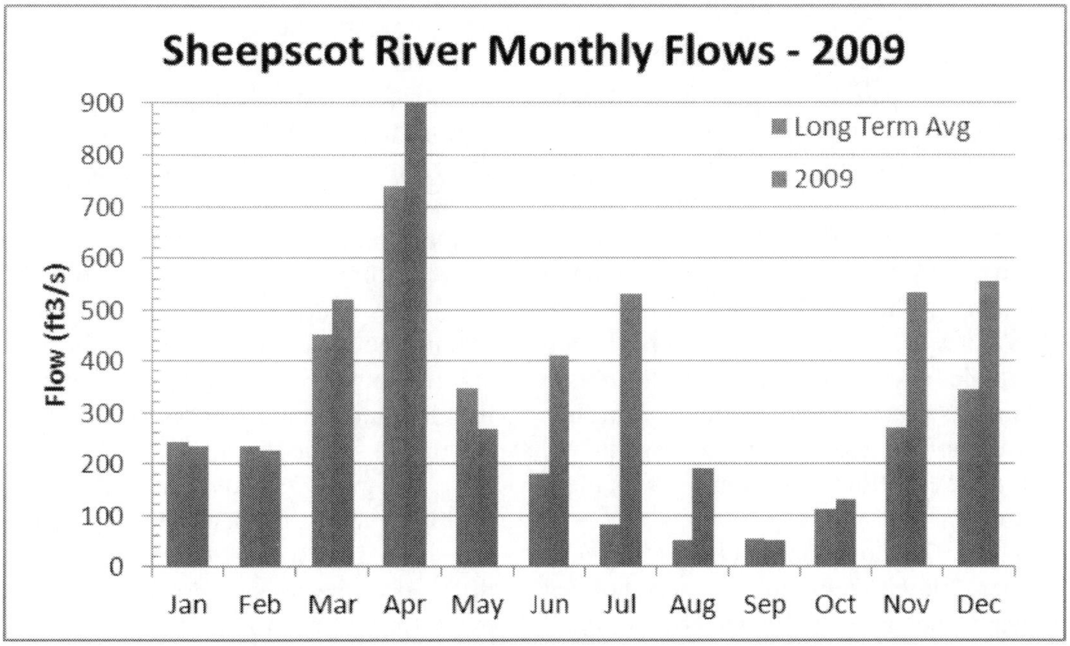

Figure 5. Sheepscot River monthly average flow – 2009.

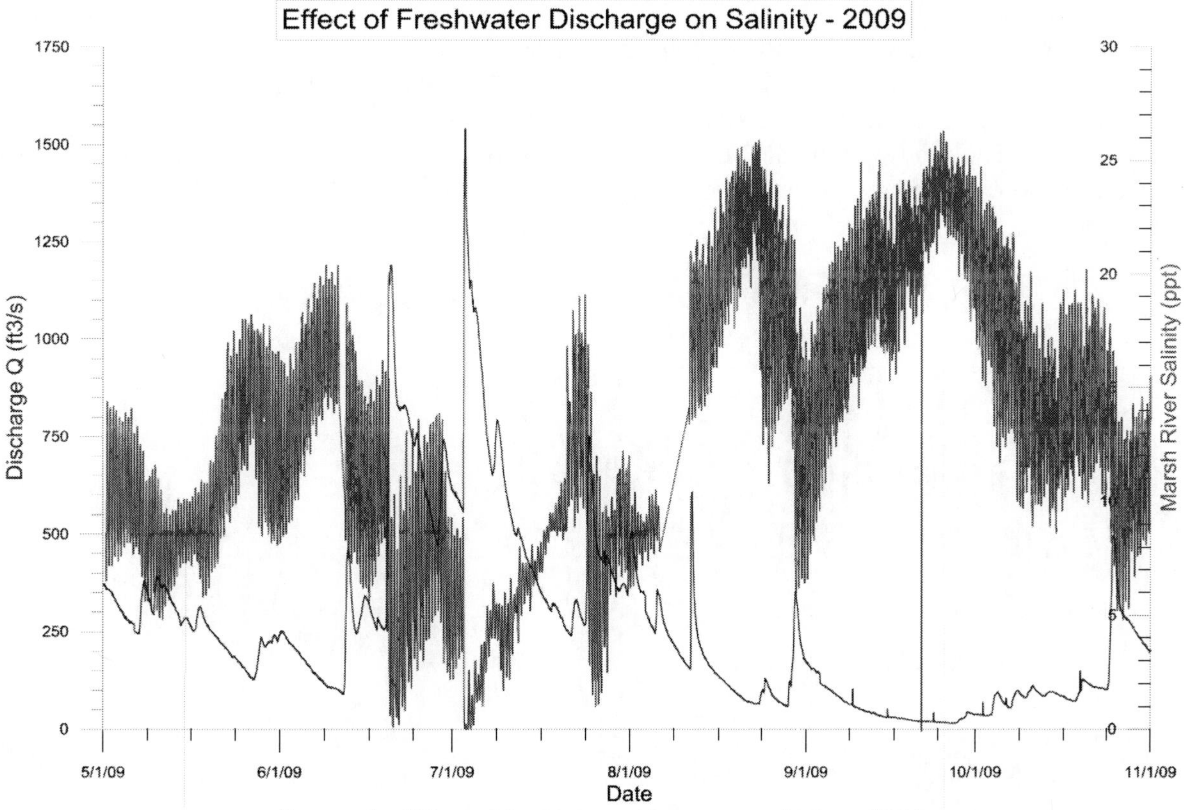

Figure 6. Effect of freshwater dilution on salinity – 2009.

69th Highway Geology Symposium, pages 1-18, 2018

Difference Between Mid-Marsh and Marsh River Below US1

Over the course of three seasons, we attempted to log salinity at the "Mid Marsh" (upstream) and "Marsh River" (downstream) locations. "Mid-Marsh" is in the marsh channel, between the former island (downstream) and the channel junction (upstream). "Marsh River" is in the channel downstream of the US-1 bridge, away from entrance/exit effects. The equipment proved to be unreliable and it was difficult to obtain coincident upstream and downstream data sets. The few complete data sets that were obtained did not always allow for simple and clear conclusions.

Figure 7a shows one such data segment in October 2009; it contains a few days of quasi-steady behavior followed by a small dilution event. A couple of simple conclusions can be drawn, consistent with what had been expected. Once a quasi-steady regime has been reached, there is essentially no significant difference over most of the tidal cycle. After a dilution event, the difference is more pronounced. The recovery in the Mid-Marsh appears to be even slower than in the Marsh River.

Figure 7b shows the quasi-steady period in more detail. Low tide corresponds to low salinity. At low tides and for a certain period following, salinities are lower in the Mid-Marsh. At these times the channels upstream have mostly drained and flow is more fresh. At the Marsh River logger, with more channel storage upstream than Mid-Marsh logger, salinities are somewhat higher than at Mid-Marsh during these periods. With regards to surface vegetation, this difference is not significant since water levels are well down in the channel and this low salinity water is not on the marsh surface.

Figure 7c shows the effect of a small dilution event. The peak Marsh River salinities are higher than in the Mid-Marsh. The differences between the salinity low points is even more pronounced, probably due to watershed drainage.

There are undoubtedly systematic differences between Marsh River and Mid-Marsh salinities. However, these differences are probably less important than the limiting factor on attainable salinities exercised by the Sheepscot River, both during quasi-steady higher salinity periods as well as during runoff events.

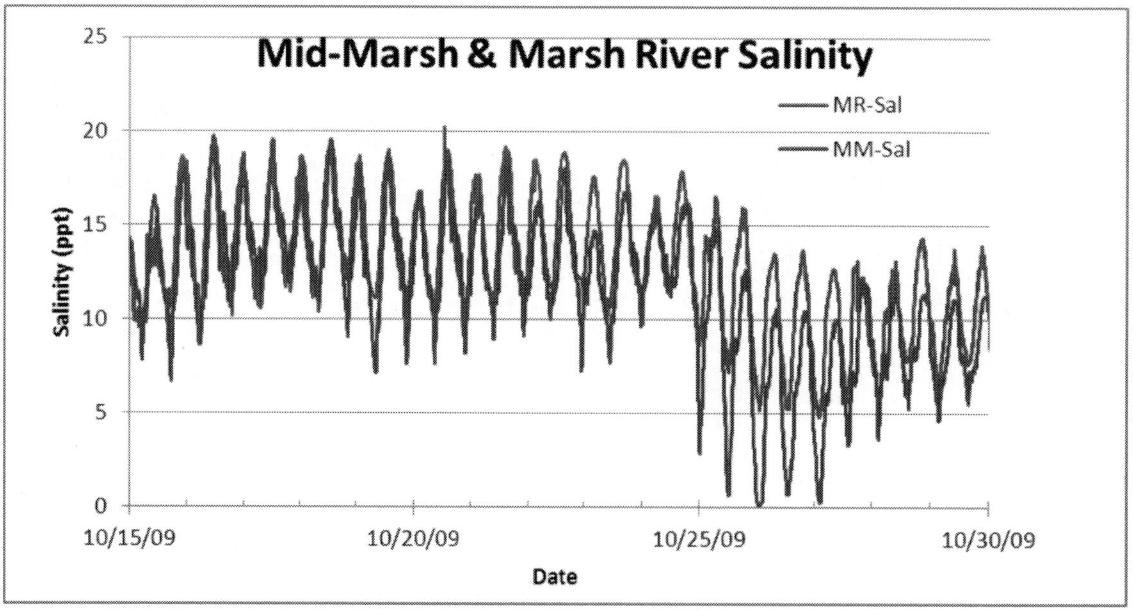

Figure 7a. Difference between Mid-Marsh and Marsh River salinity.

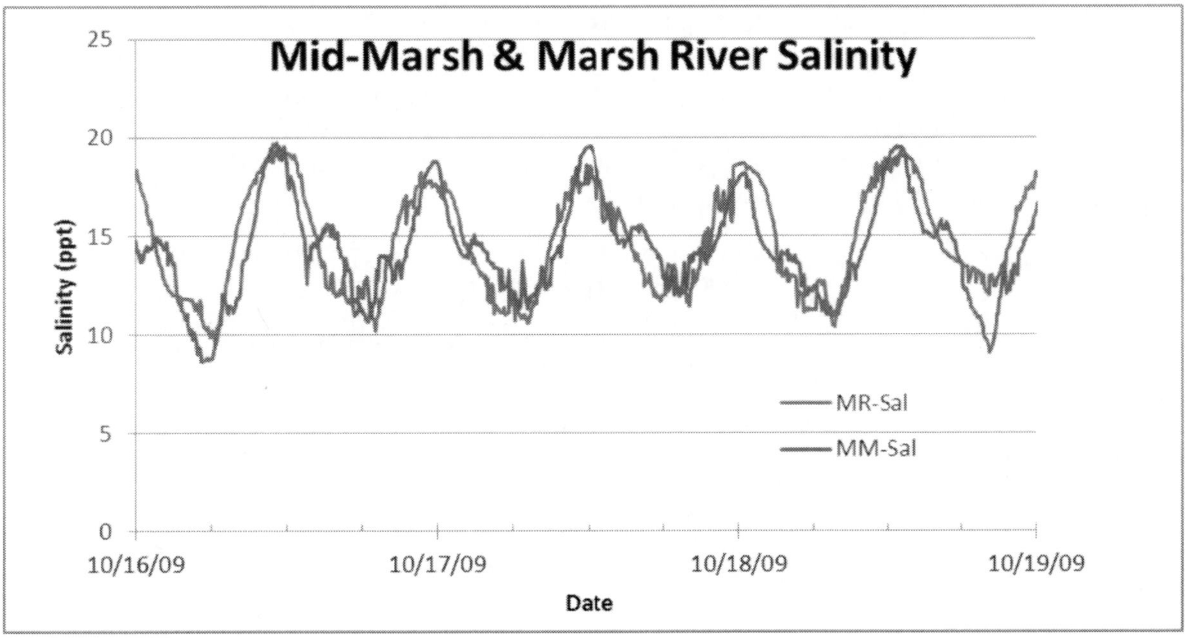

Figure 7b. Difference between Mid-Marsh and Marsh River salinity: quasi-steady period.

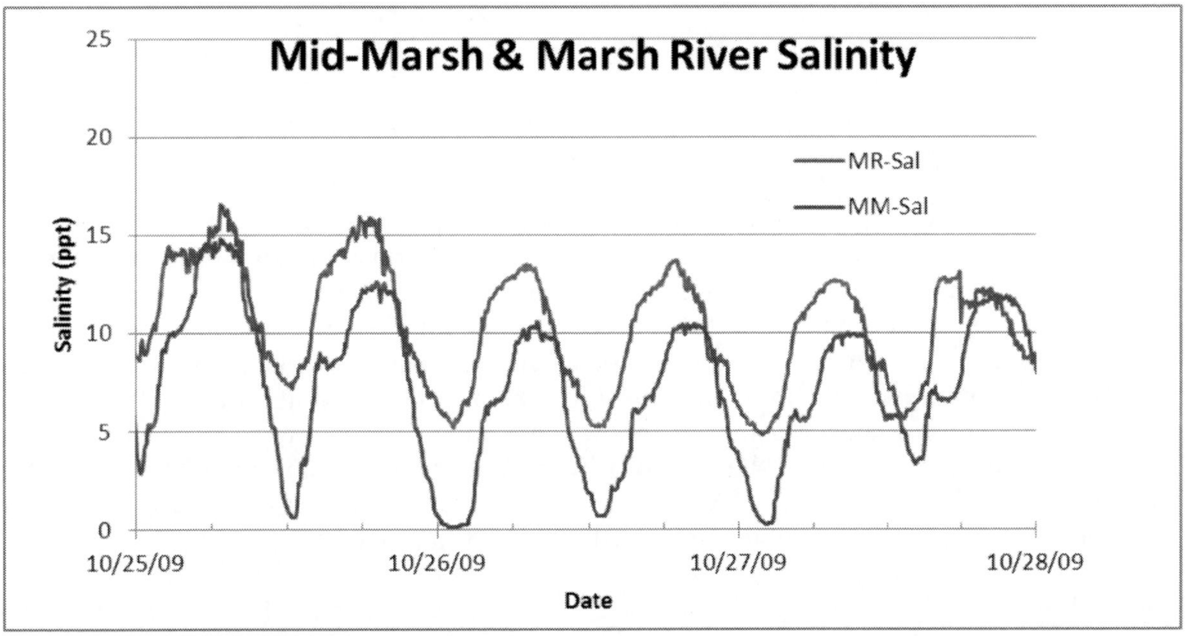

Figure 7c. Difference between Mid-Marsh and Marsh River salinity: small dilution event.

Relation of Salinity to Tidal Stage

The relation of salinity to stage is as expected. Peak salinity corresponds to high tide. The recession limb of the salinity curve can resemble a drainage curve and the salinity minimum lags the tidal minimum. Once the tide turns, the initial slug of incoming water is predominantly water that passed the logger on the way out. It takes some time for higher salinity water to makes its way back upriver, accounting for the observed lag. A small data sample is shown in Figure 8. Similar patterns were observed at the Mid Marsh logger.

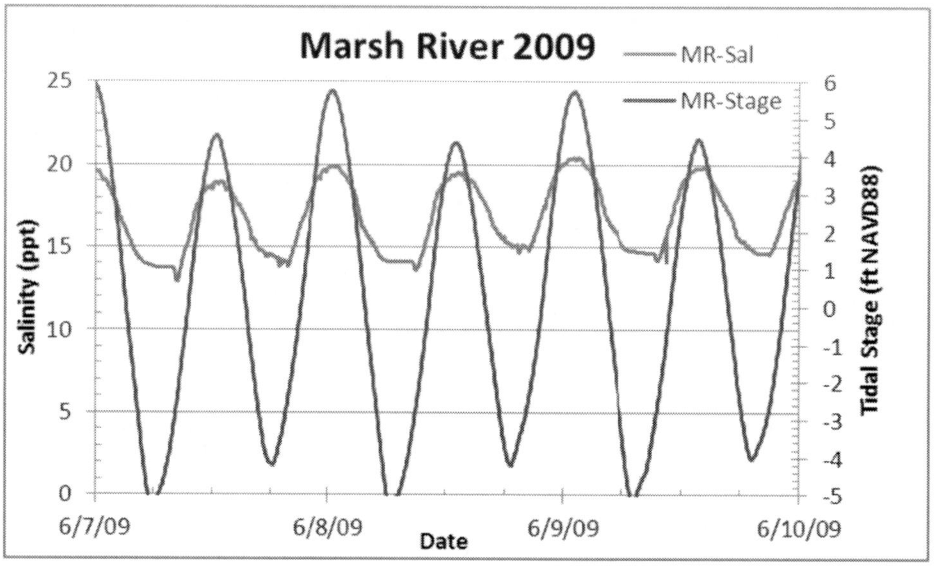

Figure 8. Relation of tidal and salinity cycles.

ADDENDUM 2: SHERMAN MARSH TIDAL HYDROLOGY OVERVIEW

Introduction

Tidal data have been collected at Sherman Marsh every summer, starting with 2006. Most years, the efforts have also encompassed spring and fall. Locations always included the "Lower Marsh" (just upstream of the US1 bridge) and the "Marsh River" (downstream of the bridge). "Mid" and "Upper" Marsh locations were also observed during several of the years; however, they added little in the way of useful additional information. Similar to the salinity data collection, the Lower Marsh tidal stage data capture the marsh tidal regime, while the Marsh River data represent the hydrologic driving force as well as the "natural" tidal regime that would presumably prevail in a completely open marsh.

Sherman Marsh is the uppermost marsh on the Marsh River, a tributary to the tidal Sheepscot River. Marsh River joins the Sheepscot just below head of tide. Thus, Sherman Marsh is significantly removed from the direct Gulf of Maine tides and is strongly influenced by freshwater discharge from the Sheepscot River as well from the Sherman Marsh watershed.

In this overview, three tidal data sources are utilized. The Portland tide station is the primary station for secondary stations in the midcoast area, so Portland data are presented as reference and are utilized to estimate long-term tidal datums at the marsh. The 2006 data are presented because they best represent the tidal regime created by the temporary emergency stabilization after the

October 2005 causeway failure. Data from 2010 are presented because they are the most consistent data set that captures the tidal regime created by the permanent stabilization and improvement constructed in winter 2008/2009.

Tidal Datums

Table 1 shows the tidal datums for the 1983 – 2001 tidal epoch for Portland (NOAA Tides and Currents / Bench Marks web page) as well as the corresponding datums estimated for Marsh River using the Modified Range Ratio Method (Computational Techniques for Tidal Datums Handbook, NOAA Special Publication NOS CO-OPS 2, September 2003). Typical marsh surface elevations are in the range 5.25-ft to 5.75-ft (Laura Jones, USM Thesis, 2007, Figure 3, p. 44) with lower elevations along the channel banks. The data period 3 July – 2 October 2010 was used to estimate the Marsh River long-term values, because this was a relatively dry period and Marsh River tides were not excessively influenced by Sheepscot River flows. The corresponding datum values for the data period are shown in Table 2; Table 2 also includes calculations for the lower marsh. However, long-term tidal datums were not transferred to the Lower Marsh station because the falling stage is still limited by a hard control elevation and also displays a residual drainage recession curve behavior.

Datum (1983-2001 epoch)	Portland	Marsh River (est)
Highest observed water level	8.869 (2Jul78)	--
Mean Higher High Water	4.651	5.31
Mean High Water	4.215	4.90
Mean Sea Level	-0.315	-0.13
Mean Tide Level	-0.348	-0.23
Mean Low Water	-4.907	-5.36
Mean Lower Low Water	-5.251	-5.86
Lowest Observed Water Level	-8.705 (30Nov55)	--
Marsh Surface	5.25' – 5.75' and lower along banks	

Table 1: Portland and Marsh River tidal datums (ft NAVD) for 1983-2001 epoch.

Datum (3 July – 2 Oct 2010)	Portland	Marsh River	Lower Marsh
Highest observed water level	6.84 (9 Sep)	7.34 (9 Sep)	7.23 (9 Sep)
Mean Higher High Water	5.02	5.67	5.61
Mean High Water	4.52	5.21	5.15
Mean Sea Level	0.05	0.23	0.47
Mean Tide Level	0.00	0.12	0.76
Mean Low Water	-4.52	-4.92	-3.58
Mean Lower Low Water	-4.67	-5.27	-3.63
Lowest Observed Water Level	-6.46 (11 Aug)	-6.48 (10 Sep)	-3.78 (8 Aug)
Marsh Surface	5.25' – 5.75' and lower along banks		

Table 2: Portland and Marsh River tidal datums (ft NAVD) for 3 July – 2 October 2010

Sherman Marsh Tides and Implications for Marsh Restoration

The purpose of collecting tidal data were several-fold. Initially, the goal was simply to develop a general understanding of tides in the newly reopened marsh. This goal was quickly refined to that of determining whether the new tidal regime was sufficient to maintain a healthy marsh. Figure 1 shows a sample trace from July 2006 data (red = Portland, + = Marsh River, blue = Lower Marsh). The flow and drainage restriction between the Marsh and Marsh River is obvious. Head losses between the two bodies of water are on the order of 1-ft and the effective control elevation is about 2.5-ft. This severely limits drainage of the marsh on the outgoing tide. The Marsh falling limb is nothing like a sinusoidal falling tidal stage; rather, it exhibits all the traits of classic reservoir drainage. These results pointed the way towards design of a permanent stabilization and improvement of the marsh outlet that make the marsh tides more nearly like those in the Marsh River just downstream. The final design was a combination of a significant enlargement of the opening as well as lowering of the outlet control elevation.

Figure 2 shows a sample data trace from August 2010, showing the dramatic effects of the stabilization and improvement work of winter 2008/2009. The head loss through the bridge has been largely eliminated and the effective control elevation has been lowered by about 5.5-ft, greatly improving marsh drainage.

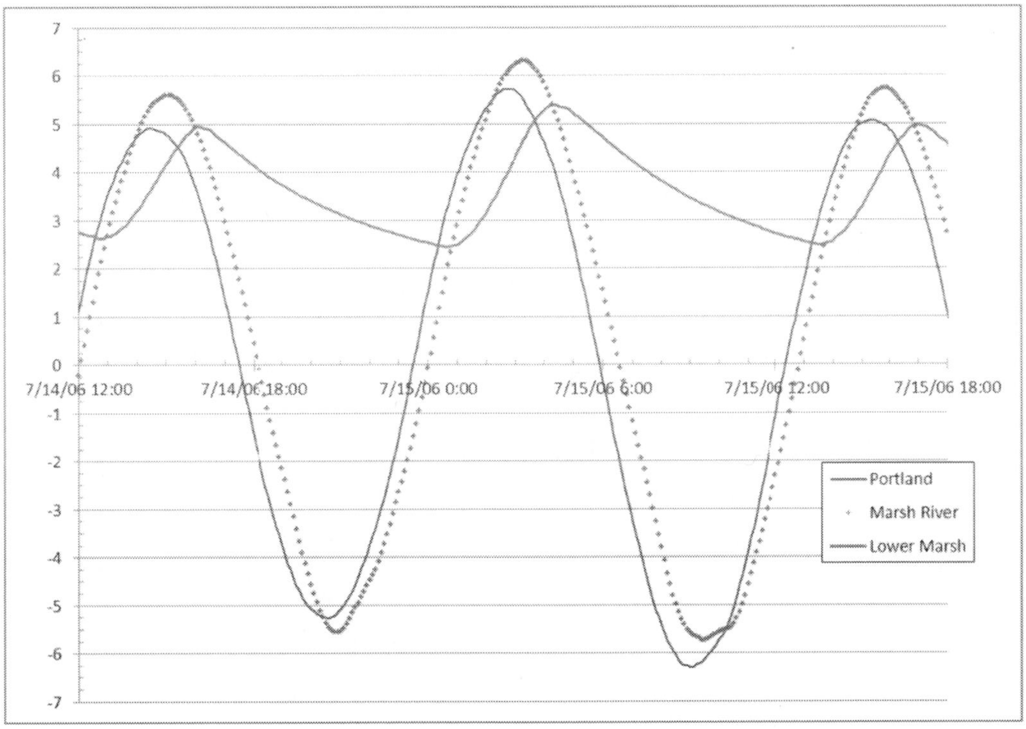

Figure 1: Sample tide trace, July 2006

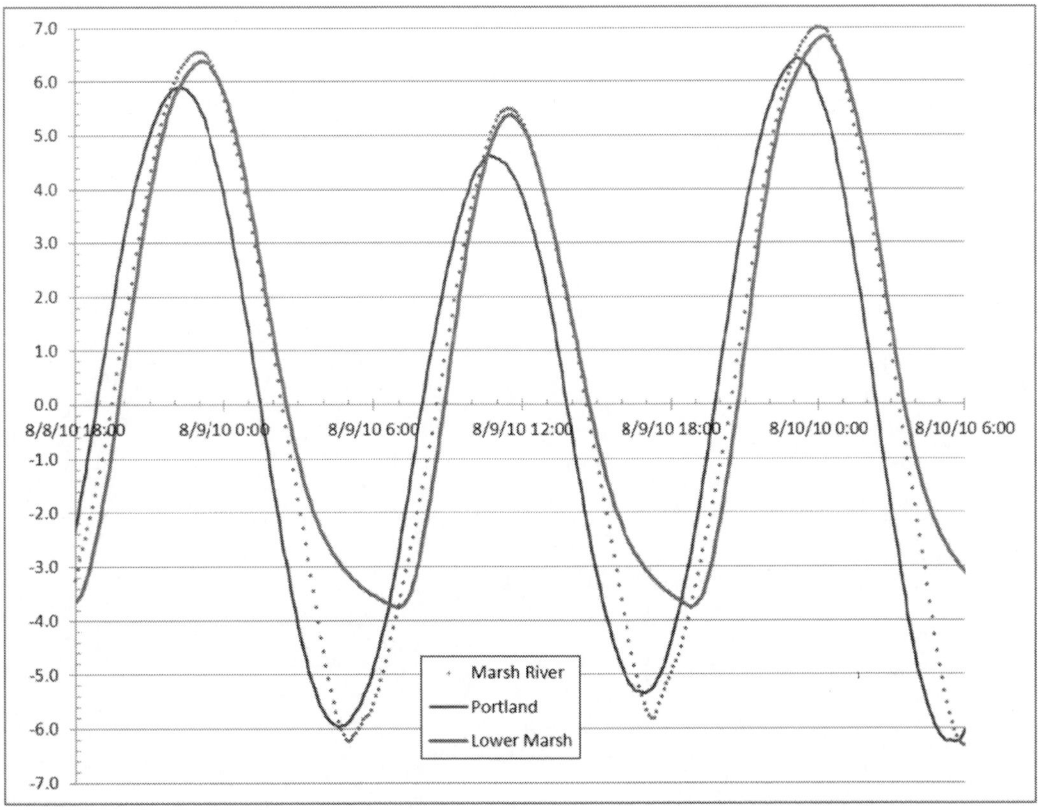

Figure 2: Sample tide trace after permanent improvements.

Formatting Issue - Best Available Paper/Graphic

69th Highway Geology Symposium, pages 19-30, 2018

1996 Rockland Harbor Landslide Site

Lindsay J. Spigel
Stephen M. Dickson
Maine Geological Survey
93 State House Station
Augusta, Maine 04333

Introduction:

Landslides are a known hazard in Maine, especially in southern Maine, where events have been recorded from early post-glacial to modern times (Morse, 1869; Novak, 1987; Berry and others, 1996; Thompson and others, 2011; Dickson and Johnston, 2015). The vast majority of Maine landslides occur in areas underlain by the Presumpscot Formation, which is a glaciomarine mud (silt and clay) deposited in the late-glacial sea that was in contact with the retreating Laurentide Ice Sheet. Isostatic rebound after deglaciation subsequently exposed the Presumpscot Formation above sea level, with upper portions weathering into a hard, tan crust while deeper unweathered portions appear blue-gray and are stiff to very soft. Informally referred to as "blue clay," the Presumpscot Formation is actually a combination of silt, clay, and fine sand, and can be massive to well-stratified (Thompson, 2015). Unweathered Presumpscot Formation creates the risk of unstable terrain and a challenge for engineers due to its low strength, compressibility, sensitivity, and unpredictable characteristics even within one work site (Andrews, 1987; Landon and Nickerson, 2015).

This stop provides an opportunity to observe a recent landslide site related to the Presumpscot Formation, how it was remediated, how it has changed since remediation, and a view of how other landowners in the harbor are dealing with coastal bluff erosion. Much of the information summarized in this guide (unless otherwise noted) is from a detailed report of the event (Berry and others, 1996), which is available for free download from the Maine Geological Survey website.

Summary of the 1996 Rockland Harbor Landslide:

Early in the morning of April 16, 1996, a 50-foot (15 m) coastal bluff failed between Samoset Road and Rockland Harbor (Figure 1). The initial movement woke up a nearby resident, Mrs. Gerrish, who looked out a window to find that her backyard had disappeared. She called the local authorities and evacuated her house along with her husband. Neighboring houses were evacuated and not long after, portions of the Gerrish home and the entirety of a neighboring house gave way as the landslide proceeded back into to the bluff, the scarp just tens of feet away from Samoset Road and city utility lines (Figure 2). Luckily, no one was injured.

About one-half acre of the bluff top was displaced, with the landslide toe moving 300-400 ft (90-120 m) out onto the intertidal mud flats. The total disturbed area was about 3.5 ac (1.4 ha), with about 60,000 yd^3 (45,800 m^3) of material displaced. The landslide was categorized as a retrogressive rotational slide – initial failure of the bluff face exposed a new unstable area which, in turn failed, allowing the slide to progress farther inland towards Samoset Road (Figures 3 and 4). After the main event, small slumps continued to occur along the scarp for about one month, making it difficult to recover materials from the destroyed homes and proceed with remediation. Figure 5 illustrates conditions before the event, while Figures 6 and 7 illustrate conditions after the event.

The potential for coastal bluff erosion, slumping, and sliding in the Rockland Harbor was known prior to the 1996 event. A similar landslide took place just 1,000 ft (305 m) to the northwest in 1973, and a small slump took place adjacent to the 1996 slide area on February 16, 1995 (Figure 1). No people or structures were harmed in the 1973 landslide and the area was left to natural processes, with exception of the west scarp. The toes of both landslides have been reworked and eroded by coastal processes.

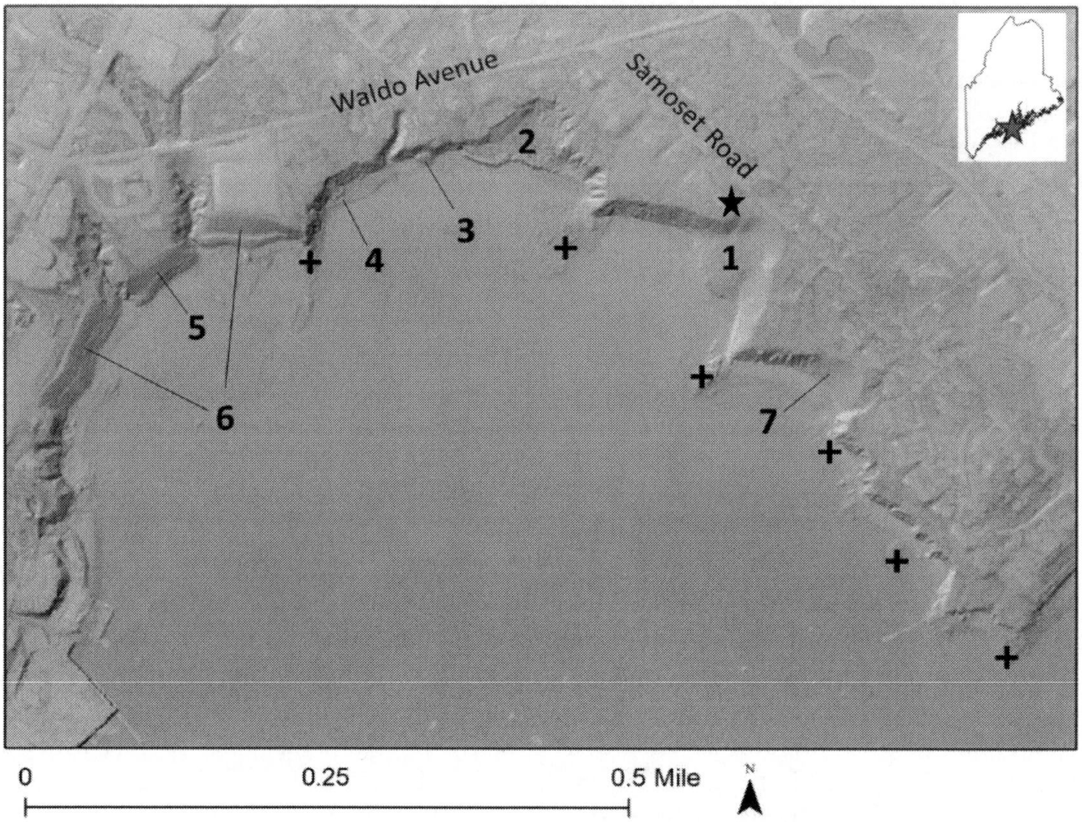

Figure 1: Locator map of Rockland Harbor, with 2010 lidar hillshade over 2009 aerial imagery. Key to labels is as follows: + = bedrock outcrop areas; 1 = 1995 slump/1996 landslide area (remediated); 2 = 1973 landslide (portions of west scarp altered, but otherwise un-remediated); 3 = slump that occurred between 1998-2003 (no remediation); 4 = slump that occurred in April 2010 (graded and rip-rap added between 2011-2013); 5 = section graded and rip-rap added prior to 1996; 6 = sections graded and rip-rap added between 1998-2003; 7 = section graded and rip-rap added between 1998-1999; star = location of house referenced in Figures 5, 6, and 7.

Figure 2. Gerrish house shortly after the landslide. View is to the south from the landslide headwall at Samoset Road toward Rockland Harbor in the background. MGS file photo.

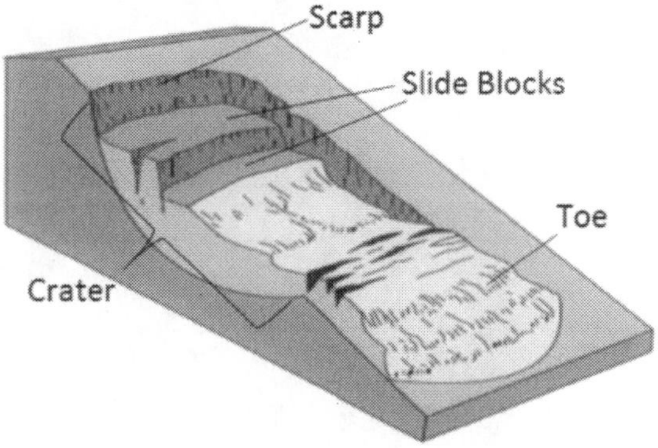

Figure 3. Cartoon illustration of a rotational landslide (modified from Highland and Bobrowsky, 2008). The 1996 Rockland landslide is an excellent example of this landslide morphology, also visible in Figures 4 and 5. A slide is considered "retrogressive" if it continues to migrate back into the slope after the initial failure. Retrogression usually ceases when enough failed material accumulates to buttress the slope.

Figure 4. Rotational slump blocks in the mid-slide area. Trees tilt up slope. MGS file photo.

Formatting Issue - Best Available Paper/Graphic

Figure 5. Aerial photo of the 1996 landslide location prior to the event (October 4, 1992). Star placed on house is the same as in Figures 6 and 7 for reference. Photo: J.W. Sewall Company.

Figure 6. Aerial photo of the 1996 landslide location after the event (May 8, 1996). Photo: J.W. Sewall Company.

Figure 7. Oblique aerial view of the 1995 slump and 1996 landslide, looking east.

The Rockland Harbor Landscape:

Bedrock geology:

The Rockland Harbor area is underlain by metamorphic rocks the Ojier Point and North Haven Formations that are likely of Silurian age. The Ojier Point Formation consists of mica schist and quartzite that are thinly interbedded and deformed. The North Haven Formation consists of greenstone and feldspar-rich gneiss that were formed from mafic and felsic volcanic rocks. In general, the bedrock structure is complex with varied foliation and folding. Fractures in the area are oriented west-northwest to north-northeast. There are no active faults in the region, and the was no seismic activity around the time of the 1996 landslide.

Depth to bedrock in the harbor area is irregular, with outcrops occurring at various locations. Seismic profiling in the harbor near the 1996 landslide site (about 1,500 ft (4,570 m) offshore) indicates an undulating bedrock surface between 16 and 65 ft (5-20 m) that is overlain by glacial and post-glacial sediments. Seismic profiling along Samoset Road and Waldo Avenue also revealed an irregular bedrock surface, with depths ranging from 0-40 ft (0-12 m).

Surficial geology:

There are two major surficial deposits in the Rockland Harbor area: glacial till and the Presumpscot Formation. Glacial till overlies bedrock in much of the area, but is not ubiquitous with thickness ranging from 0-10 ft (0-3 m). The Presumpscot Formation overlies bedrock and/or bedrock and till with thickness ranging from 35-45 ft (11-14 m) in the vicinity of the landslide. The till may have lenses of sand and gravel, while the Presumpscot Formation may have sand seams and dropstones. The Presumpscot Formation was weathered to depths of about 6-10 ft (2-3 m) in the bluff areas near the 1996 landslide site.

Groundwater:

Seismic lines and test borings in the area revealed that the local water table roughly coincided with the transition from weathered to unweathered Presumpscot Formation at depths of about 4-7 ft (1-2 m). Slight artesian conditions may exist in the glacial till and/or at the interface between bedrock and overlying glacial materials, but a longer study was needed to confirm (Jacques Whitford Consulting Engineers, 1997). Groundwater movement in the bluff areas is very localized, shallow, and slow, with water percolating down through fissures in the weathered Presumpscot Formation and then out to the harbor bluff face, or to one of a few gully drainages in the bluff face. Areas outside of 500-1,000 ft (150-300 m) of the bluff face are thought to have little influence on groundwater dynamics in the bluffs, unless they are one of the few possible areas that may recharge the till/bedrock interface (Jacques Whitford Consulting Engineers, 1997).

Geomorphology/Coastal Processes:

Maine has over 1,400 miles (2,500 km) of sedimentary coastal bluffs that make up about 40% of the coast. This shoreline type is typically found along inner bays and estuaries where wave heights are less than 6 feet (2 m) under storm conditions. Tides range from 9 feet along the southwest coast to over 20 feet in eastern Maine near the Canadian border. In Rockland Harbor, the mean tide range is 9.8 feet (3.0 m). The Highest Astronomical Tide (HAT) is 12.8 feet (3.9 m) above MLLW (mean lower low water, the "zero" tidal datum). Tides, waves, and coastal flooding all play important roles in bluff-toe erosion. Tides on this trip are large and called a "spring tide" even in the fall season. Low tide at 7:24 a.m. is -1.2 feet (below MLLW) and high tide at is 11.7 feet, a foot below the HAT. The tide range on September 12, 2018 is 12.9 feet (3.9 m).

The semidiurnal tide plays an important role in sorting sediment across often expansive intertidal areas. More extreme or "King Tides" allow non-storm marine erosion at the toe of a coastal bluff a dozen or so times a year. Storm surge, created by onshore wind, can add two feet of water to the tides a few times a year. Surges of 3.4 feet (1 m) occur with an annual probability of 10% and a 1% surge can add 4.8 feet (1.5 m) to the tides. Storm tides are often accompanied by larger waves that then can run up on a bluff and scour sediment during the time of high tide.

Fortunately, coastal flooding is moderated by the limited duration of high tide and the 1- to 3-day duration of most major storms. Historically, record-setting surges have not coincided with high tide. Moderate surges can cause significant bluff erosion if they occur during astronomically high tides. For example, the 2007 Patriots' Day Storm had 7 elevated storm tides that induced significant coastal erosion and requests for bluff stabilization with engineering structures. The storm of record was the Blizzard of February 6, 1978 when the storm tide in Portland Harbor reached 14.1 feet (4.3 m). The worst-case "superstorm" condition, which the Maine coast has yet to experience, would be with a 5-foot (1.5 m) surge on a King Tide that is accompanied by offshore waves of 30 feet (9.1 m). This scenario could produce a 16.8-foot (5.1 m) storm tide, and cause bluff erosion unlike any experienced in the last 100 years along the Maine coast.

Another factor causing bluff erosion is the gradual rise of the tides. Over the last century, sea level has risen at a rate of about 0.8 in (2 mm) per year along the coast. Since 1993, when satellite altimetry began, tide gauges in the Gulf of Maine show a 0.12 in (3 mm) per year rate of rise. This recent rate of about a foot per century rise is similar to that of the world ocean. There is no consequential vertical crustal motion contributing to Maine sea level rise. While this rate seems relatively slow compared to other coastal processes and forces, over time the reach of storm tides has moved inland. Since the early 1900's the tides have risen about a foot. This rise can translate quite far inland based on low intertidal slopes projected ashore. For example, a 1:40 erosional surface below a bluff could result in 40 feet of shoreline retreat through gradual sea level rise.

Bluff retreat tends to be episodic. As with the history of landslides in Rockland Harbor, there are periods of years with no significant changes to the shoreline. Periods of quiescence are punctuated by slope failures that are shallow and localized or large and deep-seated (as in Rockland). Sometimes the shallow failures lead to bank steepening that can then result in the more abrupt and extensive landslides. In general, bluffs composed of the Presumpscot Formation become susceptible to deep-seated failures when the top of the bluff is 20 feet (7 m) or more above the reach of the tides. In a few other Maine locations, bluff erosion has contributed to about a foot per year of upland land loss.

Groundwater, surface water, and land use all contribute to processes that can affect bluff erosion. The elevation of the water table varies seasonally but is generally highest in spring as snow pack melts and the ground thaws. In late winter, groundwater discharge sites on a bluff become visible as seeps refreeze and create concentrations of ice on the bluff face. These discharge points often coincide with small gullies where erosion is most pronounced.

Surface water discharge also tends to peak in spring for the same reason that groundwater does. A high water table and upland watersheds can both contribute to localized runoff on a bluff face. Once gullies get established, subsequent discharge tends to be concentrated on the bluff and cause landward incision with sediment deposition that gets reworked at the base of the slope near high tide.

Coastal development also affects surface and groundwater discharge. Impervious surfaces, road drainage, roof discharge of rainwater, and septic system leaching can concentrate and direct flow to coastal bluffs. Even land clearing is thought to contribute to higher bluff erosion rates. Erosion hot spots are often filled with yard debris that inhibits vegetation growth, soil stabilization with roots, and dewatering through transpiration.

Rockland Harbor bluff retreat rates from 1952 to 1996 were estimated from air photo analysis after the 1996 landslide. Including the 1973, 1995, and 1996 landslides, average retreat ranged from <1 in (<0.025 m) per year to 20 in (0.5 m) per year. When major landslide areas were excluded, average retreat was <1 in (<0.025 m) per year to 4 in (0.1 m) per year (Jacques Whitford Consulting Engineers, 1997).

Ground cover:

With exception of the bare area created by the 1995 slump, the bluff face in the vicinity of the 1996 landslide was vegetated with small trees and shrubs (Figure 3). No shore stabilization structures were in place prior to the landslide.

Local factors that contributed to the 1996 landslide:

There are many factors that can contribute to and trigger a landslide. The local factors deemed most significant by the two investigations of the 1996 landslide (Berry and others, 1996; Jacques Whitford Consulting Engineers, 1997) are:

- **Local geology and geomorphology:** The presence of unweathered Presumpscot Formation at depth, combined with a steep bluff slope certainly made the area susceptible to failure. The thickness of the Presumpscot Formation is also an important factor – areas in the harbor with less than 5 ft (1.5 m) of Presumpscot Formation in a slope toe area were found to be more stable, while areas with 20-25 ft (6-8 m) of Presumpscot Formation were much more likely to fail. A back-analysis of the landslide area resulted in a modeled factor of safety equal to 1.14 – close to failure, but not quite, which engineers attributed to a lack of completely accurate Presumpscot Formation characterization at the location.

- **Water:** A tall coastal bluff comprised of Presumpscot Formation that is exposed to coastal processes may be relatively stable without the influence of other factors. In the case of the 1996 landslide, the actual failure was likely triggered by conditions related to water. As previously mentioned, surface water drainage from the flat bluff top and groundwater movement in the bluff area is very slow due to the low permeability of the Presumpscot Formation. Therefore, any water added to the bluff area through snowmelt, rainfall, and septic leach fields drains very slowly and adds weight to the unsupported bluff slope. Added weight then leads to slope failure.

  In addition, groundwater seeping onto the bluff face from the weathered/unweathered Presumpscot Formation interface and from local, unarmored drains created consistently wet bluff slope conditions, which were found to be less stable than consistently dry slopes in the harbor regardless of coastal processes. Consistently wet bluff slopes had lower shear strength and were more susceptible to sloughing and slumping, which could lead to larger failures. Landslides could also be triggered by increased porewater pressure in deeper, unweathered Presumpscot Formation or in the underlying till, but this was determined to be less significant than the generally poor drainage.

Site remediation:

Although it is very fortunate that no one was hurt by this catastrophe, the financial impact was estimated to exceed $750,000. After the landslide area was deemed safe enough, a crane was used to salvage belongings from the two destroyed homes. The City of Rockland used emergency funds to remediate the properties and prevent further failure (Berry and others, 1997). The scarp and crater areas were graded with a combination of heavy fill near the toe and lightweight kiln dust fill in the crater (Figure 8). The area was seeded and low-growing vegetation is maintained to prevent addition of weight to the slope (Figure 9). Local drains to the shore were armored with rip-rap and an inclinometer was installed in the slope. The toe area was left to erode naturally, but rip-rap and landscape fabric were installed within the toe at the natural shoreline, which is now being exposed through erosion (Figure 10).

Formatting Issue - Best Available Paper/Graphic

Figure 8. Oblique aerial image of remediated 1996 landslide site, looking roughly east (from Berry and others, 1997).

Figure 9. Seeded, remediated slope looking northward from the slide toe to the headwall. MGS file photo.

Figure 10: Photo of eroded 1996 landslide toe (taken 11/20/17, looking northwest), which has retreated to the rip-rap barrier that was installed within the toe during remediation. A tree that was swept up in the landslide is also visible in the bank, as well as rip-rap armor that extends beyond the landslide area in the distance. Photo: Lindsay Spigel.

Natural erosion of the toe was estimated by the guide authors from aerial and lidar imagery. While data was not available for a few years immediately after the landslide, it does seem that the toe was quickly reworked by coastal processes and continues to steadily erode (Table 1).

Rip-rap was also placed at the base of the bluffs that flank the landslide (visible in Figure 7). Since the 1996 landslide, two smaller slumps occurred in the harbor and many bluff sections have been graded and armored (Figure 1).

Time Interval	Toe Erosion (m^2)	Erosion Rate (m^2/yr)
1996-2003	3420	490
2003-2004	70	70
2004-2006	300	150
2006-2007	110	110
2007-2009	160	80
2009-2010	110	110
2010-2011	100	100
2011-2013	170	80
2013-2015	150	80

Table 1: 1996 landslide toe erosion estimates.

Engineering recommendations:

Several recommendations were made by Jaques Whitford Consulting Engineers (1997) to prevent future landslides in Rockland Harbor. The most important was carefully managed drainage of bluff areas. Proper drainage of the bluff tops reduces water weight loading which likely triggered the 1996 landslide. Maintaining dry bluff slopes was also deemed important, so drainage needs to be routed off-site or to an armored outlet. Groundwater should be monitored in the area, and any development that could affect groundwater contributions to bluff areas should be restricted. A pond installed at the Samoset Resort was initially blamed for adding groundwater flow to the bluff areas, but studies confirmed that the pond drained to the east, away from the bluff areas. Development on the bluffs should be carefully regulated, as well, but ocean views remain in high demand, and a large house was built in 2007 just two doors down from the 1996 site.

There are multiple approaches to shoreline stabilization along bluff coasts. Traditional methods include rock riprap revetments backed by filter fabric and a gravel base on the slope. Hybrid shoreline stabilization combines the traditional rock works with vegetation on the bluff slope to create more opportunities for groundwater and surface water management as well as to mimic the visual character of natural slopes. Some bluff slopes can be re-graded to a lower angle and successfully planted with natural vegetation and avoid extensive rock placement. In the last few years Maine has begun to explore "living shorelines" that are seaward of the bluff toe and extend into the intertidal zone. The living shoreline concept is one that mimics a natural geologic setting and may be sacrificial over time as erosion removes material installed, instead of the natural slope retreating. So living shorelines can preempt land loss and allow a natural ecological setting of the high intertidal zone to be preserved.

There is no single solution for living shorelines since each section of coast is unique geologically and in terms of coastal processes and sediment budgets. In many sheltered locations slope failures lead to an abrupt sediment injection to the upper intertidal zone. Over time this deposit is colonized by saltmarsh cordgrass (*Spartina alterniflora*) up to mean high water and saltmeadow cordgrass (*Spartina patens*) from mean high water up to the highest astronomical tide level. Fringing salt marshes are one living shoreline type that might be suitable to slow erosion along many bluffs. Living shorelines, however, do not provide much reduction in the risk of a landslide as in Rockland Harbor. Nevertheless, fringing marshes could reduce toe erosion, delay bank over steepening, and perhaps postpone conditions that lead to landslides.

References:

Andrews, D. W., 1987, The engineering aspects of the Presumpscot Formation: in Andrews, David W., Thompson, Woodrow B., Sandford, Thomas C., and Novak, Irwin D. (editors), Geologic and geotechnical characteristics of the Presumpscot Formation, Maine's glaciomarine 'clay': unpublished proceedings of a symposium sponsored by the Maine Geological Survey, Morrison Geotechnical Engineering, University of Maine, and University of Southern Maine, March 20, 1987, variously paginated (17 p.).

Berry, H. N., IV, Dickson, S. M., Kelley, J. T., Locke, D. B., Marvinney, R. G., Thompson, W. B., Weddle, T. K., Reynolds, R. T., and Belknap, D. F., 1996, The April 1996 Rockland landslide: Maine Geological Survey, Open-File Report, 55 p., 13 figs., 3 apps., 1 plate.

Berry, H. N., IV, Dickson, S. M., Kelley, J. T., Locke, D. B., Marvinney, R. G., Thompson, W. B., Weddle, T. K., Reynolds, R. T., and Belknap, D. F., 1997, Aftermath of the 1996 Rockland Landslide: Maine Geological Survey, Geologic Facts and Localities, Circular GFL-9, 8 p.

Dickson, S. M. and Johnston, R. A., 2015, Geomorphology of Presumpscot Formation Landslides: 2nd Symposium on the Presumpscot Formation: Advances in Geotechnical, Geologic, and Construction Practice, Landon, M. E. and Nickerson, C. (Editors), Portland, Maine, 28 October 2015, 18p.

http://umaine.edu/presumpscot-symposium/2015-symposium-proceedings/

Highland, L.M., and Bobrowsky,P., 2008, The landslide handbook—A guide to understanding landslides: Reston, Virginia, U.S. Geological Survey Circular 1325, 129 p.

Jacques Whitford Consulting Engineers, 1997, Rockland Harbor Landslide Study: Jacques Whitford Consulting Engineers, 24 p., 8 figs.

Landon, M. E. and Nickerson, C. (Editors), 2nd Symposium on the Presumpscot Formation: Advances in Geotechnical, Geologic, and Construction Practice, Portland, Maine, 28 October 2015.

http://umaine.edu/presumpscot-symposium/2015-symposium-proceedings/

Morse, E. S., 1869, On the landslides in the vicinity of Portland, Me.: Boston Society of Natural History, Proceedings, v. 12. 235-244, map.

Novak, I. D., 1987, Geology of the September 1983 landslide at Gorham, Maine: in Andrews, David W., Thompson, Woodrow B., Sandford, Thomas C., and Novak, Irwin D. (editors), Geologic and geotechnical characteristics of the Presumpscot Formation, Maine's glaciomarine 'clay': unpublished proceedings of a symposium sponsored by the Maine Geological Survey, Morrison Geotechnical Engineering, University of Maine, and University of Southern Maine, March 20, 1987, variously paginated (14 p.).

Thompson, W. B., 2015, Surficial geology handbook for southern Maine: Maine Geological Survey, Bulletin 44, 97 p.

Thompson, W. B., Griggs, C. B., Miller, N. G., Nelson, R. E., Weddle, T. K. and Kilian, T. M., 2011, Associated terrestrial and marine fossils in the late-glacial Presumpscot Formation, southern Maine, USA, and the marine reservoir effect on radiocarbon ages: Quaternary Research, Vol. 75, p. 552-565.

Whiteman, N., Kelley, J. T., Belknap, D. F., and Dickson, S. M., 2016, Coastal bluff erosion, landslides and associated salt marsh environments in northern Casco Bay, Maine: in Berry, H. N., IV, and West, D. P., Jr., (editors), Guidebook for field trips along the Maine coast from Maquoit Bay to Muscongus Bay: New England Intercollegiate Geological Conference, p. 95-106, Maine Geological Survey Publications 25. http://digitalmaine.com/mgs_publications/25.

Formatting Issue - Best Available Paper/Graphic

69th Highway Geology Symposium, pages 31-40, 2018

Rock Slope Remediation at Penobscot Narrows Bridge Prospect, Maine

Marleigh L. Snow, Amber B. Granger, Bradford A. Miller, and Bryan C. Steinert
Haley & Aldrich, Inc.
75 Washington Avenue, Suite 1
Portland, Maine 04101

Introduction

Site History

Beginning in 1931, all traffic heading up U.S. Route 1 (Route 1) along the coast of Maine crossed the historic Waldo-Hancock suspension bridge to access the Down East Maine communities of Bar Harbor, Blue Hill, Castine, and Eastport. The narrow, two-lane, steel bridge spanned the Penobscot River, providing views of the Civil War-era Fort Knox and the town of Bucksport to the north, and Penobscot Bay to the south.

During the spring of 2003, engineers performing an ongoing evaluation of the main-span suspension cables found that the 75-year-old cables were more deteriorated and corroded than originally believed. Subsequently, the bridge was posted and access was denied for vehicles weighing over 24,000 pounds until stabilization and/or remedial repair options could be provided. The need to restrict truck traffic had significant economic impact on the local region and destinations Down East. An immediate decision was made by the Maine Department of Transportation (MaineDOT) to fast-track the replacement of the bridge with a new, modern structure and approach roadways while a stabilization contract was undertaken to strengthen the main-span cables until the new bridge and approaches could be completed.

The location of the replacement bridge (Penobscot Narrows Bridge) is parallel to and immediately downstream of the existing bridge as shown on Figure 1.

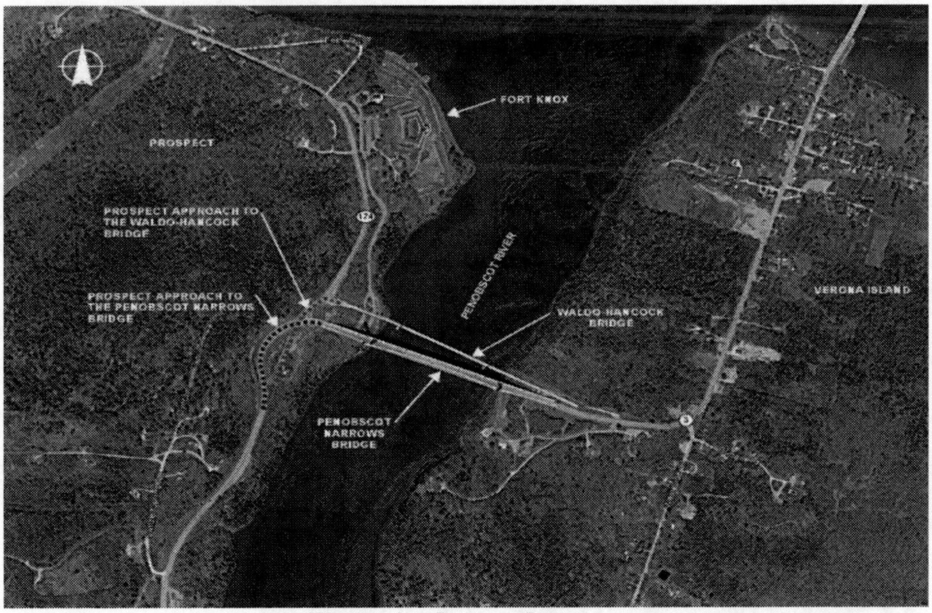

Figure 1 – Project Locus

Prospect approach

Locating the Penobscot Narrows Bridge immediately downstream of the Waldo-Hancock Bridge required realignment of an approximately 775-ft long section of the Prospect approach to the west and into a bedrock-controlled hillside to provide access to the new bridge, as illustrated by the blue dotted line in Figure 1.

The Prospect Approach roadway varies between 40 and 60-ft wide (shoulder-to-shoulder) and generally consists of two 14-ft wide travel lanes and two 8-ft wide outside shoulders. A portion of the roadway has an approximate 14-ft wide curbed median. Ground surface elevations along the concave, semi-circular approach to the new bridge ranged from approximately El. 135 to El. 140 in the vicinity of Route 1 to as high as about El. 250. The proposed grade for the new (i.e., current) roadway ranged from approximately El. 141 to El. 144. As a result, a rock cut up to approximately 100 ft was required to construct the Prospect Approach to the Penobscot Narrows Bridge.

Geologic setting

Based on the surficial geology map of the Bucksport Quadrangle (Kelley and Caron, 2013), the near surface soil conditions along the proposed roadway consists of thin drift, which is a glacial till deposit that is generally less than 10 feet thick and overlies bedrock.

Bedrock geologic maps of the region, (Wones, 1991; Stewart, 1998), indicate that the bedrock at the site consists of iron sulfide-rich schist in beds 1/32 to 2-in. thick, and graded, sandy quartz-chlorite-muscovite-plagioclase siltstone in beds 3/4 inch to 23 feet thick of the Penobscot Formation. Andalusite, corderite, and biotite are present in contact metamorphic aureoles adjacent to granitic rocks. High-grade metamorphosed rocks consist of sillimanite-cordierite-biotite gneiss. Dikes consisting of diorite and muscovite granite intruded rocks of the Penobscot Formation and became schistose when deformed within it. Immediately to the west of the site there is a mapped contact between the Ordovician age Penobscot Formation and the granite of the Mount Waldo pluton. The Mount Waldo granite, Devonian in age, is a light-gray, medium grained, equi-granular biotite granite with no apparent foliation. See Figures 2 and 3.

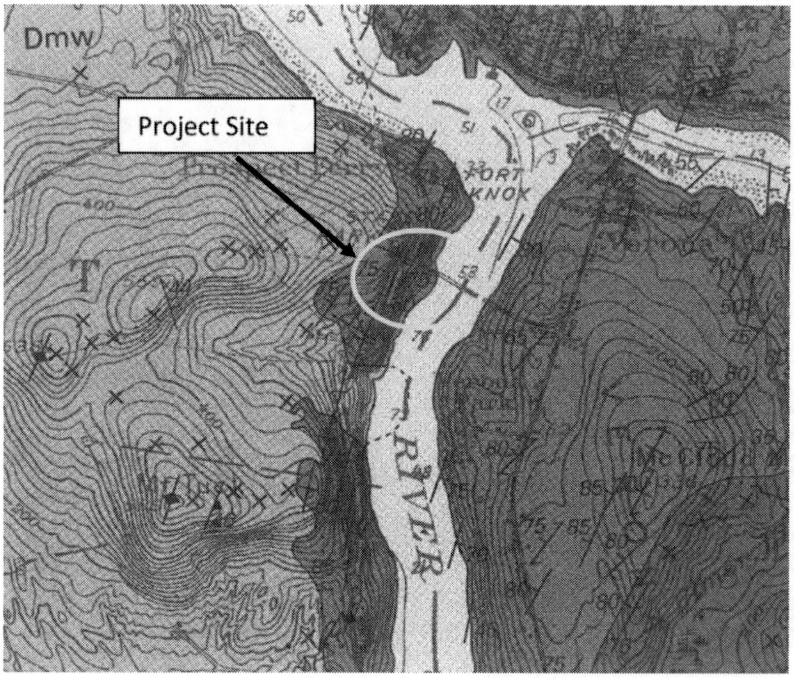

Figure 2 – Site bedrock geology. Areas shaded in light pink on the west side of the river are underlain with the Mt. Waldo granite. Areas shaded in dark pink are underlain with rocks of the Penobscot Formation. Map from Stewart (1998).

-

Figure 3 – Regional geology from Stewart (1998). The belts shaded in blue represent terranes of metamorphic and volcanic rocks that became accreted to the North American margin during the Silurian. The belts shown in shades of green represent terranes that accreted to the North American margin prior to the Silurian, and younger marine basins developed upon them. The orange areas are plutons, mostly granitic and of Silurian and Devonian ages. Site location outlined in yellow.

ORIGINAL FIELD INVESTIGATIONS AND CONDITIONS

Subsurface explorations

Design phase subsurface explorations consisting of drilling four test borings along the approach roadway was conducted at the site in 2003. The test borings were terminated at depths ranging from approximately 28 to 83 ft below the top of bedrock surface. The bedrock sampled in the test borings generally consisted of gray, fine-grained, hard, fresh, metaquartzite. Joints in the rock were described as typically low angle with steep to vertical foliation joints. The joints were generally tight and discolored, some with heavy oxidation. Veins of gray, medium to coarse grained, igneous intrusive granite were encountered in several of the test borings. Rock quality designation (RQD), a common parameter used to help assess the competency of sampled bedrock ranged from 85 to 100 percent, indicating very good to excellent rock mass quality. Localized zones of highly fractured bedrock were encountered with RQD values as low as 15 percent.

Bedrock outcrop observations

A geologic reconnaissance was conducted in August 2003. Data collected focused on structural geologic properties (e.g., strike, discontinuity dip and dip direction, infilling, visible seepage, persistence, aperture) and general rock mass properties (e.g. weathering/alteration, intact rock compressive strength).

The observed bedrock consisted of hard, gray, slightly weathered, fine-grained to aphanitic quartzite with occasional pyrite mineralization and a few calcite veins up to 2-in. thick. The rock mass contained three main joint sets. One set parallel to foliation dipping steeply to the northwest. A second set dipping steeply to the northeast, and the third set low angle to nearly horizontal. The combined orientation of the joint sets results in a blocky structure. Typical block sizes ranged from about 2 to 5 feet.

Design Rock Slope Geometry

Based on rock engineering analyses of the data collected and the conditions present along the proposed roadway, it was recommended that the proposed rock cut be sloped at a nominal 4 vertical to 1 horizontal (4V:1H). It was identified the potential for localized geologic features with adverse orientations may exist and become apparent during rock slope excavation. As a result, stability assessments were made during rock slope excavation and construction.

Rockfall analyses were completed to determine catchment area geometry at the toe of the rock slope. A catchment area is intended to retain rock blocks that may become detached from the rock slope and would otherwise enter the roadway, creating a hazard. The recommended rock slope geometry and catchment area are shown on Figure 4.

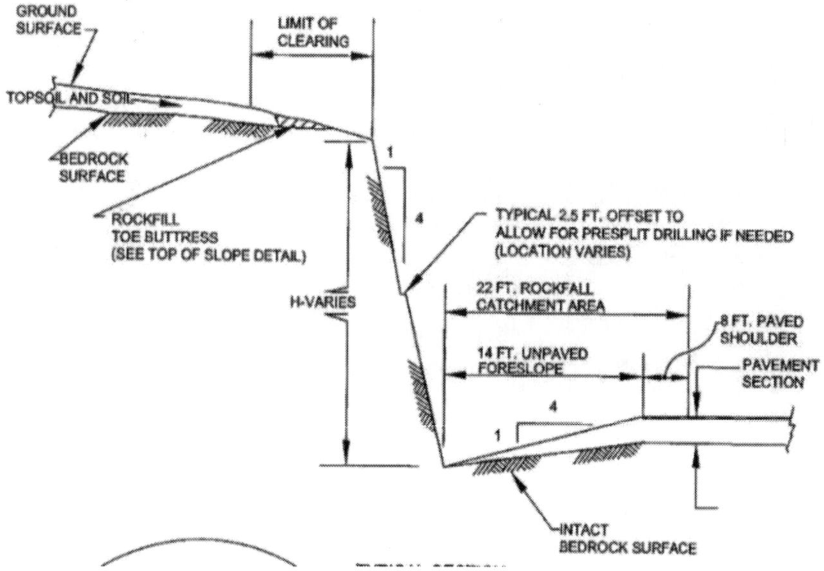

Figure 4 – Design Rock Slope and Catchment Area Geometry

ROCK SLOPE CONSTRUCTION AND INSPECTION

Blasting and excavation of the rock slope began in late 2004/early 2005 and was substantially complete by June 2005. Construction progress photographs are shown below on Figure 5.

During construction and excavation of the rock slope, areas in need of remediation were identified for stabilization. Draft sketches, details and/or specifications for remedial measures were prepared during construction and again in 2005/2006 in an effort to stabilize the identified areas. MaineDOT elected not to perform the recommended rock slope remedial work during the original bridge and approach roadway construction due to project-specific constraints at the time of the work.

In 2009 and again in 2012, further evaluations of the condition of the rock slope were completed in an effort to stabilize identified Areas along the rock slope during demolition of the Waldo-Hancock Bridge. MaineDOT elected to temporarily delay proposed rock slope remedial work until after the completion of the bridge demolition. As a result, the 2012 Haley & Aldrich work plan was modified to include recommendations for a long-term rock slope maintenance and monitoring (M&M) program. The condition of the rock slope was monitored and documented by MaineDOT in 2014.

Figure 5 – Blasting and rock slope excavation

Rock slope assessment areas

Between 2005 and 2015 a total of 23 areas along the rock slope, designated Area 1 through Area 19 (including 1A, 4A through 4C and 5A), were judged to pose potential safety and long-term maintenance issues of varying degree. Final ratings were assigned to each area after completion of the October 2015 site inspection and are summarized in Figure 6. Individual areas are shown on Figures 7 and 8.

Considering that MaineDOT had limited available funding, a rating scale was developed for specified areas in need of remediation. The scale ranged from "Most

Critical" to "Least Critical". Areas identified as "Most Critical" and "Moderately Critical" would be remediated first, based on funding. Less critical areas (i.e., "Marginally Critical" to "Least Critical" Areas) will continue to be monitored during future Maintenance and Monitoring inspections and potentially remediated as additional funding allows.

Area No.[1]	Approximate Location (Sta.)	Year Identified	Original Stabilization Recommendations	2016 Final Rating[2]	Preliminary Stabilization Recommendations[3]	Remediation in Initial Phase of Work[4]
1	360+50	2005-2009	Type B Rock Dowels		2 Rock Dowels	Yes
1A		2015	NA		NA	No
2	361+50	2005-2009	Type B Rock Dowels, Shotcrete, Shotcrete Rock Drains		NA	No
3	363+00	2005-2009	Type B Rock Dowels, Shotcrete, Shotcrete Rock Drains		NA	No
4A	364+00	2005-2009	Type B Rock Dowels, Rock Drains		2 Rock Dowels	Yes
4B			Type B Rock Dowels		2 Rock Dowels	Yes
4C			Type B Rock Dowels		3 Rock Dowels	No
5	365+00	2005-2009	Type A and B Rock Dowels, Wire Mesh		Anchored Mesh, 32 rock dowels	Yes
5A			Type B Rock Dowels		Check Scaling	Yes
6	365+50	2005-2009	Type B Rock Dowels		Cable Lashing, 4 rock dowels	Yes
7	366+00	2005-2009	Type A Rock Dowels, Shotcrete, Shotcrete Rock Drains		NA	No
8	366+50	2005-2009	Shotcrete, Shotcrete Rock Drains	S / N	NA	No
9	366+75	2005-2009	Type B Rock Dowels, Shotcrete, Shotcrete Rock Drains		Scaling, 6 Rock Dowels	Yes
10	359+50 to 360+00	2005-2009	Wire Rope		1 Vertical Rock Dowel	Yes
11	358+75	2005-2009	NA		NA	No
12	358+75	2012	NA		NA	No
13	359+75	2012	NA		NA	No
14	364+50	2012	NA		Check Scaling	Yes
15	364+50	2012	NA		Check Scaling	Yes
16	364+50	2012	NA		2 Vertical Rock Dowels, 4 Sub-Horizontal Rock Dowels	Yes
17	359+75	2015	NA		NA	No
18	362+50	2015	NA		NA	No
19	360+50	2015	NA		NA	No

Figure 6 – Final rock slope remediation area assessment

Legend:

"MOST CRITICAL"

"MODERATELY CRITICAL"

"MARGINALLY CRITICAL"

"SLIGHTLY CRITICAL"

"LEAST CRITICAL"

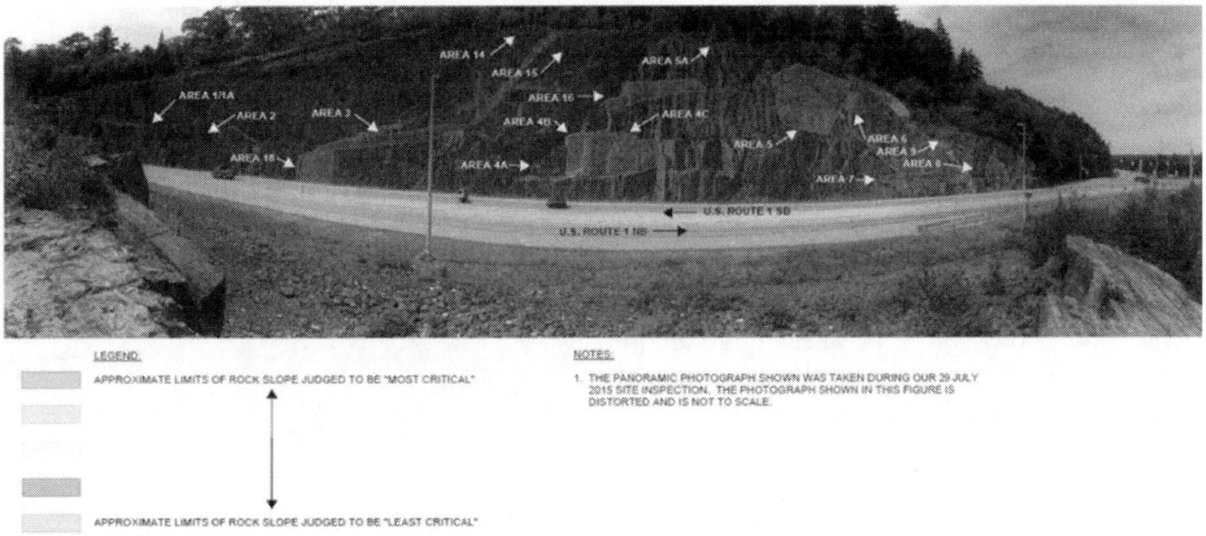

Figure 7.-Rock slope remediation area (Sheet 1 of 2). This view shows the northern portion of the rock cut.

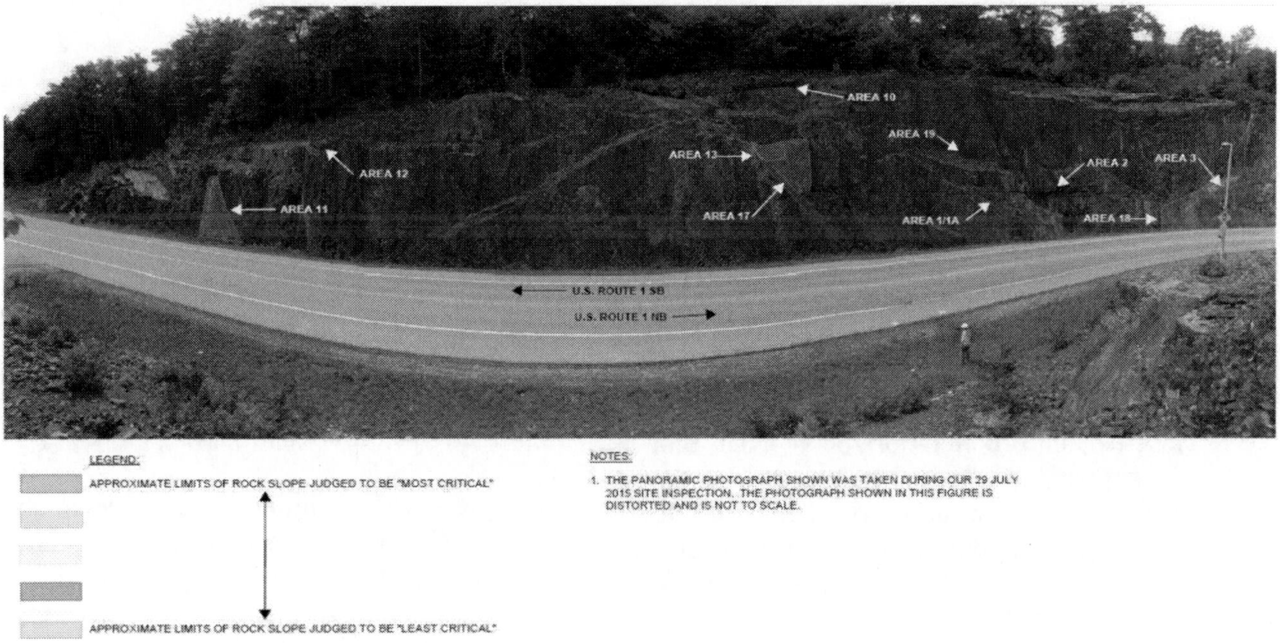

Figure 8-Rock Slope Remediation Area (Sheet 2 of 2). This view shows the southern portion of the rock cut.

ROCK SLOPE REMEDIATION

Rock slope remediation efforts were completed in the Fall of 2016 and included scaling and vegetation removal, installation of rock dowels which included a total of 67 rock dowels installed within seven different areas of the rock slope. Remedial efforts also included the installation of anchored wire mesh and cable lashing systems. Wire mesh with enhanced corrosion protection was selected given the project's proximity to a marine environment, and was powder coated to blend into the natural color of the rock slope.

Rock slope remediation elements

Rock dowels

Figure 9. Rock dowels

A rock dowel (Figure 9) is a passive (either no design load or very light lock-off load applied after installation) rock reinforcement element that typically consists of a solid steel bar that is inserted into a sub-horizontal hole that is drilled beyond potential failure surfaces in the rock mass and the bar is grouted in-place. Rock dowels can be loaded in shear and/or tension, based on desired function and physical situation in the field.

Anchored wire mesh netting

Figure 10. Wire mesh

This rock remediation measure (Figure 10) generally consists of a high tensile strength galvanized steel wire netting that is anchored to the face of the rock slope by rock dowels installed on a pre-determined grid pattern.

Wire rope cable lashing

Figure 11. Wire rope cable lashing

This relatively simple rock remediation measure (Figure 11) consists of using one or more high tensile strength galvanized wire ropes (similar to that used in the wire mesh netting) that are wrapped around loose rock blocks to secure them in place. The wire rope is typically connected to rock dowels installed in stable rock, adjacent to or behind the loose rock block.

Scaling

Rock scaling (Figure 12) is used to remove loose rock fragments/blocks, soil and vegetation from slopes that pose a rockfall hazard both during and after construction. Scaling is completed by removing the loose material using hand tools

Figure 12. Rock scaling

consisting of pry-bar, picks, and/or shovels. Where large rock blocks require removal, several different techniques can be used including inflatable air bags, hydraulic splitters, and winching as well as mechanical methods using conventional excavation equipment.

Since 2005, MaineDOT has effectively managed and implemented rock slope assessment, design and construction, along the westerly approach to the Penobscot Narrows Bridge as shown in Figure 13. Through their continued persistent efforts to secure funding and in implementing an annual Maintenance and Monitoring program, MaineDOT has reduced risk and successfully maintained the condition of the infrastructure while maximizing the available funding.

REFERENCES

Kelley, Alice R., and Caron, Lynn, 2013, Surficial geology of the Bucksport quadrangle, Maine: Maine Geological Survey, Open-File Map 13-14, map, scale 1:24,000.

MaineDOT, Preliminary Design Report, Rock Slope Remediation at Penobscot Narrows Bridge, Prospect, Maine, NHPP-1879(200) WIN 018792.00, Maine Department of Transportation Bridge Program, 2016.

Stewart, David B., 1998, Geology of Northern Penobscot Bay, Maine. U.S. Geological Survey Miscellaneous Investigations Series Map I-2551 (2 Sheets).

Wones, David R., 1991, Bedrock Geologic Map of the Bucksport [15-minute] Quadrangle, Waldo, Hancock, and Penobscot Counties, Maine: U. S. Geological Survey, Geologic Quadrangle Map GQ-1692, color map, scale 1:62,500.

Formatting Issue - Best Available Paper/Graphic

Figure 13 – Final Rock Slope Condition

Formatting Issue - Best Available Paper/Graphic

69th Highway Geology Symposium, pages 41-42, 2018

Hard Bedrock, Marine mud, and High Tidal Range: Highway Geology and Engineering Challenges along Maine's Mid-coast

Field Trip Roadlog

Wednesday, September 12, 2018

MILEAGE	*DESCRIPTION*
0.0/0.0	Begin at entrance to Holiday Inn by the Bay, 88 Spring Street, Portland, Maine and turn right onto Spring Street.
0.3/0.3	Continue straight onto Middle Street.
0.6/0.3	Turn left onto Franklin Street.
1.2/0.6	Turn right onto I-295 ramp.
23.5/22.3	Use the right two lanes to take exit 28 onto U.S. 1 North toward Brunswick.
25.9/2.4	Turn left at the light to continue on U.S. 1/Mill Street.
34.4/8.5	Bath Iron Works on the right. Continue north on U.S. 1 over the Sagadahoc Bridge spanning the Kennebec River.
45.1/10.7	Wiscasset. Continue north on U.S. 1.
49.0/3.0	**STOP 1**. Sherman Lake Rest Stop. Overview of salt marsh restoration. Rest rooms available.
	Continue north on U.S. 1.
74.9/25.9	Montpelier Mansion, home of General Henry Knox Mansion on the right. General Knox, the brilliant Revolutionary War military strategist, and first Secretary of War in Washington's cabinet, lived here from 1795 until his death in 1806.
79.0/4.1	Turn left onto Broadway.
81.3/1.3	Turn right onto Maverick Street.
81.6/0.3	Turn left onto Camden Street.
82.0/0.4	Turn right onto Waldo Ave.
82.5/0.5	Turn right onto Samoset Road.
82.7/0.2	**STOP 2**. Rockland landslide site. We will discuss the landslide event, remediation efforts, and if time and conditions permit, examine current erosion at the base of the slide.
	Turn around to continue the trip.
82.9/0.2	Turn left on Waldo Ave.
83.4/0.5	Turn right on U.S. 1/Camden Street.
108.3/24.9	Keep left to stay on U.S. 1. Shortly, cross the Passagassawakeag River at Belfast.

69th Highway Geology Symposium, pages 41-42, 2018

110.3/2.0 Turn right to Young's Lobster Pound. **LUNCH**.

Continue North on U.S. 1/ME-3.

125.4/15.1 Penobscot Narrows Bridge. Turn left on ME-174.

125.8/0.4 Turn right into Fort Knox State Park and proceed to Observatory

STOP 3. Penobscot Narrows Bridge approach and Observatory.

Note: We have very specific logistics at this site the ensure that all field trip participants get the full benefit of this stop.

We will divide into three groups by busload.

Group 1 will proceed up the short access road to the viewing point for the large roadcut on the western approach to the bridge.

Group 2 will view the example cross section of the bridge just uphill from the parking area.

Group 3 will proceed to the Observatory for the 50-second elevator ride to the top of the Observatory. The elevator has a capacity of 8, so it will take 5-6 round trips to get everyone to the top of the Observatory. The Observatory itself has a capacity of 49, so please limit your viewing to 10 minutes so that others will have an opportunity to enjoy the view.

After ½ hour, the groups switch.

Group 1. To the Obsevatory.

Group 2. Walk the short access road to view the western approach to the bridge.

Group 3. Examine the example cross section.

After ½ hour, switch again.

Group 1. Examine the bridge cross section.

Group 2. To the Observatory.

Group 3. Walk the short access road to view the western approach to the bridge.

Return to buses for trip back to Holiday Inn By the Bay.

Turn left out of Fort Knox State Park onto ME-174.

125.2/0.4 Turn right onto U.S. 1/ME-3.

142.0/16.8 Turn right on ME-3 at Belfast.

186.8/44.8 Take ramp onto I-95 South.

197.2/10.4 Keep right at fork to continue on I-295 South.

244.5/47.3 Take exit 6A for Forest Ave.

246.0/1.5 Turn left onto Congress Street.

246.1/0.1 Turn right onto Oak Street.

246.2/0.2 Turn right onto Spring Street. Holiday Inn on left.

Highway Geology Symposium
C/O John Pilipchuk
1589 Mail Service Center
Raleigh, NC, 27699-1589

ISBN 978-1-5108-8659-9

9 781510 886599

CURRAN ASSOCIATES INC.
proceedings.com